Protein-Carbohydrate Interactions in Biological Systems

The Proceedings of a symposium held under the auspices of the Federation of European Microbiological Societies and the Swedish Society for Microbiology in Luleå, Sweden, 17th–20th June 1985.

Protein-Carbohydrate Interactions in Biological Systems

The Molecular Biology of Microbial Pathogenicity

Editor

D. L. Lark

President, Syn-Tek AB, Umeå, Sweden

Associate Editors

S. Normark

Department of Microbiology, Umeå University, Umeå, Sweden

B.-E. Uhlin

Department of Microbiology, Umeå University, Umeå, Sweden

H. Wolf-Watz

Department of Microbiology, National Defence Research Institute, Umeå, Sweden

1986

ACADEMIC PRESS

Harcourt Brace Jovanovich, Publishers

London Orlando San Diego New York Austin
Boston Sydney Tokyo Toronto

ACADEMIC PRESS INC. (LONDON) LTD.
24/28 Oval Road,
London NW1 7DX

United States Edition published by
ACADEMIC PRESS INC.
Orlando, Florida 32887

British Library Cataloguing in Publication Data

Protein-carbohydrate interactions in biological systems: the molecular biology of microbial pathogenicity.—(FEMS symposia)
1. Proteins 2. Carbohydrates
I. Lark, D. L. II. Normark, S.
III. Uhlin, B.-E. IV. Wolf-Watz, H.
V. Series
574.19′245 QP551

ISBN 0-12-436665-1

Phototypeset by Dobbie Typesetting Service
Plymouth, Devon

Printed by St Edmundsbury Press,
Bury St Edmunds, England

Contributors

Adlam, C. Wellcome Biotechnology Ltd, Langley Court, Beckenham, Kent BR3 3BS, UK

Åhrén, C. Department of Medical Microbiology, University of Göteborg, Guldhedsgaton 10, S-413 46 Göteborg, Sweden

Allan, I. Infectious Diseases Unit, University Department of Paediatrics, John Radcliffe Hospital, Oxford OX3 9DU, UK

Amstrup Pedersen, P. Laboratory for Microbiology, Building 221, Technical University of Denmark, DK-2800 Lyngby, Denmark

Amyes, S. G. B. Department of Bacteriology, University Medical School, Teviot Place, Edinburgh EH8 9AG, UK

Andersson, B. Department of Clinical Immunology and Medical Biochemistry, University of Göteborg, S-413 46 Göteborg, Sweden

Andrews, G. P. Department of Gastroenterology, Division of Medicine, Walter Reed Army Institute of Research, Washington, DC 20307, USA

Ashworth, L. A. E. Experimental Pathology and Vaccine Research and Production Laboratories, PHLS Centre for Applied Microbiology and Research, Porton, Salisbury SP4 0JG, UK

Båga, M. Department of Microbiology, University of Umeå, S-901 87 Umeå, Sweden

Baudry, B. Service des Entérobactéries, U. 199 INSERM, Institut Pasteur, Paris, France

Beesley, J. E. Wellcome Biotechnology Ltd, Langley Court, Beckenham, Kent BR3 3BS, UK

Bergmans, H. Department of Molecular Cell Biology, State University of Utrecht, Padualaan 8, 3584 CH Utrecht, The Netherlands

Bitter-Suermann, D. Institute of Medical Microbiology, Johannes-Gutenserg-University, Augustusplatz/Hochhaus, D-6500 Mainz, FRG

Blackwell, C. C. Department of Bacteriology, University Medical School, Teviot Place, Edinburgh EH8 9AG, UK

Bock, K. Department of Organic Chemistry, The Technical University of Denmark, DK-2800 Lyngby, Denmark

Boedeker, E. C. Department of Gastroenterology, Division of Medicine, Walter Reed Army Institute of Research, Washington, DC 20307-5100, USA

Bölin, I. Division of Microbiology, National Defence Research Institute, S-901 82 Umeå, Sweden

Boylan, M. Department of Microbiology, Moyne Institute, Trinity College, Dublin 2, Republic of Ireland

Brée, A. Institut National de la Recherche Agronomique, Station de Pathologie Aviaire et de Parasitologie, C.R. de Tours-Nouzilly, 37380 Monnaie, France

Brendle, J. J. Department of Biological Chemistry, Walter Reed Army Institute of Research, Washington, DC 20307-5100, USA

Brettle, R. P. Pyelonephritis Clinic and Division of Infectious Diseases, City Hospital, 51 Greenbank Drive, Edinburgh, UK

Brown, J. E. Division of Biochemistry, Walter Reed Army Institute of Research, Washington, DC 20012, USA
Bundle, D. R. Division of Biological Sciences, National Research Council of Canada, Ottawa, Ontario K1A OR6, Canada
Carlström, A.-S. Swedish Sugar Company, R&D, Box 6, S-232 00 Arlöv, Sweden
Caugant, D. Department of Clinical Immunology and Medical Microbiology, University of Göteborg, S-413 46 Göteborg, Sweden.
Collier, R. J. Microbiology and Molecular Genetics, Harvard Medical School and The Shipley Institute of Medicine, Boston, MA 02181, USA
Chenoweth, D. E. Veterans Medical Center, 3550 La Jolla Village Drive, San Diego, CA 92161, USA
Clerc, P. Service des Entérobactéries, U. 199 INSERM, Institut Pasteur, Paris, France
Cluzel, R. Service de Bactériologie, Faculté de Pharmacie, 28, Place Henri Dunant, 63001 Clermont Ferrand Cedex, France
Coleman, D. Department of Microbiology, Moyne Institute, Trinity College, Dublin 2, Republic of Ireland
Danielsson, D. Department of Clinical Microbiology and Immunology, Örebro Medical Centre Hospital, S-701 85 Örebro, Sweden
Darfeuille-Michaud, A. Service de Bactériologie, Faculté de Pharmacie, 28, Place Henri Dunant, 63001 Clermont Ferrand Cedex, France
Davis, B. W. Anti-infective Research Department, Norwich Eaton Pharmaceuticals, Inc., P.O.Box 191, Norwich, NY 13815, USA
Deal, C. Department of Molecular Biology, Scripps Clinic and Research Foundation, 10666 N. Torrey Pines Rd, La Jolla, CA 92037, USA
de Cock, H. Department of Molecular Cell Biology, State University of Utrecht, Padualaan 8, 3584 CH Utrecht, The Netherlands
de Graaf, F. K. Department of Molecular Microbiology, Free University, De Boelalaan 1087, 1081 HV Amsterdam, The Netherlands
de Ree, J. M. Department of Medical Microbiology, University of Limburg, P.O.Box 616, 6200 MD Maastricht, The Netherlands
Dho, M. Institut National de la Recherche Agronomique, Station de Pathologie Aviaire et de Parasitologie, C.R. de Tours-Nouzilly, 37380 Monnaie, France
Dijksterhuis, M. Department of Molecular Cell Biology, State University of Utrecht, Padualaan 8, 3584 CH Utrecht, The Netherlands
Dinari, G. Department of Bacterial Diseases, Walter Reed Army Institute of Research, Washington, DC 20307-5100, USA
Dowsett, A. B. Experimental Pathology and Vaccine Research and Production Laboratories, PHLS Centre for Applied Microbiology and Research, Porton, Salisbury SP4 0JG, UK
Edwards, M. F. Department of Medical Microbiology, Stanford University School of Medicine, Stanford, CA 94305, USA
Ekbäck, G. Department of Microbiology, University of Umeå, S-901 87 Umeå, Sweden
Ely, S. Infectious Diseases Unit, University Department of Paediatrics, John Radcliffe Hospital, Oxford OX3 9DU, UK
Falkow, S. Department of Medical Microbiology, Stanford University, Stanford, CA 94305, USA
Fanning, G. R. Department of Biological Chemistry, Walter Reed Army Institute of Research, Washington, DC 20307-5100, USA
Felmlee, T. Department of Medical Microbiology, University of Wisconsin, Madison, WI 53706, USA
Feutrier, J. Department of Biochemistry and Microbiology, University of Victoria, Victoria, British Columbia V8W 2Y2, Canada

Fletcher, J. N. Department of Microbiology, University of Southampton Medical School, Southampton General Hospital, Southampton SO9 4XY, UK
Fliesler, S. J. Department of Ophthalmology and Biochemistry, University of Miami School of Medicine, Miami, FL 33101, USA
Forestier, C. Service de Bactériologie, Faculté de Pharmacie, 28, Place Henri Dunant, 63001 Clermont Ferrand Cedex, France
Formal, S. B. Department of Bacterial Diseases, Walter Reed Army Institute of Research, Washington, DC 20307-5100, USA
Forsberg, Å. Division of Microbiology, National Defence Research Institute, S-901 82 Umeå, Sweden
Forsman, K. Department of Microbiology, University of Umeå, S-901 87 Umeå, Sweden
Fröman, G. Department of Bacteriology and Epizootology, Swedish University of Agricultural Sciences, Biomedicum, Box 583, S-751 23 Uppsala, Sweden
Frosch, M. Institute of Medical Microbiology, Johannes-Gutenberg-University, Augustusplatz/Hochhaus, D-6500 Mainz, FRG
Fulford, S. Infectious Diseases Unit, University Department of Paediatrics, John Radcliffe Hospital, Oxford OX3 9DU, UK
Gaastra, W. Laboratory for Microbiology, Building 221, Technical University of Denmark, DK-2800 Lyngby, Denmark
Gariépy, J. Department of Medicine, Stanford University School of Medicine, Stanford, CA 94305, USA
Gemski, P. Department of Biological Chemistry, Walter Reed Army Institute of Research, Washington, DC 20307-5100, USA
Getzoff, E. Department of Molecular Biology, Scripps Clinic and Research Foundation, 10666 N. Torrey Pines Rd, La Jolla, CA 92037, USA
Giannini, G. Sclavo Research Centre, Siena, Italy
Goebel, W. Institut für Genetik und Mikrobiologie, Röntgenring 11, D-8700 Würzburg, FRG
Goergen, I. Institute of Medical Microbiology, Johannes-Gutenberg-University, Augustusplatz/Hochhaus, D-6500 Mainz, FRG
Goguen, J. D. Department of Microbiology and Immunology, University of Tennessee Center for the Health Sciences, Memphis, TN 38163, USA
Goldberg, I. Department of Microbiology and Molecular Genetics, Harvard Medical School, Boston, MA 02115, USA
Göransson, M. Department of Microbiology, University of Umeå, S-901 87 Umeå, Sweden
Gorringe, A. Experimental Pathology and Vaccine Research and Production Laboratories, PHLS Centre for Applied Microbiology and Research, Porton, Salisbury SP4 0JG, UK
Gross, U. Institute of Medical Microbiology and Immunology, University of Hamburg, Martinistrasse 52, D-2000 Hamburg 20, FRG
Guss, B. Department of Microbiology, University of Uppsala, Biomedicum, Box 581, S-751 23 Uppsala, Sweden
Haas, R. Max-Planck-Institut für Biologie, Spemannstrasse 34, D-74 Tübingen, FRG
Hacker, J. Institut für Genetik und Mikrobiologie, Röntgenring 11, D-8700 Würzburg, FRG
Hagberg, L. Department of Clinical Immunology and Medical Biochemistry, University of Göteborg, S-413 46 Göteborg, Sweden
Hale, T. L. Department of Bacterial Diseases, Walter Reed Army Institute of Research, Washington, DC 20307-5100, USA

Hales, B. A. Department of Bacteriology, University Medical School, Teviot Place, Edinburgh EH8 9AG, Scotland

Halula, M. C. Department of Microbiology and Immunology, MCV-Station, Box 678, Virginia Commonwealth University, Richmond, VA 23298-0001, USA

Hanson, L. Å. Department of Clinical Immunology and Medical Biochemistry, University of Göteborg, S-413 46 Göteborg, Sweden

Heckels, J. E. Department of Microbiology, University of Southampton Medical School, Southampton General Hospital, Southampton SO9 4XY, UK

Heesemann, J. Institute of Medical Microbiology and Immunology, University of Hamburg, Martinistrasse 52, D-2000 Hamburg 20, FRG

Herrman, R. Mikrobiologie, University of Heidelberg, 69 Heidelberg, FRG

Hinson, G. Department of Genetics, University of Leicester, Leicester LE1 7RH, UK

Hirst, T. R. Department of Medical Microbiology, University of Göteborg, Guldhedsgatan 10, S-413 46 Göteborg, Sweden

Hoekstra, W. Department Molecular Cell Biology, State University of Utrecht, Padualaan 8, 3584 CH Utrecht, The Netherlands

Hoiseth, S. K. Office of Biologics, National Center for Drugs and Biologics, Food and Drug Administration, Bethesda, MD 20205, USA

Hollyfield, J. G. Cullen Eye Institute, Baylor College of Medicine, Houston, TX 77030, USA

Holmgren, J. Department of Medical Microbiology, University of Göteborg, Guldhedsgatan 10, S-413 46 Göteborg, Sweden

Höök, M. Connective Tissue Laboratory, Diabetes Hospital, University of Alabama in Birmingham, Birmingham, AL 35294, USA

Houston, W. L. Department of Gastroenterology, Division of Medicine, Walter Reed Army Institute of Research, Washington, DC 20307-5100, USA

Hull, R. Department of Microbiology, Baylor School of Medicine, Houston, TX 77030, USA

Hull, S. Department of Microbiology, Baylor School of Medicine, Houston, TX 77030, USA

Irons, L. I. Experimental Pathology and Vaccine Research and Production Laboratories, PHLS Centre for Applied Microbiology and Research, Porton, Salisbury SP4 0JG, UK

Jarchau, T. Institut für Genetik und Mikrobiologie, Röntgenring 11, D-8700 Würzburg, FRG

Johnstone, F. D. Department of Obstetrics and Gynaecology, University Medical School, Teviot Place, Edinburgh EH8 9AG, UK

Joly, B. Service de Bactériologie, Faculté de Pharmacie, 28, Place Henri Dunant, 63001 Clermont Ferrand Cedex, France

Kamata, K. Institute of Medical Science, University of Tokyo, 4-6-1 Shiroganedai-machi, Minato-ku, Tokyo 108, Japan

Karlsson, K.-A. Department of Medical Biochemistry, University of Göteborg, P.O.Box 33031, S-400 33 Göteborg, Sweden

Kay, W. W. Department of Biochemistry and Microbiology, University of Victoria, Victoria, British Columbia V8W 2Y2, Canada

Kelly, E. P. Department of Gastroenterology, Division of Medicine, Walter Reed Army Institute of Research, Washington, DC 20307-5100, USA

Kihlström, E. Department of Medical Microbiology, University of Linköping, S-581 85 Linköping, Sweden

Kijne, J. W. Department of Plant Molecular Biology, Botanical Laboratory, University of Leiden, Nonnensteeg 3, 2311 VJ Leiden, The Netherlands

Kinane, D. F. Department of Peridontology, The Dental School, University of Dundee, Dundee, UK

Klemm, P. Department of Microbiology, Technical University of Denmark, DK-2800 Lyngby, Denmark
Klinkert, M. Mikrobiologie, University of Heidelberg, 69 Heidelberg, FRG
Knapp, S. Institut für Genetik und Mikrobiologie, Röntgenring 11, D-8700 Würzburg, FRG
Knights, J. M. Wellcome Biotechnology Ltd, Langley Court, Beckenham, Kent BR3 3BS, UK
Knoppers, M. Cambridge Research Laboratory, 195 Albany Street, Cambridge, MA 02139, USA
Korhonen, T. K. Department of General Microbiology, University of Helsinki, SF-00280, Helsinki, Finland
Kotarski, S. F. Department of Biological Chemistry, Walter Reed Army Institute of Research, Washington, DC 20307-5100, USA
Kroll, J. S. Infectious Diseases Unit, University Department of Paediatrics, John Radcliffe Hospital, Oxford OX3 9DU, UK
Lafont, J.-P. Institut National de la Recherche Agronomique, Station de Pathologie Aviaire et de Parasitologie, C.R. de Tours-Nouzilly, 37380 Monnaie, France
Lagergård, T. Department of Clinical Immunology and Medical Microbiology, University of Göteborg, S-413 46 Göteborg, Sweden
Lance, L. L. Anti-infective Research Department, Norwich Eaton Pharmaceuticals, Inc. P.O.Box 191, Norwich, NY 13815, USA
Lark, D. L. Department of Medical Microbiology, Stanford University Medical School, Stanford, CA 94305, USA
Laufs, R. Institute of Medical Microbiology and Immunology, University of Hamburg, Martinistrasse 52, D-2000 Hamburg 20, FRG
Leffler, H. Department of Clinical Immunology and Medical Biochemistry, University of Göteborg, S-413 46 Göteborg, Sweden
Levine, M. Center for Vaccine Development, Department of Medicine, University of Maryland, Baltimore, MD 21201, USA
Lindberg, A. A. Karolinska Institutet, Department of Clinical Bacteriology, Huddinge University Hospital, S-141 86 Huddinge, Sweden
Lindberg, F. Department of Microbiology, University of Umeå, S-901 87 Umeå, Sweden
Lindberg, M. Department of Microbiology, University of Uppsala, Biomedicum, Box 581, S-751 23 Uppsala, Sweden
Lindon, J. C. Wellcome Biotechnology Ltd, Langley Court, Beckenham, Kent BR3 3BS, UK
Lomberg, H. Department of Clinical Immunology and Medical Biochemistry, University of Göteborg, S-413 46 Göteborg, Sweden
Lopez-Vidal, Y. Department of Medical Microbiology, University of Göteborg, Guldhedsgatan 10, S-413 46 Göteborg, Sweden
Losonsky, G. Center for Vaccine Development, Department of Medicine, University of Maryland, Baltimore, MD 21201, USA
Lugtenberg, B. J. J. Department of Plant Molecular Biology, Botanical Laboratory, University of Leiden, Nonnensteeg 3, 2311 VJ Leiden, The Netherlands
Lund, B. Department of Microbiology, University of Umeå, S-901 87 Umeå, Sweden
Lycke, N. Department of Medical Microbiology, University of Göteborg, Guldhedsgatan 10, S-413 46 Göteborg, Sweden
MacCallum, C. J. Department of Bacteriology, University Medical School, Teviot Place, Edinburgh EH8 9AG, UK
MacLaren, D. M. Department of Medical Microbiology, School of Medicine, Free University, P.O.Box 7161, 1007 MC Amsterdam, The Netherlands

Magnusson, G. Organic Chemistry 2, Chemical Center, The Lund Institute of Technology, P.O.Box 124, 221 00 Lund, Sweden
Magnusson, K.-E. Department of Medical Microbiology, University of Linköping, S-581 85 Linköping, Sweden
Makino, S. Institute of Medical Science, University of Tokyo, 4-6-1 Shiroganedai-machi, Minato-ku, Tokyo 108, Japan
Maluszynska, G. Department of Medical Microbiology, University of Linköping, S-581 85 Linköping, Sweden
Marklund, B.-I. Department of Microbiology, University of Umeå, S-901 87 Umeå, Sweden
Marre, R. Institut für Medizinische Mikrobiologie, Ratzeburger Allee 160, D-2400 Lübeck, FRG
Maurelli, A. T. Service des Entérobactéries, U. 199 INSERM, Institut Pasteur, Paris, France
May, S. J. Department of Bacteriology, University Medical School, Teviot Place, Edinburgh EG8 9AG, UK
Mekalanos, J. Department of Microbiology and Molecular Genetics, Harvard Medical School, Boston, MA 02115, USA
Meldal, M. The Technical University of Denmark, Building 201, DK-2800, Lyngby, Denmark
Merriwether, T. L. Department of Biological Chemistry, Walter Reed Army Institute of Research, Washington, DC 20307-5100, USA
Meyer, T. F. Max-Planck-Institut für Biologie, Spemannstrasse 34, D-74 Tübingen, FRG
Miller, V. Department of Microbiology and Molecular Genetics, Harvard Medical School, Boston, MA 02115, USA
Mooi, F. R. Department of Molecular Microbiology, Free University, De Boelelaan 1087, 1081 HV Amsterdam, The Netherlands
Mörner, S. Department of Microbiology, University of Umeå, S-901 87 Umeå, Sweden
Morris J. G. Center for Vaccine Development, Department of Medicine, University of Maryland, Baltimore, MD 21201, USA
Moxon, E. R. Infectious Diseases Unit, University Department of Paediatrics, John Radcliffe Hospital, Oxford OX3 9DU, UK
Mugridge, A. Wellcome Biotechnology Ltd, Langley Court, Beckenham, Kent BR3 3BS, UK
Murphy, J. R. Section of Biomolecular Medicine, University Hospital, Boston University Medical Centre, 75E Newton Street, Boston, MA 02139, USA
Neill, R. J. Department of Biological Chemistry, Walter Reed Army Institute of Research, Washington, DC 20307-5100, USA
Nickel, P. Max-Planck-Institut für Biologie, Spemannstrasse 34, D-74 Tübingen, FRG
Nilsson, B. Swedish Sugar Company, R&D, Box 6, S232 00 Arlöv, Sweden
Nnalue, N. A. Department of Medical Microbiology, Stanford University School of Medicine, Stanford, CA 94305, USA
Norgren, M. Department of Microbiology, University of Umeå, S-901 87 Umeå, Sweden
Norlander, L. Division of Microbiology, National Defence Research Institute, S-901 82 Umeå, Sweden
Normark, S. Department of Microbiology, University of Umeå, S-901 87 Umeå, Sweden
Nowicki, B. Department of General Microbiology, University of Helsinki, SF-00280, Helsinki, Finland
Nyberg, G. Department of Microbiology, University of Umeå, S-901 87 Umeå, Sweden

Oaks, E. V. Department of Bacterial Diseases, Walter Reed Army Institute of Research, Washington, DC 20307-5100, USA

O'Hanley, P. Department of Medical Microbiology, Stanford University Medical School, Stanford, CA 94305, USA

Ohman, L. Department of Medical Microbiology, University of Linköping, S-581 85, Linköping, Sweden

Oudega, B. Department of Molecular Microbiology, Free University, De Boelelaan 1087, 1081 HV Amsterdam, The Netherlands

Orndorff, P. E. Department of Microbiology, Pathology and Parasitology, School of Veterinary Medicine, North Carolina State University, Raleigh, NC 27606, USA

Palmer, S. Department of Microbiology, Medical School, Framlington Place, University of Newcastle upon Tyne, NE2 4HH, UK

Parry, S. H. Immunology Department, Unilever Research, Colworth House, Sharnbrook, Bedfordshire MK44 1LQ, UK

Pedersen, H. Department of Organic Chemistry, The Technical University of Denmark, DK-2800, Lyngby, Denmark

Pellett, S. Department of Medical Microbiology, University of Wisconsin, Madison, WI 53706, USA

Pere, A. Department of General Microbiology, University of Helsinki, SF-00280, Helsinki, Finland

Perugini, M. Sclavo Research Centre, Siena, Italy

Poikonen, K. Department of Medical Microbiology, University of Oulu, SF-90220 Oulu 22, Finland

Porras, O. Department of Clinical Immunology and Medical Biochemistry, University of Göteborg, S-413 46 Göteborg, Sweden

Quentin-Millet, M.-J. Institute Merieux, Marcy L'Etoile, B.P.No. 3-69752, Charbonnières/Bains Cedex, France

Quiocho, F. A. Department of Biochemistry, Rice University Houston, TX 77251, USA

Rappuoli, R. Sclavo Research Centre, Siena, Italy

Ratti, G. Sclavo Research Centre, Siena, Italy

Rayborn, M. E. Cullen Eye Institute, Baylor College of Medicine, Houston, TX 77030, USA

Rhen, M. Department of General Microbiology, University of Helsinki, SF-00280, Helsinki, Finland

Robinson, A. Experimental Pathology and Vaccine Research and Production Laboratories, PHLS Centre for Applied Microbiology and Research, Porton, Salisbury SP4 0JG, England

Rooke, D. M. Department of Microbiology, Medical School, Framlington Place, University of Newcastle upon Tyne, NE2 4HH, UK

Roosendaal, B. Department of Molecular Microbiology, Free University, De Boelelaan 1087, 1081 HV Amsterdam, The Netherlands

Rosqvist, R. Division of Experimental Medicine, National Defence Research Institute, Department 4, S-901 82 Umeå, Sweden

Rothman, S. W. Division of Biochemistry, Walter Reed Army Institute of Research, Washington, DC 20012, USA

Rowe, B. The Central Public Health Laboratory, 61 Colindale Avenue, London NW9, UK

Ryter, A. Unité de Microscopie Electronique, Département de Biologie Moléculaire, Institut Pasteur, Paris, France

Sakai, T. Institute of Medical Science, University of Tokyo, 4-6-1 Shiroganedai-machi, Minato-ku, Tokyo 108, Japan

Sansonetti, P. J. Service des Entérobactéries, U. 199 INSERM, Institut Pasteur, Paris, France

Sasakawa, C. Institute of Medical Science, University of Tokyo, 4-6-1 Shiroganedai-machi, Minato-ku, Tokyo 108, Japan

Saukkonen, K. National Public Health Institute, SF-00280, Helsinki, Finland

Savelkoul, P. H. M. Department of Medical Microbiology, University of Limburg, P.O.Box 616, 6200 MD Maastricht, The Netherlands

Schaller, H. Mikrobiologie, University of Heidelberg, 69 Heidelberg, FRG

Schauer, D. Department of Microbiology, Pathology and Parasitology, School of Veterinary Medicine, North Carolina State University, Raleigh, NC 27606, USA

Schmidt, G. Institut für Experimentelle Biologie und Medizin, 2061 Borstel, FRG

Schmidt, M. A. Zentrum für Molekulare Biologie der Universität Heidelberg, Im Neuenheimer Feld 282, D-6900 Heidelberg, FRG

Schmoll, T. Institut für Genetik und Mikrobiologie, Röntgenring 11, D-8700 Würzburg, FRG

Schoolnik, G. K. Department of Medical Microbiology, Stanford University Medical School, Stanford, CA 94305, USA

Schröder, J. Institute of Medical Microbiology and Immunology, University of Hamburg, Martinistrasse 52, D-2000 Hamburg 20, FRG

Schultz, J. E. Karolinska Institutet, Department of Clinical Bacteriology, Huddinge University Hospital, S-141 86 Huddinge, Sweden

Schwillens, P. Department of Medical Microbiology, University of Limburg, P.O.Box 616, 6200 MD Maastricht, The Netherlands

Segal, E. Department of Molecular Biology, Scripps Clinic and Research Foundation, 10666 N. Torrey Pines Rd, La Jolla, CA 92037, USA

Sekura, R. D. Laboratory of Developmental and Molecular Immunity, National Institute of Child Health and Human Development, National Institutes of Health, Bethesda, MD 20205, USA

Serrano, N. S. Department of Biological Chemistry, Walter Reed Army Institute of Research, Washington, DC 20307-5100, USA

Sherman, P. M. Department of Gastroenterology, Division of Medicine, Walter Reed Army Institute of Research, Washington, DC 20307-5100, USA

Skurnik, M. Department of Medical Microbiology, University of Oulu, SF-90220 Oulu 22, Finland

Smit, G. Department of Plant Molecular Biology, Botanical Laboratory, University of Leiden, Nonnensteeg 3, 2311 VJ Leiden, The Netherlands

Smyth, C. J. Department of Microbiology, Moyne Institute, Trinity College, Dublin 2, Republic of Ireland

So, M. Department of Molecular Biology, Scripps Clinic and Research Foundation, 10666 N. Torrey Pines Rd, La Jolla, CA 92037, USA

Sodd, M. A. Department of Biological Chemistry, Walter Reed Army Institute of Research, Washington, DC 20307-5100, USA

Spears, P. A. Department of Microbiology, Pathology and Parasitology, School of Veterinary Medicine, North Carolina State University, Raleigh, NC 27606, USA

Stendahl, O. Department of Medical Microbiology, University of Linköping, S-581 85 Linköping, Sweden

Stern, A. Max-Planck-Institut für Biologie, Spemannstrasse 34, D-74 Tübingen, FRG

Stocker, B. A. D. Department of Medical Microbiology, Stanford University School of Medicine, Stanford, CA 94305, USA

Strömberg, N. Department of Medical Biochemistry, University of Göteborg, P.O.Box 33031, S-400 33 Göteborg, Sweden

Svanborg Edén, C. Department of Clinical Immunology and Medical Biochemistry, University of Göteborg, S-413 46 Göteborg, Sweden
Svennerholm, A.-M. Department of Medical Microbiology, University of Göteborg, Guldhedsgatan 10, S-413 46 Göteborg, Sweden
Swindlehurst, J. Cambridge Research Laboratory, 195 Albany Street, Cambridge, MA 02139, USA
Switalski, L. Connective Tissue Laboratory, Diabetes Hospital, University of Alabama in Birmingham, Birmingham, AL 35294, USA
Tachovsky, T. G. Cambridge Research Laboratory, 195 Albany Street, Cambridge, MA 02139, USA
Tainer, J. Department of Molecular Biology, Scripps Clinic and Research Foundation, 10666 N. Torrey Pines Rd, La Jolla, CA 92037, USA
Taschke, C. Mikrobiologie, University of Heidelberg, 69 Heidelberg, FRG
Taylor, R. Department of Microbiology and Molecular Genetics, Harvard Medical School, Boston, MA 02115, USA
Teneberg, S. Department of Medical Biochemistry, University of Göteborg, P.O.Box 33031, S-400 33 Göteborg, Sweden
Tehunen, J. Department of General Microbiology, University of Helsinki, SF-00280, Helsinki, Finland
Thom, S. M. Department of Bacteriology, University Medical School, Teviot Place, Edinburgh EH8 9AG, UK
Tippett, J. Infectious Diseases Unit, University Department of Paediatrics, John Radcliffe Hospital, Oxford OX3 9DU, UK
Trust, T. J. Department of Biochemistry and Microbiology, University of Victoria, Victoria, British Columbia V8W 2Y2, Canada
Uhlin, B. E. Department of Microbiology, University of Umeå, S-901 87 Umeå, Sweden
Väisänen-Rhen, V. Department of General Microbiology, University of Helsinki, SF-00280, Helsinki, Finland
van Bergen en Henegouwen, P. M. P. Department of Molecular Cell Biology (EMSA), State University of Utrecht, Padualaan 8, 3584 CH Utrecht, The Netherlands
van den Bosch, J. F. Department of Medical Microbiology, University of Limburg, P.O.Box 616, 6200 MD Maastricht, The Netherlands
van Die, I. Department of Molecular Cell Biology, State University of Utrecht, Padualaan 8, 3584 CH Utrecht, The Netherlands
van Doorn, J. Department of Medical Microbiology, School of Medicine, Free University, P.O.Box 7161, 1007 MC Amsterdam, The Netherlands
Virji, M. Department of Microbiology, University of Southampton Medical School, Southampton General Hospital, Southampton SO9 4XY, UK
Virkola, R. Department of General Microbiology, University of Helsinki, SF-00280, Helsinki, Finland
Vuopio-Varkila, J. National Public Health Institute, SF-00280, Helsinki, Finland
Vyas, N. K. Department of Biochemistry, Rice University Houston, TX 77251, USA
Wadström, T. Department of Bacteriology and Epizootology, Swedish University of Agricultural Sciences, Biomedicum, Box 583, S-751 23 Uppsala, Sweden
Walan, Å. Department of Medical Microbiology, University of Linköping, S-581 85 Linköping, Sweden
Walz, W. Zentrum für Molekulare Biologie der Universität Heidelberg, Im Neuenheimer Feld 282, D-6900 Heidelberg, FRG
Weir, D. M. Department of Bacteriology, University Medical School, Teviot Place, Edinburgh EH8 9AG, UK

Welch, R. A. Department of Medical Microbiology, University of Wisconsin, Madison, WI 53706, USA

Wells, N. S. Department of Biological Chemistry, Walter Reed Army Institute of Research, Washington, DC 20307-5100, USA

Westling, M. Karolinska Institutet, Department of Clinical Bacteriology, Huddinge University Hospital, S-141 86 Huddinge, Sweden

Williams, J. M. Wellcome Biotechnology Ltd, Langley Court, Beckenham, Kent BR3 3BS, UK

Williams, P. Department of Genetics, University of Leicester, Leicester LE1 7RH, UK

Wilton-Smith, P. Experimental Pathology and Vaccine Research and Production Laboratories, PHLS Centre for Applied Microbiology and Research, Porton, Salisbury SP4 0JG, UK

Wolhieter, J. A. Department of Bacterial Immunology, Walter Reed Army Institute of Research, Washington, DC 20307-5100, USA

Wolf, M. K. Department of Gastroenterology, Division of Medicine, Walter Reed Army Institute of Research, Washington, DC 20307-5100, USA

Wolf-Watz, H. Division of Microbiology, National Defence Research Institute, S-901 82 Umeå, Sweden

Wood, L. V. Anti-infective Research Department, Norwich Eaton Pharmaceuticals, Inc. P.O.Box 191, Norwich, NY 13815, USA

Yamagata-Murayama, S. Institute of Medical Science, University of Tokyo, 4-6-1 Shiroganedai-machi, Minato-ku, Tokyo 108, Japan

Yoshikawa, M. Institute of Medical Science, University of Tokyo, 4-6-1 Shiroganedai-machi, Minato-ku, Tokyo 108, Japan

Yother, J. Department of Microbiology and Immunology, University of Alabama in Birmingham, Birmingham, AL 35294, USA

Zak, K. Department of Microbiology, University of Southampton Medical School, Southampton General Hospital, Southampton, SO9 4XY, UK

Zamze, S. Infectious Diseases Unit, University Department of Paediatrics, John Radcliffe Hospital, Oxford OX3 9DU, UK

Zuidweg, E. Department of Molecular Cell Biology, State University of Utrecht, Padualaan 8, 3584 CH Utrecht, The Netherlands

Preface

This first FEMS Conference in Sweden "Molecular Biology of Microbial Pathogenicity: The Role of Protein-Carbohydrate Interactions" was held in Luleå, Sweden, 17-20 June 1985, under the sponsorship of the Swedish Society for Microbiology. Originally conceived as a meeting to examine the genetic approach to microbial pathogenicity, the conference developed into an interdisciplinary programme combining molecular biologists, structural biologists and chemists, carbohydrate chemists and protein chemists. The meeting size was limited and the meeting place was selected to ensure maximum opportunities for interactions among the participants. The meeting—featuring 52 speakers—examined the molecular and chemical aspects of the known microbial pathogenicity determinants. The programme was heavily scheduled but every session, including an additional poster session, was enthusiastically and well attended.

As the meeting progressed it became obvious to most participants that complex carbohydrates are going to have an increasing impact on biology. Evidence is accumulating that these protein-carbohydrate interactions are important in both species and organ parasite tropism and eukaryotic cell differentiation and growth. Thus, these biologically active compounds may modulate organ embryogenesis, differentiation and oncogenesis, and may determine host susceptibility to various infectious diseases encompassing pathogenic bacteria, viruses and protozoons.

Because this first interdisciplinary conference was so successful, we assume there will be future meetings that will examine in many additional biological systems the role of protein-sugar interactions.

We are grateful to the speakers and participants at this conference whose enthusiasm and support contributed to its success and to this published volume.

David L. Lark
Staffan Normark
Bernt Eric Uhlin
Hans Wolf-Watz

The organizers wish to express their gratitude to the following companies and organizations for providing generous support.

Alfa Laval Agri International Corp.
Bayer Corp., Germany
BioCarb Corp.
Ewos Corp.
Federation of European Microbiological Societies
Glaxo Medicine
Kabi Vitrum Corp.
The Community of Luleå
LKB Sweden Corp.
Marcus Wallenberg Foundation for International Cooperation in Science
Norrlandsfonden
Pharmacia Corp.
Rank Xerox
Skandinaviska Enskilda Banken
Swedish Sugar Company
Sparbanken
Biotechnology Research Foundation, Sweden
Swedish National Board for Technical Development
Seth M. Kempes Memory Foundation
Syn-Tek Corp.
University of Umeå
Uminova

Acknowledgements

The conference was supported by FEMS and the Swedish Society for Microbiology. In addition the organizers wish to express their gratitude to the companies and organizations listed on the opposite page for providing generous support.

Syn-Tek AB supported Carin Bergman and myself in the organization of the conference and we were also assisted by the Department of Microbiology, Umeå University.

The publication of this book was also made possible through the donations of these companies and organizations and the efficient assistance of Academic Press (London).

Carin Bergman provided the principal support during the many months prior to the conference. She coordinated the activities, organized the accommodation and ensured a smooth and efficient meeting. Maj-Lis Enwall provided efficient secretarial support and looked after many of the details to ensure the book was edited properly.

Finally, to all the conference participants, authors and all other individuals and groups, I thank you for all your support in making this an exciting and successful meeting.

David L. Lark
Umeå, Sweden
1985

Contents

II. ANTIGENIC VARIATION OF BACTERIAL ADHESINS

Co-Chairmen: Dan Danielsson and Gary Schoolnik

Poster Session

III. ROLE OF PILI AND ADHESINS IN PATHOGENICITY

Co-Chairpersons: David Lark and Catharina Svanborg-Edén

Poster Session

IV. CARBOHYDRATE RECEPTORS: IDENTIFICATION, ANALYSIS AND THEIR BIOLOGICAL ROLE

Co-Chairmen: Karl-Anders Karlsson and Klaus Bock

Poster Session

V. ADHESINS: THEIR RECEPTORS AND INTERACTIONS

Co-Chairmen: Alf Lindberg and Timo Korhonen

Poster Session

VI. BACTERIAL INVASION

Co-Chairpersons: Helena Mäkelä and Hans Wolf-Watz

Poster Session

VII. BACTERIAL SURFACE COMPONENTS AND THEIR IMPORTANCE FOR VIRULENCE

Co-Chairpersons: Frederick Sparling and Maggie So

Poster Session

VIII. BACTERIAL TOXINS AND RECEPTORS

Co-Chairmen: R. John Collier and Torkel Wadström

Introduction

David L. Lark

SYN-TEK AB, P.O. Box 1010, S-90120 Umeå, Sweden

Host-parasite interactions, cellular differentiation and dedifferentiation, and intercellular communication are major topics in biology. Disturbances in these biological processes usually manifest themselves microscopically as cellular distortion, or if extensive enough as organ malfunction and in certain cases death. Until 5-10 years ago we could only explain these phenomena by the ubiquitous biological "black box". However, with the advent and the maturation of molecular biology—stimulated by molecular genetics—it has provided a rational means to probe these complex phenomena by identifying key molecules (principally proteins), isolating them, and examining their effect in defined systems. As a result of these investigations, it appears that many of these complex biological observations are the result of a disturbance in key "regulator" molecules. So far most effort has focused investigating biologically active proteins since they are the principal direct gene by-products. However, recent studies have included carbohydrates that may be defined as "biologically active" in a more strict sense. These carbohydrates are most often recognized specifically by proteins and this interaction may mediate a particular biological response. These protein-carbohydrate interactions will confer specificity, and then it is possible to localize a specific response in a complex biological system. A well known example is the blood group recognition by complementary antibodies [1]. These blood group carbohydrates differ between individuals and inappropriate mixing may result in blood transfusion complications or tissue graft rejection. Other examples of biologically interesting protein–carbohydrate interactions include the evaluation of carbohydrates as tumour antigens by use of monoclonal antibodies. There are cell membrane proteins mediating carbohydrate uptake (high-affinity transport of monosaccharides, clearing of glycoproteins from blood plasma by the liver). Carbohydrates are also of decisive importance for membrane biogenesis and assembly. There is even evidence that plant lectins (protein) may, by specific binding to cell surface carbohydrate, induce a selective behaviour of animal cells (protein synthesis, growth, motility, death). A more spectacular aspect is the assumed role of protein-carbohydrate interactions in specific cell-to-cell adhesion during growth and precisely programmed tissue formation. It may be that cell membrane defects result in tumour cells with their unregulated growth and spread (metastasis). These aspects are covered by review articles in a series of monographs [2].

However, in spite of great efforts, there is still no precise carbohydrate receptor identified on animal cells mediating a *normal* physiological function, although a recent report claims the presence of a minor glycolipid of chicken hepatocytes essential for adhesion [3]. In contrast, there is a rapidly increasing volume of evidence for specific carbohydrate receptors on animal cells mediating *pathological* processes. These receptors are attachment sites for viruses, bacteria and bacterial toxins, which are required for the essential colonization or membrane penetration by these parasites or their products. This is one main theme of the present conference.

Although microbiological ligands may bind to peptides, it appears that carbohydrate receptors are preferentially selected. This is probably due to the abundance of carbohydrates on the animal cell surface, since they are found both as strictly membrane-bound glycoconjugates (glycoproteins, glycolipids, proteoglycans) and as more loosely associated material (proteoglycans, mucins), most of which have more unspecific roles for the target cell (e.g. structural, protective against proteolytic damage). This implies, that as a parasite containing a binding lectin approaches an animal target cell, it is much more likely to collide with carbohydrate than peptide. The classical example is the influenza virus and its strict dependence on surface sialic acid.

Technical facilities are now rapidly improving for studying carbohydrate receptors and their biological role. These include methods of detection, isolation, identification, chemical synthesis, conformation analysis, and interaction of protein with carbohydrate. In spite of the greater structural complexity of carbohydrate compared to peptide, modern high-resolution isolation methods and refined analysis methods including high-field NMR and mass spectrometry, have gained access to complex oligosaccharides at the milligram level. Biological reagents, such as monoclonal antibodies, improve both mapping in tissues and isolation. Synthesis of complex sequences, with natural or non-natural structure, is not only useful in biological test systems to prove or disprove theories about the forces involved in the recognition, but also for potential applications in biology and medicine. Very importantly, the three-dimensional structure of complex oligosaccharides in aqueous solution is now possible to define through analysis by very high-field proton and carbon-13 NMR spectroscopy combined with simple computer calculations. This and the few examples of refined X-ray data of protein-carbohydrate complexes are providing insight into the specific forces involved in the binding between carbohydrates and proteins (where the latter compounds are solving the almost impossible physical chemical problem of extracting a sugar molecule out of water). Electrostatic interactions are probably the most important in determining the specificity of the overall driving force for association in water, whereas the non-bonded van der Waal interactions and hydrophobic effects are important for the overall strength of the established binding. Most of the above-mentioned points will be covered in the following contributions.

The realization that biologically active proteins and carbohydrates complement one another and those interactions have a general biological importance justifies the effort to formulate a multi-disciplinary approach to examine all aspects regarding these protein-carbohydrate interactions. The meeting and this publication reflect my attempt to initiate this process. This specific interplay of biologists, carbohydrate and protein chemists, geneticists, and physicians was very exciting. If this continues the perspectives for biology and medicine are fascinating.

REFERENCES

1. Lemieux, R. U., D. R. Bundle and D. A. Baker (1975). *J. Am. Chem. Soc.* **97**, 4076.
2. The Glycoconjugates (M. I. Horowitz, ed.), Vols 1-4. Academic Press, New York and London.
3. Roseman, S. (1985). *J. Biochem.* **97**, 709.

I. Genetics of Bacterial Adhesins

Co-Chairmen: Staffan Normark
Bernt-Eric Uhlin

Minor Pilus Components Acting as Adhesins

Staffan Normark[a], Frederik Lindberg, Björn Lund, Monica Båga, Gustav Ekbäck, Mikael Göransson, Stellan Mörner, Mari Norgren, Britt-Inger Marklund and Bernt Eric Uhlin

Department of Microbiology, University of Umeå, Umeå, Sweden

INTRODUCTION

Bacterial attachment to eukaryotic cells is believed to be mediated by adhesin proteins on the bacterial surface that bind to specific glycolipid or glycoprotein receptors on the host cells. In Gram-negative bacteria pili (fimbriae) of various classes have been shown to mediate receptor specific binding to target cells.

Purified pili are composed of one major subunit, the pilin, which has led to the assumption that the pilin protein itself binds to the receptor. It has, however, been difficult to understand how pili can vary so extensively in their antigenic properties, while maintaining the same receptor binding specificity. Sequencing studies of a number of antigenic variants of *Escherichia coli* (see van Die *et al.*, this volume) as well as *Neisseria gonorrhoeae* pilins [1] has revealed regions within the respective proteins that vary little or not at all. It is possible that such conserved regions are involved in binding to the receptor, as has been concluded for the *N. gonorrhoeae* pili [2]. It is probable, however, that some of these conserved regions are a result of structural constraints on the pilin for its ability to be exposed and assembled into a pilus fibre. A conceptually different approach for the bacterium would be to vary antigenically a major pilin subunit not involved in receptor binding, maintaining the structure of a minor receptor binding pilin more constant. This approach has been used in the case of the digalactoside binding *E. coli* P-fimbriae or Pap-pili [3-6]. Recent data suggest that a similar situation may prevail also for S-fimbriae [7] and for mannose-binding pili [8].

[a] *Telephone:* 46-90-101774

Protein-Carbohydrate Interactions in Biological Systems. ISBN 0 12 436665 1

GENETIC AND FUNCTIONAL ORGANIZATION OF THE *pap* GENE CLUSTER OF UROPATHOGENIC *E. COLI*

Uropathogenic *E. coli* frequently contain chromosomal gene clusters that encode pilus formation and an adhesin binding to a Galα(1$\rightarrow$4)βGal moiety present in the globo series of glycolipids to which the P-blood group antigens belong [9]. One such gene cluster was cloned from the UTI strain J96 (0:4, K:6, F13) [10] and subjected to genetic analysis [11].

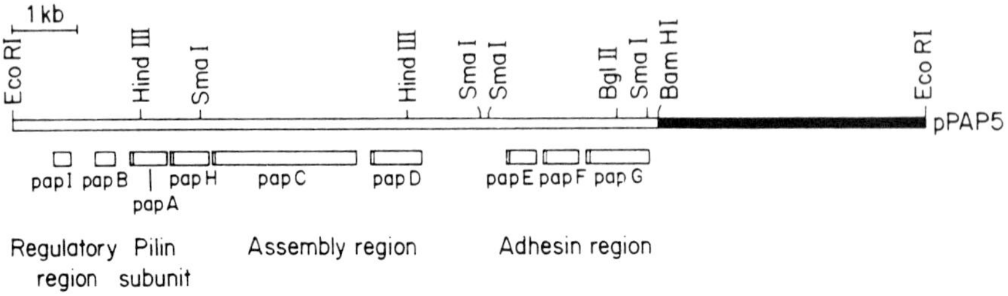

Figure 1. Gene organization and physical map of the Pap pilus-adhesin gene cluster carried by pPAP5. The white horizontal bar represents a chromosomal section cloned from the uropathogenic strain *E. coli* J96. The black horizontal bar represents vector (pBR322) DNA. The boxes indicates positions of individual *pap* genes (A–I), all genes except *papI* and *papB* encode signal peptides. For the *papH* gene, the existence of a signal peptide has been postulated from the DNA sequence. The position of the different functional units are indicated at the bottom of the figure.

Nine *pap* genes have been identified on a chromosomal section of roughly 9.5 kb (Fig. 1). Two of these genes, *papI* and *papB* are involved in the transcriptional regulation of the *papA* pilin gene ([12]; see also Uhlin *et al.*, this volume). DNA linker insertions early in the *papA* gene or deletions of the entire gene abolish piliation as determined by electron microscopy [3,5]. Mutants with such lesions were, however, still proficient for binding. Transcriptionally distal to the *papA* gene, an open reading frame is located that corresponds to a weakly expressed 22 kD (kilodalton) polypeptide (Båga, unpubl. obs.). This gene, *papH*, is not required for binding or piliation. The amino acid sequence of the PapH protein as deduced from the DNA sequence of the gene shows similarities to PapA and other *E. coli* pilins (Båga, unpubl. obs.). The reason for the poor expression of *papH* is probably the strong transcriptional terminator present between *papA* and *papH* [12].

papC encodes an 81 kD protein [4]. Genes for similarly sized proteins are found in all *E. coli* pili gene clusters so far analysed. In the K88 system it has been shown that the corresponding protein migrates in the outer membrane fraction in sucrose gradients (see Mooi *et al.*, this volume). In analogy, we believe PapC has the same localization as its K88 counterpart, although this remains to be experimentally investigated. Mutational inactivation of *papC* results in loss of both piliation and the binding phenotype. Pilin antigen can, however, be detected after disruption of the cells [4] and by Western blot analysis [5]. Probably PapC acts as a nucleation centre for polymerization.

The gene *papD* directs the synthesis of a 28.5 kD large protein which is likewise essential for pilus biogenesis. Mutants in *papD* do not express pili and do not bind. In contrast to *papC* mutants, pilin antigen can not be detected even after disruption of the cells [4]. Pulse-chase experiments reveal that the PapA protein (the major pilin subunit) is made in a *papD* mutant, but it is rapidly degraded (Norgren, unpubl. obs.). Thus the PapD protein seems to be necessary to stabilize the pilin after its synthesis and before its assembly into a pilus.

A subclone carrying the *papA*, *papH*, *papC* and *papD* genes express visible pili on *E. coli* HB101 [5]. No digalactoside-specific binding is expressed from this clone. Tn*5* insertions in a region transcriptionally downstream of the *papD* gene abolished binding but not piliation [4]. This region was found to carry the two genes *papF* and *papG*, both of which are essential for the expression of digalactoside binding pili [3,4]. Immediately upstream from *papF* we identified a gene, *papE*, that is not involved in receptor binding *per se* since *papE* mutants still bind the receptor. However, if Pap-pili are purified from such a mutant the preparation showed a much reduced ability to agglutinate erythrocytes in a P-specific manner [3]. This suggests that the *papE* gene product might be involved in associating the adhesin to the pilus.

PRIMARY STRUCTURE OF THE *pap* ADHESIN GENE PRODUCTS

The primary sequences of PapE, PapF, and PapG proteins have been deduced from the DNA sequence of their respective genes. Both *papE* and *papF* have features in common with known *E. coli* pilins [6]. Thus both proteins have a size (149 and 148 amino acids respectively) similar to PapA and other pilins. They contain two cysteine residues in the amino terminal half. The penultimate amino acid is a tyrosine residue, as is the case for most *E. coli* pilins. Amino acids in the amino and carboxy termini which are conserved in most *E. coli* pilins are also present in PapE and PapF. Furthermore the use of two different algorithms to compare the primary sequences of PapE, PapF and the *E. coli* pilins indicated that these proteins have tertiary structure similarity. Based on these data we suggest that both PapE and PapF are pilin-like proteins and that they might polymerize either alone or together with the major pilin.

Thus the *pap* gene cluster contains four genes coding for pilins or pilin-like proteins (*papA*, *papH*, *papE*, *papF*). The proximity of *papA-papH*, and *papE-papF* suggest they may have arisen by a duplication event. Of these genes, only *papA* is highly expressed.

The primary sequence of PapG showed no significant resemblance to that of other *E. coli* pilins (Lund, unpubl. obs.). This protein is fairly large 35.5 kD. Only the K88 pilin antigen approaches this size, but also this protein lacks amino acid homology with PapG. Unlike the PapA, PapE and PapF proteins the gene product of *papG* is not dependent on PapD for stability.

A *pap* PILUS PREPARATION CONTAINS MINOR PILIN COMPONENTS

To understand the molecular basis for attachment, it is pertinent to know whether or not PapE, PapF and PapG are part of the Pap pilin structure. Pap pili consist of more than 99% of the major pilin, PapA [11]. To detect possible minor constituents in the preparation, purified Pap pili were radioiodinated [6]. Two heavily labelled proteins were seen, one corresponding to PapA and the other to PapE. The PapE protein was also seen after silver staining of the SDS-gels and was not detected in pili purified from a *papE* mutant.

MINOR PILIN COMPONENTS ARE IMMUNOLOGICALLY ACTIVE

From a medical standpoint, it is important to know if minor pilus constituents are immunogenic. A radioimmunoassay was therefore developed using the host strain HB101 containing wild-type or mutant *pap* clones as the solid phase. Antibodies raised against wild-type Pap-pili were allowed to bind to the bacteria and then detected with iodinated Protein A. As shown in Table 1, the antiserum contained antibodies that bound to a *papA* mutant expressing all Pap proteins except PapA. This reactivity was lost when the antiserum was preabsorbed with HB101 cells harbouring the *papA* mutation mutant clone. If the antiserum was preabsorbed with either a *papE* or a *papF* mutant a significant reactivity to the *papA* mutant remained in each case. Since these preabsorbed antisera contained antibodies that bound equally well to wild-type and a *papA* mutant they cannot represent cross-reacting PapA antibodies. We believe instead that the remaining

Table 1. Reaction of absorbed antisera with *pap* mutants

	Serum absorbed with:				
Solid phase	*host*	*wt*	*papA*	*papE*	*papF*
host	–	–	–	–	–
wt	+ +	–	+ +	+	+
papA	+	–	–	+	+
papE	+ +	–	+ +	–	+
papF	+ +	–	+ +	+	–

E. coli K12 HB101 was used as a host in all experiments. HB101 carrying the *pap* gene cluster cloned on pPAP5 is denoted wt.
Titres are given as follows: + + = $\approx 1\times10^4$; + = $\approx 3\times10^3$; – = $<1\times10^1$.

Female white rabbits were immunized with purified Pap pili. Antiserum was serially diluted and incubated with the different mutants. After incubation, cells were washed and incubated with 50 000 cpm of ^{125}I protein A. After 1 h of incubation, the cells were washed, and the radioactivity in the pellet determined. Titres are determined as the dilution at which the bound ^{125}I protein A corresponds to 50% of maximum activity bound, using non-absorbed serum with the wild-type in the solid phase. Absorbed sera were prepared by incubating serum overnight with 10^9 bacteria, and repeating this process until no binding activity to the absorbing strain remained. The mutants used are all derivatives of pPAP5 with *Xho*I linker insertions in the corresponding genes [3].

reactivity represents antibodies directed against epitopes on PapE and PapF, respectively. This also means that these proteins must be present in the pilus preparation as well as on the surface of a *papA* mutant.

SEQUENCE VARIABILITY IN THE *pap* ADHESIN GENES

A number of gene clusters encoding digalactoside binding P-fimbriae of different serotypes have been cloned and characterized. We have compared two such gene clusters cloned from two independent *E. coli* 0:6 isolates [14,15] with the gene cluster from the 0:4 strain J96 [13]. The overall gene organization was similar for all three clones. The *papH*, *papC* and *papD* genes appeared well conserved as judged from restriction endonuclease mapping, partial DNA sequencing and Southern blot hybridization. In the construction of one of the 0:6 clones (pDC5) the pilin gene was deleted. Nevertheless, this clone expressed digalactoside-specific binding in the absence of pilus formation. Thus, it is likely that the adhesin genes are distinct from the major pilin gene in all digalactoside binding gene clusters. The "adhesin genes" (*papF*, *papG* and the adhesin anchoring gene *papE*) were sequenced from one of the 0:6 clones (pDC5) and compared to the equivalent of J96. Figure 2 outlines in a schematic fashion the amino acid sequence differences between the *papE* and *papF* variants in pPAP5 (from J96) and pDC5 (from IA2). Thirty-five per cent of the amino acid positions differ between the two PapE variants. These sequence differences are clustered in the central part of the proteins. Analogous to different major pilin proteins, the amino and carboxy termini of the PapE variants are very similar. PapF, on the other hand, differed between the two clones in only 9% of the amino acid positions (Lund,

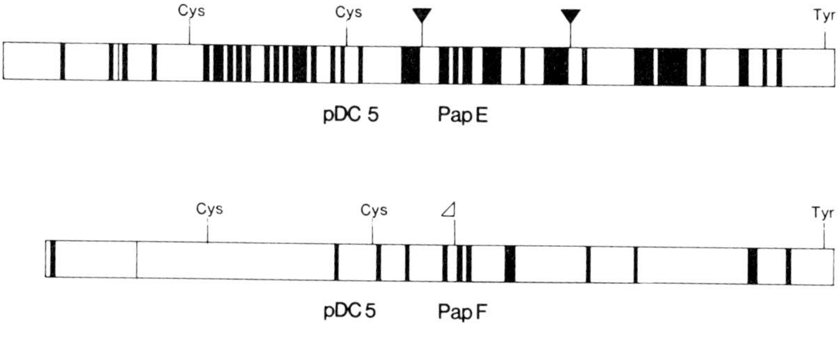

Figure 2. Structural variation of the PapE and PapF proteins. The *papE* and *papF* genes from the 0:4 clone pPAP5 and the 0:6 clone pDC5 were DNA sequenced and the deduced amino acid compared. The proteins coded for by pDC5 are shown. Thick vertical bars indicate amino acid differences as compared to the corresponding genes of pPAP5. The thin vertical line found in the left part of the proteins represent the postulated cleavage site of the signal peptide. The width of the vertical bars shown in the signal peptides indicates one amino acid shift, double width indicates two amino acid shifts in a row, etc. The *papE* gene of pDC5 have two amino acid insertions (▲) as compared to pPAP5, *papF* has one deletion (△). The positions of the cysteines and the penultimate tyrosine are shown.

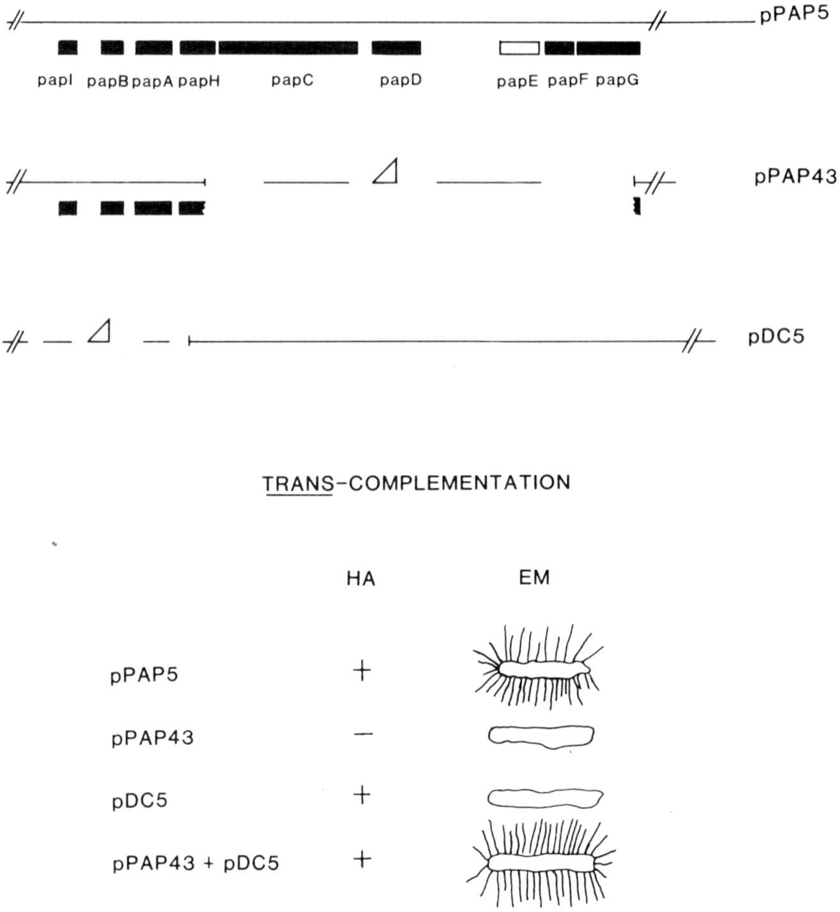

Figure 3. Gene organization of plasmid constructs used for *trans*-complementation. The upper panels show the 0:4 clone pPAP5 and its deletion derivative pPAP43. The 0:6 clone pDC5 lacks DNA encoding the major pilin subunit (*papA*) and *papIB*. Plasmids PAP5 and PAP43 have a pBR322 (pMBI) replicon, whereas pDC5 have a pACYC184 (p15A) replicon. The result of the *trans*-complementation experiments are outlined at the bottom part of the figure. HA indicates haemagglutination, an assay for adhesin expression. EM shows, in a schematic fashion, the appearance of bacteria harbouring the different plasmids.

unpubl. obs.). It is yet to be determined if these differences affect antigenic epitopes within the proteins.

A *papG* probe from pPAP5 did not hybridize to pDC5 or pPIL110-35, showing that the *papG* allele of pPAP5 must differ significantly from the others [13]. This was confirmed by DNA sequencing (Lund, unpubl. obs.). DNA corresponding to *papE*, *papF*, the intercistronic region between *papF* and *papG*

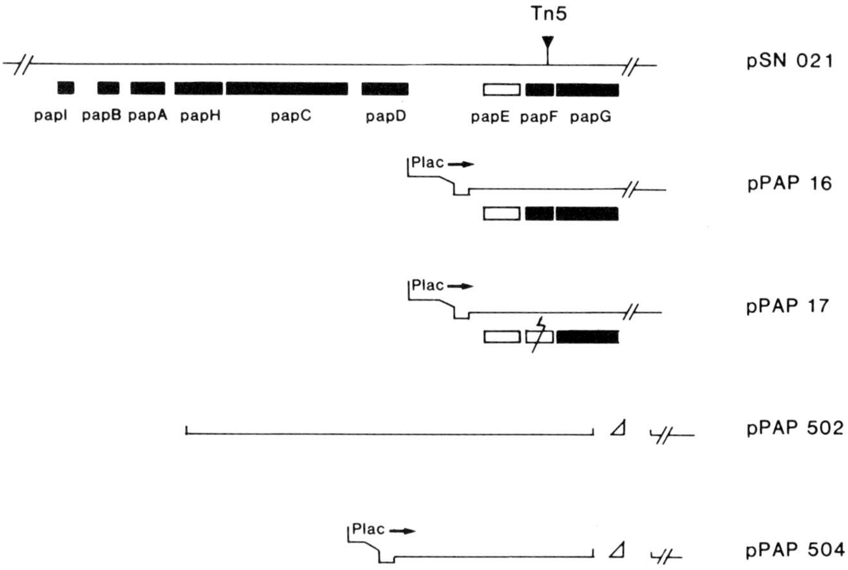

TRANS-COMPLEMENTATION

	HA
pPAP 16 + pPAP 502	+
pPAP 17 + pPAP 502	−
pSN 021 + pPAP 504	+

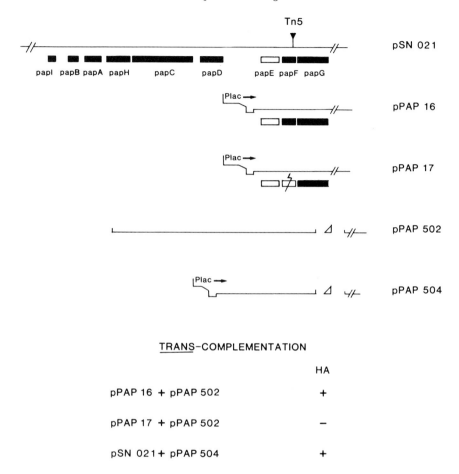

Figure 4. *Trans*-complementation of the adhesin function. The gene organization of the different plasmid constructs is shown. The pDC5 derivative pPAP504 as well as the pPAP5 derivatives pPAP16 and pPAP17 have a pBR322 (pMBI) replicon, whereas the pPAP5 derivative pSN021 and the pDC5 derivative pPAP502 have a pACYC184 (p15A) replicon. Slide haemagglutination assays (HA) were performed to test for expression of adhesin. The result is outlined in the figure.

and the very 5′-end of the *papG* gene was very homologous between pPAP5 compared to pDC5 and pPIL110-35. However, from the centre of the signal sequence gene and downstream the two *papG* genes revealed only patchy homologies. Despite these marked sequence differences, PapG proteins are absolutely essential for digalactoside-specific binding in both clones [3,13].

The *papG* allele of pPAP5 seems to be very different from most *papG* alleles in gene clusters encoding digalactoside-specific adhesin. An internal *papG* probe from pPAP5 did not hybridize to any of 66 UTI strains tested, whereas a corresponding *papG* probe from pDC5 hybridized to 23 out of the 24 UTI strains expressing digalactoside-specific binding as well as to a few other strains [16]. Thus the *papG* gene of pDC5 is structurally more similar to most other *papG* alleles than is the *papG* allele of pPAP5. Only DNA sequencing of a number of *papG* alleles will tell us how conserved this gene actually is.

Trans-complementation analyses were performed to test for functional similarities between the gene clusters present on pPAP5 and pDC5, respectively. A derivative of pPAP5 carrying the papA pilin gene (pPAP43) complemented pDC5 to pili formation ([13] and Fig. 3). Likewise, a *papG* deletion of pDC5 (pPAP502) was complemented in *trans* to express binding by a pPAP5 derivative (pPAP16) carrying *papE*, *papF* and *papG* under the transcriptional control of the *lac* promoter ([13], Fig. 4). Thus, the adhesion function can be complemented by the intact *papEFG* region. The three proteins of one clone may interact with the highly conserved PapC and PapD protein to express the adhesion phenotype.

Attempts to complement individual mutations in the *papEFG* region with the *papG* deletion of pDC5 have failed (pPAP17 and pPAP502 in Fig. 4). However, the *papF* protein encoded by pDC5 can complement a mutation in the equivalent gene of pPAP5 as shown by pPAP504 and pSNO21 in Fig. 4. Hence, the *papF* protein can be *trans*-complemented between the clones whereas the *papG* protein can not.

In *Neisseria gonorrhoeae*, the basis for structural diversity of pilin proteins is based on DNA rearrangements between silent copies of the pilin genes and one or two expression-linked copies [17]. *E. coli* may contain several complete gene clusters coding for P-fimbriae of different serotypes (see van Die *et al.*, this volume). It is also possible that incomplete or non-expressing gene clusters are present in the genome. Chromosomal DNA of *E. coli* J96, for example, contains four *Sma*I fragments hybridizing to a *papE-papF*, or a *papG* probe (Lund, unpubl. obs.). A temperature-sensitive replicon carrying the entire *pap* gene cluster with a linker insertion in the *papG* gene was constructed and transformed into *E. coli* J96. Chromosomal plasmid integrates were isolated by selecting for the vector resistance marker (tetracycline) at the restrictive temperature. After growth under non-selective conditions, tetracycline-sensitive clones which had deleted the plasmid DNA and one copy of the gene cluster were isolated. The clones that carried the linker insertion did not agglutinate digalactoside-coated latex beads. These clones in Southern blot analysis still carry multiple *Sma*I fragments hybridizing to a probe representing the adhesin genes, suggesting that the multiple hybridization bands are caused by multiple gene copies, and not by microheterogenicity among the bacteria. Only one of these copies seems to specify digalactoside binding.

SUMMARY

Digalactoside-specific binding of uropathogenic *E. coli* expressing P-fimbriae or Pap pili is mediated by minor components present in a pilus preparation. Of the two proteins associated with the binding phenotype, PapF appears from its primary sequence to be a minor pilin subunit, whereas PapG shows no resemblance to other *E. coli* pilins. The PapE protein that appears to link the adhesin to the pilus also revealed pilin-like features. In two sequenced clones from independent isolates, the PapF protein was structurally very conserved; whereas PapE and especially papG differed significantly in their amino acid sequence. The presence of antibodies specific to PapE and PapF in an antiserum raised against wild-type pili will encourage us to test the possible immunoprophylactic effect of these adhesin proteins against upper urinary tract infections.

ACKNOWLEDGEMENTS

This work was supported by grants from the Swedish Medical Research Council (Dnr 505504769-9), Swedish Natural Research Council (Dnr B-BU 3373-110 and BU 1670), and the Board for Technological Development (Dnr 81-3384B). B.L. was in part supported by the Lennander Foundation. M.B. was supported by grant B85-16P-6893-02B from the Swedish Medical Research Council.

REFERENCES

1. Hagblom, P., E. Segal, E. Billyard and M. So (1985). Intragenic recombination leads to pilus antigenic variation in *Neisseria gonorrhoeae. Nature* **315**, 156-159.
2. Rothbard, J., R. Fernandez, L. Wang, N. Teng and G. Schoolnik (1985). Antibodies to peptides corresponding to a conserved sequence of gonococcal pilins block bacterial adhesion. *Proc. Natl. Acad. Sci. USA* **82**, 915-919.
3. Lindberg, F. P., B. Lund and S. Normark (1984). Genes of pyelonephritic *E. coli* required for digalactoside-specific agglutination of human cells. *EMBO J.* **3**, 1167-1173.
4. Norgren, M., S. Normark, D. Lark, P. O'Hanley, G. Schoolnik, S. Falkow, C. Svanborg Edén, M. Båga and B.-E. Uhlin (1984). Mutations in *E. coli* cistrons affecting adhesion to human cells do not abolish Pap pili fiber formation. *EMBO J.* **3**, 1159-1165.
5. Uhlin, B. E., M. Norgren, M. Båga and S. Normark (1985). Adhesion to human cells by *Escherichia coli* lacking the major subunit of a digalactoside-specific pilus adhesin. *Proc. Natl. Acad. Sci. USA* **82**, 1800-1804.
6. Lundberg, F. P., B. Lund and S. Normark (1986). Gene products specifying adhesion of uropathogenic *Escherichia coli* are minor components of pili. *Proc. Natl. Acad. Sci. USA* **83**, in press.
7. Hacker, J., G. Schmidt, C. Hughes, S. Knapp, M. Marget and W. Goebel (1985). Cloning and characterization of genes involved in production of mannose-resistant, neuraminidase-susceptible (X) fimbriae from a uropathogenic 06:K15:H31 *Escherichia coli* strain. *Infect. Immun.* **47**, 434-440.
8. Maurer, L. and P. Orndorff (1985). A new locus, pil E, required for the binding of Type 1 piliated *Escherichia coli* to erythrocytes. *FEMS Microbiol. Lett.* **30**, 59-66.
9. Svensson, S. B., H. Hultberg, G. Källenius, T. K. Korhonen, R. Möllby and J. Winberg (1983). P-fimbriae of pyelonephritic *Escherichia coli*: Identification and chemical characterization of receptors. *Infection* **II**, 61-67.
10. Hull, R., R. Gill, P. Hsu, B. Minshew and S. Falkow (1981). Construction and expression of recombinant plasmids encoding type 1 or D-mannose-resistant pili from a urinary tract infection *Escherichia coli* isolate. *Infect. Immun.* **33**, 933-938.

11. Normark, S., D. Lark, M. Norgren, M. Båga, P. O'Hanley, G. Schoolnik and S. Falkow (1983). Genetics of digalactoside-binding adhesin from a uropathogenic *Escherichia coli* strain. *Infect. Immun.* **41**, 942-949.
12. Båga, M., M. Görannsson, S. Normark and B.-E. Uhlin (1985). Transcriptional activation of a Pap pilus virulence operon from uropathogenic *Escherichia coli. EMBO J.* **4**, 3887-3893.
13. Lund, B., F. Lindberg, M. Båga and S. Normark (1985). Globoside-specific adhesins of uropathogenic *E. coli* are encoded by similar *trans*-complementable gene clusters. *J. Bacteriol.* **162**, 1293-1301.
14. Clegg, S. (1982). Cloning of genes determining the production of mannose-resistant fimbriae in a uropathogenic strain of *Escherichia coli* belonging to serogroup 06. *Infect. Immun.* **38**, 739-744.
15. van Die, I., C. van den Hondel, H.-J. Hamstra, W. Hoekstra and H. Bergmans (1983). Studies on the fimbriae of an *Escherichia coli* 06:K2:H1:F7 strain: molecular cloning of a DNA fragment encoding a fimbrial antigen responsible for mannose-resistant hemagglutination of human erythrocytes. *FEMS Microbiol. Lett.* **19**, 77-82.
16. Ekbäck, G., S. Mörner, B. Lund and S. Normark (1986). Correlation of genes in the *pap* gene cluster to expression of globoside-specific adhesion by uropathogenic *E. coli. FEMS Microbial. Lett.*, in press.
17. Meyer, T., M. Mlawer and M. So (1982). Pilus expression in *Neisseria gonorrhoeae* involved chromosomal rearrangement. *Cell* **30**, 45-52.

Regulation and Biogenesis of Digalactoside-Binding Pili

Bernt Eric Uhlin[a], Monica Båga, Kristina Forsman, Mikael Göransson, Frederik Lindberg, Björn Lund, Mari Norgren and Staffan Normark

Department of Microbiology, University of Umeå, Umeå, Sweden

INTRODUCTION

Uropathogenic *E. coli* isolates frequently express pili-adhesins that mediate binding to a digalactoside moiety present in glycoconjugates (e.g., the globo series of glycolipids) on human cells. Molecular genetic studies have shown that formation of these pili (Pap pili or P-fimbriae) is mediated by gene clusters containing cistrons for pilus and adhesin components as well as gene products involved in compartmentalization and assembly of subunits [1-3]. Regulation of pili formation probably occurs at different levels. It may be anticipated that gene expression in pilus-adhesin gene clusters is regulated such that production of catalytic and structural components is coordinated. Characterization of the regulatory mechanisms governing pili-adhesin formation is important for our understanding of the biogenesis. We will here discuss regulatory features of the determinant for digalactoside-binding pili in the *E. coli* isolate J96. Nine different cistrons (*papA-I*) have been identified in a gene cluster cloned from this strain (Fig. 1A). Molecular genetic analysis has been used to assess the roles of the gene products in adhesion and pili formation ([1,4,5], see also Normark *et al.*, this volume). Two of the cistrons in the central region of the gene cluster are essential for the appearance of subunit proteins on the bacterial cell surface. The PapC and PapD proteins can presumably interact with different sets of pili and adhesin subunits. Biogenesis of Pap pili and biogenesis of Pap adhesin can be independently manipulated, but in both cases expression of the *papC* and *papD* gene products is required [1,5].

[a] *Telephone:* 46-90-101292

Protein-Carbohydrate Interactions in Biological Systems. ISBN 0 12 436665 1

REGULATORY REGION OF THE *pap* GENE CLUSTER

To determine the level of expression of the different *pap* genes we constructed several operon fusions with the *E. coli lacZ* gene. Since the *papA-G* genes are all transcribed in the same direction (as discussed below, *papI* has the opposite orientation), they could theoretically be expressed as a multicistronic operon. However, the operon fusion studies indicated that there may be internal promoters responsible for at least low-level expression of the *papC-D* and *papE-G* genes (M. Göransson, K. Forsman and B. E. Uhlin, unpubl. obs.). The dominant transcriptional activity was found with fusions of the gene for the major pilin subunit, *papA*. That the region upstream of *papA* contains some regulatory function became evident with the discovery that transcription in *papA-lacZ* fusions is thermoregulated [6]. A detailed characterization of the region upstream of *papA* was therefore initiated and it is clear that it contains several regulatory determinants involved in control of Pap pili synthesis [7].

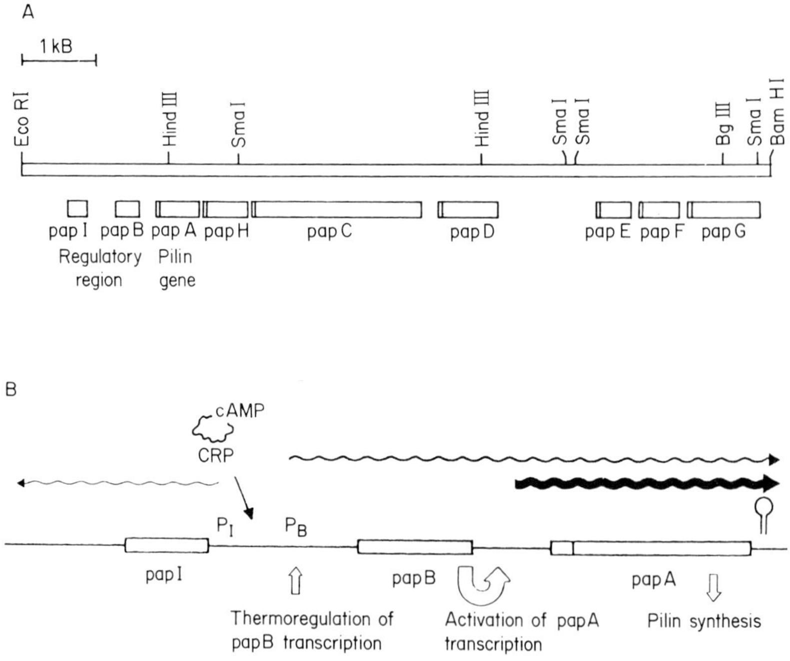

Figure 1. (A) Physical and genetic map of the Pap pili-adhesin determinant cloned from *E. coli* J96. The positions of the nine *pap* cistrons are indicated by horizontal boxes. (B) Transcriptional organization and regulatory features of the region determining PapA pilin synthesis. The horizontal wavy arrows indicate the length and origin of mRNA molecules detected by Northern blot hybridizations. Proposed locations of promoters for the *papI* and *papB* transcripts are indicated by P_I and P_B, respectively. A stem-loop structure represents the transcriptional terminator identified downstream of the *papA* gene. The postulated site of action of the cAMP-CRP complex is shown by the small black arrow.

The transcripts of the *papI*, *papB*, and *papA* genes were identified by Northern blot hybridization with *in vivo* synthesized mRNA and cistron-specific DNA probes [7]. The *papA* gene was primarily represented by an 800-nucleotide-long transcript but appeared also to be cotranscribed with *papB* as a less abundant 1300-nucleotide-long mRNA. Both transcripts presumably terminate at the same site downstream of the *papA* coding sequence. A transcriptional terminator was postulated from DNA sequence data to be located immediately distal to the 3′-end of *papA*. The weakly expressed *papI* gene was transcribed as a 600-nucleotide-long mRNA in the opposite direction to that of *papB* and *papA*. The transcriptional organization of the *papA* regulatory region is outlined in Fig. 1B. From DNA sequencing data we could postulate potential promoter sequences in the intercistronic region between *papI* and *papB*. However, despite the fact that the *papA* gene is the most efficiently transcribed gene, we found no obvious promoter sequence in the region between *papB* and *papA*.

POSITIVE REGULATION OF *papA* TRANSCRIPTION

Inactivation of the *papB* gene by transposon Tn*5* mutagenesis results in a markedly decreased expression of Pap pili antigen [1]. By utilizing *pap-lacZ* operon fusions we could study the role of the *papB* and *papI* genes in expression of the major Pap pili subunit. The transcriptional activity of *papA* as measured by *papA-lacZ* fusions was 50 times greater in the presence of the *papI-papB* region than in its absence. The *del(papI, papB)* construct pHMG11 contains the entire intercistronic region between *papB* and *papA*. Despite the fact that it does carry the region in which the 800-nucleotide-long major *papA* transcript was found to originate, the pHMG11 construct showed virtually no *papA-lacZ* expression. A fusion derivative (pHMG15), with a mutation abolishing the *papB* gene function due to a small insertion in the structural gene, also showed reduced *papA-lacZ* expression. The effect of such mutations therefore suggested that *papB-papI* have some positive regulatory role in the transcription of *papA*. A series of complementation tests provided evidence for this hypothesis. DNA fragments encoding *papB* and/or *papI* were cloned on separate plasmids and subsequently introduced into bacteria carrying the above mentioned *papA-lacZ* fusion

Table 1. Complementation tests

pap *genes on complementing plasmid*	*Relative expression of* β-galactosidase from papA-lacZ *fusion:*[a] *pHMG1* (papI$^+$,papB$^+$)	*pHMG15* (papI$^+$,papB1)	*pHMG11* (del(papI,papB))
None (vector only)	1.0	0.13	0.03
papI$^+$	0.88	0.22	0.03
papB$^+$	0.62	0.69	0.02
papI$^+$, *papB*$^+$	1.42	1.14	0.03

[a]Values are expressed relative to the level of β-galactosidase specific activity from cells carrying pHMG1 and the vector control which was set to 1.0 [7].

constructs. As shown in Table 1, the complementing *papB*$^+$ plasmids could restore expression from the *papB1* mutant pHMG15. Also, in the case of the non-mutant fusion derivative pHMG1 there was a stimulatory effect on expression by the plasmid which carries the regulatory region. The results clearly indicate that the *papB* gene encodes a *trans*-acting positive effector for *papA* transcription. However, a full stimulatory effect seemed to require that the *papI* gene product was also present *in trans*. Similar experiments with *papB-lacZ* operon fusions indicate that both the *papB* and the *papI* gene products can stimulate *papB* transcription. The *del(papI, papB)* derivative pHMG11, on the other hand, did not show any increase in expression upon the introduction of the complementing plasmids. The positively acting *papA* regulatory mechanism thereby seems to require not only the *trans*-acting gene products, but also the *papI-papB* region *in cis*. A possible explanation for the apparent *cis*-dependence could be that regulatory proteins bind to two separate DNA regions and that an interaction between such sites would be required for activation of *papA* transcription. One site would then be located in the *papB-papA* intercistronic region (i.e., where the 800-nucleotide-long *papA* mRNA originates) and the other presumably upstream or within the *papB* gene. Interactions of this kind between separate binding sites for regulatory proteins have recently been proposed in regulation of expression of the *ara* and *gal* operons in *E. coli* [8,9]. However, other explanations for the observed features in the *papA* expression are also plausible. At present, it is not ruled out that the 800-nucleotide-long *papA* transcript could be a processing product coming from partial degradation of the 1300-nucleotide-long mRNA. If so, fully active transcription of *papA* would be *cis*-dependent upon the *papB* promoter region.

REGULATION OF TRANSCRIPTION IN RESPONSE TO GROWTH CONDITIONS

In several cases the expression of pili-adhesins seems to be affected by growth conditions such as temperature and composition of the growth media [10]. Expression of pili and adhesive properties by digalactoside-binding *E. coli* is reduced if glucose is present in the growth medium [11]. To determine if such glucose effects are related to the catabolite repression which regulates bacterial metabolic activity the cloned Pap determinant was analysed with respect to possible involvement of cAMP and its receptor protein (CRP) in *papA* expression. Expression of β-galactosidase by both *papA-lacZ* and *papB-lacZ* transcriptional fusion constructs was clearly reduced if the *crp* or *cya* genes of *E. coli* were inactive (Table 2). In the case of the adenylate cyclase mutant (*cya*$^-$), externally added cAMP could restore expression of the *pap* genes. A binding site for the cAMP-CRP complex could be postulated from DNA sequence data to be localized in the *papI-papB* intercistronic region (Fig. 1B). Presumably cAMP-CRP primarily stimulates transcription of the *papB* gene and thereby indirectly also stimulates transcription of *papA*, since the latter gene requires *papB* for its expression. Through this regulatory circuit the production of pili is coupled to the systems that respond to changes in the metabolic load and energy state of the bacterial cell.

Table 2. Effect of *crp* and *cya* mutations on *pap* gene expression

pap *genotype*	Relative expression of β-galactosidase:[a]		
	crp^-/crp^+	cya^-/cya^+	cya^- (+cAMP)[b]/cya$^+$
$papI^+$,$papB^+$,*papA-lacZ*	0.03	0.05	0.56
$papI^+$, *papB-lacZ*	–	0.29	0.96
papB-lacZ	0.31	0.51	1.06

[a] Ratio between mutant and non-mutant hosts [7].
[b] Addition of 1 mM cAMP to the medium.

For the majority of pili-adhesins there is a temperature-dependent mode of expression such that production is optimal at physiological temperatures (i.e. 37°C) but decreases drastically at lower temperatures. Expression therefore seems optimized under the condition found inside an infected host. Studies with the cloned *pap* DNA established that transcription of the *papA* pilin gene is thermoregulated [6]. Further analysis of the regulatory region by the use of *lacZ* fusions has shown that also the *papB* gene transcription is temperature-dependent (M. Göransson, K. Forsman and B. E. Uhlin, unpubl. obs.). The results suggest that the *papB* promoter region contains the primary determinant involved in this thermoregulation. Similar to the cAMP-CRP stimulation, the temperature effect is therefore presumably transmitted to the level of pili synthesis by virtue of the *papB* activator function.

SUMMARY

Expression of the major Pap pili subunit gene, *papA*, is dependent on several regulatory determinants. The *papB* gene product acts *in trans* in stimulating *papA* transcription and can therefore be regarded as a positive regulator of pili synthesis. Transcription of the *papB* gene is stimulated *in trans* by *papB* itself and by the *papI* gene. Synthesis of Pap pili subunits is influenced by the cellular level of cAMP through cAMP-CRP-mediated stimulation of *papB* transcription. Similarly temperature conditions influence pili synthesis through regulation via the *papB* gene. A thermoregulatory determinant has been localized within the *papB* promoter region. Since clinical isolates often appear to carry more than one gene cluster for pili-adhesins one might ask if there are independent control systems for expression. The detailed characterization of how *papA* gene transcription is regulated will facilitate studies of possible regulatory interactions between separate pili-adhesin gene clusters in a given strain.

ACKNOWLEDGEMENTS

The excellent secretarial assistance of Britt-Inger Strömberg is gratefully acknowledged. The authors' work was supported by grants from the Swedish Natural Science Research Council (project nos. BU 1670 and BU 3373), the Swedish Medical Research Council (project nos. 5428 and

B85-16P-6893), and the Board for Technological Development (project nos. 81-3384B and 84-5463).

REFERENCES

1. Norgren, M., S. Normark, D. Lark, P. O'Hanley, G. Schoolnik, S. Falkow, C. Svanborg-Edén, Båga, M. and B. E. Uhlin (1984). Mutations in *E. coli* cistrons affecting adhesion to human cells do not abolish Pap pili formation. *EMBO J.* **3**, 1159-1165.
2. Uhlin, B. E., M. Båga, M. Göransson, F. Lindberg, B. Lund, M. Norgren and S. Normark (1985). Genes determining adhesin formation in uropathogenic *E. coli*. *Curr. Top. Microbiol. Immunol.* **188**, 163-178.
3. Normark, S., M. Båga, M. Göransson, F. P. Lindberg, B. Lund, M. Norgren and B. E. Uhlin (1986). Genetics and biogenesis of *Escherichia coli* adhesins. *In* "Microbial Lectins and Agglutinins: Properties and Biological Activity (Ed. D. Mirelmann). John Wiley & Sons, Inc. New York. In press.
4. Lindberg, F. P., B. Lund and S. Normark (1984). Genes of pyelonephritic *E. coli* required for digalactoside-specific agglutination of human cells. *EMBO J.* **3**, 1167-1173.
5. Uhlin, B. E., M. Norgren, M. Båga and S. Normark (1985). Adhesion to human cells by *Escherichia coli* lacking the major subunit of a digalactoside-specific pilus adhesin. *Proc. Natl. Acad. Sci. USA* **82**, 1800-1804.
6. Göransson, M. and B. E. Uhlin (1984). Environmental temperature regulates transcription of a virulence pili operon in *E. coli*. *EMBO J.* **3**, 2885-2888.
7. Båga, M., M. Göransson, S. Normark and B. E. Uhlin (1985). Transcriptional activation of a Pap pilus virulence operon from uropathogenic *E. coli*. *EMBO J.* **4**, 3887-3893.
8. Dunn, T. M., S. Hahn, S. Ogden and R. F. Schleif (1984). An operator at −280 base pairs that is required for repression of *araBAD* operon promoter: Addition of DNA helical turns between the operator and promoter cyclically hinders repression. *Proc. Natl. Acad. Sci. USA.* **81**, 5017-5020.
9. Majumbar, A. and S. Adhya (1984). Demonstration of two operator elements in *gal*: *In vitro* repressor binding studies. *Proc. Natl. Acad. Sci. USA* **81**, 6100-6104.
10. Gaastra, W. and F. K. de Graaf (1982). Host-specific fimbrial adhesins of noninvasive enterotoxigenic *Escherichia coli* strains. *Microbiol. Rev.* **46**, 129-161.
11. Svanborg-Edén, C. and H. A. Hansson (1978). *Escherichia coli* pili as possible mediators of attachment to human urinary tract epithelial cells. *Infect. Immun.* **21**, 229-237.

Genetics and Biogenesis of the K88ab and K99 Fimbrial Adhesins

Frits R. Mooi[a], Bert Roosendaal,
Bauke Oudega and Frits K. de Graaf

Department of Molecular Microbiology, Biology Laboratory, Free University, Amsterdam, The Netherlands

INTRODUCTION

The K88ab and K99 fimbrial adhesins are extracellular filamentous proteins found on enterotoxigenic *Escherichia coli* strains, and involved in the colonization of the small intestine of domestic animals (see [1] for a recent review). Both adhesins are encoded by plasmids, and their production is dependent on the growth temperature. The two fimbrial adhesins differ in a number of features. The K88ab adhesin consists of very thin and flexible structures, while the K99 adhesin is thicker and more rigid. Furthermore, the molecular weight of the subunit constituting the major component of the K88ab adhesin (i.e. 27 564) is much larger than the molecular weight of the major subunit of the K99 adhesin (i.e. 16 703). The K88ab adhesin is involved in colonization of the small intestine of piglets, while the K99 adhesin is less host-specific, and involved in colonization of the intestine of piglets, calves and lambs. K88 adhesins have been found to exist in multiple antigenic variants (i.e. K88ab, K88ac and K88ad), which have probably evolved to escape the immune response of the host. No antigenic variants of the K99 adhesin have been described.

STRUCTURE AND REGULATION OF THE K88ab and K99 GENE CLUSTERS

The K88ab and K99 gene clusters have been isolated by molecular cloning, and genetic maps have been derived for the cloned DNA fragments (Fig. 1) [2,3].

[a]*Telephone:* 31-20-54470

Protein-Carbohydrate Interactions in Biological Systems. ISBN 0 12 436665 1

These genetic maps revealed that, like most adhesin gene clusters analysed to date, the K88ab and K99 gene clusters contain three consecutive, similar, structural genes. The first gene (*C*) codes for a fimbrial subunit, the second (*D*) for a large outer membrane protein, and the third (*E*) for a polypeptide located in the periplasmic space. Analyses of the amino acid sequences of the gene *D* and *E* products revealed that the corresponding K88ab and K99 polypeptides show significant homology (unpubl. res.). Another similarity observed between the two gene clusters is the occurrence of a region of dyad symmetry located between the genes *C* and *D* (Figs 1 and 2).

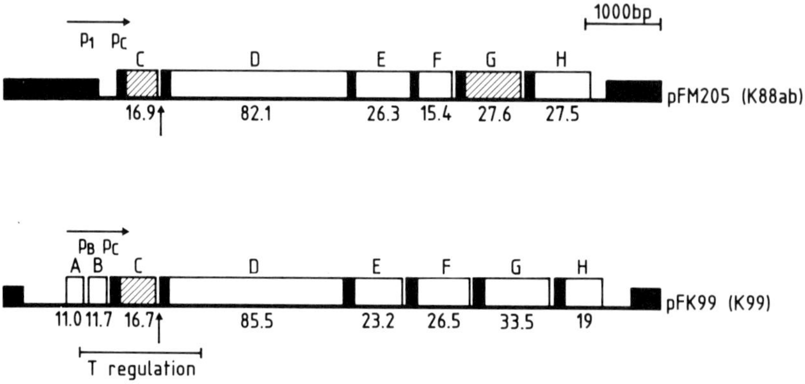

Figure 1. Genetic organization of DNA regions involved in biogenesis of the K88ab and K99 fimbrial adhesins. The locations of the various structural genes have been indicated by boxes. The black ends of boxes indicate parts of genes coding for signal peptides. Fimbrial subunit genes have been shaded. The numbers below the structural genes refer to the molecular weights ($\times 10^{-3}$) of the corresponding polypeptides. The vertical arrows indicate regions with dyad symmetry. A region involved in temperature-dependent regulation of K99 fimbrial adhesin production has been indicated. Horizontal arrows indicate the direction of transcription. The black boxes indicate pBR322 DNA. P, promoter sequence; bp, base pair.

The K88ab gene cluster contains two fimbrial subunit genes *faeC* and *faeG* [4] (fae, *f*imbrial *a*dhesin *e*ighty-eight). One is located at the 5′-end of the cluster, at the same relative position as the K99 fimbrial subunit gene *fanC* (fan, *f*imbrial *a*dhesin *n*inety-nine), and is expressed at a very low level. The other K88ab fimbrial subunit gene (*faeG*) is located at the 3′-end of the cluster, expressed at a high level, and codes for the major component of the K88ab fimbrial adhesin. There is evidence that *faeC* is preceded by a weak promoter. No other promoters have been located on the cloned K88ab DNA.

Two possible promoters have been identified in the K99 gene cluster (Pb and Pc, Fig. 1) ([5], and unpubl. res.). Pb probably represents an efficient promoter, because its sequence complies well with the consensus sequence of strong *E. coli* promoters, while the sequence of Pc suggests it is a weak promoter. The *fanA*

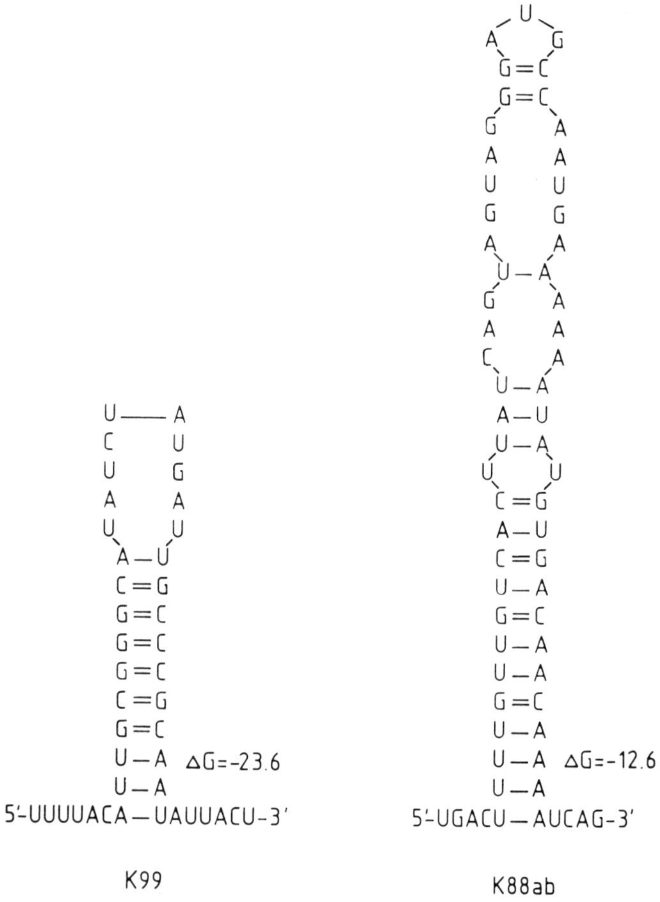

Figure 2. Possible secondary structures formed by K99 and K88ab mRNA. The DNA regions coding for these structures are indicated by vertical arrows in Fig. 1.

and *fanB* products have not yet been detected in minicells, presumably because they are produced in very low amounts. The primary structure of these polypeptides, deduced from the DNA sequence, reveals that they are the only K99 polypeptides not synthesized with a signal peptide (unpubl. res.). Furthermore, the two polypeptides are predicted to be relatively hydrophilic. Taken together, these observations suggest that the two polypeptides are located in the cytoplasmic space. The DNA region involved in temperature-dependent regulation of K99 adhesin production has been located at the 5′-end of the gene cluster (Fig. 1) (unpubl. res.). In addition to the fimbrial subunit gene, this DNA region contains *fanB*, the promoters Pb and Pc and the region of dyad symmetry. Operon fusions have shown that the temperature-dependent regulation within this region occurs at the level of transcription. The *fanA* product is apparently not required for temperature-dependent regulation. It probably acts as a positive regulator, since mutations in *fanA* result in a three-fold reduction in K99 adhesin production.

ROLE OF THE GENE *D* AND *E* PRODUCTS IN THE BIOGENESIS OF THE K88ab AND K99 FIMBRIAL ADHESINS

Fimbrial subunits belong to the very few proteins of *E. coli* that are transported across both the inner and outer membranes. We have shown that fimbrial subunits accumulate transiently in the periplasmic space before they are transported across the outer membrane [6] (Fig. 3). No evidence has been obtained that K88ab or K99 encoded polypeptides are required to transport fimbrial subunits across the inner membrane, and we presume that this transport occurs via the normal secretory pathway of periplasmic proteins. Transport of K88ab fimbrial subunits

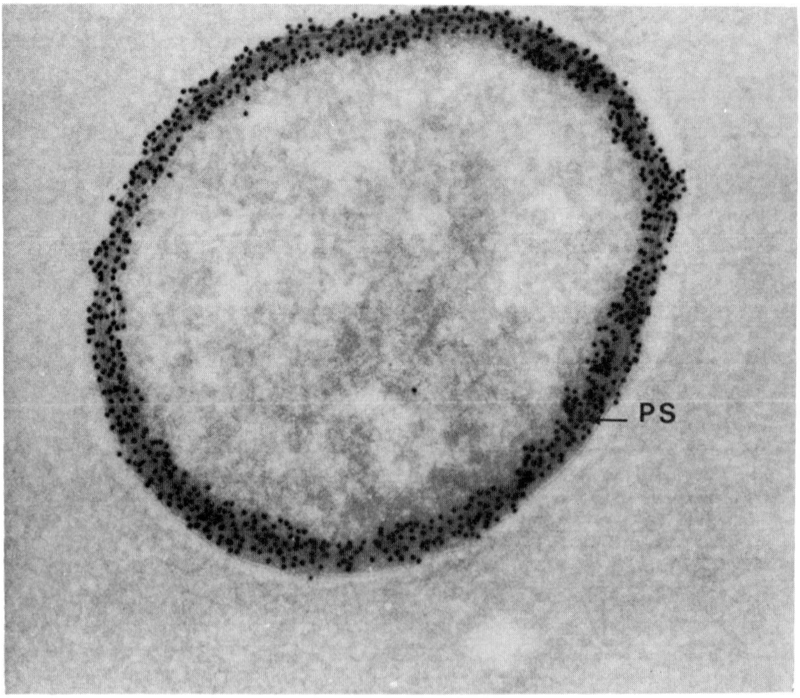

Figure 3. Localization of the K99 fimbrial subunit by immuno-electron microscopy. The electron micrograph shows a cross-section of an *E. coli* cell, in which K99 fimbrial subunits have been visualized with colloidal gold attacked to IgG. K99 fimbrial subunits appear as black spots on the electron micrograph. Note the presence of fimbrial subunits in the periplasmic space (PS).

across the outer membrane requires the presence of at least two K88ab DNA encoded polypeptides, encoded by the *D* and *E* genes. As mentioned above, similar, probably homologous, genes are observed in all fimbrial gene clusters analysed to date, and it seems likely that they code for the polypeptides with identical functions.

The K88ab and K99 gene *D* products are located in the outer membrane ([7], unpubl. res.), and it has been shown that mutations in *faeD* result in the accumulation of the large K88ab fimbrial subunit in the periplasmic space where it is found associated with the *faeE* product [6]. We presume that the gene *D* product is a trans-membrane protein that recognizes fimbrial subunits associated with the gene *E* product at the periplasmic side of the outer membrane, transports the subunits across the outer membrane, possibly through pores specific for fimbrial subunits, subsequently initiates polymerization of the subunits, and finally anchors the fimbria to the cell (Fig. 4).

The K88ab and K99 gene *E* products are located in the periplasmic space, and it appears that the large K88ab fimbrial subunit associates with the *faeE* product before it is transported across the outer membrane (Fig. 4) [6]. Mutations in the *E* genes generally result in the degradation of fimbrial subunits, suggesting that association with the gene *E* product affects the conformation of the fimbrial subunit, rendering it resistant to proteolysis. In view of the instability of the fimbrial subunit in the absence of the gene *E* product, it seems likely that the association between the two polypeptides occurs within a very short span of time after the fimbrial subunit has passed the inner membrane. This may be accomplished by a high concentration of the gene *E* product in the proximity of the inner membrane. In this context an interesting feature of both the K88ab and K99 gene *E* products is worth mentioning. The primary structure of these polypeptides has revealed the presence of regions enriched for basic amino acids and devoid of acidic amino acids (unpubl. res.). It seems plausible that these positively charged regions may facilitate the association of the gene *E* products with the negatively charged phospholipids of the inner (and outer) membrane. Association of the fimbrial subunit with the gene *E* product could decrease its affinity for the inner membrane, allowing the complexes composed of the fimbrial subunit and the gene *E* product to diffuse to the outer membrane, where they may attach to the gene *D* product. What functions for the gene *E* product may be envisaged? It seems unlikely that the gene *E* product functions only as a shuttle for fimbrial subunits between the inner and outer membranes. We suppose that the gene *E* product induces or stabilizes a particular conformation of the fimbrial subunit, which might be required for several reasons. First, as is shown for phage tail proteins and flagellin subunits [8], structural proteins are often synthesized in a form that does not spontaneously assemble to prevent the formation of a polymeric structure at the wrong site. The function of the gene *E* products could be to induce or stabilize such a form of the subunit, to prevent the fimbrial subunits from premature polymerization in the periplasmic space. Second, a particular conformation of the fimbrial subunit might be required to transport it through pores formed by the gene *D* product. For example, this transport might require a relatively unstructured conformation of the fimbrial subunit. Third,

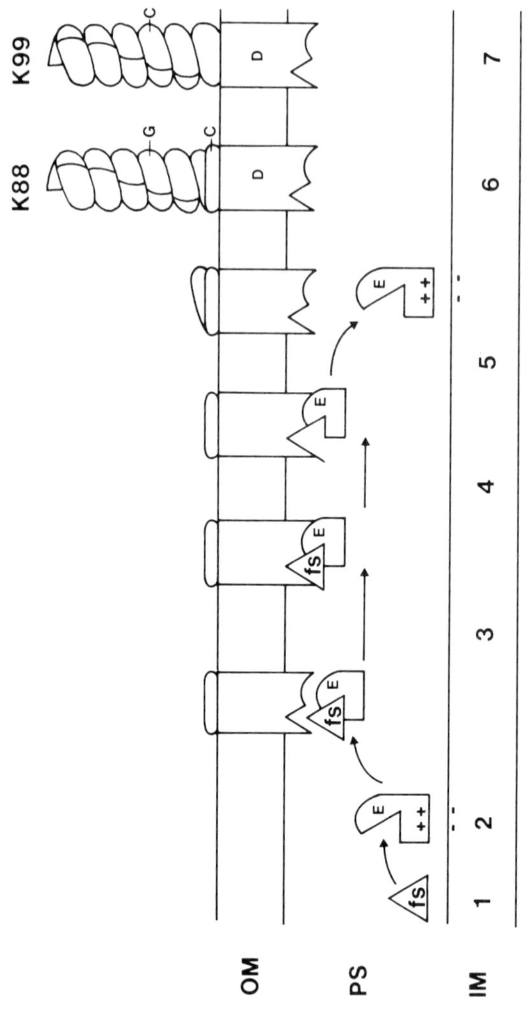

Fig. 4. Subcellular location and putative functions of some K88ab and K99 gene products. (1) Fimbrial subunits are transported across the inner membrane following the normal secretory pathway of periplasmic proteins. (2) Within a very short span of time after entering the periplasmic space, the fimbrial subunit associates with the gene *E* product. The latter contains stretches of positively charged amino acid residues which may function to attach the polypeptide to the negatively charged phospholipids of the inner membrane. (3) Association of the fimbrial subunit with the gene *E* product results in its release from the inner membrane, and the complex composed of the two proteins diffuses to the outer membrane where it is bound by the gene *D* product. (4) and (5) After the fimbrial subunit is translocated across the outer membrane, the gene *E* product is released from the outer

this particular conformation might be necessary to deliver the energy for transport and assembly. If the fimbrial subunit folds into a strained and energetically unfavourable conformation in the periplasmic space, some of the energy built into the protein during the polymerization of amino acids might be retained. The conformational energy thus built into the complexes composed of the fimbrial subunits and the gene *E* products might be used to drive transport and assembly.

IS THE K88ab FIMBRIAL ADHESIN COMPOSED OF TWO FIMBRIAL SUBUNITS?

The K88ab gene cluster contains two fimbrial subunit genes, *faeC* and *faeG*. We have not been able to locate the *faeC* product because it is produced in very small amounts. However, the fact that mutations in *faeC* abolish fimbria formation, and result in the intracellular accumulation of the *faeG* product [4] suggests that both fimbrial subunits are part of the same structure. As mentioned above, we presume that the *faeD* product serves to anchor the fimbria to the cell, and it is interesting to note that *faeC* and *faeD* form a translationally coupled gene pair (i.e. efficient translation of *faeD* requires prior translation of *faeC*: unpubl. res.). Since it is thought that translational coupling is a means to ensure a proper molar production of polypeptides that are part of the same structure [9], it seems likely that the *faeC* and *faeD* products form a complex in the outer membrane (Fig. 3).

SUMMARY

Like most fimbrial gene clusters analysed to date, the K88ab and K99 gene clusters carry three consecutive, similar, structural genes. The first (gene *C*) codes for a fimbrial subunit, the second (gene *D*) for an outer membrane protein, and the third (gene *E*) for a periplasmic protein. The primary structure of the polypeptides encoded by the *D* and *E* genes has revealed that the corresponding K88ab and K99 polypeptides share significant homology. The two polypeptides are probably required to transport fimbrial subunits across the outer membrane, and anchor the fimbriae to the cell.

The K88ab gene cluster contains two fimbrial subunit genes. One (*faeC*) is located at the 5′-end of the cluster, at the same relative position as the K99 fimbrial subunit gene, and is expressed at a very low level. The other fimbrial subunit gene (*faeG*) is located at the 3′-end of the gene cluster, and codes for the major

---

membrane, and regains its affinity for the inner membrane. (6) The K88ab fimbrial adhesin is composed of two fimbrial subunits, encoded by *faeC* and *faeG*, respectively. The *faeC* product might be part of the basal structure of the fimbria, whereas the *faeG* product constitutes the major component of the filamentous structure, and probably contains the receptor binding domain. (7) No evidence has yet been found for the presence of more than one subunit in the K99 fimbrial adhesin. Letters used to label the various polypeptides correspond to those used in Fig. 1. fs, fimbrial subunit; OM, outer membrane; IM, inner membrane; PS, periplasmic space.

subunit of the K88ab fimbrial adhesin. There is evidence that the *faeC* product is also a (minor) component of this structure.

REFERENCES

1. Mooi, F. R. and F. K. de Graaf (1985). Molecular biology of fimbriae of enterotoxigenic *Escherichia coli*. *Current Topics Microbiol. Immunol.* **118**, 119-138.
2. Mooi, F. R., C. Wouters, A. Wijfjes and F. K. de Graaf (1982). Construction and characterization of mutants impaired in the biosynthesis of the K88ab anigen. *J. Bacteriol.* **150**, 512-521.
3. De Graaf, F. K., B. E. Krenn and P. Klaasen (1984). Organization and expression of genes involved in the biosynthesis of K99 fimbriae. *Infect. Immun.* **43**, 508-514.
4. Mooi, F. R., M. van Buuren, E. Roosendaal and F. K. de Graaf (1984). A K88ab gene of *Escherichia coli* encodes a fimbria-like protein distinct from the K88ab fimbrial adhesin. *J. Bacteriol.* **159**, 482-487.
5. Roosendaal, E., W. Gaastra and F. K. de Graaf (1984). The nucleotide sequence of the gene encoding the K99 subunit of enterotoxigenic *Escherichia coli*. *FEMS Microbiol. Lett.* **22**, 253-258.
6. Mooi, F. R., A. Wijfjes and F. K. de Graaf (1983). Identification and characterization of mutants impaired in the biosynthesis of the K88ab fimbria of *Escherichia coli*. *J. Bacteriol.* **154**, 41-49.
7. Van Doorn, J., B. Oudega, F. R. Mooi and F. K. de Graaf (1982). Subcellular location of polypeptides involved in the biosynthesis of K88ab fimbriae. *FEMS Microbiol. Lett.* **13**, 99-104.
8. King, J. (1980). Regulation of structural protein interactions as revealed in phage morphogenesis. In "Biological Regulation and Development" (Ed. R. F. Goldberger), Vol. 1, pp. 101-128. Plenum Press, New York.
9. Das, A. and C. Yanofsky (1984). A ribosome binding site is necessary for efficient expression of the distal gene of a translationally-coupled gene pair. *Nucl. Acids. Res.* **12**, 4757-4768.

Two Modes of Control of *pilA*, the Gene Encoding Type 1 Pilin in *Escherichia coli*

Paul E. Orndorff[a], P. A. Spears,
David Schauer and Stanley Falkow

Department of Microbiology, Pathology and Parasitology, School of Veterinary Medicine, North Carolina State University, Raleigh, NC 27606, USA, and Department of Medical Microbiology, Stanford University, Stanford, CA 94305, USA

Type 1 pili of *Escherichia coli* are filamentous, proteinaceous structures about 5 nm wide and 2 μm long composed of a repeating 17 kilodalton (kD) monomer, pilin [1]. Type 1 pili are required for the binding of *E. coli* to a variety of eukaryotic cells [2] via a mannose-sensitive interaction with a receptor on the surface of the eukaryotic cell.

The comparatively few genes involved in type 1 pili production [3] and the facility with which pilus polymerization takes place under defined conditions [3] make type 1 pili a useful model for studying the assembly of extracellular supramolecular structures. Also, the genetic regulation of type 1 pili expression poses interesting problems, both in the area of metastable gene expression [4-6] and in the area of coordinated regulation of gene expression [6,7]. In this report, we describe two regulatory mechanisms involved in the control of transcription of *pilA*, the gene encoding the structural subunit of type 1 pili.

GENETIC BACKGROUND

The genes responsible for the production of type 1 pili are located at approximately 98 min on the *E. coli* chromosomal map [6,8]. Six of the genes involved in piliation (Fig. 1) have been identified [3], mapped [3,6] and general functions ascribed to individual gene products [3].

[a] *Telephone:* 1-919-829-4200 Ext. 343

Protein-Carbohydrate Interactions in Biological Systems. ISBN 0 12 436665 1

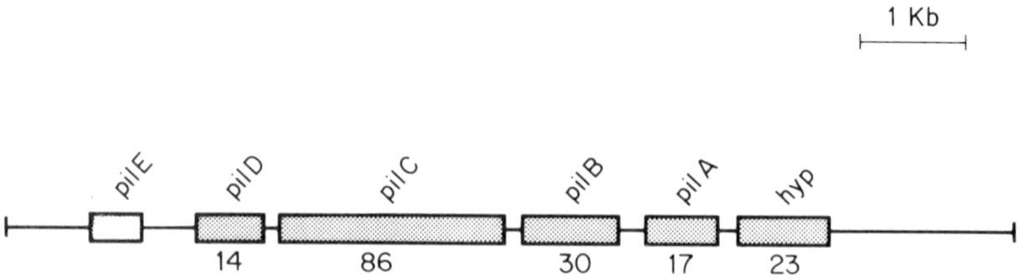

Figure 1. Genetic organization of the Pil region. The boxes represent genes involved in type 1 piliation. Below each box is listed the molecular weight of its protein product in kilodaltons (kd). An open box designates *pilE* locus. The open box is strictly representational; no gene product has been associated with this region.

The products of genes *pilA*, *pilB* and *pilC* are involved in pilus polymerization; the products of *pilB* and *pilC* are required for the polymerization of the *pilA* gene product, pilin, to form pili. Mutants that have lesions in either *pilB* or *pilC* abolish the formation of polymerized pili in minicells but do not effect synthesis or processing of pilin [3]. Mutations in *pilA* completely abolish pilus formation [9] and the *pilA* gene has been sequenced and shown to encode the pilin polypeptide [9].

Lesions which define *pilE* and map to *pilD* result in cells that produce pili that are morphologically and antigenically identical to the parental pili but fail to agglutinate guinea pig erythrocytes (Maurer and Orndorff, submitted). The *pilE* locus, currently defined by a single mutation, may specify a trans-acting product required for haemagglutination or may be required for *pilD* expression. The *pilD* gene product is a 14 kd envelope protein [3] that is immunoprecipitated with antibody to purified pili suggesting a possible association of the 14 kd protein with pili [3]. At present it is not clear how the *pilD* gene product and the *pilE* locus function to facilitate haemagglutination. However, since pili are also required for haemagglutination [3; Maurer and Orndorff, submitted], we assume that the product of the *pilD* gene and the putative product of the *pilE* locus act either directly, to mediate binding of piliated cells to erythrocytes or indirectly, to modify pilin in such a way so as to make pili capable of binding.

The *hyp* gene produces a 23 kD protein that is not required for pilus assembly or receptor specificity but is involved in regulating the level of piliation [6,7]. Mutants having lesions in *hyp* can have as much as 40-fold more pili than parental strains [7]. Some of our experiments designed to discover how the *hyp* gene product acts to repress piliation are discussed below. In addition, a second mechanism of control of piliation is described. The latter control mechanism is a metastable (on or off) type of regulation. Experiments designed to determine possible relationships between the two mechanisms are also described.

EFFECT OF THE *hyp* GENE PRODUCT ON *pilA* TRANSCRIPTION

Mutants that have Tn5 insertions in the *hyp* gene are hyper-piliated [7]. To better define how and where the *hyp* gene product acted, we constructed two subclones of the Pil region. Both subclones contained the putative promotor region for *pilA* and the sequence encoding the amino-terminal portion of the *pilA* gene with the genes for lactose utilization (*lacZYA*) inserted into the *Pvu*II site in the *pilA* gene (Fig. 2). However, the plasmids differed in the orientation of the promotorless *lacZ* gene. In pORN116, *lacZ* was oriented in the reading frame of *pilA*, in pORN121 it was oriented in the opposite direction (toward the β-lactamase gene in pBR322) (Fig. 2). The level of β-galactosidase activity conferred by the plasmids was measured in cells harbouring a compatible plasmid that contained the Pil region with or without a functional *hyp* gene (Fig. 2). The *hyp* gene product repressed the transcription of the *pilA* gene as evidenced by the approximately five-fold lower level of β-galactosidase activity in strains containing

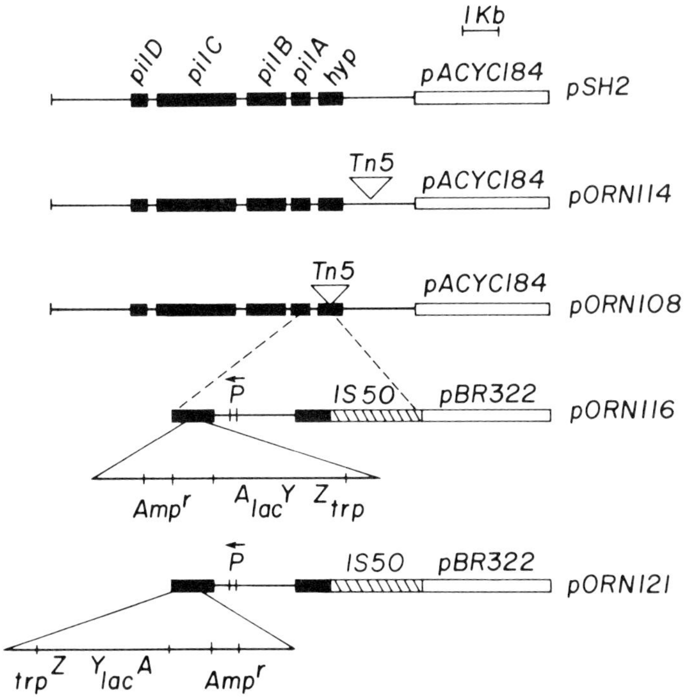

Figure 2. Plasmids used to determine the effect of the *hyp* gene product on *pilA* expression. pSH2 is the parental plasmid; the dark boxes represent the genes involved in pilus biosynthesis. pORN114 has the parental phenotype but has a Tn*5* insertion outside the Pil region. pORN108 contains a Tn*5* insertion mutation in *hyp*. pORN116 and pORN121 were constructed from pORN108 and contain the amp^r lacZYA region from pMC81 [10] introduced into the *pilA* coding region *in vitro* by taking advantage of a *Pvu*II site in *pilA*. pORN116 and pORN121 are identical except for the orientation of the *lac* region. The putative promotor region of *pilA* and the direction of transcription of *pilA* are designated by a "P" and a small arrow respectively.

Table 1. Effect of a *hyp* mutation on *pilA* transcription[a]

Plasmid[c]	Beta-galactosidase activity[b] in combination with: pORN116	pORN121
pACYC184 (hyp^-)	1.97	2.20
pORN114 hyp^+	0.35	1.86
pORN103 hyp^-	1.73	2.07

[a] Beta-galactosidase assays were done using logarithmic phase cells containing the plasmid combinations shown. The plasmids were introduced into strain ORN103 (Δ*pil* Δ(*argF-lac*)U169 *recA*13) by transformation with selection for the appropriate antibiotic resistances to keep both plasmids resident.

[b] Beta-galactosidase activity is shown as relative activity per cell as measured against a reference population of ORN103 containing pACYC184 and pBR322. pORN116 and pORN121 are shown diagrammatically in Figure 1. One unit of beta-galactosidase activity in this table is equivalent to *c*. 167 Miller units.

[c] The *hyp* genotype is listed next to each plasmid. The pACYC184 plasmid is the vector used for pORN114 and pORN103. Consequently, the *hyp* genotype for pACYC184 is listed parenthetically. pORN114 contains a Tn*5* insertion to the right of *hyp* and confers a parental (Pil^+) phenotype. pORN103 contains a Tn*5* insertion in *hyp* and also contains a *c*. 246 bp deletion in the putative *pilA* promotor region.

a functional *hyp* gene (Table 1). Also, we tentatively infer from the results that the repression effected by *hyp* was specific to the putative *pilA* promotor region since the *hyp* gene product had no effect on transcription promoted in the direction opposite to the *pilA* promotor (Table 1).

THE METASTABLE REGULATION OF *pilA* EXPRESSION

The expression of type 1 pili in *E. coli* is metastable [11]. Piliated cells give rise to non-piliated cells, which give rise to piliated cells. The switch from piliated to non-piliated states, and vice versa, occurs at a frequency seemingly too high to be accounted for by spontaneous mutation. Recent experiments by Eisenstein [4], using Mu(Ap^r *lac*) fusion methodology, suggested that at least one of the genes involved in piliation is transcriptionally regulated in cis [5] in a metastable manner. However, the gene (or genes) subject to this regulation was not identified. We have constructed, an *in vitro*, *pilA'-lacZ* fusion and found that after introduction of this fusion into the chromosome, *pilA* is subject to metastable expression. Our experiments also show that the element responsible for the metastable expression of *pilA* is closely linked to *pilA*.

CONSTRUCTION AND INTRODUCTION OF *pilA'-kan^r-lacZYA* FUSIONS INTO THE Pil REGION OF THE *E. COLI* K-12 CHROMOSOME

The *pilA'-kan^r-lacZYA* fusions described here were constructed on recombinant plasmids as shown in Fig. 3. *Sal*I-cleaved pORN129 and pORN130 were each

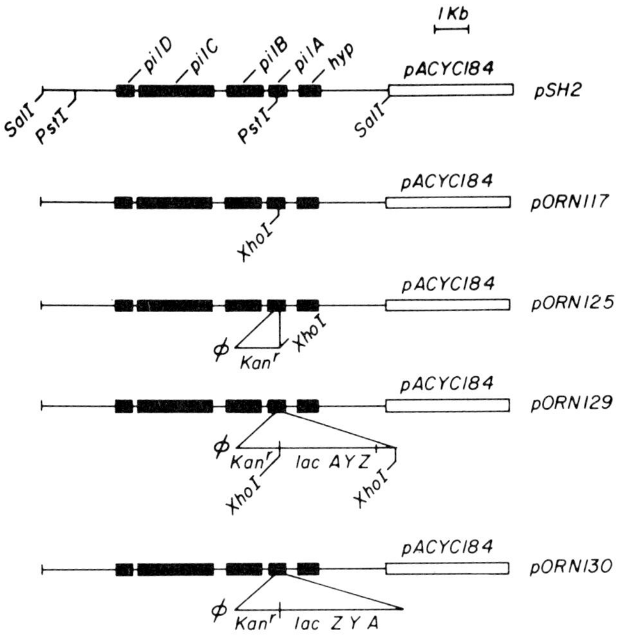

Figure 3. Plasmids used to construct *pilA'-kan^r-lacZYA* fusions. pSH2 is the parental plasmid. pORN117 was constructed by partial digestion of pSH2 with *Pst*I and the addition of a synthetic *Xho*I linker at the *Pst*I site in *pilA*, as described previously [9]. pORN125 was constructed by inserting the *c.* 2.4 kb *Xho*I-*Sal*I fragment of Tn*5* containing the kanamycin (neomycin) phosphotransferase gene into *Xho*I-cut ORN117. The *Xho*I-*Sal*I ligation used to construct pORN125 reproduced the original *Xho*I restriction endonuclease site at one end and a null site at the other end (*Xho*I and *Sal*I-cut DNA have identical overlapping ends). pORN129 was constructed by inserting a *Hin*dIII-*Bam*H1 DNA fragment containing the *lacZYA region* [10]. The *Hin*dIII-*Bam*H1 segment was given *Xho*I ends prior to ligation with *Xho*I-cut pORN125. pORN129 contains the *lacZYA* region inserted into *pilA* as shown (with *lacZ* oriented in the direction of transcription of *pilA*). pORN130 is the same as pORN129, except that the *lacZYA* region is oriented in the opposite direction.

introduced, by transformation, into strain ORN105 (Fig. 4). The *recBC sbcB* phenotype of ORN105 allowed transformation with linear DNA (as described in [6]) and the chromosomal Pil region of ORN105 provided homology for the introduction of the Pil region contained on the linear plasmid DNA. The double cross-over event depicted in Fig. 4 resulted in Kan^r Cam^s Pil$^-$ transformants which were shown, by a variety of genetic [6] and physical techniques (our unpubl. res.), to contain the *kan^r-lacZYA* region inserted as shown for strain ORN110 in Fig. 4.

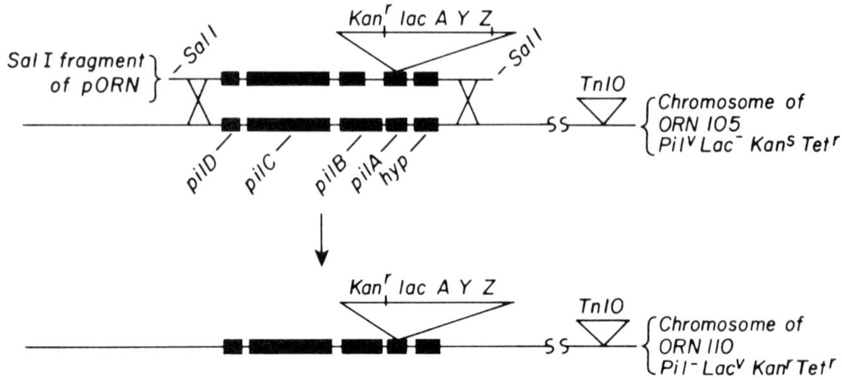

Figure 4. The incorporation of the *pilA'-kan^r-lacZYA* fusion in pORN129 into the chromosomal Pil region. The double cross-over event depicts the formation of strain ORN110 and the resultant phenotype. The phenotypic designation Pil^v and Lac^v are used in this figure to denote variability of expression and are used to emphasize that the variable character of wild-type piliation in ORN105 was observed for the Lac phenotype of the transformant ORN110. The location of Tn*10* is between *hsd* and *serB* on the *E. coli* K-12 chromosome (*c.* 99 min). The directional order of the Pil genes relative to the chromosome is unknown and depicted arbitrarily as going counter-clockwise, *pilA* to *pilD*. The *pilA'-kan^r-lacZYA* fusion of pORN130 was introduced into the ORN105 chromosome in the same manner as for pORN129.

SPONTANEOUS VARIATION IN *pilA* EXPRESSION

Transductants receiving the fusions derived from either pORN129 or pORN130 were found to have different Lac phenotypes. Transductants containing the *pilA'-kan^r-lacZYA* fusion in the chromosome, where the *lacZ* gene was in the same orientation as the *pilA* gene (e.g., strain ORN110, Fig. 4) were Lac variable. That is, a given Lac$^+$ clone of ORN110 could give rise to progeny clones having a Lac$^-$ phenotype, and those Lac$^-$ clones could give rise to clones having a Lac$^+$ phenotype (Fig. 5). The frequency and magnitude of the change from Lac$^+$ to Lac$^-$ phenotypes and vice versa for strain ORN110 is shown in Table 2. In contrast, strain ORN111, which had the *lacZ* reading frame oriented in the opposite direction from *pilA*, showed very low beta-galactosidase activity and exhibited no switching to Lac$^+$ (Table 2).

The putative genetic element responsible for the switching from *pilA* "ON" (transcribed and therefore Lac$^+$) to *pilA* "OFF" (untranscribed, Lac$^-$) and vice versa was greater than 90% cotransducable with *pilA* [6]. Phage P1 grown on a *pilA* ON (Lac$^+$) variant population of ORN110 or a *pilA* OFF (Lac$^-$) variant population of ORN110 conferred the same ON or OFF phenotype to transductants regardless of the phenotype (Pil$^+$ or Pil$^-$) of the recipient strain.

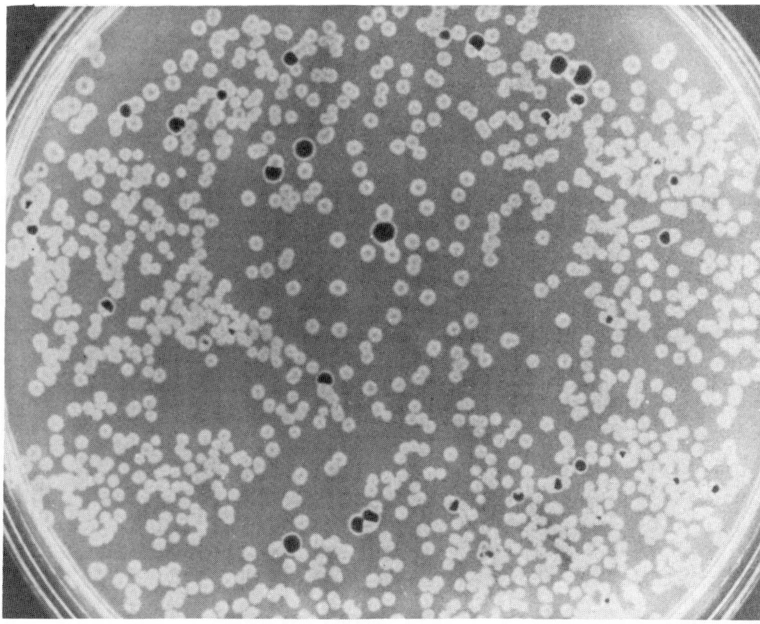

Figure 5. Lac phenotypic variation in ORN110. A single Lac^+ colony of ORN110 was picked and grown for *c.* 6 h in L-broth and dilutions plated on lactose tetrazolium agar containing kanamycin. The Lac^- colonies are dark, and the Lac^+ colonies are light.

One genetic element closely linked to *pilA* and known to affect the level of piliation was *hyp* (see above). We asked if a lesion in *hyp* influenced the metastable transcription of *pilA*. (Cloned segments of the Pil region on multicopy plasmids do not exhibit metastable regulation [6,7]). Experimentally, this question was posed by introducing a mutation into the chromosomal *hyp* gene and observing the effect on *pilA* transcription, as measured by using a chromosomal *pilA'-kan^r-lacZYA* fusion.

THE EFFECT OF A *hyp* MUTATION ON THE METASTABLE REGULATION OF *pilA* EXPRESSION

Mutants containing a *pilA'-kan^r-lacZYA* fusion and a *hyp*::Tn*5*-132 mutation were constructed by exchanging the Tn*5* insertion in the *hyp* gene in strain ORN113 [6] for Tn*5*-132 [12]. Tn*5*-132 contains the tetracycline resistance gene in place of the central *Bgl*II fragment of Tn*5* [12], thus replacing the gene encoding kanamycin phosphotransferase. The exchange of Tn*5* for Tn*5*-132 produced strain ORN114, which was then used to introduce the *hyp*::Tn*5*-132 mutation into strain ORN117 by P1 transduction, producing strain ORN118 (*pilA'-kan^r-lacZYA hyp*::Tn*5*-132) (Fig. 6).

Table 2. Effect of the orientation of *lacZ* in *pilA* on the observed metastable expression of β-galactosidase activity[a]

Strain[b]	*Rate of Lac⁺ to Lac⁻ variation*[c]	*Rate of Lac⁻ to Lac⁺ variation*[d]	*β-galactosidase activity in phenotypically Lac⁺ and Lac⁻ populations*[e]	
			Lac⁺	*Lac⁻*
ORN110	4.0×10^{-4}	6.2×10^{-4}	0.917	0.067
ORN111	no variants observed	no variants observed	0.004	none observed

[a] Metastable expression of the Lac⁺ phenotype was observed by plating strain ORN110 or ORN111 on lactose MacConkey plates and observing colony colour. Single Lac⁺ or Lac⁻ colonies were inoculated into tubes containing L-broth and the initial ratios of Lac⁺ or Lac⁻ phenotypes determined by plate counts on indicator agar. Cells were diluted to give 0 to 1 cell per tube and grown for 8 h, at which time viable counts were determined on indicator agar. These determinations were repeated over 40 times to obtain the rates shown.

[b] ORN110 contains the *lacZ* gene oriented in the direction of transcription of *pilA*. ORN111 contains the *lacZ* gene oriented opposite to the direction of transcription of *pilA*.

[c] The rate of Lac⁺ to Lac⁻ variation is expressed as Lac⁻ variants per cell per generation and represents an average value taken over 367 generations.

[d] The rate of Lac⁻ to Lac⁺ variation is expressed as Lac⁺ variants per cell per generation and represents an average value taken over 454 generations.

[e] Beta-galactosidase activity was measured on logarithmic phase cells as described in the text. The values represent a relative activity per cell measured against strain ORN105 Δ(*argF-lac*)U169. One unit of β-galactosidase activity in this table is equivalent to *c.* 167 Miller units.

Strain ORN118 was found to undergo metastable variation from Lac⁺ to Lac⁻ at a frequency indistinguishable from the parent. However, strain ORN118 exhibited an approximately two-fold increase in β-galactosidase activity in the "ON" (Lac⁺) mode compared to a control strain containing Tn*5*-132 unlinked to *hyp*. We conclude from this that *hyp* does not effect metastable regulation of *pilA* expression but does effect a repression of *pilA* transcription. Hence, two modes of control appear to be active in the regulation of *pilA* transcription.

The results described above indicate that *pilA* is subject to metastable regulation at the transcriptional level. However, the element (or elements) involved in effecting metastable expression of *pilA* remain a mystery. A possible candidate for effecting the metastable regulation of *pilA* expression is the *pilA* promotor which we have found to undergo a high frequency of precise excision [7,9]. However, at present, we cannot reconcile the instability of the promotor region with the observed metastability of *pilA* expression. Presumably, examination of DNA from *pilA* "ON" and *pilA* "OFF" clones using a probe consisting of the *pilA* promotor region should reveal any differences in the *pilA* promotor between these two populations. The function of the *hyp* gene product appears to be one of repression although the repression effected, at least in the chromosome, appears to be rather modest. Whereas this may be due to technical reasons, the low level of repression in the chromosome may reflect a more complex type of repressor activity. For example, the *hyp* gene product may serve as a coordinator of pilus assembly with repressor activity dependent upon other products of the Pil region.

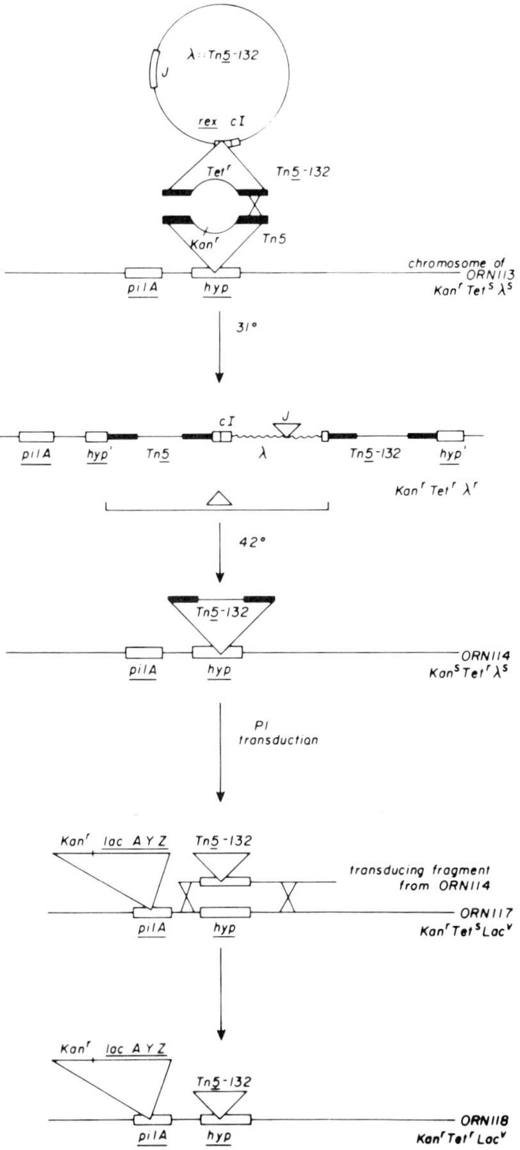

Figure 6. Construction of strains ORN114 and ORN118. Strain ORN113 was infected with λ::Tn*5*-132 and lysogens (Tet^r Kan^r λ^r) selected at 31°. Homology between IS50 in Tn*5* and Tn*5*-132 is indicated by the bold bars; a single cross-over event is depicted. Mutants having undergone spontaneous excision of λ::Tn*5* at 31° (denoted by the delta) were selected at 42° and checked for the predicted Tet^r Kan^r λ^s phenotype. One such clone was subsequently called ORN114. P1 phage were grown on ORN114 and used to transduce strain ORN117 to tetracycline resistance, forming strain ORN118.

One might predict from this that the repressor activity of the *hyp* gene product would be allosterically regulated. This possibility is currently being considered in our laboratory.

SUMMARY

We present evidence that the *pilA* gene, encoding the structural subunit of type 1 pili, is subject to metastable transcriptional regulation. A *pilA'-lacZ* fusion, constructed *in vitro* on a recombinant plasmid, was used in conjunction with a *recBC sbcB* mutant of *E. coli* K-12 to introduce the fusion into the chromosomal Pil region. This fusion was found to be subject to metastable transcriptional control. The rate of switching from the Lac^+ to the Lac^- phenotype was 4×10^{-4} per cell per generation and 6.2×10^{-4} in the opposite direction. An approximately ten-fold difference in beta-galactosidase activity was observed between phenotypically "ON" (Lac^+) and "OFF" (Lac^-) populations. P1 transduction experiments showed that the element determining the ON or OFF phenotype was tightly linked to *pilA*. In addition to the metastable regulation of *pilA*, a second type of transcriptional regulation was effected by the product of a gene, *hyp*, adjacent to *pilA*. Using a recombinant plasmid containing just a *pilA'-lacZ* fusion and the putative *pilA* promotor, we found that a lesion in *hyp* conferred an approximately five-fold β-galactosidase activity when compared to a strain possessing the parental *hyp* gene. Mutants constructed to have a *pilA'-lacZ* fusion and a *hyp*::Tn*5*-132 mutation in the chromosome exhibited a frequency of switching from Lac^+ to Lac^- and vice versa indistinguishable from the parental strain. However, in the ON (Lac^+) mode, *hyp*::Tn*5*-132 mutants showed a two-fold higher β-galactosidase activity. Thus, *hyp* does not appear to effect metastable variation but does affect the level of transcription of the *pilA* gene in the ON (transcribed) mode.

ACKNOWLEDGEMENTS

This work was supported by North Carolina fund project number 197227, and by grant 1RO1 AI22223-01 from the National Institute of Allergy and Infectious Diseases.

REFERENCES

1. Brinton, C. C. Jr (1965). The structure, function, synthesis and genetic control of bacterial pili and a molecular model of DNA and RNA transport in Gram-negative bacteria. *Trans. N.Y. Acad. Sci.* **27**, 1003–1054.
2. Pearce, W. A. and T. M. Buchanan (1980). Structure and cell membrane-binding properties of bacterial fimbriae. *In* "Bacterial Adherence" (Ed. E. H. Beachey), Receptors and Recognition, Ser. B, Vol. 6, pp. 289–344. Chapman and Hall, London.
3. Orndorff, P. E. and S. Falkow (1984). Organization and expression of genes responsible for type 1 piliation in *Escherichia coli*. *J. Bacteriol.* **159**, 736–744.
4. Eisenstein, B. I. (1981). Phase variation of type 1 fimbriae in *Escherichia coli* is under transcriptional control. *Science* **214**, 337–339.

5. Freitag, C. S., J. M. Abraham, J. R. Clements and B. I. Eisenstein (1985). Genetic analysis of the phase variation control of expression of type 1 fimbriae in *Escherichia coli. J. Bacteriol.* **162**, 668-675.
6. Orndorff, P. E., P. A. Spears, D. Schauer and S. Falkow (1985). Two modes of control of *pilA*, the gene encoding type 1 pilin in *Escherichia coli. J. Bacteriol.* **164**, 321-330.
7. Orndorff, P. E. and S. Falkow (1984). Identification and characterization of a gene product that regulates type 1 piliation in *Escherichia coli. J. Bacteriol.* **160**, 61-66.
8. Freitag, C. S. and B. I. Eisenstein (1983). Genetic mapping and transcriptional orientation of the *fimD* gene. *J. Bacteriol.* **156**, 1052-1058.
9. Orndorff, P. E. and S. Falkow (1985). The nucleotide sequence of *pilA*, the gene encoding the structural component of type 1 pili in *Escherichia coli. J. Bacteriol.* **162**, 454-457.
10. Casadaban, M. and S. N. Cohen (1980). Analysis of gene control signals by DNA fusion and cloning in *Escherichia coli. J. Mol. Biol.* **138**, 179-207.
11. Brinton, C. C. Jr (1959). Nonflagellar appendages of bacteria. *Nature (Lond.)* **183**, 782-786.
12. Berg, D. E., C. Egner, B. J. Hirschel, H. Howard, L. Johnsrud, R. Jorgensen and T. Tlsty (1980). Insertion, excision and inversion of transposon Tn5. *Cold Spring Harbor Symp. Quant. Biol.* **45**, 115-132.

Structural Variation of P-Fimbriae from Uropathogenic *Escherichia coli*

Irma van Die[a], Marja Dijksterhuis, Hans de Cock, Wiel Hoekstra and Hans Bergmans

Department of Molecular Cell Biology, State University of Utrecht, Utrecht, The Netherlands

The ability to adhere to uroepithelial cells appears to be an important virulence factor for uropathogenic *Escherichia coli* [1]. In the majority of pyelonephritogenic strains this adherence is mediated by P-fimbriae, which specifically recognize the α-D-Gal(1→4) β-D-Gal moiety of P blood group antigens [2]. P-fimbriae also mediate mannose resistant haemagglutination (MRHA) of human erythrocytes. Serological studies have revealed the immunogenic diversity of these fimbriae; at least 7 serotypes (F7-F12, Pap) are distinguished [3,4].

Recent interest in the molecular biology of P-fimbriae has led to the cloning of a number of DNA fragments encoding the expression of P-fimbriae from different uropathogenic *E. coli* strains into *E. coli* K12. In this paper we will compare a number of these clones isolated in our and collaborating laboratories (Table 1, Fig. 1).

COMPARISON OF THE CLONED FRAGMENTS (Fig. 1)

The DNA fragments needed for expression of P-fimbriae encompass about 9 kb in all cases. The restriction maps show a large degree of homology, but all have some particular differences. In three cases ($F7_1$, $F7_2$ and Pap) the genetic organization has been studied in detail. The three gene clusters have a similar organization; one gene (A) encoding the fimbrial subunit, and a number of genes encoding accessory proteins, some of which will be involved in the processing of the fimbrial subunit into fimbriae. One of the accessory proteins appears to be the lectin involved in MRHA [10]. The proteins encoded by the accessory

[a]*Telephone:* 31-30-533184

Protein-Carbohydrate Interactions in Biological Systems. ISBN 0 12 436665 1

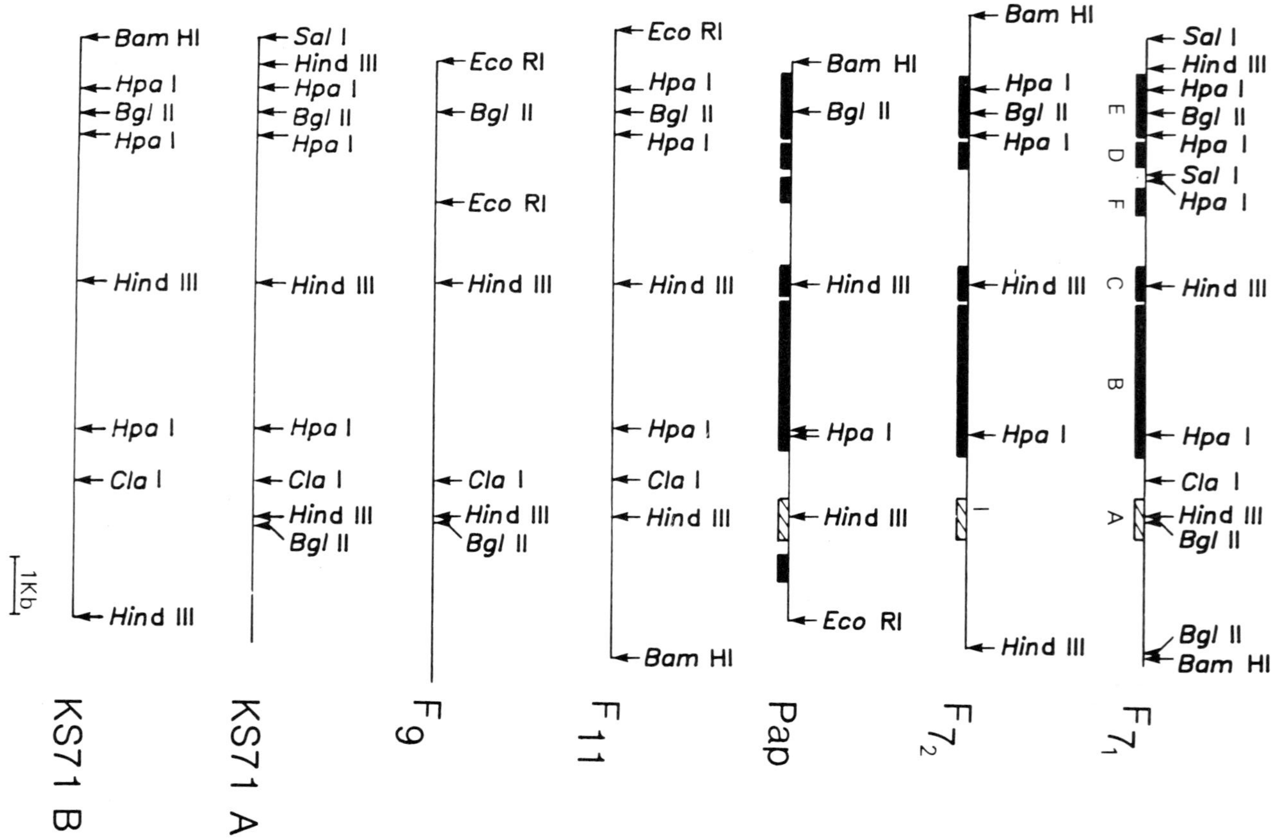

Figure 1. Comparison of the restriction maps of DNA fragments encoding the expression of P-fimbriae mentioned in Table 1. Shaded bars show the location of the fimbrial subunit genes; solid bars show the location of accessory proteins. Nomenclature of genes according to [10].

Table 1. Characteristics of bacterial strains and fimbrial subunits

Parental strain	*Serotype*	*Serotype fimbriae*	*Apparent MW of fimbrial subunit (KD)*	*Computed MW of fimbrial subunit (KD)*	*Reference*
AD110	06:K2:H1	$F7_1$	20.0	17.2	5
AD110	06:K2:H1	$F7_2$	17.0	18.1	6
KS71	04:K12	KS71A	22.0	17.2	5
KS71	04:K12	KS71B	20.0		5
J96	04:K6:H+	Pap	19.5	16.5	7
C1018	02:K5:H4	F9	21.0		8
C1976	01:K1:H7	F11	18.0	16.4	9

genes have a similar apparent molecular weight in SDS-PAGE. Clear differences are found in the apparent molecular weights of the fimbrial subunit proteins. The homology between the restriction maps suggests that all clones contain a similar gene cluster. This is further corroborated because in all cases the expression of P-fimbriae could be separated genetically from the expression of MRHA [10, Van Die *et al.*, submitted]. However, the fimbrial subunits have evolved to be serologically distinct proteins, probably due to strong selective pressure to escape the immune system of the host. This raises the question as to how far the accessory proteins of different gene clusters are able to process fimbrial subunits of other gene clusters into fimbriae.

FUNCTIONAL HOMOLOGY BETWEEN THE GENE CLUSTERS

Complementation studies have shown that transposon insertion mutations in genes A–E of the $F7_2$ gene cluster can be complemented in *trans* by appropriate subclones of the $F7_1$ gene cluster [11]. This complementation usually results in weak expression of fimbriae. This weak expression may be due to several artefacts, for example polar effects of the transposon mutations or copy number effects of the complementing plasmids.

Such artefacts cannot occur in hybrid plasmids where parts of different gene clusters are fused at homologous restriction sites. Using the *Cla*I restriction site between the subunit and the accessory proteins it was shown that the F11 subunit is processed efficiently into fimbriae by the $F7_1$ and F9 accessory proteins and vice versa. Also, gene fusions in gene C (*Hind*III site) and gene E (*Bgl*II site) between $F7_1$, $F7_2$ and F11 gene clusters resulted in efficient processing of the fimbrial subunits.

No hybrid subunit proteins have been constructed as yet, because the *Hind*III site in the $F7_1$, F9 and F11 A genes is not homologous in all cases.

These results indicate that evolutionary changes in the subunit proteins have not influenced the regions essential for interaction with the accessory proteins.

STRUCTURAL COMPARISON OF THE $F7_1$, $F7_2$, F11, Pap AND KS71A SUBUNIT PROTEINS

The nucleotide sequences of the genes encoding these subunit proteins and the primary structures of these proteins have been published before (Table 1), except for the sequence of the F11 subunit gene, shown in Fig. 2. The nucleotide sequence of the $F7_1$ and KS71A genes are completely homologous. Figure 3 shows a comparison of the primary structures of the KS71A/$F7_1$, $F7_2$, Pap and F11 subunit proteins, aligned for maximal homology.

The computed molecular weights of the proteins cannot be compared directly to the apparent molecular weights estimated in SDS-PAGE (Table 1). This may be due to the influence of the protein conformation on the apparent molecular weights. Five regions can be discriminated in Fig. 3: the signal sequence (before the alanine residue in position 1), the *N*-terminal part (residues 1–21), the cys–cys loop (residues 22–61), the main nonconserved region (residues 62–153) and the *C*-terminal part (residues 154–168).

The signal sequences and the *N*- and *C*-terminal regions are highly homologous. Less homology is found in the region between the two cysteine residues and even less homology is found between the second cysteine residue and the conserved *C*-terminal part. Only in this "main non-conserved" region are gaps needed for maximal alignment of the sequences. The two largest non-homologous stretches are also located in this region (residues 87–95) and 121–128).

The conserved regions are much less hydrophilic than the non-conserved regions: 37% of the conserved amino acids are hydrophobic (hydrophilicity value < -1.0), 11.6% are hydrophilic (hydrophilicity value > 1.0), and for the non-conserved regions these values are 16% and 24% respectively. This could mean that the non-conserved regions are located at the outside of the fimbrial subunit proteins, and also at the outside of the fimbriae. This would account for the immunogenic diversity of these proteins. Indeed, most of the monoclonal antibodies isolated by De Ree *et al.* against these subunit proteins [12] are specific; only three appear to recognize a common epitope in the $F7_2$, Pap and F11 subunit proteins.

Homology between the nucleotide sequences of the genes is shown in Table 2. The five genes can be grouped into two families: $F7_1$/KS71A-$F7_2$ and Pap-F11. Although the original strains from which KS71A and F_1 fimbrial subunits have been cloned have a different serotype, and consequently may have had a considerable separate history, the nucleotide sequences of the two genes are completely identical. The nucleotide sequences upstream of all 5 genes show absolute homology (or 1 base pair inhomology in two cases) over the entire regions where sequences are available (90–180 base pairs). This homology is continued between Pap and F11 over 207 base pairs in the gene, except for a strange duplication of one codon at the end of the signal sequence of Pap.

Homology between the five genes is high again at the *C*-terminus (up to 100%), but is lost behind the *C*-terminus, although only short sequences could be compared in some cases. Only in the case of KS71A and $F7_1$ is considerable homology (95% over 81 base pairs) found in this region.

```
                                                                    TGTATTGATCT

                                     50
GGTTATTAAAGGTAATCGGGTCATTTTAAATTGCCAGATATCTCTGGTGTGTTCAGTAATGAAAAAGAGGTTGTTATTT


          100                                                                 150
ATG ATT AAG TCG GTT ATT GCC GGT GCG GTA GCT ATG GCA GTG GTG TCT TTT GGT GTA AAT
Met Ile Lys Ser Val Ile Ala Gly Ala Val Ala Met Ala Val Val Ser Phe Gly Val Asn
-21

                                                                200
GCT GCT CCA ACT ATT CCA CAG GGG CAG GGT AAA GTA ACT TTT AAC GGA ACT GTT GTT GAT
Ala Ala Pro Thr Ile Pro Gln Gly Gln Gly Lys Val Thr Phe Asn Gly Thr Val Val Asp
 -1  +1                                                      15

                                                   250
GCT CCA TGC AGC ATT TCT CAG AAA TCA GCT GAT CAG TCT ATT GAT TTT GGA CAG CTT TCA
Ala Pro Cys Ser Ile Ser Gln Lys Ser Ala Asp Gln Ser Ile Asp Phe Gly Gln Leu Ser
                                         30

                                    300
AAA AGC TTC CTT GAG GCA GGA GGT ACA TCC AAG CCA ATG GAT TTA GAT ATT GAG TTA GTA
Lys Ser Phe Leu Glu Ala Gly Gly Thr Ser Lys Pro Met Asp Leu Asp Ile Glu Leu Val
                     45

                       350
AAT TGC GAT ATT ACA GCT TTT AAA CAA GGT CAA GCA GCT AAA AAC GGT AAG GTT CAA TTA
Asn Cys Asp Ile Thr Ala Phe Lys Gln Gly Gln Ala Ala Lys Asn Gly Lys Val Gln Leu
 60                                                          75

          400                                                                 450
TCT TTT ACT GGG CCA CAA GTA ACG GGG CAG GCT GAA GAA TTA GCA ACT AAC GGC GGT ACG
Ser Phe Thr Gly Pro Gln Val Thr Gly Gln Ala Glu Glu Leu Ala Thr Asn Gly Gly Thr
                                         90


                                                               500
GGC ACA GCT ATT GTA GTT CAG GCT GCA GGC AAA AAC GTT TCT TTC GAT GGG ACT GCA GGT
Gly Thr Ala Ile Val Val Gln Ala Ala Gly Lys Asn Val Ser Phe Asp Gly Thr Ala Gly
                    105

                                                  550
GAC GCT TAT CCC CTG AAA GAT GGT GAT AAT GTT CTT CAT TAT ACA GCT CTT GTG AAA AAA
Asp Ala Tyr Pro Leu Lys Asp Gly Asp Asn Val Leu His Tyr Thr Ala Leu Val Lys Lys
120                                                          135

                                    600
GCG AAT GGC GGT ACT GTT TCT GAA GGT GCT TTT TCT GCA GTT GCA ACC TTT AAT TTA AGT
Ala Asn Gly Gly Thr Val Ser Glu Gly Ala Phe Ser Ala Val Ala Thr Phe Asn Leu Ser
                                        150

                    650
TAC CAG TAA TTCATCTGATTTTTACAGGGTGAGG
Tyr Gln Stop
```

Figure 2. Nucleotide sequence of F11 fimbrial subunit gene and predicted amino acid sequence of the encoded subunit protein. The alanine residue in position 1 is supposed to be the first amino acid of the mature protein. The nucleotide sequence shown was obtained by analysis of both DNA strands except the part between base pairs 501-541, which was obtained by analysis of only one DNA strand.

KS71A/F7 1	MET ILE LYS SER VAL ILE ALA GLY ALA VAL ALA MET ALA VAL VAL SER PHE GLY ALA ASN ALA
F7 2	MET ILE LYS SER VAL ILE ALA GLY ALA VAL ALA MET ALA VAL VAL SER PHE GLY ALA TYR ALA
Pap	MET ILE LYS SER VAL ILE ALA GLY ALA VAL ALA MET ALA VAL VAL SER PHE GLY VAL ASN ASN ALA
F11	MET ILE LYS SER VAL ILE ALA GLY ALA VAL ALA MET ALA VAL VAL SER PHE GLY VAL ASN ALA

1

ALA ALA SER ILE PRO GLN GLY GLN GLY GLU VAL SER PHE LYS GLY THR VAL VAL ASP ALA PRO CYS GLY ILE GLU THR GLN SER ALA LY
ALA PRO THR ILE PRO GLN GLY GLN GLY LYS VAL THR PHE ASN GLY THR VAL VAL ASP ALA PRO CYS GLY ILE ASP ALA GLN SER ALA AS
ALA PRO THR ILE PRO GLN GLY GLN GLY LYS VAL THR PHE ASN GLY THR VAL VAL ASP ALA PRO CYS SER ILE SER GLN LYS SER ALA AS
ALA PRO THR ILE PRO GLN GLY GLN GLY LYS VAL THR PHE ASN GLY THR VAL VAL ASP ALA PRO CYS SER ILE SER GLN LYS SER ALA AS

31

GLN GLU ILE ASP PHE GLY GLN ILE SER LYS SER PHE LEU GLN GLU GLY GLY GLU THR GLN PRO LYS ASP LEU ASN ILE LYS LEU VAL ASN
GLN SER ILE ASP PHE GLY GLN VAL SER LYS LEU PHE LEU GLU ASN ASP GLY GLU SER GLN PRO LYS SER PHE ASP ILE LYS LEU ILE ASN
GLN SER ILE ASP PHE GLY GLN LEU SER LYS SER PHE LEU GLU ALA GLY GLY VAL SER LYS PRO MET ASP LEU ASP ILE GLU LEU VAL ASN
GLN SER ILE ASP PHE GLY GLN LEU SER LYS SER PHE LEU GLU ALA GLY GLY THR SER LYS PRO MET ASP LEU ASP ILE GLU LEU VAL ASN

61

CYS ASP ILE THR ASN LEU LYS GLN LEU GLN GLY GLY ALA ALA LYS LYS GLY THR VAL SER LEU THR PHE SER GLY VAL PRO ALA GLU
CYS ASP ILE THR ASN PHE LYS LYS ALA ALA GLY GLY GLY GLY ALA LYS THR GLY THR VAL SER LEU THR PHE SER GLY VAL PRO SER GLY
CYS ASP ILE THR ALA PHE LYS GLY GLY ASN GLY ALA LYS LYS GLY THR VAL LYS LEU ALA PHE THR GLY PRO ILE VAL ASN
CYS ASP ILE THR ALA PHE LYS GLN GLY GLN ALA ALA LYS ASN GLY LYS VAL GLN LEU SER PHE THR GLY PRO GLN VAL THR

91

ASN ALA ASP ASP MET LEU GLN THR VAL GLY ASP THR ASN THR ALA ILE VAL VAL THR ASP SER SER GLY LYS ARG VAL LYS PHE ASP GLY
PRO GLN SER ASP MET LEU GLN THR VAL GLY ALA THR ASN THR ALA ILE VAL VAL THR ASP PRO HIS GLY LYS ARG VAL LYS PHE ASP GLY
GLY HIS SER ASP GLU LEU ASP THR ASN GLY GLY THR GLY THR ALA ILE VAL VAL GLN GLY ALA GLY LYS ASN VAL VAL PHE ASP GLY
GLY GLN ALA GLU GLU LEU ALA THR ASN GLY GLY THR GLY THR ALA ILE VAL VAL GLN ALA ALA GLY LYS ASN VAL SER PHE ASP GLY

121

ALA THR GLU THR GLY ALA SER ASN LEU ILE ASN GLY ASP ASN THR ILE HIS PHE THR ALA PHE VAL LYS LYS ASP ASN SER GLY LYS
ALA THR ALA THR GLY VAL SER TYR LEU VAL ASP GLY ASP ASN THR ILE HIS PHE THR ALA ALA VAL ARG LYS ASP GLY SER GLY ASN
SER GLU GLY ASP ALA ASN THR LEU LYS ASP GLY GLU ASN VAL LEU HIS TYR THR ALA VAL VAL LYS LYS SER SER ALA VAL GLY ALA
THR ALA GLY ASP ALA TYR PRO LEU LYS ASP GLY ASP ASN VAL LEU HIS TYR THR ALA LEU VAL LYS LYS ALA ASN GLY GLY

151

ASN VAL ALA GLU GLY ALA PHE SER ALA VAL ALA ASN PHE ASN LEU THR TYR GLN
PRO VAL THR GLU GLY ALA PHE SER ALA VAL ALA ASN PHE ASN LEU THR TYR GLN
ALA VAL THR GLU GLY ALA PHE SER ALA VAL ALA ASN PHE ASN LEU THR TYR GLN
THR VAL SER GLU GLY ALA PHE SER ALA VAL ALA THR PHE ASN LEU SER TYR GLN

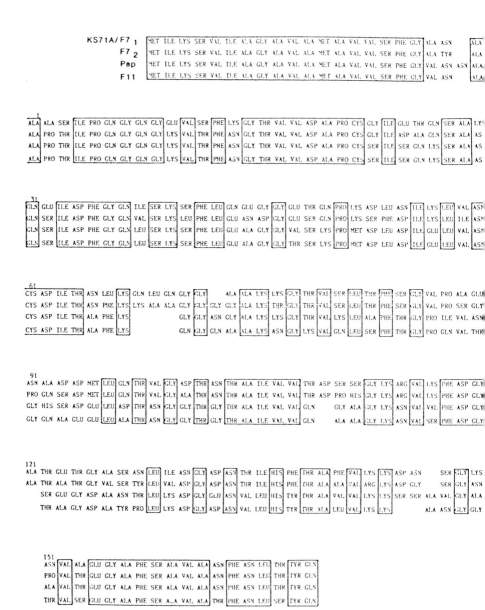

Figure 3. Comparison of the amino acid sequences of $F7_1$, KS71A, $F7_2$, Pap and F11 fimbrial subunit proteins. The alanine residues in position 1 are the first amino acids of the mature proteins. Homologous amino acids are boxed.

Table 2. Homology between the nucleotide sequences encoding fimbrial subunit proteins[a]

	Percentage homology between					
Region[b]	*$F7_1$-$F7_2$*	*$F7_1$-Pap*	*$F7_1$-F11*	*$F7_2$-Pap*	*$F7_2$-F11*	*Pap-F11*
Complete gene	74.8	64.2	63.0	66.8	64.7	82.6
Signal sequence	96.8	92.4	96.8	92.4	96.8	95.5
N-terminal	76.6	79.7	79.7	93.8	93.8	100.0
cyc-cys loop	71.9	61.2	60.3	64.5	62.8	93.4
main non-conserved	66.8	49.5	50.9	50.5	49.8	71.8
C-terminal	97.8	100.0	73.3	97.8	75.6	73.3

[a]For this comparison nucleotide sequences have been aligned according to the amino acid alignment shown in Fig. 3.
[b]Regions of the subunit proteins are defined in the text.

The strong conservation of the region upstream of the *N*-terminal part of the genes suggests that these regions have some important function. No large open reading frames are found in this region, so we suppose that this region has a regulatory function.

The strong conservation and large variations that are observed in some cases in different regions within one pair of genes (Pap/F11 and Pap/$F7_1$) can hardly be explained by random mutation of the two subunit genes alone. Some other mechanism, e.g. incorporation of blocks of genetic information from other fimbrial subunit genes by recombination, might be responsible for the observed variability.

In summary, we conclude that the fimbrial subunits have a number of non-conserved hydrophilic regions that may function as antigen determinants. These regions may offer interesting possibilities for introducing further changes in the subunit proteins by means of genetic engineering. The conserved hydrophobic regions may have a role in biogenesis of fimbriae, either by determining the shape of the fimbrillin, or because the accessory proteins recognize these regions during biogenesis. Further experiments, for example localized mutagenesis of these regions, are needed to elucidate this point.

REFERENCES

1. Svanborg-Edén, C., L. A. Hanson, U. Jodal, U. Lindberg and A. SohlÅkerlund (1976). Variable adhesion to normal urinary tract epithelial cells of *Escherichia coli* strains associated with various forms of urinary tract infection. *Lancet* **ii**, 490–492.
2. Källenius, G., R. Möllby, S. B. Svenson, J. Winberg, A. Lundblad, S. Svensson and B. Cedergren (1980). The p^k antigen as receptor for the hemagglutinin of pyelonephritic *Escherichia coli*. *FEMS Microbiol. Lett.* **7**, 297–302.
3. Ørskov, I. and F. Ørskov (1983). Serology of *Escherichia coli* fimbriae. *Prog. Allergy* **33**, 80–105.
4. Hull, R. A., R. E. Gill, P. Hsu, B. H. Minshew and S. Falkow (1981). Construction and expression of recombinant plasmids encoding type 1 or D-mannose-resistant pili from a urinary tract infection *Escherichia coli* isolate. *Infect. Immun.* **33**, 933–938.
5. Rhen, M., I. van Die, V. Rhen and H. Bergmans. Comparison of the nucleotide sequences of the genes encoding the KS71A and $F7_1$ fimbrial antigens of uropathogenic *Escherichia coli*. Submitted for publication.

6. Van Die, I. and H. Bergmans (1984). Nucleotide sequence of the gene encoding the $F7_2$ fimbrial subunit of a uropathogenic *Escherichia coli* strain. *Gene* **32**, 83-90.
7. Båga, M., S. Normark, J. Hardy, P. O'Hanley, D. Lark, O. Olsson, G. Schoolnik and S. Falkow (1984). Nucleotide sequence of the *papA* gene encoding the *papA* pilus subunit of human uropathogenic *Escherichia coli*. *J. Bacteriol.* **157**, 330-333.
8. De Ree, J. M., P. Schwillens, L., Promes, I. van Die and H. Bergmans (1985). Molecular cloning and characterization of F9 fimbriae from a uropathogenic *Escherichia coli*. *FEMS Microbiol. Lett.* **26**, 163-169.
9. De Ree, J. M., P. Schwillens and J. F. van den Bosch (1985). Molecular cloning of F11 fimbriae from a uropathogenic *Escherichia coli* and characterization of fimbriae with polyclonal and monoclonal antibodies. *Infect. Immun.* **29**, 91-97.
10. Lindberg, F. P., B. Lund and S. Normark (1984). Genes of pyelonephritogenic *E. coli* required for digalactoside-specific agglutination of human cells. *EMBO J.* **3**, 1167-1173.
11. Van Die, I., G. Spierings, I. van Megen, E. Zuidweg, W. Hoekstra and H. Bergmans (1985). Cloning and genetic organization of the gene cluster encoding $F7_1$ fimbriae of a uropathogenic *Escherichia coli* and comparison with the $F7_2$ gene cluster. *FEMS Microbiol. Lett.* **28**, 329-334.
12. De Ree, J. M., P. Schwillens and J. F. van den Bosch (1985). Monoclonal antibodies that recognize the P-fimbriae $F7_1$, $F7_2$, F9 and F11 from uropathogenic *Escherichia coli*. *Infect. Immun.* **50**, 900-904.

The *fim* Genes of *Escherichia coli* K-12: Aspects of Structure, Organization and Expression

Per Klemm[a]

Department of Microbiology, Technical University of Denmark, Lyngby, Denmark

Type 1 fimbriae are produced by the majority of *Escherichia coli* strains. These fimbriae confer adhesion to D-mannose containing structures on a variety of eukaryotic cells. Consequently such adhesion can be inhibited by the monosaccharide D-mannose. The localization of the actual binding sites on type 1 fimbriae has been reported to be lateral rather than terminal. This could indicate that each subunit provides a binding site [1]. No conclusive function has yet been assigned type 1 fimbriae. They do not seem to be implicated in the disease-causing potential of pathogenic strains, since no correlation is apparent between the presence of these structures and virulence (see [2]). However, since type 1 fimbriae mediate attachment to urinary mucus, it is plausible that their biological role is to adhere to mucus in the large intestine, which make up the natural habitat of *E. coli*.

A single type 1 fimbria, with a diameter of 7 nm and a length of 1 μm, is composed of around 1000 identical subunits. The expression of type 1 fimbriae is subject to phase variation, and in a given cell population both fimbriated and nonfimbriated cells are present. The actual shift from one phase to the other is an all-or-none kind of event reported to be regulated on the transcriptional level [3]. The rate of change is in the order of 10^{-3} per generation [3,4].

The *fim* genes responsible for the expression of type 1 fimbriae in *E. coli* K-12 have been located at 98 min on the linkage map [5,6]. Recently these genes were cloned [7]. The *fim* gene cluster encompasses at least five genes that seem to be necessary for the regulation and expression of functional type 1 fimbriae. They

[a] *Telephone:* 45-2-884066

Protein-Carbohydrate Interactions in Biological Systems. ISBN 0 12 436665 1

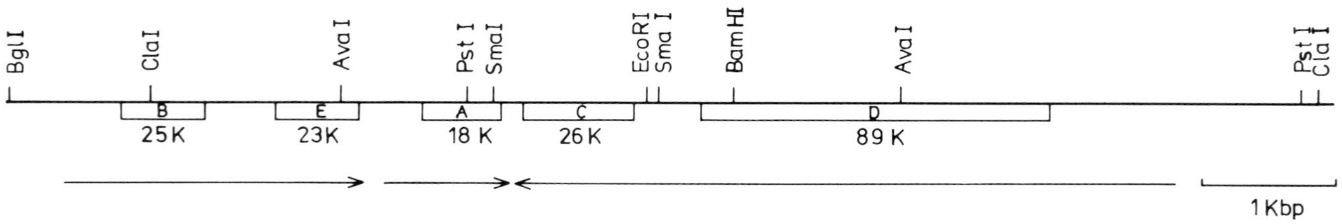

Figure 1. Organization of the *fim* genes of *Escherichia coli* K-12. Sizes of the corresponding proteins are given. Transcriptional units are indicated by arrows.

encode proteins with apparent molecular weights of 16.5 kD, 23 kD, 25 kD, 26 kD and 89 kD. The order of the *fim* genes as well as their organization into three transcriptional units is outlined in Fig. 1. The *fimA* gene encodes an 18 kD protein, which in its mature form appears as the 16.5 kD protein, and forms the structural protein of the type 1 fimbriae. The other genes in the *fim* gene cluster encode several auxillary proteins necessary for translocation, anchorage and probably regulation of expression of the fimbrial subunit protein.

The *fimA* gene, encoding the structural protein of the type 1 fimbriae has been sequenced [8]. There exists extensive homology between this protein and other fimbrial proteins, such as fimbriae from uropathogenic strains and the K99 antigen. These different fimbriae exhibit no immunological relationship, but their primary structures are clearly indicative of a group of evolutionarily related proteins that have "drifted apart". The ubiquitous presence of the type 1 fimbrial protein could indicate this fimbrial species to be the ancestor of the whole group. Fimbriae are excellent immunogens and *in vivo* continually exposed to the immune defences of the host organisms. These proteins must therefore be prone to strong antigenic pressures leading to a fine structural balance between the optimal immunological diversity and the conservation of the necessary structural and functional domains.

Recently the region of 2.5 kb upstream of the *fimA* has been sequenced (P. Klemm, unpubl. data). This region harbours two open reading frames for two proteins with molecular weight of 23 kD. The corresponding gene product can be visualized by minicell analysis with apparent molecular weights of 25 kD and 23 kD. A promoter has been shown to be located upstream (to the left in Fig. 1) of the *fimB* gene. Both proteins have a very large content of basic amino acid residues and high isoelectric points. They could be envisaged to be DNA-interacting proteins, being positive or negative regulators or perhaps playing a role in the phase variation of type 1 fimbriae.

During DNA-sequence work of the region upstream of the *fimA* gene a region of 300 bases was found in two configurations; the second being an inversion of the first. The invertible region is flanked by two inverted repeats. From DNA-sequence data the promotor for the *fimA* gene was suggested to be located in this region [8]. The insertion of a 1 kb DNA fragment harbouring this segment in front of a promotor-deficient beta-lactamase gene resulted in ampicillin resistance of the transformed host (P. Klemm, unpubl.). The invertible DNA-segment does not contain any substantial open reading frame. If its inversion is indeed responsible for the phase variation of type 1 fimbriae, the organization is different from those of the DNA-inversion systems represented by the site specific recombinases *hin*, *gin* and *pin* [9,10].

REFERENCES

1. Sweeney, G. and J. H. Freer (1979). Location of binding sites on common type 1 fimbriae of *Escherichia coli*. *J. Gen. Microbiol.* **112**, 321-328.
2. Klemm, P. (1985). Fimbrial adhesins of *Escherichia coli*. *Rev. Infect. Dis.* **7**, 321-340.
3. Eisenstein, B. I. (1981). Phase variation of type 1 fimbriae in *Escherichia coli* is under transcriptional control. *Science* **214**, 337-339.

4. Swaney, L. M., Y.-P. Liu, K. Ippen-Ihler and C. C. Brinton Jr (1977). Genetic complementation analysis of *Escherichia coli* type 1 somatic pilus mutants. *J. Bacteriol.* **130**, 506–511.
5. Brinton, Jr, C. C., P. Gemski, S. Falkow and S. Baron (1961). Location of the piliation factor on the chromosome of *Escherichia coli. Biochem. Biophys. Res. Commun.* **5**, 293–298.
6. Freitag, C. S. and B. I. Eisenstein (1983). Genetic mapping and transcriptional orientation of the *fimD* gene. *J. Bacteriol.* **156**, 1052–1058.
7. Klemm, P., B. J. Jørgensen, I. van Die, H. de Ree, and H. Bergmans (1985). The *fim* genes responsible for synthesis of type 1 fimbriae in *Escherichia coli*, cloning and genetic organization. *Mol. Gen. Genet.* **199**, 410–414.
8. Klemm, P. (1984). The *fimA* gene encoding the type 1 fimbrial subunit of *Escherichia coli*, nucleotide sequence and primary structure of the protein. *Eur. J. Biochem.* **143**, 395–399.
9. Zieg, J. and M. Simon (1980). Analysis of the nucleotide sequence of an invertible controlling element. *Proc. Natl. Acad. Sci. USA* **77**, 4196–4201.
10. Plasterk, R. H. A., A. D. Brinkman and P. van de Putte (1983). DNA inversions in the chromosome of *Escherichia coli* and in bacteriophage Mu: Relationship to other site-specific recombination systems. *Proc. Natl. Acad. Sci. USA.* **80**, 5355–5358.

Poster Session

Transfer of the Plasmid-Mediating AF/R1 Pili from Enteropathogenic *Escherichia coli* Strain RDEC-1

Marcia K. Wolf[a], Gerard P. Andrews, Wendy L. Houston and Edgar C. Boedeker

Department of Gastroenterology, Division of Medicine, Walter Reed Army Institute of Research, Washington, DC 20307, USA

RDEC-1 is an enteropathogenic strain of *Escherichia coli* (EPEC) which causes diarrhoea in rabbits. It adheres to the proximal small intestine but does not invade mucosal cells and does not produce HS or LT enterotoxin. Adherence to intestinal epithelium correlates with the presence of AF/R1 pili on the surface of RDEC-1 [1]. Piliation and adherence were demonstrated in *Shigella flexneri* after transfer of an 85 MD plasmid from RDEC-1 [2].

We have tagged the pilus plasmid, which we designate pG1, with Tn5 which encodes resistance to kanamycin. This was mated with *E. coli* HB101 and kanamycin-resistant derivatives of HB101 were obtained. One was selected which expressed pili as determined by agglutination of anti-AF/R1 antisera. This derivative was named M5. M5 adheres to brush borders prepared from rabbits which confirms that functional pili are expressed. An 86 MD plasmid, which we call pG1::Tn5, is present in M5. Two other exconjugants were named M6 and M7. Both are resistant to kanamycin and have plasmids similar in size to those in RDEC-1 (40 MD and 86 MD, respectively), but neither expressed AF/R1 pili as assayed by slide agglutination and attachment to brush borders. By DNA restriction digests M7 seems to contain a deletion derivative of pG1::Tn5.

M5 was negative in both assays for pilus expression when grown in media known to suppress pilus expression in RDEC-1. Therefore expression of AF/R1 from pG1::Tn5 in HB101 is under similar control to that in RDEC-1.

[a] *Telephone:* 1-202-576 2582

Protein-Carbohydrate Interactions in Biological Systems. ISBN 0 12 436665 1

Crossed-line immunoelectrophoresis of M5 with antibody raised against RDEC-1 was performed. The only evident plasmid-mediated antigen in M5 was AF/R1.

pG1::Tn5 has been transferred into two *E. coli* strains isolated from healthy rabbits. Both derivatives express AF/R1 pili when assayed by slide agglutination using antisera raised against AF/R1 pili. Again pili were not expressed in media which suppress pilus expression in RDEC-1 and HB101.

These data confirm that pG1 mediates expression of RDEC-1 pili. Regulation of pilus expression by media is not peculiar to RDEC-1; similar expression was observed in three different backgrounds. Therefore it is likely that the structural gene and regulatory regions for pili are present on pG1. Restriction digests of plasmids from M5 and M7 suggest that genes necessary for pilus expression are present on a specific fragment of pG1. The genes for pilus expression will be cloned from pG1::Tn5. It should now be possible to investigate the contribution of pG1 and pilus expression to virulence.

REFERENCES

1. Cheney, C. P., S. B. Formal, P. A. Schad and E. C. Boedeker (1983). Genetic transfer of a mucosal adherence factor (R1) from an enteropathogenic *Escherichia coli* strain and the phenotypic suppression of this adherence factor. *J. Infect. Dis.* **147**, 711-723.
2. Berendson, R., C. P. Cheney, P. A. Schad and E. C. Boedeker (1983). Species-specific binding of purified pili (AF/R1) from the *Escherichia coli* RDEC-1 to rabbit intestinal mucosa. *Gastroenterology* **85**, 837-845.

The views of the authors do not purport to reflect the positions of the Department of the Army or the Department of Defense. (Para. 4-3, AR 360-5).

Molecular Cloning of the CS3 Fimbriae Determinant of Enterotoxigenic *Escherichia coli* of Serotype 06:K15:H16 or H−

Maire Boylan[a], David Coleman and Cyril J. Smyth

Department of Microbiology, Moyne Institute, Trinity College, Dublin 2, Republic of Ireland

Enterotoxigenic *Escherichia coli* of serotype 06:K15:H16 or H− produce three antigenically distinct types of fimbriae according to the biotypes of strains [1-3]. These fimbriae had previously been termed coli-surface (CS) antigens, namely CS1, CS2 and CS3. Most strains express two types of fimbriae, either CS1 or CS2 together with CS3 fimbriae. The expression of these fimbriae has been shown to be plasmid mediated [2,4]. CS1 and CS2 fimbriae are rigid, rod-like structures of 7 nm diameter, morphologically indistinguishable from common type fimbriae, whereas CS3 fimbriae are thin, wiry, flexible structures of 2-3 nm diameter. When plasmids of differing molecular weights are mobilized from wild-type 06:H16 strains into laboratory K-12 strains, only CS3 fimbriae are expressed in the K-12 host. Upon mobilization back into CSFim$^-$ variants of the same wild-type 06:H16 strains which originally lost the plasmids, expression of CS1 or CS2 and CS3 fimbriae is regained [2,4]. Thus, although expression of CS fimbriae is associated with the presence of a plasmid, it appears to be influenced by host-related factors. The genetic control of CS fimbriae expression is not understood.

To investigate whether or not expression of CS fimbriae is controlled by separate genes, molecular cloning was undertaken. Complete *Hin*dIII digests of a CS fimbriae associated plasmid designated pCS001 [4] yielded 12 fragments which were ligated to *Hin*dIII digested vector plasmid pBR322. The recombinant plasmids were transformed into *E. coli* K-12 strain C600. Ampicillin-resistant,

[a]*Telephone:* 44-1-772941 ext. 1195

Protein-Carbohydrate Interactions in Biological Systems. ISBN 0 12 436665 1

tetracycline-sensitive transformants were screened for CS fimbriae expression by colony immunoblotting; 13% expressed CS3 fimbriae but none CS1 or CS2 fimbriae. The recombinant plasmid containing the smallest DNA insert was designated pCS100 [pBR322Ω(0 kb::pCS001 CSFim 5.1 kb)]. A restriction endonuclease map of pCS100 revealed *Eco*RI, *Pst*I and *Hin*cII sites within the DNA insert, whereas *Hin*dIII, *Cla*I, *Pvu*II, *Sal*I and *Bam*HI did not cut within the 5.1 kb insert.

Transfer of pCS100 into wild-type hosts of serotype 06:K15:H16 or H− by direct transformation or mobilization using plasmids R64 and ColK yielded transformants and transconjugants, respectively, which only expressed CS3 fimbriae. Repeated tests for heat-stable enterotoxin production by strain C600 bearing pCS100, using heat-treated culture supernatant fractions in the infant suckling mouse model, were negative.

A physical map of pCS100 was obtained using transposon Tn*5* insertion mutagenesis. Ten independent Tn*5* insertions were mapped. This has defined a minimum region of approximately 3.2 kb necessary for the expression of CS3 fimbriae on the cell surface.

It is concluded that the recombinant plasmid pCS100 does not contain the determinants required for expression of heat-stable enterotoxin and CS1 and CS2 fimbriae. The size of the minimum region of the DNA insert necessary for CS3 expression implies the presence of a cluster of genes as has been described for other fimbriae such as the K88 and K99 antigens of enterotoxigenic *E. coli*.

ACKNOWLEDGEMENT

This work was supported by a postgraduate studentship to M.B. and a grant from the Medical Research Council of Ireland.

REFERENCES

1. Smyth, C. J. (1984). Serologically distinct fimbriae on enterotoxigenic *Escherichia coli* of serotype 06:K15:H16 or H−. *FEMS Microbiol. Lett.* **21**, 51-57.
2. Mullany, P., A. M. Field, M. M. McConnell, S. M. Scotland, H. R. Smith and B. Rowe (1983). Expression of plasmids coding for colonization factor antigen II (CFA/II) and enterotoxin production in *Escherichia coli*. *J. Gen. Microbiol.* **129**, 3591-3601.
3. Levine, M. M., P. Ristaino, G. Marley, C. Smyth, S. Knutton, E. Boedeker, R. Black, C. Young, M. L. Clements, C. Cheney and R. Patnaik (1984). Coli surface antigens 1 and 3 of colonization factor antigen II-positive enterotoxigenic *Escherichia coli*: Morphology, purification, and immune responses in humans. *Infect. Immun.* **44**, 409-420.
4. Boylan, M. B. and C. J. Smyth (1985). Mobilization of CS fimbriae associated plasmids of enterotoxigenic *Escherichia coli* of serotype 06:K15:H16 or H− into various wild-type hosts. *FEMS Microbiol. Lett.* **29**, 83-89.

Mannose-Resistant Haemagglutination Gene(s) of *Salmonella typhimurium*

Madelon Halula[a]* and B. A. D. Stocker

Department of Medical Microbiology, Stanford University, Stanford, CA 94305, USA

The initial step in Salmonella enteritis is thought to be association of the organism with and invasion of intestinal epithelial cells. The presence of mannose-resistant haemagglutination (MRHA) activity in *Salmonella typhimurium* has been correlated with the ability of bacteria to adhere to and invade HeLa cells [1]. We initiated studies to further characterize the gene(s) responsible for this phenotype in *S. typhimurium*. Our survey demonstrated MRHA production by 90% of the 70 clinical and laboratory strains we have examined to date.

We have previously reported isolation of a 5.8 kb DNA fragment from *S. typhimurium* strain SL5166 which causes MRHA of sheep erythrocytes and adherence to HeLa cells when present on a plasmid in *Escherichia coli* HB101 [2]. We have used Tn5 transposon mutagenesis to determine which regions of the cloned DNA are required for MRHA and adherence. In cases examined so far, Tn5 insertions which cause loss of MRHA also cause reduced adherence to HeLa cells.

To study the effect of these insertion mutations in the Salmonella background, plasmid fragments containing Tn5 insertions which cause loss of MRHA and adherence in *E. coli* were re-introduced into the *S. typhimurium* chromosome replacing the wild-type DNA segments. These strains show the loss of MRHA and reduced adherence.

[a]*Telephone:* 1-415-497-2673

*Address from June 1, 1986: Department of Microbiology and Immunology, MCV-Station, Virginia Commonwealth University, Richmond, Virginia, USA.

Protein-Carbohydrate Interactions in Biological Systems. ISBN 0 12 436665 1

Using minicell and *in vitro* translation techniques we have found the 5.8 kb DNA fragment codes for at least four peptides with molecular weights of 58, 34, 27 and 12 kD. Further studies with deletion mutants should enable us to assign specific peptides to particular regions of the cloned fragment.

ACKNOWLEDGEMENTS

This work was supported by Public Health Service research grant AI07168 from the National Institute of Health and by a donation from Johnson and Johnson, Inc.

REFERENCES

1. Jones, G. W. and Leigh A. Richardson (1981). *J. Gen. Micro.* **127**, 361–370.
2. Halula, M. and B. A. D. Stocker (1985). *Abstr. Ann. Meet. Am. Soc. Microbiol.*

Regulatory Aspects of the K99 Fimbriae Synthesis

Bert Roosendaal[a], Paul M. P. van Bergen en Henegouwen*, Frits R. Mooi and Frits K. de Graaf

*Department of Molecular Microbiology, Free University, Amsterdam, The Netherlands, and *Department of Molecular Cell Biology (EMSA), State University of Utrecht, Utrecht, The Netherlands*

K99 fimbriae are extracellular filamentous structures that enable *E. coli* strains to adhere to intestinal epithelial cells of calves, lambs and piglets. The K99 genetic determinant encodes for at least seven polypeptides involved in the biogenesis of K99 fimbriae [1]. Expression of K99 fimbriae is dependent on at least four factors, namely: (1) growth rate; (2) pH; (3) temperature; and (4) alanine concentration. The aim of our study is to unravel the mechanisms which cause the mentioned regulatory features. The experimental approach was as follows: (1) we analysed the DNA region responsible for the regulation of K99 expression by means of mutagenesis and operon fusions; and (2) an immuno-gold labelling study was carried out to study the effect of temperature on the synthesis and assembly of K99 fimbriae.

After DNA sequencing the promoter region of the K99 operon (Fig. 1) we cloned regulatory sequences into a galactokinase transcription vector [2]. With this cloning system we confirmed the existence of two promoters and a terminator, as deduced from the DNA sequence, and located two more promoters, one of which is probably involved in the expression of gene A, while the other is probably not involved in K99 expression. The gene A product is involved in K99 expression since a mutation in this gene diminished K99 expression 2-3 times.

Temperature regulates K99 synthesis on the level of transcription since an *Hpa*I fragment (Fig. 1), containing the promoters P_B and P_C as well as the genes B and C and the terminator T, cloned in a galactokinase transcription vector, showed the same temperature regulation for galactokinase. Experiments to verify which

[a]*Telephone:* 31-20-5483548

Protein-Carbohydrate Interactions in Biological Systems. ISBN 0 12 436665 1

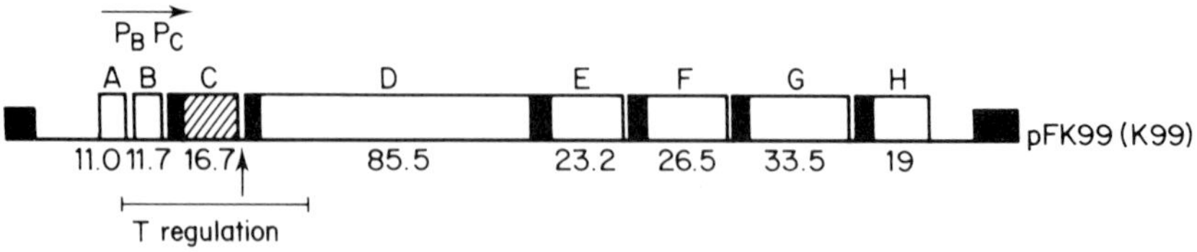

Figure 1. Genetic map of cloned K99 DNA. The thick black lines and the thin line represent pBR322 and cloned DNA, respectively. The black ends of the boxes indicate the parts of genes coding for the signal peptide. The fimbrial subunit gene is shaded. The horizontal arrow indicates the direction of transcription. The vertical arrow indicates the region of dyad symmetry (T). The numbers correspond to the molecular weights of the polypeptides in kilodaltons. P, promoter; bp, base pair.

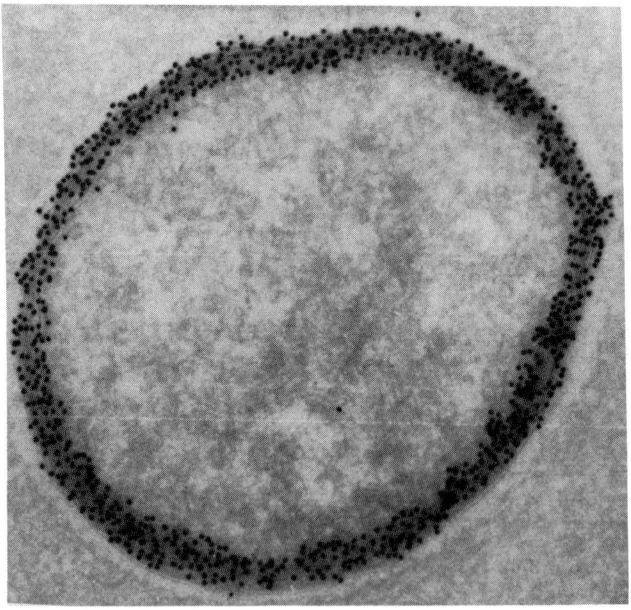

Figure 2. Cross-section of an *E. coli* cell 20 min after a temperature shift from 18°C (non-fimbriae expressing state) to 37°C (fimbriae expressing state). K99 fimbrial subunits are labelled with a K99 specific goat serum and gold (10 nm) labelled rabbit anti-goat as second antibody.

of the regulatory signal(s) is/are responsible for this feature are in progress.

The immuno-gold labelling of K99 antigens in a temperature shift (18°C to 37°C) experiment revealed that limited K99 synthesis occurred at 18°C while no assembly occurred at this temperature. After the shift K99 synthesis increased and at least 60 min later assembly of subunits into fimbriae was observed. The K99 subunits are probably cotranslationally transported across the inner membrane and accumulate transiently in the periplasmic space (Fig. 2). The subunits are subsequently transported across the outer membrane and assembled into fimbriae.

REFERENCES

1. De Graaf, F. K., B. Krenn and P. Klaassen (1984). *Infect. Immun.* **43**, 508–514.
2. McKenney, K., H. Shimatake, D. Court, U. Schmeissner, C. Brady and M. Rosenberg (1981). *In* "Gene Amplification and Analysis" (Eds J. C. Chirikjian and T. S. Papens) Vol. II, Analysis of Nucleic Acids by Enzymatic Markers. Elsevier-North Holland, Amsterdam.

Characterization and Cloning of Non-Fimbrial Protein Adhesins of Two *Escherichia coli* Strains of Human Origin

Gary Hinson[a] and Peter Williams

Department of Genetics, University of Leicester, Leicester, UK

We have previously described two facultatively enteropathogenic strains of *Escherichia coli*, 444-3 (O?:H4) and 469-3 (021:H−), isolated from human infants with symptoms of dysentery-like diarrhoea [1]. Both strains cause mannose-resistant haemagglutination (MRHA) only of human erythrocytes, and are able to adhere to and subsequently invade human epithelial cells in culture. These functions are mediated by surface proteins comprising aggregates of more than 300 identical subunit molecules, each of 14 to 14.5 kD, although electron microscopy has failed to reveal the presence of true fimbrial structures associated with these adhesins [2]. Derivatives deficient in MRHA, isolated following chemical mutagenesis of both strains, did not produce the MR adhesins and were unable to adhere to cultured human epithelial cells.

In order to study the adhesins in greater detail, we have constructed a cosmid library of total cellular DNA from 469-3 in a non-haemagglutinating *E. coli* K-12 host strain. We then screened for individual haem-adsorbing clones by a technique which involves making a nitrocellulose print of an agar plate containing a number of colonies, washing the filter with saline containing BSA, then with 5% fresh human erythrocytes. Clones expressing adhesins leave clearly visible red areas on the print, and the original colonies are picked off the plate and streaked out. Several hundred isolates may thus be screened with great ease.

The haem-adsorbing clones from the 469-3 library were able to haemagglutinate human erythrocytes and to adhere to cultured human epithelial cells. They

[a]*Telephone:* 44-533-551234

Protein-Carbohydrate Interactions in Biological Systems. ISBN 0 12 436665 1

specifically react with antiserum raised against 469-3 and absorbed with non-haemagglutinating derivatives of 469-3, and they express large, surface-located aggregates of identical protein subunits of 14 kD. Subcloning from the original cosmids is in progress, and preliminary data indicate that the system resides within a DNA fragment smaller than 12 kb. It will be interesting to determine the detailed genetic organization of the cloned adhesin determinants in comparison with those of other fimbrial and non-fimbrial adhesins that have been described. Furthermore, and perhaps more important, is the possibility of analysing regulatory mechanisms for adhesin expression. The parent strains express their MR adhesins maximally at 37°C, but to less than 1% of this level at a growth temperature of 22°C. Moreover, we have shown by immune fluorescence microscopy that less than one third of the cells in a genetically homogeneous population express the surface MR adhesins at one time, while other cells may carry type I (MS) fimbriae or glycocalyces (capsules). Our findings that haem-adsorbing clones from the 469-3 cosmid library also show temperature-dependent expression of MRHA activity and epithelial cell adhesion, and demonstrate antigenic variability within populations, indicate that regulatory regions are included in the clones and are thus amenable to molecular genetic analysis.

REFERENCES

1. Williams, P. H., S. Knutton, M. G. M. Brown, D. C. A. Candy and A. S. McNeish (1984). Characterisation of nonfimbrial mannose-resistant protein hemagglutinins of two *Escherichia coli* strains isolated from infants with enteritis. *Infect. Immun.* **44**, 592–598.
2. Knutton, S., P. H. Williams, D. R. Lloyd, D. C. A. Candy and A. S. McNeish (1984). Ultrastructural study of adherence to and penetration of cultured cells by two invasive *Escherichia coli* strains isolated from infants with enteritis. *Infect. Immun.* **44**, 599–608.

II. Antigenic Variation of Bacterial Adhesins

Co-Chairmen: Dan Danielsson
Gary Schoolnik

Regulation and Production of *N. Gonorrhoeae* Pilus Phase and Antigenic Variation

Ellyn Segal[a] and Magdalene So

Department of Molecular Biology,
Scripps Clinic and Research Foundation, La Jolla, CA 92037, USA

Neisseria gonorrhoeae is responsible for the sexually transmitted disease of gonorrhoea. Attachment to a mucosal surface is necessary for infection to occur. One of the most thoroughly examined surface structures is the gonococcal pilus. The pilus protein is a homopolymer composed of identical subunits (pilins) of ~18 kilodaltons [1]. The pilus undergoes phase variation: *in vitro*, piliated (P+) cells can produced non-piliated (P−) derivatives at a frequence of $\sim 10^{-3}$, depending on strain and culture conditions. The frequency of a P− to P+ switch is $\sim 10^{-5}$. P+ cells are infectious while P− are not [2]. A second membrane protein, PII (Op), also undergoes phase variation. Thus, there are four phenotypic combinations possible: P+PII+, P+PII−, P−PII+ and P−PII−. A single gonococcal colony of one phenotype can give rise to any of the other three. There is no correlation between the phase of piliation and PII. Pilus antigenic variation has been observed in two different strains in the laboratory [3,4] and many pilus serotypes have been identified in clinical isolates. The genetics of pilin gene phase variation and variability will be discussed below.

PHASE VARIATION

A piliated transparent (P+PII−) isolate of gonococcal strain MS11 was used as the progenitor of lines of pilus phase variants (Fig. 1) [5]. A single P+ colony (A series) was passaged daily until a P− variant (B) was isolated. P+ revertants

[a] *Telephone:* 1-619-455-8005

Protein-Carbohydrate Interactions in Biological Systems. ISBN 0 12 436665 1

	A		B		C		D		E		F		G		H		I	
Piliation	+	→	−	→	+	→	−	→	+	→	−	→	+	→	−	→	+	→

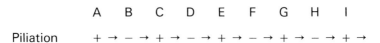

Figure 1. Derivation of lines of phase variants. A represents the original MS11 P+ isolate. The state of piliation is indicated by a (+) for piliated and a (−) for non-piliated.

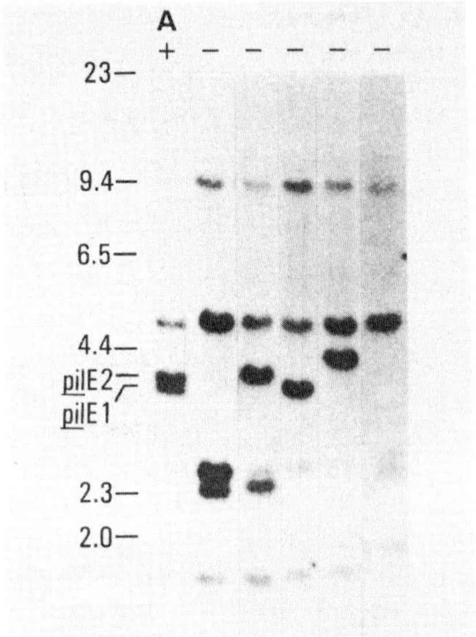

Figure 2. Southern blot analysis of the original (A) P+ isolate and five P− (B) phase variants derived from it. *Cla*I-digested total genomic DNA was probed with an expression site pilin gene cloned from the A isolate. The locations of *pil*E1 and *pil*E2 are indicated.

(C) were isolated from the B (P−) series. The D series (P−) were then isolated from C. All isolates were in a PII− background. Using this protocol, 35 lines of phase variants were derived, with some lines having undergone phase variation up to eight times. These lines were used to study the genetics of pili phase variation and variability.

Numerous regions of the gonococcal genome contain pilin sequences. However only two regions of the chromosome function as pilin expression site [6]. These two loci, *pil*E1 and *pil*E2, are separated from each other by ~20 kb. *pil*E1 is located within a *Cla*I fragment of 4.0 kb, while *pil*E2 is within a 4.1 kb *Cla*I fragment. *pil*S1, a region which contains silent pilin sequences, is ~15 kb away from *pil*E1.

A cloned expressed pilin gene from the original (A) MS11 P+ isolate was used as a probe in Southern blot analysis of phase variants. Figure 2 shows that a P+ to P− switch can result from a deletion event at either one or both expression sites. Sequence data from several cloned B series *pil*E1 and *pil*E2 loci showed that the deletions were of varying sizes and often resulted from single or multiple recombination events between directly repeated sequences within and flanking the expression sites [5]. Although the deletion event can be of variable length, a region 3′ of the gene is never affected. This region is ~70 bp downstream from the pilin stop codon and is termed the *Sma* Repeat.

Many of the P− variants studied have one expression locus intact [5] and ~10% of all P− variants have neither expression site deleted. These intact P− expression sites contain the entire pilin gene and promoter sequence [4]. Seven such P− variants were examined and found to make very little or no pilin (E. Billyard, E. Segal and M. So, unpubl. obs.). Clones of these intact P− expression sites are capable of pilin synthesis in *Escherichia coli* (E. Billyard and M. So, unpubl. obs.). Therefore the P+ to P− switch results from one of two mechanisms. In one, the pilin structural gene is removed from the expression loci by a deletion event. Secondly, a *trans*-acting repressor regulates pilus expression; this factor is itself regulated by the phase switch. Support for the existence of this repressor has recently become available. Crude extracts made from P− cells will specifically bind a DNA fragment containing the pilin promoter (E. Billyard and M. So, unpubl. obs.). Additionally, a recombinant clone from a P− gene bank has been identified which acts negatively on the pilin gene promoter present on another plasmid in *E. coli* (C. Marchal and M. So, unpubl. obs.).

ANTIGENIC VARIATION

Expressed pilin genes

Data obtained from primer extension of mRNA from the original (A) MS11 isolate and its piliated derivatives [4] has provided detailed information on the production of pili variability. A single P+ progenitor can give rise to piliated derivatives which express pilin different in sequence. The pilin gene deleted from an expression site during a P+ to P− switch can be re-expressed. In strain MS11, certain pilin genes are preferentially expressed. The expressed pilin gene is divided into three regions: constant (C), semivariable (SV), and hypervariable (HV)

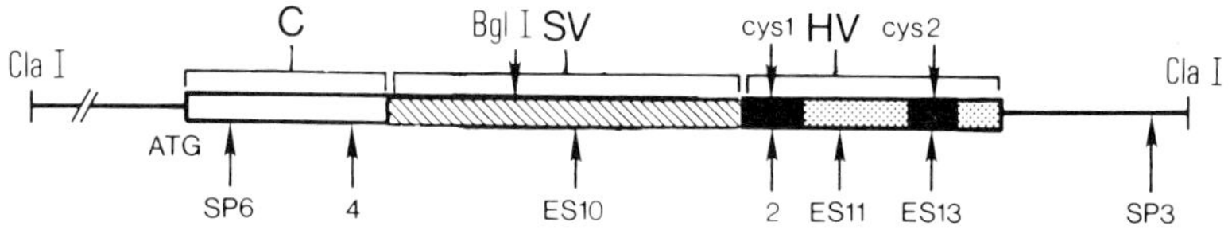

Figure 3. Map of the expression site pilin gene. The C region is represented by the clear box, SV by the hatched box and HV by the stippled box. Solid boxes show the cys1 and cys2 regions. The locations of oligonucleotides synthesized are indicated below.

(Fig. 3). The SV and HV gene segments can be mixed and matched in an expressed pilin gene. The C region comprises amino acids 1-53. The SV region, extending from amino acids 54-114, is characterized by single amino acid substitutions, resulting from single base changes. The HV carboxy-terminal portion of the protein is characterized by in-frame insertions and deletions of one to four codons, along with amino acid substitutions [4]. cys1 and cys2, two segments surrounding the HV region, are always conserved. Centring around the cysteine residues at positions 121 and 151, respectively they are 13 and 10 amino acids long.

To access the similarity between laboratory derived variants and *in vivo* isolates, mRNA primer extension analysis was carried out on clinical isolates from three patients whose origin of infection could be traced to a single source [4]. The isolates were shown to be members of the same strain. Sequence data showed that pilin antigenic variation did occur *in vivo*. The C, SV and HV regions defined by the MS11 sequences were observed in these variants, and variation resulted from nucleotide changes mechanistically similar to those seen for MS11.

The relationship between sequence change and antigenic variation was investigated through the use of double-diffusion assays. Rabbit IgG made against the original MS11 pilin was tested with pilus prepared from P+ derivatives. Changes in sequence correlated to a difference in antigen type in all variants tested [4].

P− derivatives with intact pilin expression sites synthesize very little or no pilin. In three such P− variants studied (4.1B, 5.1B and 6.1B), examination of pilin transcripts produced in *E. coli* showed that variation has occurred during the P+ to P− switch [4]. The pilin genes encoded by these P− variants have all the characteristics observed for P+ switchers. These observations suggest that pilus phase and antigenic variation are closely linked, though presently it cannot be determined if the linkage is obligatory.

Silent loci

Oligonucleotides were synthesized to sequences specific to several regions of the expressed pilin gene in the original MS11 isolate and the phase variants (Fig. 3) and used in Southern hybridizations against total genomic DNA of phase variant line members. For each line, a single filter was used for all probes. The data obtained indicates that copies of a complete pilin gene are only present at the two expression sites. A complete pilin gene is assembled by recombination events involving at least three separate DNA segments.

Figure 4 shows the results of using C (probe no. 4) and HV (probe ES11) specific probes against *Cla*I digested DNA from the 3 line of switchers and an *Sma* Repeat probe (SP3) against DNA from the 2 line. The C probe hybridizes to silent loci of 3.1 and 1.2 kb, while the HV probe, expressed in variants 3C, 3G and 3I, hybridized to a *Cla*I silent locus of 1.8 kb. ES1a, a second HV probe, hybridized to a *Cla*I fragment of 6.8 kb, while ES12, a third HV specific probe, to *Cla*I fragments of 6.4 and 1.8 kb. These results imply that C and HV pilin gene regions do not reside together in the silent loci of the genome. A SV specific probe (ES10) hybridized to all the *Cla*I fragments recognized by a complete pilin gene except

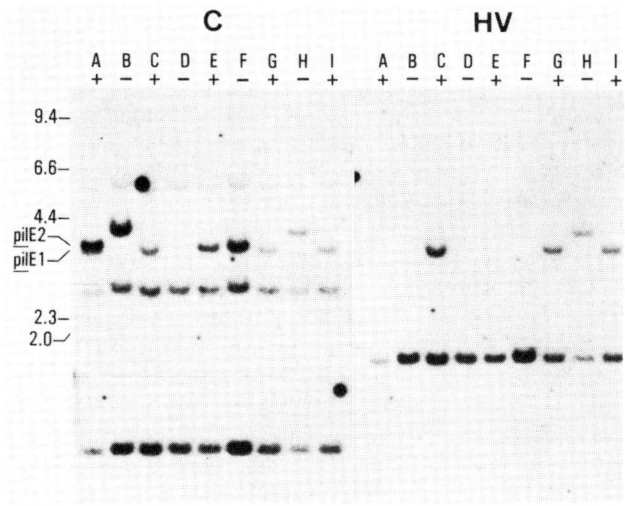

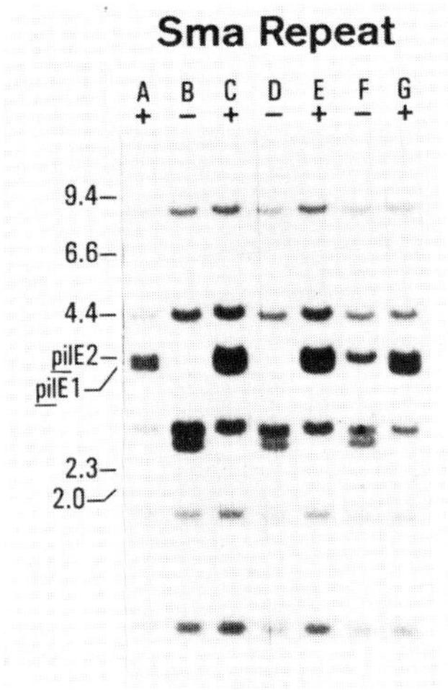

Figure 4. Southern blot analysis of silent C, HV, and *Sma* Repeat pilin sequences within the chromosome of line 3 and line 2 phase variants. Total genomic DNA was digested with *Cla*I and probed with an oligonucleotide specific for the designated region.

the 3.1 kb fragment encoding the silent C region and the 6.4 kb fragment containing silent HV sequences [7]. Thus silent C region and HV sequences do not always have silent SV region sequences associated with them.

The hybridization pattern resulting from the use of the *Sma* Repeat specific probe (SP3) is the same as seen upon using an entire pilin gene as a probe [5]. The identical results were observed upon using cys1 and cys2 specific probes (unpubl. obs.).

A pilin signal sequence probe (SP6), when hybridized to phase variant DNA under high stringency conditions, recognized only the expression loci. A deletion of the pilin genes from the expression loci results in the absence of the signal sequence from the genome. Only under reduced hybridization stringency conditions does the probe recognize additional restriction fragments in both P+ and P− variants (Segal *et al.*, submitted). These results imply that the 21 bp signal sequence is split into at least two sections in the silent loci and is only complete when present at a pilin expression site.

DISCUSSION

The genetic control of *N. gonorrhoeae* pilus phase and antigenic variation is novel and complex. A switch from P+ to P− can result from the deletion of the expressed pilin gene(s) from the expression sites. Additionally a *trans*-acting repressor controls pilin gene expression through action at the pilin expression locus. The repressor itself must be switch-inducible, implying a series of steps being involved in pilin gene control.

A complete copy of the pilin gene is present only in the expression sites. Silent constant and hypervariable pilin gene sequences occur in separate genomic regions. Silent semivariable sequences are not always located in the silent loci which contain constant and hypervariable sequences. Additionally, the pilin signal sequence is present in its entirety only in the expression sites. Providing homology needed for initial base pairing are the cys1, cys2, and *Sma* Repeat sequences, which are present in all silent loci. The *Sma* Repeat sequence is present in the 3′ part of the expression sites of both P+ and P− variants, regardless of whether the pilin gene is deleted, and may act as the initiation site for gene conversion.

The formation of a complete pilin gene in a deleted expression site requires multiple recombination steps involving several silent loci and the expression locus. The study of *N. gonorrhoeae* pilus phase and antigenic variation provides an insight into a novel method of prokaryotic gene regulation and rearrangement.

ACKNOWLEDGEMENTS

This work was supported by National Science Foundation grant PCM8340588. We wish to thank Emily Chen and Marc Nasoff for their help in synthesizing oligonucleotides.

REFERENCES

1. Salit, I. E., M. Blake and E. Gotschlich (1980). Intra-strain heterogeneity of gonococcal pili is related to opacity colony variance. *J. Exp. Med.* **151**, 716–725.
2. Kellogg, D. S., W. L. Peacock, W. E. Dacon, L. Brown and C. I. Pirkle (1963). *Neisseria gonorrhoeae* I. Virulence genetically linked to clonal variation. *J. Bacteriol.* **85**, 1274–1279.
3. Buchannan, T. M. (1975). Antigenic heterogeneity of gonococcal pili. *J. Exp. Med.* **141**, 1470–1475.

4. Hagblom, P., E. Segal, E. Billyard and M. So (1985). Intragenic recombination leads to pilus antigenic variation in *Neisseria gonorrhoeae. Nature* **315**, 156–158.
5. Segal, E., E. Billyard, M. So, S. Storzbach and T. F. Meyer (1985). Role of chromosomal rearrangement in *N. gonorrhoeae. Cell* **40**, 293-300.
6. Meyer, T. F., E. Billyard, R. Haas, S. Storzbach and M. So (1984). Pilus genes of *Neisseria gonorrhoeae*: Chromosomal organization and DNA sequence. *Proc. Natl. Acad. Sci. USA* **81**, 6110-6114.

Structural Model for *Neisseria gonorrhoeae* Pilin and Identification of a Non-Pilin-Mediated Glycolipid Binding Activity

Carolyn Deal[a], Elizabeth Getzoff, Nicklas Strömberg*, Gunilla Nyberg‡, Staffan Normark‡, Magdalene So, John Tainer and Karl Karlsson*

*Department of Molecular Biology, Scripps Clinic and Research Foundation, La Jolla, CA 92037, USA, *Department of Medical Biochemistry, Göteborg University, Göteborg, Sweden, and ‡Department of Microbiology, Umeå University, Umeå, Sweden*

The *Neisseria gonorrhoeae* pilin presents several unique problems to investigators in the field of protein structure determination. The pilin undergoes extensive antigenic variation [1], yet at the same time must form a physical structure, the pilus, which does not discernibly vary in structure. At the same time it must conserve other cellular functions, such as binding to DNA during the initial step of transformation and binding to eukaryotic cells [2]. Since the protein folding problem remains unsolved, the wealth of primary sequence data of pilin variants now available presents a challenge to the pursuit of the protein structure. The addressing of this problem in the gonococcal pilin system has yielded results which have been useful in suggesting immunological experiments and in advancing X-ray crystallographic studies.

STRUCTURAL PREDICTIONS

The 150–200 crystallographic three-dimensional protein structures provide a database for understanding the relationship between amino acid sequence and

[a]*Telephone:* 1-619-455-8005 *Present address of all authors.

Protein-Carbohydrate Interactions in Biological Systems. ISBN 0 12 436665 1

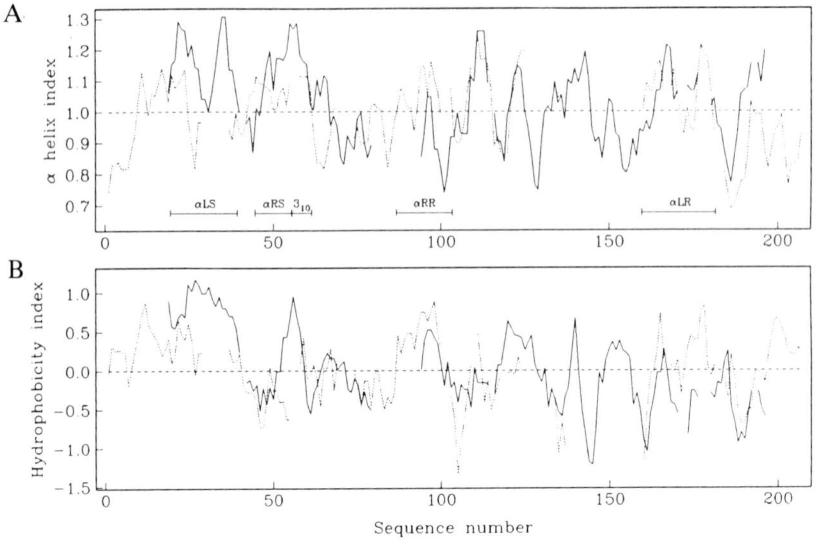

Figure 1. (a) Alpha-helix conformational prediction plotted against sequence number for MS11 gonococcal pilin (solid line) and TMV coat protein (dotted line). Numbering is defined by position in the sequence alignment (see Fig. 2). Breaks in each curve indicate deletions relative to the other sequence. Values above the indicated cutoff of 1.0 on the vertical axis define probable alpha-helical secondary structure. Bars at the bottom represent TMV alpha-helics and are labelled by the nomenclature used in describing the crystallographic structure [2]. (b) Comparison of hydrophobicity profiles for MS11 gonococcal pilin (solid line) and TMV coat protein (dotted line).

protein structure. The folding domains in the known protein structures can be classified into approximately fifteen groups, each of which forms a stable structural unit. These groups are defined by the identity and arrangements of core secondary structural elements such as alpha-helices, beta sheets and turns. The position of secondary structural elements in the sequence can be predicted with approximately 60% accuracy from empirically determined parameters reflecting the preferred occurrence of each amino acid in each type of structure. A secondary structure prediction for the gonococcal pilin (Fig. 1) was generated from a secondary structure prediction program written by Duncan McRee and Elizabeth Getzoff based on Chou-Fasman parameters [3] and the Eisenberg consensus hydrophobicity scale [4]. This predicted an overall secondary structure dominated by alpha-helix, but containing interspersed beta strands and regions dominated by turns, as a possible folding motif. Proteins with known alpha-helical domains in the same size range are the oxygen binding proteins, small cytochromes and the Tobacco Mosaic Virus (TMV) coat protein. Of these, the topology of the

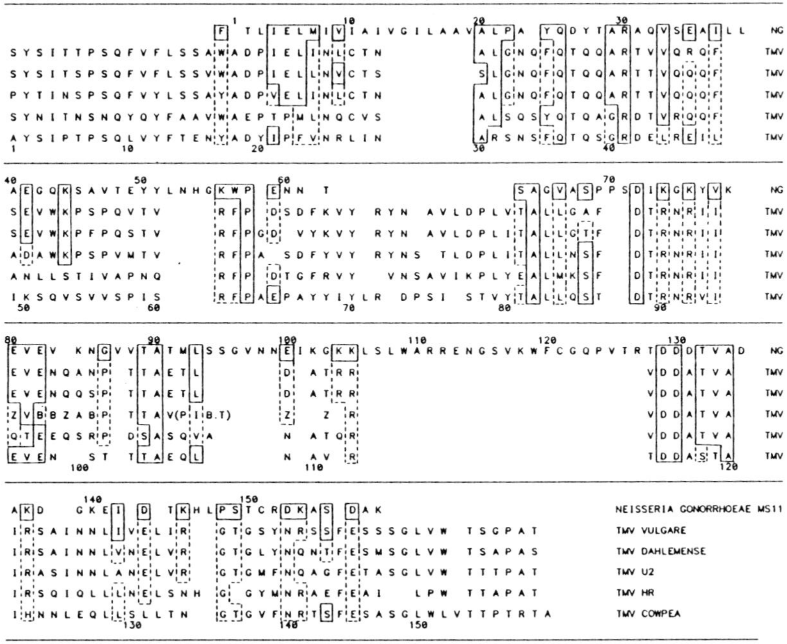

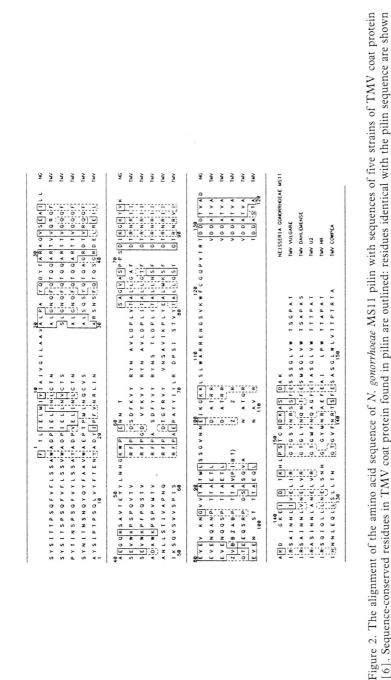

Figure 2. The alignment of the amino acid sequence of *N. gonorrhoeae* MS11 pilin with sequences of five strains of TMV coat protein [6]. Sequence-conserved residues in TMV coat protein found in pilin are outlined: residues identical with the pilin sequence are shown with a solid outline; similar residues are shown with a dashed outline. If identical amino acids could not be aligned, residues with similar physical and chemical properties were aligned to conserve charge, hydrogen-bonding abilities, aromatic rings, hydrophobicity, and size. The sets within which residues were considered similar are: (F,W,Y), (F,I,L,M,V), (A,G), (G,P), (D,N), (Q,N), (E,Q,Z), (B,D,E,Z), (H,K,R), and (S,T). The MS11 pilin sequence numbers are given at the top, and the TMV at the bottom.

TMV coat protein domain [5], an antiparallel helix bundle, is the most compatible with the predicted pilin secondary structure.

The pilin sequence was aligned with five sequences of TMV coat protein (Fig. 2) obtained from the Protein Sequence Database [6]. Amino acid identities between pilin and only one TMV were ignored, while amino acids present in the five TMV strains were assumed to have functional or structural significance. In this alignment amino acid identity was not necessary, but pattern matching was conserved; that is, residues with similar physical and chemical properties were aligned, conserving charge, hydrogen-bonding ability, hydrophobicity, and size. The similarities identified were among regions in the TMV variants which are important for intra-subunit folding [5], inter-subunit contacts, and binding of nucleic acid [7]. Additionally a detailed comparison of the predicted pilin secondary structure with the secondary structure of TMV [5] determined from the X-ray crystal structure shows good agreement of the helical regions. As indicated in Fig.1, of the four TMV helices, the first three helices align with helically predicted regions of pilin. The fourth aligns with a shorter region in the pilin predicted to consist of turns which could interlock to form a 3-10 helix. The two turns in TMV align with predicted turns in pilin. An alignment of the hydrophobicity profiles predicts similar folding motifs, and indicates a probable similarity in presentation to the environment of inside/outside components of the molecule.

Based on this analogy, the phase diagram of TMV solubility [8] was used to predict conditions for the dissociation of pilus filaments into subunits, and their subsequent reassembly into filaments was found to be indistinguishable by electron microscopy from the original filaments (Fig. 3). The subunits obtained by this dissociation were used to grow block-like crystals measuring approximately 50 μm in each direction. Electron microscopy of the crystals revealed a lattice spacing of 32 nm by 20 nm by 3 nm. This packing is compatible with that predicted from the deduced subunit structure based on the analogy with TMV. Therefore the prediction has been useful in determining conditions which allow the growth of crystals of a previously difficult to crystallize molecule. Solving of the pilin crystal structure is the first step in solving the supramolecular structure of the pilus filament.

CARBOHYDRATE DIRECTED RECOGNITION

Adherence of bacteria to eukaryotic cell surfaces is an important initial event in bacterial-host cell interactions [9]. For *N. gonorrhoeae*, several surface molecules, pilin and protein II for example, have been implicated as playing an important role in this interaction [2]. In the case of pilin, which exhibits extensive antigenic and sequence variation in the carboxy-terminal region of the protein [1], binding is suggested to be mediated by the constant amino terminal region [10]. Due to the postulated role of pilin in mediating binding, it was decided to couple structural studies with an investigation of the function of the pilin in binding. Of additional interest was the nature of the receptor on eukaryotic cells which *N. gonorrhoeae* recognizes. Cell surface glycoconjugates, glycolipids and

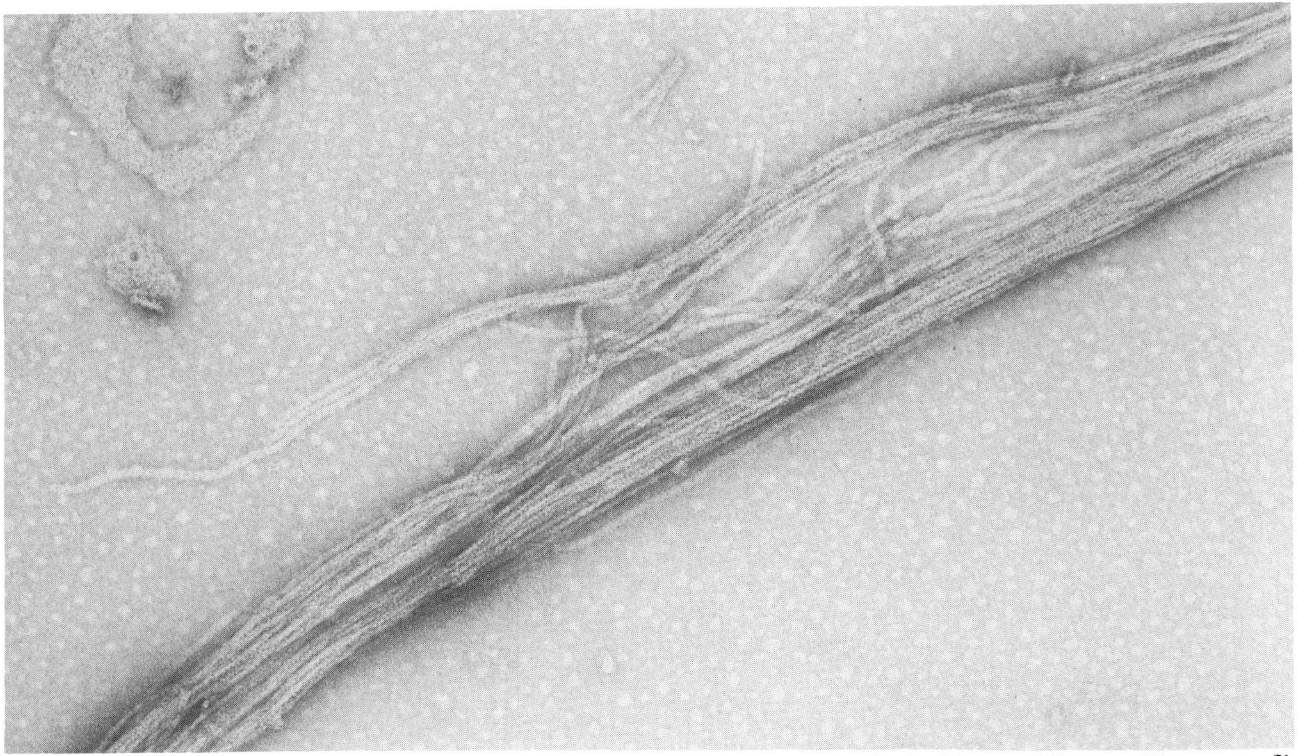

Figure 3. Electron micrograph (168 000 ×) of reassembled pilin filaments negatively stained with uranyl formate. The narrow fibres (about 6 nm width) are indistinguishable from negatively stained native pili. In the pilin sheets (lower right) which appear in solution at pH 8 to 9, the longitudinal filaments 6 nm apart are clearly distinguished and regular fine structure can be seen that is not present in individual fibres.

A

B

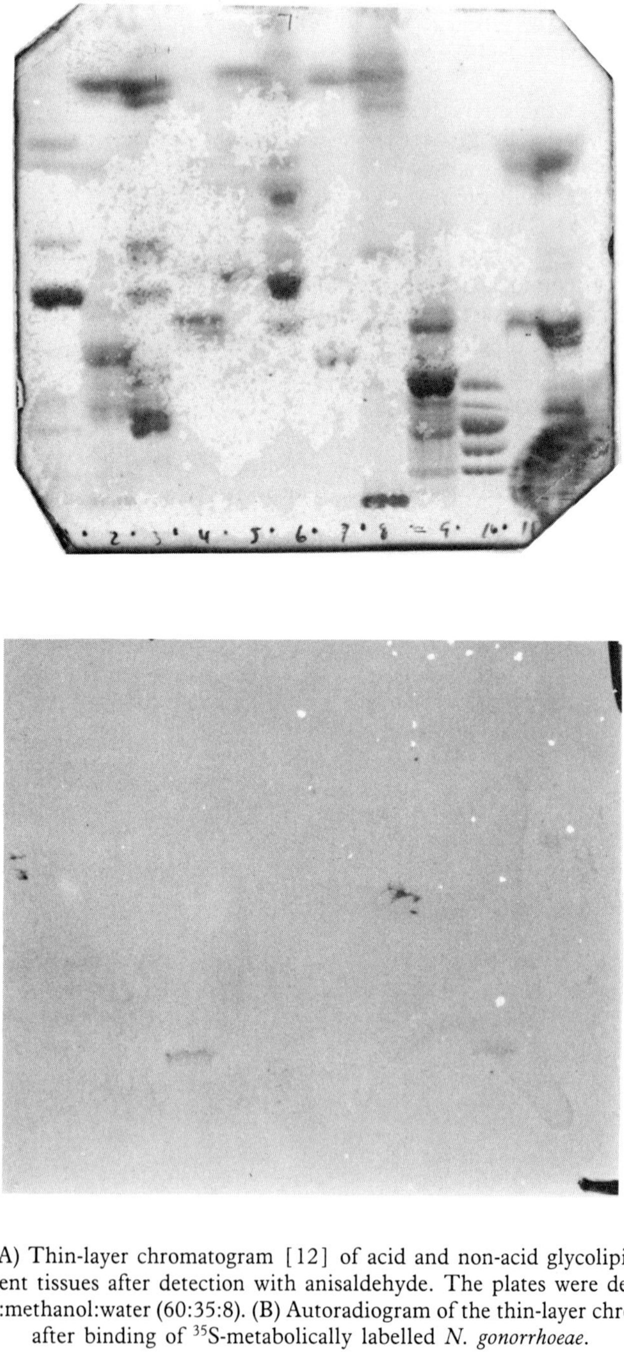

Figure 4. (A) Thin-layer chromatogram [12] of acid and non-acid glycolipid fractions from different tissues after detection with anisaldehyde. The plates were developed in chloroform:methanol:water (60:35:8). (B) Autoradiogram of the thin-layer chromatogram after binding of ^{35}S-metabolically labelled *N. gonorrhoeae*.

glycoproteins, have been suggested to function as specific receptors for bacteria, viruses, and toxins [9,11], therefore these molecules were the subject of our investigation.

To investigate the interaction of *Neisseria* with potential receptor candidates on the eukaryotic cell, gonococcal cells were exposed to glycolipids separated on thin layer chromatography plates [12]. As seen in Fig. 4, there was a specific recognition of a small number of bands from the larger library of glycolipids present on the TLC plate, indicating a specific carbohydrate-directed recognition and adhesion. Detailed analysis by mass spectrometry and N.M.R. spectroscopy of the carbohydrate moieties represented by these bands will allow the determination of the exact epitope recognized by the gonococcus and the further elucidation of its biological role.

To determine the moiety on the gonococcal cell necessary for this binding, several genetically characterized strains [1,13] of *Neisseria* were tested. Analysis of the strains in Table 1 reveals that the binding is not dependent on the state

Table 1. Genetically defined strains of *Neisseria gonorrhoeae* which demonstrated the thin-layer chromatogram binding pattern illustrated in Fig. 4. Isogenetic piliated and non-piliated strains of MS11 were isolated as described [1,13]. Protein II variants were kindly provided by Dr Janne Cannon. Protein I serotypes were provided by Dr Dan Daniellson. UM and the plasmid-free strains were obtained from the Microbiology Department of Umeå University. ML1 and LP1 are clinical isolates obtained from Dr Fred Sparling [1]

Strain	*Phenotype*	*Binding to thin layer plate*
MS11	P+ PII− PIBahjk	+
MS11-B4	P− PII−	+
MS11-B2	P− PII−	+
MS11-B1	P− PII−	+
MS11-C2	P+ PII−	+
MS11-C10	P+ PII−	+
UM	P+ PIAedghih	+
UM	P−	+
FA1090	P− PII−	+
FA1090	P− PIIa	+
FA1090	P− PIIb	+
FA1090	P− PIIc	+
FA1090	P− PIId	+
FA1090	P− PIIe	+
FA1090	P− PIIf	+
MLI	P+	+
LPI	P+	+
KH4318	P+ PII+ PIBack Cryptic plasmid-free)	+

of piliation of the gonococcal cell. Additionally, neither protein II nor any function encoded by the cryptic plasmid appears to be involved (Table 1). Therefore this assay identifies a non-pilin lectin activity present on the gonococcal cell surface which recognizes the receptor. Present activities are directed towards characterizing the exact molecular nature of this carbohydrate-directed recognition both from the receptor and bacterial points of view.

DISCUSSION

Our findings suggest analogies to the *Escherichia coli pap* pilus [14], where a minor adhesin component present in the pilus mediates binding to eukaryotic cells. It is possible that the pilus itself serves an architectural role, incorporating the adhesin into the pilus supramolecular structure in order to correctly present the adhesin to the eukaryotic cell surface. Alternatively, common epitopes on antigenic variants of pilin could mediate one mode of binding and other adhesins present on the cell surface could mediate other types of binding or trigger invasion events. Determination of these various possibilities will be possible following purification and genetic analysis of the adhesin proteins involved, coupled with structure determinations of the pilin and any associated proteins.

REFERENCES

1. Hagblom, P., E. Segal, E. Billyard and M. So (1985). *Nature* **315**, 156-158.
2. Cannon, J. G. and P. F. Sparling (1984). *Ann. Rev. Microbiol.* **38**, 111-133.
3. Chou, P. Y. and G. D. Fasman (1978). *Adv. Enzymol.* **47**, 45-148.
4. Eisenberg, D., R. M. Weiss, T. C. Terwilliger and W. Wilcox (1982). *Faraday Symp. Chem. Soc.* **17**, 109-120.
5. Bloomer, A. C., J. N. Champness, G. Bricogne, R. Staden and A. Klug (1978). *Nature* **276**, 362-368.
6. Barker, W. C., L. T. Hunt, B. C. Orcutt, D. G. George, L. S. Yeh, H. R. Chen, M. C. Blomquist, G. C. Johnson, E. I. Seibel-Ross and R. S. Ledley (1984). "Protein Sequence Database of the Protein Identification Resource". National Biomedical Research Foundation, Washington, DC.
7. Stubbs, G. and C. Stauffacher (1981). *J. Mol. Biol.* **152**, 387-396.
8. Durham, A. C. H., J. T. Finch and A. Klug (1971). *Nature New Biol.* **229**, 37-50.
9. Beachey, E. H. (Ed.) (1980). "Bacterial Adherence". Receptors and Recognition Series b, Vol. 8. Chapman and Hall, London.
10. Rothbard, J. B., R. Fernandez, L. Wang, N. N. H. Teng and G. K. Schoolnik (1985). *Proc. Natl. Acad. Sci. USA* **82**, 915-919.
11. Karlsson, K. A. (1982). "Biological Membranes" Vol. 4, pp. 1-74. Academic Press, London.
12. Hansson, G. C., K. A. Karlsson, G. Larson, N. Strömberg and J. Thurin (1985). *Anal. Biochem.* **146**, 158-163.
13. Segal, E., E. Billyard, M. So, S. Storzbach and T. F. Meyer (1985). *Cell* **40**, 293-300.
14. Normark, S., D. Lark, R. Hull, M. Norgren, M. Båga, P. O'Hanley, G. Schoolnik and S. Falkow (1983). *Infect. Immun.* **41**, 942-949.

Genomic Organization of Pilus and Opacity Genes in *Neisseria gonorrhoeae*

Rainer Haas, Peter Nickel, Anne Stern and Thomas F. Meyer[a]

*Max-Planck-Institut für medizinische Forschung and Zentrum für molekulare Biologie der Universität, Heidelberg, FRG, and *Max-Planck-Institut für Biologie, Tübingen, FRG*

Neisseria gonorrhoeae, a Gram-negative coccus causing a human sexually transmitted disease, has evolved an intriguing means of counteracting the immune system of its host. Several proteinaceous components of the surface of gonococci as well as an extracellularly secreted protease specific for human IgAl have been implicated with the virulence of the pathogen [1-3]. The relevant proteins, e.g. pilin, protein I (PI), opacity protein (PII or Op), and IgA protease, have been well characterized during the recent years and are now being investigated on the molecular genetics level [4-7], aiming towards a basic understanding of gonococcal pathogenesis.

Colonization of gonococci in the human mucosa appears primarily to depend on the expression of pilin, the subunit component of the pilus structure, and on the opacity protein (Op or PII) of the outer membrane. These proteins are subject to phase and antigenic variation. The production of pilus and opacity proteins can be turned on and off and a single cell can give rise to a large number of cell lines which express proteins of different structural, serological and functional characteristics. This flexibility in the expression of essential surface proteins and adhesins is thought to allow the pathogen to adapt to changes in the environments of its host organism and to evade the host immune response.

The cloning of pilus and Op genes enabled us to determine the structure and organization of these genes in the genome of *N. gonorrhoeae* and to study the molecular basis of surface antigen variation. As it turns out, a single gonococcal cell contains numerous gene copies for both pilus and Op protein, some of which

[a] *Telephone:* 49-7071-6011

*Present address of all authors.

Protein-Carbohydrate Interactions in Biological Systems. ISBN 0 12 436665 1

can be expressed and others are silent [8]. Silent and expression genes frequently undergo genetic recombination, thus causing changes in the expression of the respective proteins. These recombination events represent two classes, i.e. site-specific homologous recombination [9] and, perhaps more significant for antigenic variation in *N. gonorrhoeae*, gene conversion (R. Haas and T. F. Meyer, in prep.; also see below).

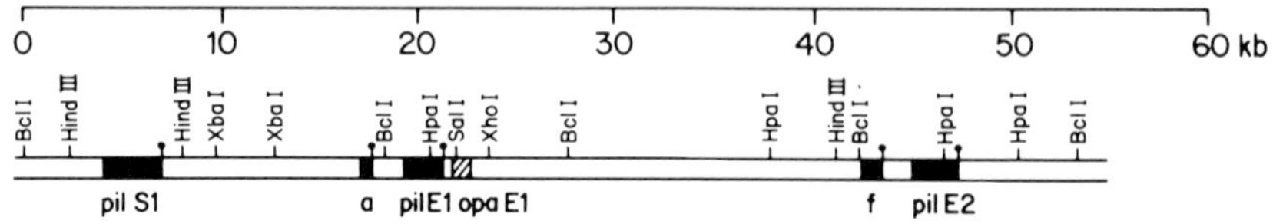

Figure 1. Genomic map of pilus and Op loci in *N. gonorrhoeae* MS11. The relative location of the pilus expression loci *pil*El and *pil*E2, the silent locus *pil*Sl, and the Op expression locus *opa*El have been reported, previously [5,8]. Two further silent pilus loci (a) and (f) have been mapped proximal to the pilus expression loci (R. Haas, unpub.). All mapped pilus loci are arranged in direct orientation as indicated by vertical lines with dots.

THE PILUS GENE SYSTEM

For the pilus genes, we identified a chromosomal section extending over 45 kb which contains the sole active expression loci *pil*El and *pil*E2 of *N. gonorrhoeae* MS11, and three further loci, among them the major silent locus *pil*Sl (Fig. 1). Physical analysis indicates that the relative orientation of the pilus loci mapped in this chromosomal section is identical with respect to pilus gene sequences and certain characteristic restriction sites. Hence, the pilus gene system provides a structural basis much different from the invertible segments controlling flagella phase variation in *Salmonella* [10] or expression of common pili in *Escherichia coli* [11].

The DNA sequence analysis of the expression loci [8] and, more recently, of the silent locus *pil*Sl (R. Haas and T. F. Meyer, submitted) provides further clues on the structure of gonococcal pilus genes. The expression loci, *pil*El and *pil*E2, both harbour complete genes capable to express pilin when cloned in *E. coli* [4]. The DNA sequence of both expression genes is identical, coding for a pilin molecule of 159 amino acids, plus a seven amino acid *N*-terminal leader peptide [8].

The silent locus *pil*Sl carries six pilus gene copies all lacking *N*-terminal sequences coding for the common determinants of gonococcal pilin (Fig. 2). The truncated gene copies are tandemly arranged and connected via a short repetitive segment. Interestingly, this striking repeat is also present in the expression loci upstream of the pilus promoter. Together with other repetitive segments, such as the *Sma*I/*Cla*I fragments at the 3′-terminal end of all pilus loci, it is assumed to be involved in the transfer of variant sequence information between *pil*Sl and pilus expression loci.

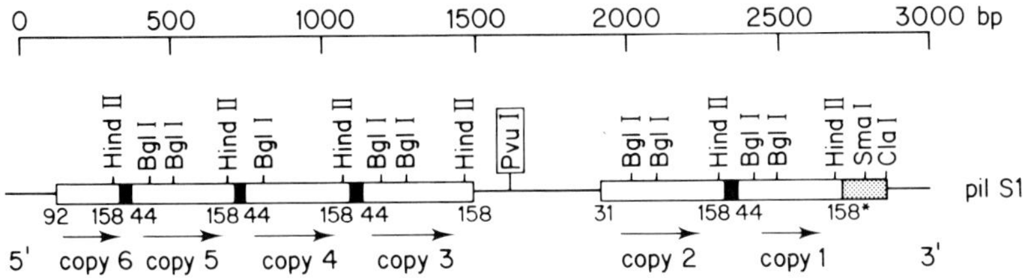

Figure 2. Detailed physical map of the major silent plus locus *pil*Sl. As determined by DNA sequencing (R. Haas and T. F. Meyer submitted), *pil*Sl contains six tandemly arranged pilus gene copies (copies 6 to 1) which are linked together by repetitive segments (solid boxes). At their 5′-terminal ends the gene copies are truncated, the numbers indicating the first and last amino acids present in each copy. Copy 1 is affiliated with an TGA stop codon (⋆) after positive 158 in the correct frame, which is followed by a conserved sequence with *Sma*I/*Cla*I restriction sites (dotted box).

The truncated pilus gene copies of *pil*Sl indeed harbour variant sequence information: While most silent copy sequences start at codon position 44 of pilin, this position also marks the beginning of a semi-variable (from positions 47 to 114) and a hyper-variable section (from positions 126 to 159) of the silent gene copies of *pil*Sl, as compared to the pilus expression gene. The variant sequences reveal numerous nucleotide exchanges as well as insertions and/or deletions of whole triplet codons with an increasing incidence of alterations towards the 3′-terminal ends of the silent gene copies. At distinct positions, the semi- and hyper-variable sections are interrupted by short, strictly conserved regions, up to 33 bp in size. These conserved regions, which are also found in the expression loci [12], may be essential to maintain the principal features of pili, but in addition to this, to allow sequence specific interaction between variant pilus genes. In this latter respect the intragenic conserved sequences and those located outside of the structural genes are supposed to fulfill similar functions.

THE OPACITY GENE FAMILY

As for the pilus, *N. gonorrhoeae* possesses a family of genes for opacity protein (Op). In order to investigate the mechanisms associated with the variable expression of Op proteins, we isolated several Op genes from different locations in the gonococcal genome. Strikingly, in three cases the cloned Op loci were located proximal to one of the pilus loci. The Op expression site *opa*El, which has been studied in greater detail is located approximately 600 bp downstream of *pil*El, and is transcribed in the same direction (see Fig. 1). In the light of the functional relatedness of both proteins this observation may be an indication for a subtle, as yet unknown, common control mechanism of both gene families. This hypothesis is confirmed by findings of Salit and coworkers who observed an apparent linkage in the expression of pilin and Op proteins in several gonococcal strains [13].

When we cloned the *opa*El locus from two isogenic variants of strain MS11 producing two different Op proteins, the *E. coli* clones, too, expressed the Op proteins made in the respective *N. gonorrhoeae* variants. A detailed sequence analysis of the coding regions of both variant genes, both derived from the *opa*El locus, revealed extensive homology in both sequences. This homology, however, is disturbed by a number of nucleotide exchanges and small deletions/insertions, both having no effect on the continuity of the open reading frame. Most of these small sequence alterations are clustered in two distinct areas within the Op gene

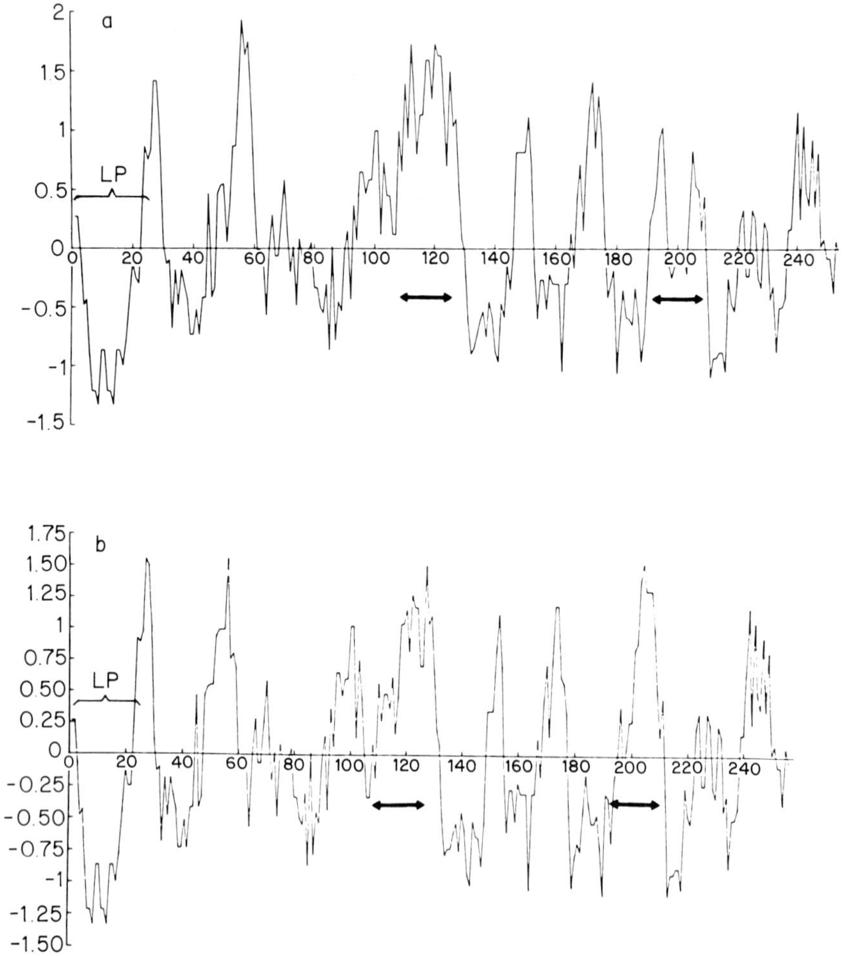

Figure 3. Hydropathy blot of two variant Op proteins. Both proteins are encoded by the *opa*El locus cloned from two different Op variants, (a) V28, and (b) VO [5]. LP indicates the hydrophobic leader peptide common to both proteins. The two directional arrows refer to the two hyper-variable regions of the Op proteins. These regions are strongly hydrophilic.

locus. Interestingly, the heterologous sequences code for hydrophilic amino acids (Fig. 3). We therefore assume that the hyper-variable sequences code for surface epitopes exposing the immunogenic parts of the Op proteins.

By comparison with several known *N*-terminal amino acid sequences of several Op proteins [14], the amino acid sequence predicted from our DNA sequence reveals an amino-terminal leader peptide (LP) in the precurser of Op protein (Fig. 3). This leader peptide carries positively charged residues followed by a strongly hydrophobic section, thus comprising the characteristics of a typical prokaryotic leader peptide. The leader is supposed to guide the Op protein to

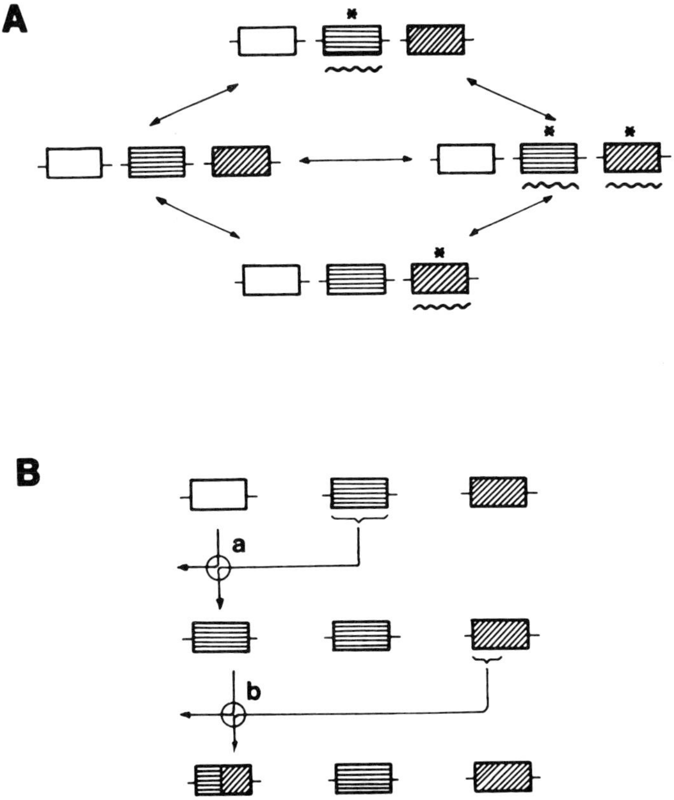

Figure 4. Two mechanisms proposed for phase and antigenic variation of Op protein. Boxes represent independent *opa* expression loci. (A) Phase switch mechanism. Each *opa* locus can be turned on and off, independently, causing the production of none, one, or two different Op proteins in a single cell. Active loci are marked with an asterisk and a wavy line. (B) Gene conversion mechanism. As supported by hybridization experiments with synthetic oligonucleotides (A. Stern and T. F. Meyer in prep.) complete (a) or partial (b) Op gene sequences can be duplicated, replacing analogous gene sequences in a second *opa* locus. Such an event, for example, has been observed for the *opa*El locus causing the expression of variant Op proteins (see Fig. 3).

the outer membrane of gonococci. Whether the Op protein is also transported to the cell surface in the *E. coli* clones remains to be seen.

To analyse the organization of Op loci in the chromosome of gonococci, and also to determine the origin of variant nucleotide sequences detected in the *opa*El locus of isogenic variants, we used specific DNA probes for genomic hybridization. With a complete Op gene as probe, the experiment revealed a large number of cross-hybridizing DNA fragments rendering a complex hybridization pattern which was identical for the different Op colony variants tested. Therefore no major chromosomal rearrangements were found to be associated with Op antigenic variation. However, using a small oligonucleotide probe specific for variable sequences in the *opa*El locus, only few chromosomal loci will cross-hybridize and we can demonstrate the exchange of variable sequence information between the different Op gene loci. These sequence variations do not affect the sizes of the respective fragments. The data suggest that gene conversion is one of the principal mechanisms responsible for antigenic variation in the opacity system.

The cloning of Op genes distinct from the *opa*El locus reveals that the other Op loci also contain complete structural genes having their own promoters. We therefore assume that two mechanisms may contribute to antigenic variation of Op protein: first, the phase switch (i.e. the "on" and "off" switch), an unknown control mechanism which regulates the expression of individual Op gene loci (Fig. 4A) and, second, a gene conversion event which can occur between variant Op genes (Fig. 4B), thus increasing the repertoire of variant Op genes by intragenic recombination.

ACKNOWLEDGEMENTS

We thank Sabine Stoerzbach for assistance during these studies and Melissa Brown for comments on the manuscript. This work was supported by the Deutsche Forschungsgemeinschaft grant number Me 705/2-2.

REFERENCES

1. Blake, S. M. and E. C. Gotschlich (1983). Gonococcal membrane proteins: speculation on their role in pathogenesis. *Progr. Allergy* **33**, 298–313.
2. Schoolnik, G. K., J. Y. Tai, and E. C. Gotschlich (1983). A pilus peptide vaccine for the prevention of gonorrhoeae. *Progr. Allergy* **33**, 314–331.
3. Plaut, A. G. (1983). The IgAl proteases of pathogenic bacteria. *Ann. Rev. Microbiol.* **37**, 603–622.
4. Meyer, T. F., N. Mlawer and M. So (1982). Pilus expression in *Neisseria gonorrhoeae* involves chromosomal rearrangement. *Cell* **30**, 45–52.
5. Stern, A., P. Nickel, T. F. Meyer and M. So (1984). Opacity determinants of *Neisseria gonorrhoeae*: Gene expression and chromosomal linkage to the gonococcal pilus gene. *Cell* **37**, 447–456.
6. Halter, R., J. Pohlner and T. F. Meyer (1984). IgA protease of *Neisseria gonorrhoeae*: Isolation and characterization of the gene and its extracellular product. *EMBO J.* **3**, 1595–1601.

7. Ganss, M. T., T. M. Buchanan, P. K. Kohl and T. F. Meyer (1985). Cloning of *Neisseria gonorrhoeae* protein IB in *Escherichia coli. In* "The Pathogenic Neisseriae" (Eds H. Dale *et al.*). American Society for Microbiology, in press.
8. Meyer, T. F., E. Billyard, R. Haas, S. Stoerzbach and M. So (1984). Pilus genes of *Neisseria gonorrhoeae*: Chromosomal organization and DNA sequence. *Proc. Natl. Acad. Sci. USA* **81**, 6110-6114.
9. Segal, E., E. Billyard, M. So, S. Stoerzbach and T. F. Meyer (1985). Role of chromosomal rearrangement in *N. gonorrhoeae* pilus phase variation. *Cell* **40**, 293-300.
10. Zieg, J. and M. Simon (1980). Analysis of the nucleotide sequence of an invertible controlling element. *Proc. Natl. Acad. Sci. USA* **77**, 4196-4200.
11. Freitag, C. S., J. M. Abraham, J. R. Clements and B. I. Eisenstein (1985). Genetic analysis of the phase variation control of expression of type 1 pimbriae in *Escherichia coli. J. Bacteriol.* **162**, 668-675.
12. Hagblom, P., E. Segal, E. Billyard and M. So (1985). Intragenic recombination leads to pilus antigenic variation in *Neisseria gonorrhoeae. Nature* **315**, 156-158.
13. Salit, I. E., M. Blake and E. C. Gotschlich (1980). Intrastrain heterogeneity of gonococcal pili is related to opacity colony variance. *J. Exp. Med.* **151**, 716-725.
14. Blake, M. S. and E. C. Gotschlich (1984). Purification and partial characterization of the opacity-associated proteins of *Neisseria gonorrhoeae. J. Exp. Med.* **159**, 716-725.

Antigenic Variation of Gonococcal Surface Proteins: Effect on Virulence

John E. Heckels[a] and Mumtaz Virji

Department of Microbiology, University of Southampton Medical School, Southampton General Hospital, Southampton, UK

The surface proteins of gonococci exhibit considerable antigenic variation. Of three major protein antigens which have been identified, namely pili and outer membrane proteins I and II (PI, PII), both pili and PII may differ between colonial variants of the same strain. In studies with a single strain, P9, we have identified variants which produce one of four pilus types α, β, γ, δ with subunit molecular weight of 19.5 kD, 20.5 kD, 21 kD and 18.5 kD respectively [1]. Variants of the same strain have also been isolated which express zero, one or two different molecular species of PII from a possible repertoire of at least six [2]. Since pilus and PII expression apparently vary independently, a large number of distinct variants can be generated by laboratory subculture of a single strain. Similar variations occur during the course of the natural infection so that variants of a strain growing at different anatomical sites express distinct pili and/or PII. To examine the effect of antigenic variation on virulence and immunity to gonococcal infection we have utilized variants of strain P9 with defined differences in surface antigen expression.

ANTIGENIC SPECIFICITY OF VARIANT PROTEINS

Antisera were raised by immunization of rabbits with purified pili. The degree of antigenic cross-reactivity between the different pili was determined by ELISA inhibition. In each case the antibodies produced were highly specific for the homologous pilus type [3]. Similar antigenic specificity was seen with antisera directed against PII [4]. Thus although variant proteins share considerable structural homology, type specific determinants are immunodominant.

[a] *Telephone:* 44-703-772222 ext. 3918

Protein-Carbohydrate Interactions in Biological Systems. ISBN 0 12 436665 1

Similar antigenic specificity is seen during the course of gonococcal infection. Antibodies produced are highly specific, reacting with only a single PII species produced by the infecting strain [5]. Antigenic shift in PII or pilus expression may thus enable gonococci to evade the consequences of the host immune response.

INFLUENCE OF ANTIGENIC VARIATION ON ATTACHMENT, INVASION AND CELL SPECIFICITY

A characteristic virulence attribute of gonococci which distinguishes gonococci from commensal bacteria is their ability to attach to and invade the epithelial cells of the genital tract. One model system which mimics this effect is the measurement of the cytotoxic effect of gonococci on monolayers of Chang conjunctival cells growing in tissue culture [3]. Cells are incubated with gonococci for 3 h after which non-adherent bacteria removed by washing. After a further incubation for 16 h the degree of cell killing is determined. In this system purified antigens show no effect and gonococcal cytotoxicity is abolished by the addition of cytoclasin B, demonstrating that virulence in the model is not due to the release of toxic factors but, like the natural infection, is dependent on gonococcal uptake into the cell. When variants of strain P9 are compared in this system (Table 1), considerable differences in virulence are seen. In each case the presence of pili increases cytotoxicity as does PIIb but PIIa is without effect.

Table 1. Effect of surface protein variation on the cytotoxic effect of variants of strain P9 (killing of Chang cells relative to the variant lacking Pili and PII)

Variant proteins present		*Relative virulence*
Pili	PII	
—	—	1.0
α	—	3.2
γ	—	6.2
δ	—	2.5
—	IIa	0.5
—	IIb	5.0

Although all pili increase cytotoxicity, their effect is not equal. The differences in virulence of the variants is paralleled by their relative adhesion. Thus the more toxic γ-piliated variant attaches to Chang cells much more readily than does the α-piliated variant. However, this relative difference is reversed when adhesion to buccal epithelial cells is measured. Similarly while PIIb is associated with greatest adhesion to Chang cells PIIa is most efficient in buccal cell adherence.

Thus variations in both pili and PII are associated with altered specificity of adhesion to different human epithelial types. Such specificity would facilitate colonization of different anatomic sites encountered during the natural infection.

EFFECT OF VARIATION ON PMN INTERACTION AND PHAGOCYTIC KILLING

Previous studies have suggested that both pili and protein II may play an important role in gonococcal-polymorphonuclear leukocyte (PMN) interactions. In particular pili have been associated with resistance to phagocytosis and hence killing [6] although studies by Swanson and co-workers suggest that the initial gonococcal-PMN interaction is dominated by PII [7]. We have utilized our variants of strain P9 to examine the individual effects of PII and pili on phagocytic killing and the initial gonococcal-PMN interaction as measured by luminol-dependent chemiluminescence (CL). All variants which lacked PII were resistant to killing by PMN and showed no significant differences in CL between non-piliated and piliated forms. In contrast, all variants containing PII were readily killed and showed an equivalent increase in PMN interaction as measured by CL. When piliated PII containing variants were compared with their non-piliated counterparts the species of PII present determined the degree of interaction (Fig. 1).

The similarity between CL assay and phagocytic killing suggests that killing is an inevitable consequence of the initial gonococcal-PMN interaction and that neither PII nor pili promote increased intracellular survival. Although PII causes

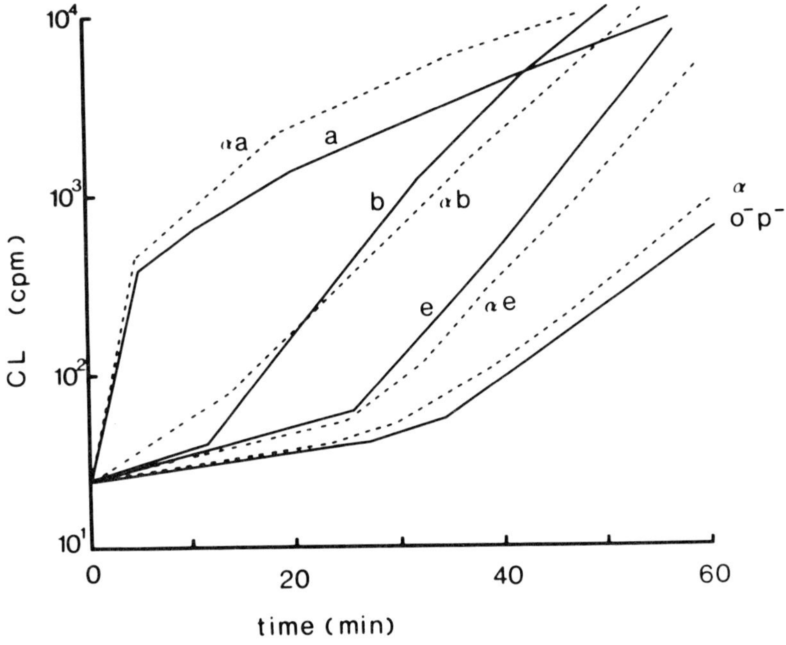

Figure 1. Effect of PII and pili on gonococcal PMN interactions. (Variants contained PII species IIa, IIb or IIe either in combination with α pili or alone.)

increased phagocytosis, most clinical isolates possess at least one PII, indicating it must confer other essential properties such as increased adhesion to mucosal surfaces. However PII-lacking variants arising during the natural infection may also play an important role escaping phagocytosis and ultimately recolonizing other niches.

BIOLOGICAL EFFECT OF ANTIBODIES DIRECTED AGAINST PILI AND PII

Since effective immunization against gonococcal infection would require production of widely cross-reacting antibodies, we have used monoclonal antibodies with defined specificity [8] to examine the potential protective effect of antibodies directed against conserved and variable pilus domains and also against a variable domain of PIIb (Table 2). No antibodies to conserved PII domains were available.

Table 2. Biological effect of monoclonal antibodies directed against variable surface proteins

Monoclonal antibody directed against:	*Inhibition of invasion*	*Bactericidal activity*	*Opsonic effect*
Pilus variable determinants	+	±	+
Pilus conserved determinants	–	–	–
PII variable determinants	+	+	+

In each case antibodies to the variable determinants exerted a protective effect. In contrast, two antibodies to distinct pilus conserved domains were without effect. The location of the epitopes recognized by these cross-reacting antibodies (SM1 and SM2) was investigated using the two fragments generated on CNBr cleavage of pili described by Schoolnik *et al.* [9].

Antibody SM1 reacted with fragment CNBr2 which contains common sequences between amino acids 9–92, while SM2 recognized a conserved epitope within the otherwise highly variable CNBr3. The location of the SM1 reactive epitope was further defined to lie within residues 48–60 by use of synthetic peptides provided by Dr G. Schoolnik.

The low levels of cross-reacting antibodies which are produced on immunization with intact pili also react with peptide 48–60. Like monoclonal SM1, antibodies produced against this synthetic peptide are unable to prevent pilus binding to epithelial cells [10]. Thus the natural immune response to pili generates only low levels of cross-reacting antibodies, and these are directed against a non-protective epitope.

SUMMARY

Antigenic shift in expression of pili and outer membrane protein II occur during gonococcal infection. This antigenic variation influences the ability to adhere to different cell types and may permit colonization of different anatomic sites. Although PII expression is associated with increased killing by human PMN most clinical isolates express PII.

With both pili and PII, the variable domains are immunodominant and little antigenic cross-reactivity is observed. Thus antigenic shift during infection may circumvent immune defences.

Antibodies to variable determinants are protective in model systems, but only low levels of cross-reacting antibodies are produced in response to pili and these are directed against nonprotective epitopes.

ACKNOWLEDGEMENTS

We are very grateful to Dr G. Schoolnik for providing synthetic pilus peptides. Portions of this work were supported by the Medical Research Council and the World Health Organization.

REFERENCES

1. Lambden, P. R., J. E. Heckels, H. McBride and P. J. Watt (1981). *FEMS Microbiol. Lett.* **10**, 339-341.
2. Lambden, P. R. and J. E. Heckels (1979). *FEMS Microbiol. Lett.* **5**, 262-265.
3. Virji, M., J. S. Everson and P. R. Lambden (1982). *J. Gen. Microbiol.* **128**, 1095-1100.
4. Diaz, J.-L. and J. E. Heckels (1982). *J. Gen. Microbiol.* **128**, 585-591.
5. Zak, K., J.-L. Diaz, D. Jackson and J. E. Heckels (1984). *J. Infect. Dis.* **149**, 166-173.
6. Dilworth, J. A., J. O. Hendley and G. L. Mandell (1975). *Infect. Immun.* **11**, 512-516.
7. King, G. J. and J. Swanson (1978). *Infect. Immun.* **21**, 575-584.
8. Virji, M. and J. E. Heckels (1983). *J. Gen. Microbiol.* **129**, 2761-2768.
9. Schoolnik, G. K., R. Fernandez, J. Y. Tai, J. Rothbard and E. C. Gotschlich (1984). *J. Exp. Med.* **159**, 1351-1370.
10. Rothbard, J. B., R. Fernandez, L. Wang, N. N. H. Teng and G. K. Schoolnik (1985). *Proc. Natl. Acad. Sci. USA* **82**, 915-919.

Serological Variants of the K88 Antigen

Wim Gaastra[a] and Per Amstrup-Pedersen

Laboratory for Microbiology, Technical University of Denmark, Lyngby, Denmark

Functional and antigenic domains can be assigned to various regions of the gonococcal fimbrial subunit protein ([1] and elsewhere in this volume). The central part of the molecule is a highly conserved region in all gonococcal fimbriae and mediates receptor binding. The *C*-terminal part of the molecule contains a variable region, which determines type-specific antigenicity and thus the serological diversity of the gonococcal fimbriae. The hydrophobic *N*-terminal of the protein may be involved in the stabilization of the polymeric structure of the fimbriae and is also highly conserved. A high antigenic variation also exists among meningococcal fimbriae [2]. Serologically at least three variants of the K88 antigen can be distinguished. The K88 antigen is involved in the pathogenesis of enterotoxigenic *Escherichia coli* strains which cause diarrhoea in pigs [3]. Functional and antigenic domains may probably also be assigned to the K88 fimbrial subunit molecule.

The amino acid sequence of the K88 fimbrial subunit protein has been determined for the three major serological variants known today, i.e. K88ab, K88ac and K88ad. The three serological variants all have a common antigenic factor consisting of one or more antigenic determinants. This common antigenic factor is termed "b", "c" or "d" in this terminology. Between the K88ab and the K88ac protein subunit there are 21 differences in the amino acid sequence and between the K88ab and the K88ad protein subunit there are 28 differences in the amino acid sequence (Fig. 1). The most striking of these differences are the insertion of a Lys residue between amino acid residues 104 and 105 and the deletion of three amino acid residues between residues 164 and 168 in the K88ac sequence which results in the fact that this protein is two amino acids shorter than the other two variants.

[a] *Telephone:* 45-2-224066 ext. 3546

Protein-Carbohydrate Interactions in Biological Systems. ISBN 0 12 436665 1

```
                       -21                   -15                 -10                 -5              -1
K88AB  GGG TTT AAT TTT ATG AAA AAG ACT CTG ATT GCA CTG GCA ATT GCT GCA TCT GCT GCA TCT GGT ATG GCA CAT GCC
                       Met Lys Lys Thr Leu Ile Ala Leu Ala Ile Ala Ala Ser Ala Ala Ser Gly Met Ala His Ala

K88AC  GGG TTT AAT CTT ATG AAA AAG ACT CTG ATT GCA CTG GCA ATT GCT GCA TCT GCT GCA TCT GGT ATG GCA CAT GCC
                       Met Lys Lys Thr Leu Ile Ala Leu Ala Ile Ala Ala Ser Ala Ala Ser Gly Met Ala His Ala

       1               5                   10                  15                  20                  25
K88AB  TGG ATG ACT GGT GAT TTC AAT GGT TCG GTC GAT ATC GGT GGT AGT ATC ACT GCA GAT GAT TAT CGT CAG AAA TGG
       Trp Met Thr Gly Asp Phe Asn Gly Ser Val Asp Ile Gly Gly Ser Ile Thr Ala Asp Asp Tyr Arg Gln Lys Trp
K88AD  TGG ATG ACT GGT GAT TTC AAT GGT TCG GTC GAT ATC GGT GGT AGT ATC ACT GCA GAT GAT TAT CGT CAG AAA TGG
       Trp Met Thr Gly Asp Phe Asn Gly Ser Val Asp Ile Gly Gly Ser Ile Thr Ala Asp Asp Tyr Arg Gln Lys Trp
K88AC  TGG ATG ACT GGT GAT TTC AAT GGT TCG GTC GAT ATC GGT GGT AGT ATC ACT GCA GAT GAT TAT CGT CAG AAA TGG
       Trp Met Thr Gly Asp Phe Asn Gly Ser Val Asp Ile Gly Gly Ser Ile Thr Ala Asp Asp Tyr Arg Gln Lys Trp

                       30                  35                  40                  45                  50
K88AB  GAA TGG AAA GTT GGT ACA GGT CTT AAT GGA TTT GGT AAT GTA TTG AAT GAC CTG ACC AAT GGT GGA ACC AAA CTG
       Glu Trp Lys Val Gly Thr Gly Leu Asn Gly Phe Gly Asn Val Leu Asn Asp Leu Thr Asn Gly Gly Thr Lys Leu
K88AD  GAA TGG AAA GTT GGT ACA GGT CTT AAT GGA TTT GGT AGT GTA TTG AAT GAC CTG ACC AAT GGT GGA ACC GAA CTG
       Glu Trp Lys Val Gly Thr Gly Leu Asn Gly Phe Gly Ser Val Leu Asn Asp Leu Thr Asn Gly Gly Thr Glu Leu
K88AC  GAA TGG AAA GTT GGT ACA GGT CTT AAT GGA TTT GGT AAT GTA TTG AAT GAC CTG ACC AAT GGT GGA ACC AAA CTG
       Glu Trp Lys Val Gly Thr Gly Leu Asn Gly Phe Gly Asn Val Leu Asn Asp Leu Thr Asn Gly Gly Thr Lys Leu

                       55                  60      ●           65      ●           70              ●   75
K88AB  ACC ATT ACT GTT ACT GGT AAT AAG CCA ATT TTG TTG GGC CGA ACC AAA GAA GCA TTT GCT ACG CCA GTA AGT GGT
       Thr Ile Thr Val Thr Gly Asn Lys Pro Ile Leu Leu Gly Arg Thr Lys Glu Ala Phe Ala Thr Pro Val Ser Gly
K88AD  ACC ATT ACT GTT ACT GGT AAT AAG CCA AGT TTG TTA GGC CGA ACC AGA GAG GCA TTT GCT ACG CCA GTG GTT GGT
       Thr Ile Thr Val Thr Gly Asn Lys Pro Ser Leu Leu Gly Arg Thr Arg Glu Ala Phe Ala Thr Pro Val Val Gly
K88AC  ACC ATT ACT GTT ACT GGT AAT AAG CCA ATT TTG TTA GGC CGA ACC AAA GAA GCA TTT GCT ACG CCA GTA ACT GGT
       Thr Ile Thr Val Thr Gly Asn Lys Pro Ile Leu Leu Gly Arg Thr Lys Glu Ala Phe Ala Thr Pro Val Thr Gly

                       80                  85                  90                  95                  100
K88AB  GGT GTA GAT GGA ATT CCT CAG ATT GCA TTT ACT GAC TAT GAA GGA GCT TCT GTA AAA CTC AGA AAC ACT GAT GGT
       Gly Val Asp Gly Ile Pro Gln Ile Ala Phe Thr Asp Tyr Glu Gly Ala Ser Val Lys Leu Arg Asn Thr Asp Gly
K88AD  GGT GTA GAT GGA ATT CCT CAT ATT GCA TTT ACT GAC TAT GAA GGA GCT TCT GTA GAA CTC AGA AAC CCT GAT GGT
       Gly Val Asp Gly Ile Pro His Ile Ala Phe Thr Asp Tyr Glu Gly Ala Ser Val Glu Leu Arg Asn Pro Asp Gly
K88AC  GGT GTA GAT GGA ATT CCT CAT ATT GCA TTT ACT GAC TAT GAA GGA GCT TCT GTA GTA CTC AGA AAC CCT GAT GGT
       Gly Val Asp Gly Ile Pro His Ile Ala Phe Thr Asp Tyr Glu Gly Ala Ser Val Val Leu Arg Asn Pro Asp Gly

                           105                 110                 115                 120                 125
K88AB  GAA ACT AAT AAA     GGT TTA GCA TAT TTT GTT CTG CCG ATG AAA AAT GCA GAG GGC ACT AAA GTT GGT TCA GTG AAA
       Glu Thr Asn Lys     Gly Leu Ala Tyr Phe Val Leu Pro Met Lys Asn Ala Glu Gly Thr Lys Val Gly Ser Val Lys
K88AD  GAA ATT GAA AAA     GGT TTA GCA TAT TTT GTT CTG CCG ATG AAA AAT GCA GAG GGC ACT AAA GTT GGT TCA GTG AAA
       Glu Ile Glu Lys     Gly Leu Ala Tyr Phe Val Leu Pro Met Lys Asn Ala Glu Gly Thr Lys Val Gly Ser Val Lys
K88AC  GAA ACT AAT AAA AAA GGT TTA GCA TAT TTT GTT CTG CCG ATG AAA AAT GCA GAG GGC ACT AAA GTT GGT TCA GTG AAA
       Glu Thr Asn Lys Lys Gly Leu Ala Tyr Phe Val Leu Pro Met Lys Asn Ala Glu Gly Thr Lys Val Gly Ser Val Lys
```

```
       Val Asn Ala Ser Tyr Ala Gly Ala Leu Gly Arg Gly Gly Val Thr Ser Ala Asp Gly Glu Leu Met Ser Leu Phe
K88AC  GTG AAT GCA TCT TAT GCC GGT GTG TTA GGG AGA GGT GGG GTT ACT TCT GCG GAC GGG GAG CTG CTT TCG CTT TTT
       Val Asn Ala Ser Tyr Ala Gly Val Leu Gly Arg Gly Gly Val Thr Ser Ala Asp Gly Glu Leu Leu Ser Leu Phe

        •              155                 160                 165                 170                 175
K88AB  GCG GAC GGG TTG CGC GCT ATC TTT TAT GGT GGT TTG ACG ACG ACT GTT TCG GGT GCT GCA CTC ACG AGT GGG AGT
       Ala Asp Gly Leu Arg Ala Ile Phe Tyr Gly Gly Leu Thr Thr Thr Val Ser Gly Ala Ala Leu Thr Ser Gly Ser
K88AD  GCC GAG GGG TCG CAC GCT ATC TTT TAT GGT GGT TTG CCG ACG AAT GTT CAG AAT TCT GCA CTC CCG GGT GGG AGT
       Ala Glu Gly Ser His Ala Ile Phe Tyr Gly Gly Leu Pro Thr Asn Val Gln Asn Ser Ala Leu Pro Gly Gly Ser
K88AC  GCC GAC GGG TTG AGC GCT ATC TTT TAT GGT GGT TTG CCG AGG                 GGT TCT GAA TCG TCG GCT GGG AGT
       Ala Asp Gly Leu Ser Ala Ile Phe Tyr Gly Gly Leu Pro Arg                 Gly Ser Glu Ser Ser Ala Gly Ser

                       180                 185                 190                 195                 200
K88AB  GCC GCA GCG GCG CGC ACA GAG TTG TTT GGA AGT CTA TCA AGA AAT GAT ATT CTC GGA CAG ATT CAA AGA GTA AAC
       Ala Ala Ala Ala Arg Thr Glu Leu Phe Gly Ser Leu Ser Arg Asn Asp Ile Leu Gly Gln Ile Gln Arg Val Asn
K88AD  GCC GCA GCG GCG CGC ACA GAG TTG TTT GGA AGT CTA TCA AAA AAT GAT ATT CTC GGA CAG ATT CAA AGA GTA AAC
       Ala Ala Ala Ala Arg Thr Glu Leu Phe Gly Ser Leu Ser Lys Asn Asp Ile Leu Gly Gln Ile Gln Arg Val Asn
K88AC  GCC GCA GCG GCG CGC ACA AAG TTG TTT GGA AGT CTA TCA AGA AAT GAT ATT CTC GGA CAG ATT CAA AGA GTA AAC
       Ala Ala Ala Ala Arg Thr Lys Leu Phe Gly Ser Leu Ser Arg Asn Asp Ile Leu Gly Gln Ile Gln Arg Val Asn

                       205                 210                 215                 220                 225
K88AB  GCA AAT ATT ACT TCT CTT GTT GAC GTC GCA GGT TCT TAC AGG GAA GAC ATG GAG TAC ACT GAT GGA ACT GTT GTT
       Ala Asn Ile Thr Ser Leu Val Asp Val Ala Gly Ser Tyr Arg Glu Asp Met Glu Tyr Thr Asp Gly Thr Val Val
K88AD  GCA AAT ATT ACT TCT CTT GTT AAC GTC CCA GGT TCT TTC AAT GAA AAC ATG GCG TAC ACT GAT GGA TCC GTT GTT
       Ala Asn Ile Thr Ser Leu Val Asn Val Pro Gly Ser Phe Asn Glu Asn Met Ala Tyr Thr Asp Gly Ser Val Val
K88AC  GCA AAT ATT ACT TCT CTT GTT GAC GTC GCA GGT TCT TAC AGG GAA AAC ATG GAG TAC ACT GAT GGA ACT GTT GTT
       Ala Asn Ile Thr Ser Leu Val Asp Val Ala Gly Ser Tyr Arg Glu Asn Met Glu Tyr Thr Asp Gly Thr Val Val

        •              230                 235                 240                 245                 250
K88AB  TCT GCT GCC TAT GCA CTG GGT ATT GCA AAC GGT CAG ACT ATT GAG GCA ACT TTT AAT CAG GCT GTA ACT ACC AGC
       Ser Ala Ala Tyr Ala Leu Gly Ile Ala Asn Gly Gln Thr Ile Glu Ala Thr Phe Asn Gln Ala Val Thr Thr Ser
K88AD  TCC GTT GCC TAT GCA CTG GGT ATT GCA AAC CGG CAG ACT ATT GAG GCA ACT TTT AAT CAG GCT GTA ACT ACC AGC
       Ser Val Ala Tyr Ala Leu Gly Ile Ala Asn Arg Gln Thr Ile Glu Ala Thr Phe Asn Gln Ala Val Thr Thr Ser
K88AC  TCT GCT GCC TAT GCA CTG GGT ATT GCA AAC GGT CAG ACT ATT GAG GCA ACT TTT AAT CAG GCT GTA ACT ACC AGC
       Ser Ala Ala Tyr Ala Leu Gly Ile Ala Asn Gly Gln Thr Ile Glu Ala Thr Phe Asn Gln Ala Val Thr Thr Ser

                       255                 260             264
K88AB  ACT CAG TGG AGC GCT CCG CTG AAC GTA GCA ATA ACT TAT TAC
       Thr Gln Trp Ser Ala Pro Leu Asn Val Ala Ile Thr Tyr Tyr
K88AD  ACT CAG TGG AGC GCT CCG CTG AAC GTA GCA ATA ACT TAT TAC
       Thr Gln Trp Ser Ala Pro Leu Asn Val Ala Ile Thr Tyr Tyr
K88AC  ACT CAG TGG AGC GCT CCG CTG AAC GTA GCA ATA ACT TAT TAC
       Thr Gln Trp Ser Ala Pro Leu Asn Val Ala Ile Thr Tyr Tyr
```

Figure 1. The nucleotide sequence of the DNA region encoding the K88ac subunit protein. The corresponding amino acid sequence is also given. The nucleotide sequences of the K88ab and K88ad have been included for comparison. Differences between the three amino acid sequences are boxed, whereas base substitutions not resulting in amino acid differences are indicated by dots.

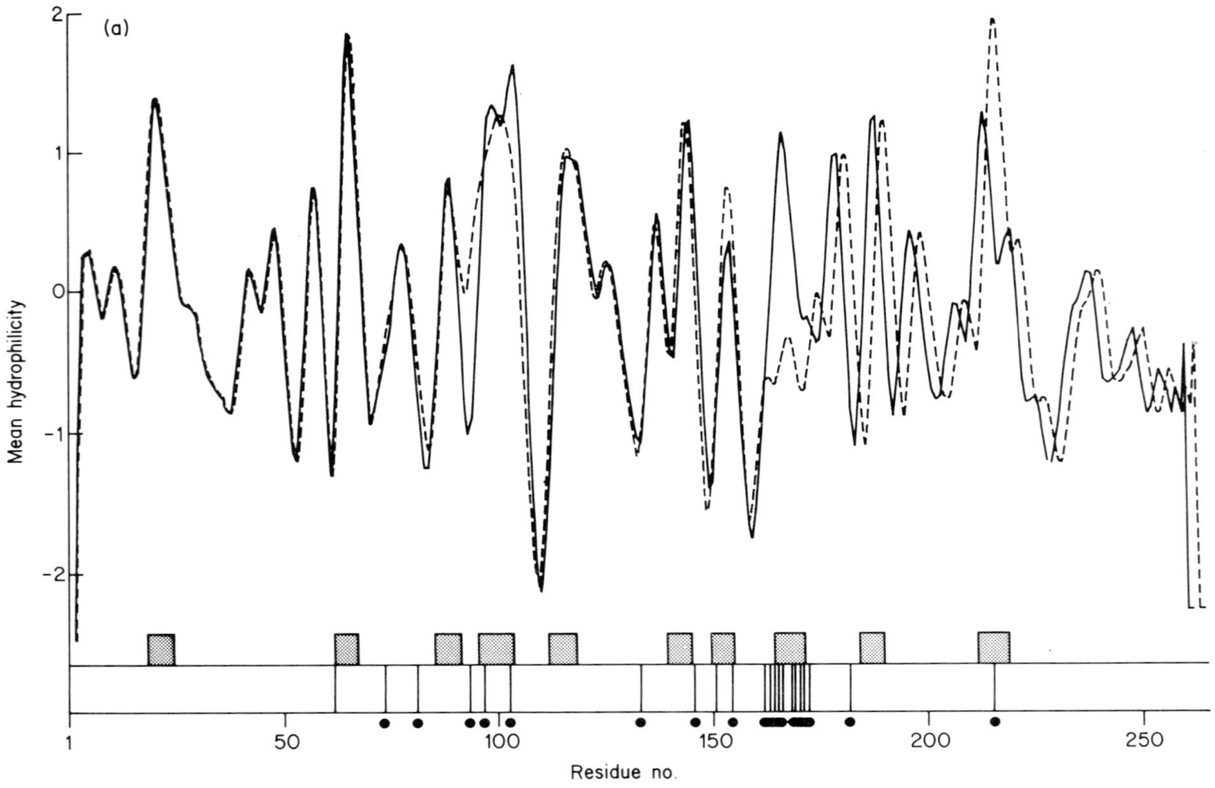
(a)
Mean hydrophilicity
2
1
0
-1
-2
1
50
100
150
200
250
Residue no.

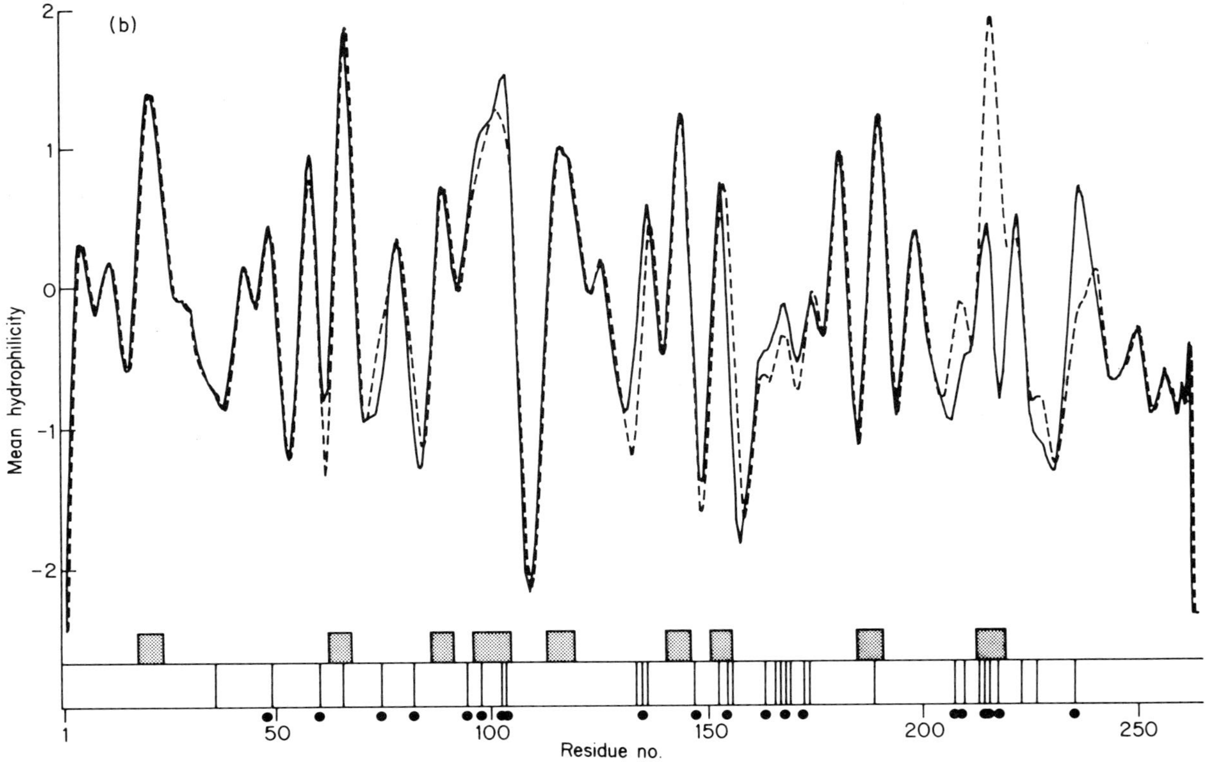

Figure 2. (a) Hydrophilicity profiles of the K88ac (solid line) and K88ab (dashed line) depicting potential antigenic determinants. (b) Hydrophilicity profiles of the K88ad (solid line) and K88ab (dashed line) depicting potential antigenic determinants. Predicted antigenic determinants are indicated by bars. Positions with sequences differences between the two proteins are shown by vertical lines. Non-conservative substitutions are marked by dots.

As can be seen from Fig. 1, the differences found between the amino acid sequences of the three variants are not restricted to a particular part of the molecule, although they tend to come in clusters. No differences are observed in the $N-$ and $C-$terminal parts of the molecule which might reflect a role for these parts of the protein in the assembly of the subunits into intact fimbriae. Using the method of Hopp and Woods [4] several regions of high antigenic potential can be predicted within the K88 molecule [5] (Fig. 2). The possible antigenic determinants are located between amino acid residues 19-24, 63-68, 87-92, 96-104, 114-120, 140-145, 151-156, 186-191 and 213-219. A number of amino acid substitutions seem to cluster within some of the predicted antigenic determinants, i.e. between amino acid residues 96-104, 151, 156 and 213-219. Therefore it might be that these regions are part of the variable antigenic factors.

As seen in Fig. 1 there is a hyper-variable region between amino acid residues 162 and 175. Since in this region a deletion of three amino acid residues is observed, it seems justified to assume that this part of the molecule forms a sort of loop which can be shorter in some K88 variants. This region, therefore, could be a good candidate for an antigenic determinant which could be part of the variable "b", "c" and "d" factors. No antigenic determinant is predicted in this region of the molecule in the K88ab and ad variants (Fig. 2). A possible antigenic determinant is, however, predicted in this region of the K88ac variant. Some of the peptides encompassing possible antigenic determinants within the K88 protein have been chemically synthesized and are tested for their ability to inhibit the binding of antibodies to the K88 fimbria in an ELISA test. The results of these experiments are presented elsewhere in this volume. Additional information concerning the location of the "a", "b", "c" and "d" antigenic factors, was obtained as described below.

Plasmids pFM 205 (encoding the K88ab variant), pBac (encoding the K88ac variant), and pBad (encoding the K88ad variant) were digested with the restriction enzyme *Eco*RI. In all three cases the *N*-terminal part of the K88 protein subunit (up to amino acid residue 80) is encoded on a 5.9 MD fragment (including the cloning vehicle). The remaining part of the molecule (residues 81-269) is encoded on an 1.1 MD *Eco*RI fragment in the case of the K88ab and K88ac variants, whereas residues 81-169 are encoded on a 0.2 MD fragment and residues 170-264 on a 0.9 MD fragment in the case of the K88ad variant. The various *Eco*RI fragments were isolated, the 5.9 MD fragments treated with alkaline phosphatase to prevent circularization and the fragments were ligated in the following combinations: AB 1.1+AC 5.9, AB 1.1+AD 5.9, AC 1.1+AB 5.9 and AC 1.1+AD 5.9 (capital letters refer to *Eco*RI DNA fragments derived from a particular serological variant). After transformation to *E. coli* K-12 C600 of the resulting recombinant DNA molecules, a number of transformants were tested for their serological characteristics. In Ouchterlony double immuno-diffusion tests with specific anti-K88b, anti-K88c and anti-K88d sera, it was found that every combination encoded a K88 protein subunit that reacted with antiserum against the variable antigenic factor, from which the 1.1 MD fragment originated.

These results therefore indicate that the variable antigenic factor is located between amino acid residues 80 and 264. These results and the deletion found in the hypervariable region between amino acid residues 162–175 in the K88ac variant show that the specific transport system, responsible for translocation of the K88 protein subunit molecules to the final destination on the outer membrane [6,7], does tolerate a number of differences in the sequence of the K88 protein to be transported.

Since it is one of the major frustrations in gene technology to get proteins transported out of *E. coli* this leads us to investigate whether a K88 fusion product is still transported to the surface of the cells.

A fusion of the first 79 amino acid residues of K88 and β-galactosidase is however not transported. The hyper-variable region between amino acid residues 162–175 is located between two unique restriction sites which are 73 nucleotides apart and can therefore easily be replaced by other sequences.

Removal of the 73 nucleotides followed by incubation with mungbean nuclease and insertion of an eight nucleotides long *Bgl*II linker results in a K88 protein molecule with a deletion of 23 amino acids. This shortened K88 molecule is, however, not transported to the surface of the *E. coli* cells. Insertion of a DNA fragment of 121 nucleotides from the DNA A protein between the two restriction sites flanking the hyper-variable region, however, resulted into the transport of a K88 molecule with a slightly higher molecular weight to the surface of the *E. coli* cells. This protein could be isolated in the same way as K88 fimbriae by heating the cells to 60°C and reacting with K88 antibodies. It is not known whether these longer proteins also assemble to form fimbriae. At present we are investigating the minimal and maximal length of DNA fragments that can be inserted in this way.

SUMMARY

The primary structure of the three major serological variants of the K88 antigen has been determined. Some possible functions for the various parts of the molecule are indicated, as well as the possible location of the various antigenic determinants. The differences between the amino acid sequences of the three K88 variants are scattered over the whole protein, but tend to come in clusters. One especially variable region is noted between amino acid residues 162–175. This region is located in the DNA between two unique restriction sites and can therefore be removed and replaced by other DNA sequences. One of these constructs results in a slightly longer K88 molecule which is still transported to the surface of the *E. coli* cells.

ACKNOWLEDGEMENTS

This work was supported by the Danish Technical and Medical Research Councils.

REFERENCES

1. Schoolnik, G. K., E. C. Gotschlich, J. Rothbard, J. Y. Tai and R. Fernandez (1984). Gonococal pili, primary structure and receptor binding domain. *J. Exp. Med.* **159**, 1351–1370.

2. Olafson, R. W., A. R. Bhatti, J. S. G. Dooley, J. E. Heckels, P. J. McCarthy and T. J. Trust (1985). Structural and antigenic analysis of meningococcal piliation. *Infect. Immun.* **48**, 336-342.
3. Ørskov, I., F. Ørskov, J. M. Leach and W. J. Sojka (1961). Simultaneous occurrence of *E. coli* B and L antigens in strains from diseased swine. *Acta Pathol. Microbiol. Scand. Sect. B.* **53**, 404-422.
4. Hopp, T. P. and K. R. Woods (1981). Prediction of protein antigenic determinants from amino acid sequences. *Proc. Natl. Acad. Sci. USA* **78**, 3824-3828.
5. Klemm, P. and L. Mikkelsen (1982). Prediction of antigenic determinants and secondary structures of the K88 and CFA/1 fimbrial proteins from enteropathogenic *Escherichia coli. Infect. Immun.* **38**, 41-45.
6. Mooi, F. R., D. Bakker, F. K. de Graaf and N. Harms (1981). Organization and expression of genes involved in the production of the K88ab antigen. *Infect. Immun.* **32**, 1155-1163.
7. Mooi, F. R., A. Wijfjes, C. Wouters and F. K. de Graaf (1982). Construction and characterization of mutants impaired in the biosynthesis of the K88ab antigen. *J. Bacteriol.* **150**, 512-521.

Novel Type 1 Fimbriae of *Salmonella Enteritidis*

Josiane Feutrier, William W. Kay, Torkel Wadström* and Trevor J. Trust[a]

*Department of Biochemistry and Microbiology, University of Victoria, Victoria, British Columbia, Canada, and *Department of Microbiology, Swedish University of Agricultural Sciences, Uppsala, Sweden*

The type 1 fimbriae of *Escherichia coli* and the haemagglutination they mediate was first reported in 1955 [1]. Since then, type 1 fimbriae have been demonstrated on a wide variety of Gram-negative bacteria, including the invasive enteropathogen *Salmonella* [2]. In this case, attention has focused almost exclusively on the type 1 fimbriae of *Salmonella typhimurium*, and the amino acid composition and *N*-terminal 28 residue amino acid sequence of the fimbrin subunits of this serovar have been reported [3,4]. There is little information available on the mannose-sensitive fimbriae produced by other Salmonellae.

One serotype commonly associated with Salmonellosis of humans is *Salmonella enteritidis*. Indeed *S. enteritidis* generally ranks among the three most commonly isolated serotypes world-wide. We have isolated the mannose-sensitive fimbriae from a human isolate of *S. enteritidis* and have biochemically characterized the polypeptide subunit of this fimbrial species. Here we report on the purification procedure and on the chemical and antigenic properties of this fimbrin.

The strain we studied, *S. enteritidis* 27655-3b was isolated from human faeces in India. The strain produced a heavy pellicle during growth in static liquid culture. The cells of strain 27655-3b were strongly hydrophobic and autoaggregated at 0.2 M $(NH_4)_2SO_4$. Electron microscopy revealed that strain 27655-3b was heavily fimbriated and the 8 nm fimbriae displayed a channelled morphology typical of type 1 fimbriae (Fig. 1a). Strain 27655-3b was able to haemagglutinate erythrocytes of a variety of species and this haemagglutination was inhibited by 0.1 M D-mannose.

[a]*Telephone:* 1-18-17 4595

Protein-Carbohydrate Interactions in Biological Systems. ISBN 0 12 436665 1

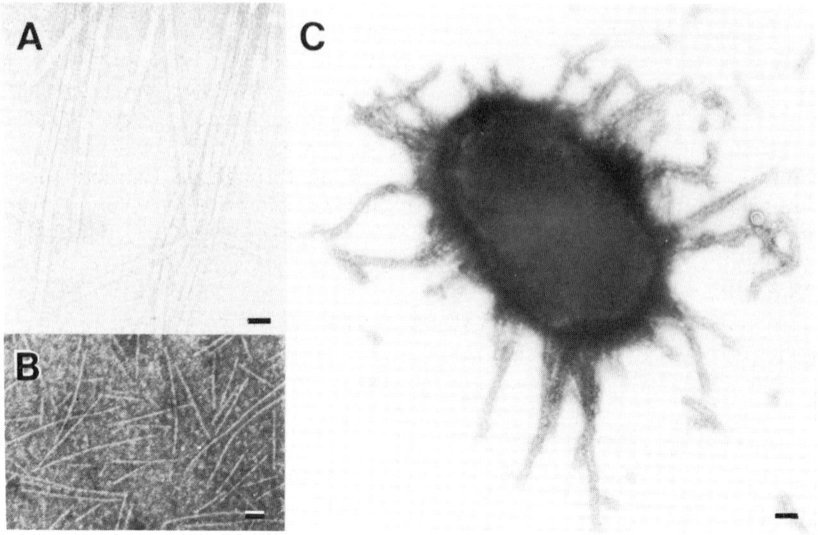

Figure 1. Electron microscopy of fimbriae of *S. enteritidis* strain 27655-3b stained by 1% uranyl acetate (pH 4.2). (A) Native fimbriae, bar = 20 nm; (B) purified fimbriae, bar = 100 nm; (C) immunolabelling with 1:32 dilution of antiserum raised in rabbits to purified pili. The fimbriae appear completely decorated and multiple fimbriae are cross-linked into aggregates extending from the cell surface. Bar = 100 nm.

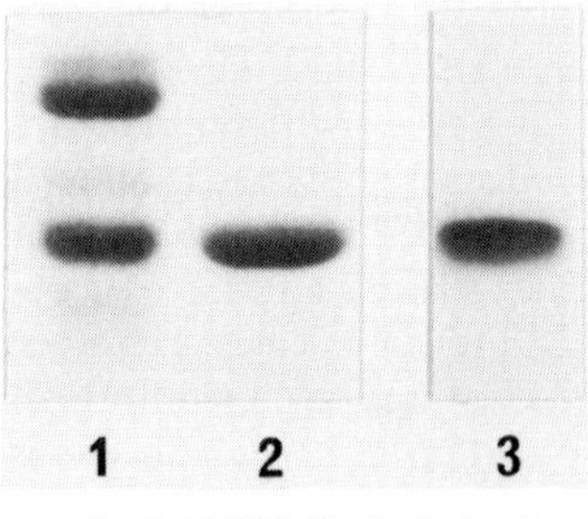

Figure 2. SDS-PAGE of *S. enteritidis* 27655-3b fimbrin. Lane 1, M_r standards at 21 500 and 14 400; Lane 2, Purified fimbrin stained by Coomassie blue, Lane 3, autoradiogram of fimbrin immunoprecipitated by a 1:1000 dilution of antiserum raised in rabbits to purified fimbriae.

Fimbriae were isolated and purified by a simple procedure. Cells were grown for 48 h in static CFA broth, harvested, suspended in 0.15 M ethanolamine buffer (pH 10.5) and fimbriae removed by blending for 2 min. After removal of cells by centrifugation, fimbriae were recovered from the supernatant by 0–10% ammonium sulphate precipitation. The fimbrial pellet was resuspended in ethanolamine and subjected to two rounds of acetone precipitation followed by

0–40% ammonium sulphate precipitation. The precipitate was then resuspended in ethanolamine and centrifuged at 115 000×*g* for 60 min to remove contaminating outer membrane material, dialysed against 20 mM Tris HCl buffer (pH 7.5) and lyophilized. The lyophilisate was then resuspended in Tris buffer containing 0.2% SDS to solubilize any flagella that may be contaminating the preparation. Fimbriae were then recovered by centrifugation at 115 000×*g* for 3 h, resuspended in ethanolamine, and the SDS removed by elution through an Extracti-Gel D column. The purified fimbriae were then dialysed extensively against distilled water. This simple protocol resulted in purification to homogeneity of a polypeptide of apparent M_r 14 500 as assessed by Coomassie blue R staining of SDS-PAGE gels (Fig. 2, lane 2).

The amino acid composition and *N*-terminal amino acid sequence of this purified *S. enteritidis* fimbrin was then determined. Based on composition (Table 1) an M_r of 14 500 was calculated. This is markedly smaller than the M_r of 22 100 reported for *S. typhimurium* type 1 fimbriae [3], and also smaller than the M_r 17 100 reported for *E. coli* type 1 fimbriae [5]. The amino acid composition of the *S. enteritidis* fimbrin was readily distinguishable from both

Table 1. Amino acid composition of type 1 fimbrins

	Residues per fimbriae subunit[a]		
Amino acid	*S. enteritidis* 27655-3b	*S. typhimurium*	*E. coli*
Asx	11	22	18
Thr	14	25	20
Ser	10	23	9
Glx	12	19	16
Pro	5	11	2
Gly	19	23	21
Ala	18	34	34
Val	11	16	14
Met	0	Tr	ND
Ile	4	7	5
Leu	4	12	14
Tyr	2	4	2
Phe	6	9	8
His	1	3	2
Lys	4	9	4
Arg	2	4	2
Cys	0	ND	2
Try	1	ND	ND
Total residues per mol	125	221	173
M_r apparent (×1000)	14.55	22.1	17.1
Reference		[1]	[4]

[a] Tr, trace; ND, not detected.

Table 2. Amino acid sequence of the *N*-terminal region of type 1 fimbrin from *S. enteritidis* 27655-3b, and type 1 fimbrin of *E. coli* and *S. typhimurium* aligned for greatest homology. Dashes infer residues homologous with *S. enteritidis* fimbrin, conservative replacements with *E. coli* type 1 fimbrin are underlined, and deletion introduced into *S. enteritidis* fimbrin to make a better fit is indicated by dot. X is not identified

Organism	*Residue*
	1 10 20 30
S. enteritidis	AGFV GNKA · VVQAA VTIAAQNTTS
E. coli	AATTVNG_T_ HF_GE__N__ CAVD_GSVDQ
S. typhimurium	ADP TPVSVSG_TI HFEGKL_N__ XAVST

other type 1 fimbrins. The *N*-terminal 30 amino acid sequence of the *S. enteritidis* fimbrin is shown in Table 2. The best fit of this sequence with the *N*-terminal sequences of the type 1 fimbrins of *E. coli* [4] and *S. typhimurium* [6] was obtained by aligning the two *ala* residues at positions 12 and 13 of the *S. enteritidis* sequence with the two *ala* residues at positions 19 and 20 of the *E. coli* sequence. When aligned in this manner the *S. enteritidis* fimbrin molecule could be seen to be lacking a 7–10 residue *N*-terminal sequence present in the *E. coli* and *S. typhimurium* molecules. This alignment provides a length of sequence between residues 7 and 25 of the *E. coli* fimbrin with a mean homology score of 91 [7] with the *S. enteritidis* sequence. In this segment of sequence, the two *Salmonella* molecules appear to be more highly related to the *E. coli* molecule than to each other. The *S. enteritidis* fimbrin shares 8 identical residues with the *E. coli* fimbrin and only 4 identical residues with the *S. typhimurium* fimbrin, while the *S. typhimurium* fimbrin shares 11 identical residues with the *E. coli* sequence. The amino acid sequence to residue 63 of the *S. enteritidis* molecule has also been determined, and when the first 60 residues were compared to the corresponding *E. coli* sequence using the log-odds score of Dayhoff [7], a mean homology score of 79 was obtained, with only 12 identical residues.

The antigenicity of the *S. enteritidis* fimbriae was then examined. Antisera raised in rabbits to purified fimbriae gave a titre of 1:10 000 when tested by ELISA. The antibodies in this antiserum did not react with the *S. enteritidis* fimbrin subunit in Western immunoblot assays, but were reactive in immunoprecipitation assays (Fig. 2, lane 3) indicating conformation-dependent epitopes. Immuno-electron microscopy (Fig. 1C) showed antibodies binding along the entire length of the native fimbrial rods, confirming that the protein which had been purified was the fimbrial protein and further showing that epitopes recognized by the antibodies were surface-exposed. This antisera also prevented haemagglutination by *S. enteritidis* 27655-3b. Immuno-dot blot assay (Fig. 3) of intact cells showed that the antibodies were strain specific, with no reaction being displayed with the type 1 fimbriae of *E. coli* and other *Salmonella* strains, or with strains producing other adhesins.

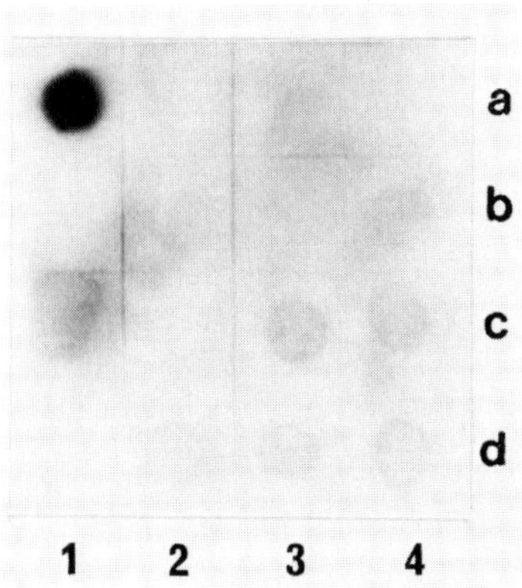

Figure 3. Autoradiogram of immuno-dot blot of bacterial cells using a 1:1000 dilution of antiserum raised in rabbits to purified fimbrin. *S. enteritidis* 27655-3b (al); *E. coli* strains with type 1 (a2-3), K88 (a4) K99 (b1), CFA I (b2) and CFA II (b3) fimbriae; *S. typhimurium* strains (b4, c1-2) and other Salmonellae (c3-4; d1-4) with type 1 fimbriae.

To provide additional evidence that the fimbriae of *S. enteritidis* 27655-3b were involved in the haemagglutinating activity, and also contributed to the surface hydrophobicity of the strain, a non-fimbriate mutant was isolated by transposon mutagenesis. Transposon Tn10 was introduced into 27655-3b by phage P22 Tc10, and Tc^r clones resulting from the transposition were screened for fimbriae production by immuno-dot blot assay. One clone exhibited no reactivity with the anti-fimbrial antiserum and electron microscopy confirmed the absence of fimbriae. This mutant was unable to haemagglutinate and displayed a marked reduction in hydrophobicity aggregating at 1.5 M $(NH_4)_2SO_4$ compared to the 0.2 M $(NH_4)_2SO_4$ of the parent strain. Preliminary experiments employing an oligonucleotide probe based on residues 25 to 30 of the *N*-terminal *S. enteritidis* amino acid sequence, and Southern blotting against restriction endonuclease digests of the DNAs from the parent and mutant strains, suggest that the mutation is not in the structural fimbrin gene but is in a gene regulating fimbrial expression. The precise nature of this mutation is currently under investigation.

ACKNOWLEDGEMENTS

This work was supported in part by Program Grant No. PG024 from the Medical Research Council of Canada. Our thanks to S. Kielland for help with sequence analysis.

REFERENCES

1. Duguid, J. P., I. W. Smith, G. Dempster and P. N. Edmunds (1955). Non-flagellar filamentous appendages ('fimbriae') and hemagglutinating activity of *Bacterium coli*. *J. Path. Bact.* **70**, 335-348.

2. Duguid, J. P. and R. R. Gillies (1958). Fimbriae and haemagglutination diversity in *Salmonella, Klebsiella, Proteus* and *Chromobacterium. J. Gen. Microbiol.* **75**, 519–520.
3. Korhonen, T. K., K. Lounatmaa, H. Ranta and N. Kuusi (1980). Characterization of type 1 pili of *Salmonella typhimurium* LT2. *J. Bacteriol.* **144**, 800–805.
4. Klemm, P. (1984). The fim A gene encoding the type-1 fimbrial subunit of *Eschericha coli.* Nucleotide sequence and primary structure of the protein. *Eur. J. Biochem.* **143**, 395–399.
5. Salit, I. E. and E. G. Gotschlich (1977). Hemagglutination by purified type 1 *Escherichia coli* pili. *J. Exp. Med.* **146**, 1169–1179.
6. Waalen, K., K. Sletten, L. O. Froholm, V. Väisänen and T. K. Korhonen (1983). The *N*-terminal amino acid sequence of type 1 fimbria (pili) of *Salmonella typhimurium* LT2. *FEMS Microbiol. Lett.* **16**, 149–151.
7. Dayhoff, M. O. (1978). "Atlas of Protein Sequence and Structure", Vol. 5, Supplement 3, pp. 345–358. National Biomedical Research, Washington, DC.

Prospects for a Gonorrhoea Vaccine Based on Gonococcal Adhesins

D. Danielsson[a]

Department of Clinical Microbiology and Immunology, Örebro Medical Centre Hospital, Örebro, Sweden

Pili are important virulent markers of many different bacteria, e.g. *Salmonella*, *Shigella*, *E. coli*, gonococci, meningococci and others, because they mediate adhesion to mucosal surfaces during the initial stages of infection. The general view is that this is mediated through the interaction between ligands, i.e. particular structures of the proteinaceous pilus subunits, and surface receptors of the host cell which are assumed to represent different carbohydrates linked to lipid or protein (glycolipids, glycoproteins) [1,2]. Pili are strong candidates for vaccines because antipilus antibodies have been shown to be effective in inhibiting the attachment of piliated bacteria to host cells [3,4].

As to gonococcal pili, these are composed of identical subunits (pilins) of relative molecular mass (M_r) between 17 000 and 22 000 with apparent variations regarding immunogenic and antigenic properties [5]. The relative M_r of a particular pilin is related to the gonococcal strain involved. The immunogenic and antigenic properties are related to particular regions of the pilin [6-8]. Three distinct regions or domains of the gonococcal pilins have been recognized as demonstrated by comparative amino acid sequence analysis of laboratory derived strains and clinical isolates in combination with serological studies with polyclonal and monoclonal antibodies. The first region at the amino-terminus of the polypeptide represented by 53 or so amino acids is the constant (C) domain with epitopes common to all gonococcal (and probably meningococcal) pili. The second and third regions represent semi-variable (SV) and hyper-variable (HV) domains. The SV and HV regions represent the remaining 100–110 amino acids of the carboxy-terminus of the protein. The epitopes of these polypeptides are responsible for inter-strain diversities, i.e. type specificities, and the intra-strain

[a] *Telephone:* 46-19-151000

Protein-Carbohydrate Interactions in Biological Systems. ISBN 0 12 436665 1

antigenic variations. In a recent study this was shown to be due to chromosomal rearrangements that result from intragenic recombinations between a repertoire of different pilin genes [9].

Polyclonal and monoclonal antibodies against the C and SV domains of the pilus subunits effectively block the attachment to host cells [7,10]. Monoclonal antibodies directed against the HV region likewise inhibit the attachment to host cells [10]. These observations indicate that both conserved and variable regions are involved in binding to cell receptors. This might also mean that there are multiple ligands with specificities for different receptors and that attachment is inhibited by antibodies reacting with epitopes for each one of these ligands. However, we still lack information about the chemical structures of the host cell receptors. Inhibition of attachment has so far not been successful with different mono- and disaccharides. It must also be kept in mind that mechanisms other than specific ligand–receptor interactions might be involved, e.g. through hydrophobic interactions, ionic bonds, van der Waals attractions, etc., or combinations thereof. Also, antibodies might react with non-ligand epitopes and accomplish inhibition of attachment just by steric hindrance or changes of surface charges.

Other adhesins than pili might also be involved in the attachment of gonococci to mucosal surfaces. There is indirect evidence that protein II (P.II), which is also called the opacity protein, might play a role because gonococci with opacity protein show a more effective attachment than those without [11]. Practically nothing is known about possible ligands and specific cell receptors responsible for these interactions. It is of great interest, however, that this protein is subjected to phase variations and antigenic variations very similar to those for pili [12,13]. Comparative amino acid sequence determinations show that many amino acids are conserved in all the proteins, but there are differences in a few amino acids suggesting that the proteins may be products of different structural genes. Genetic transformation experiments have also provided evidence that multiple genes are involved in specifying the different P.II species and that there is a heritable difference between expressed and unexpressed forms of a P.II gene [14,15].

The variability of both piliation and opacity proteins seem to involve at least two distinct genetic phenomena. In the first, pilus and opacity protein phase variations, i.e. from piliation to non-piliation and from the presence to the absence of opacity proteins or vice versa, are reversibly switched on and off at a high rate. In the second process, antigenic variation, i.e. the appearance of new epitopes on the pilin and the opacity protein, is caused by the expression of alternative pilin genes and opacity protein genes due to chromosomal rearrangements from intragenic recombinations between repertoires of different pilin genes and opacity protein genes [9,14]. The genes seem to be closely located on the chromosome suggesting common, though as yet unknown, control mechanisms for these two proteins. They illustrate the success with which gonococci escape immune defence mechanisms and also explain the ease with which this bacterium causes repeated infection in humans. The genetic mechanisms have parallels in other parasites that exhibit antigenic variation and in processes such as the generation of immunoglobulin diversity.

It is a logical consequence to develop vaccines against important surface structures mediating attachment in the initial stage of infection. Antipilus antibodies obtained from humans after gonorrhoea or after vaccination with native pili or synthetic peptides corresponding in sequence to residues in the C and SV regions, do block the attachment of homologous and heterologous gonococci to human mucosal cells [4,6,10]. These antibodies may belong to different Ig-classes. What is of particular interest in this respect is the attention that has recently been paid to the appearance of anti-idiotype antibodies which probably act as a normal ingredient or as modulators of the immune response in vertebrates [16,17]. This means that antibodies with specificity for the epitope of the ligand will induce a class 1 anti-idiotype antibody, specific for the receptor of the host cell. This antibody will then induce a class 2 anti-idiotype antibody, specific for the epitope of the ligand, etc. This means that the class 1 anti-idiotype antibodies might react with the receptor of the host cell and block the attachment, and thus the inhibition of attachment might be enhanced, which would be a highly desirable effect. However, this might also have negative side effects since the reaction of the class 1 anti-idiotype antibody with the host cell receptor might activate complement or cytotoxic cells and initiate cell damage, something which parallels possible pathogenetic mechanisms in autoimmune diseases [18].

Vaccines against gonorrhoea based on defined components of appropriate adhesins certainly seem promising and are now ready for field trials. They would work by eliciting specific antibodies and enhancing anti-idiotype antibodies that would prevent the attachment of gonococci to mucosal surfaces. However, this approach may be compromised by the demonstrated phase variations and antigenic diversities of pili and opacity proteins. Controlled field trials will elucidate these matters.

REFERENCES

1. Beachey, E. H. (Ed.) (1980). "Bacterial Adherence", Receptors and Recognition, Series B, Vol. 6. Chapman & Hall, London.
2. Hansson, G. C., K-A. Karlsson, G. Larsson *et al.* (1985). *Anal. Biochem.* **145**, 158-163.
3. Darfeuille, A., B. Lafeuille, B. Joly and R. Cluzel (1983). *Ann. Microbiol.* (*Inst. Pasteur*) **134A**, 53-64.
4. Tramont, E. C. (1976). *Infect. Immun.* **14**, 593-595.
5. Brinton, C. C., J. Bryan, J-A. Dillon *et al.* (1978). *In* "Immunobiology of Neisseria Gonorrhoeae" (Eds G. F. Brooks *et al.*), pp. 156-178. American Society for Microbiology, Washington, DC.
6. Schoolnik, G. K., J-Y. Tay and E. C. Gotschlich (1982). *In* "Seminars in Infectious Disease" (Eds L. Weinstein *et al.*) Vol IV, Bacterial Vaccines, pp. 172-188. Thieme-Stratton, New York.
7. Rothbard, J. B., R. Fernandez, L. Wary *et al.* (1985). *Proc. Natl. Acad. Sci. USA* **82**, 915-921.
8. Meyer, T. F., E. Biilyard, R. Haas *et al.* (1984). *Proc. Natl. Acad. Sci. USA* **81**, 6110-6114.
9. Hagblom, P., E. Segal, E. Biilyard and M. So (1985). *Nature* **315**, 156-158.
10. Virji, M. and J. E. Heckels (1984). *J. Gen. Microbiol.* **130**, 1089-1095.
11. Forslin, L. and D. Danielsson (1980). *Gynecol. Obstet. Invest.* **11**, 327-340.

12. Swansson, J. and O. Barrera (1983). *J. Exp. Med.* **157**, 1405–1420.
13. Heckels, J. E. (1981). *J. Bacteriol.* **145**, 736–742.
14. Stern, A., P. Nickel, T. F. Meyer and M. So (1984). *Cell* **37**, 447–454.
15. Schwalbe, R. S., W. J. Black, D. G. Klapper and J. G. Cannon (1985). *In* "The Pathogenic Neisseria: Proceedings of the Fourth International Symposium. ASM, Washington, D.C.
16. Jerne, N. K. (1974). *Ann. Immunol (Paris)* **125C**, 373–389.
17. Kohler, H., S. Muller and C. Bona (1985). *Proc. Soc. Exp. Biol. Med.* **178**, 189–195.
18. Shoenfield, Y. and R. S. Schwartz (1984). *New Engl. J. Med.* **311**, 1019–1029.

Poster Session

Characterization of Fimbriae from *Bacteroides fragilis*

J. van Doorn*[a], F. R. Mooi‡, J. de Graaff§ and D. M. MacLaren*

*Departments of *Medical Microbiology, ‡Molecular Microbiology, and §Oral Microbiology, Free University, Amsterdam, The Netherlands*

INTRODUCTION

Little is known about colonization factors of *Bacteroides fragilis*, an obligate anaerobe from the intestinal tract of healthy animals and humans. *B. fragilis* forms less than 0.5% of the faecal flora, but is the most commonly isolated bacterium in human anaerobic and intra-abdominal infections. Fimbriated *B. fragilis* cells were described for the first time by Pruzzo *et al.* in 1984 [1]. They proposed that these cells agglutinated red blood cells of many species, but it was shown that this activity is caused by capsular components of the bacteria.

No haemagglutinating activity was found in purified fimbria preparations from *Bacteroides nodosus* and *Bacteroides gingivalis*. These two species are being studied at the molecular level.

The aim of our investigation is to study the role of *B. fragilis* fimbriae in colonization and in mixed infections.

METHODS

Southern blot experiments were carried out using standard procedures. Chromosomal DNA of several *Bacteroides* species were probed with the ^{32}P-labelled plasmid 7AI. This plasmid was constructed by Elleman *et al.* [2] by cloning the chromosomal gene coding for the 18 000 D fimbrial subunit of *B. nodosus* strain 198 in the vector pBR322.

Standard procedures were used to isolate the surface components of *B. fragilis* strain BE-1. Essentially, the shear fraction of the cells was precipitated with ammonium sulphate and subjected to desoxycholate sucrose or CsCl gradient

[a] *Telephone:* 31-20-5483955

Protein-Carbohydrate Interactions in Biological Systems. ISBN 0 12 436665 1

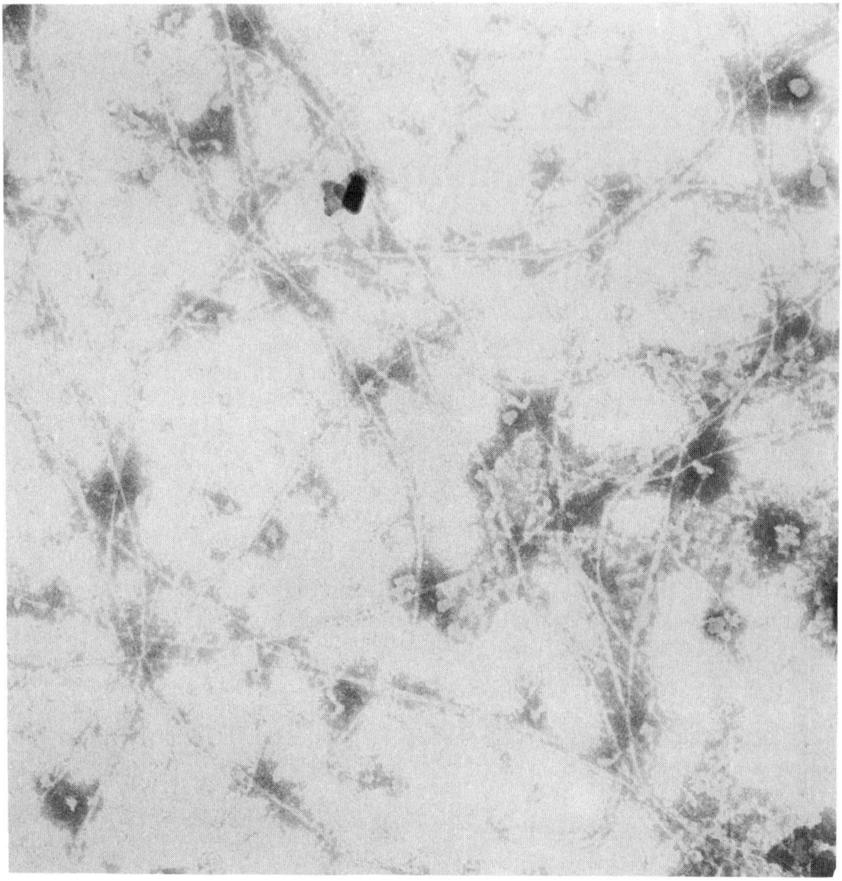

Figure 1. Electron micrograph of a partially purified fimbria preparation of *B. fragilis* strain BE-1. The filaments have a diameter of about 4 nm. Magnification ×80 500.

ultracentrifugation. Fractions containing fimbrial filaments, as determined by electron microscopy, were further purified by gel filtration in the presence of 6 M urea.

RESULTS

Homology between fimbrial subunits coding sequences of *Bacteroides* species was estimated by Southern blotting. Chromosomal DNA of several strains of *B. fragilis*, *B. gingivalis*, *Bacteroides ovatus* and *Bacteroides vulgatus* were probed with plasmid 7AI, but no hybridization was detected. The tentative molecular weight of the fimbrial subunit of *Bacteroides fragilis* was estimated. Several

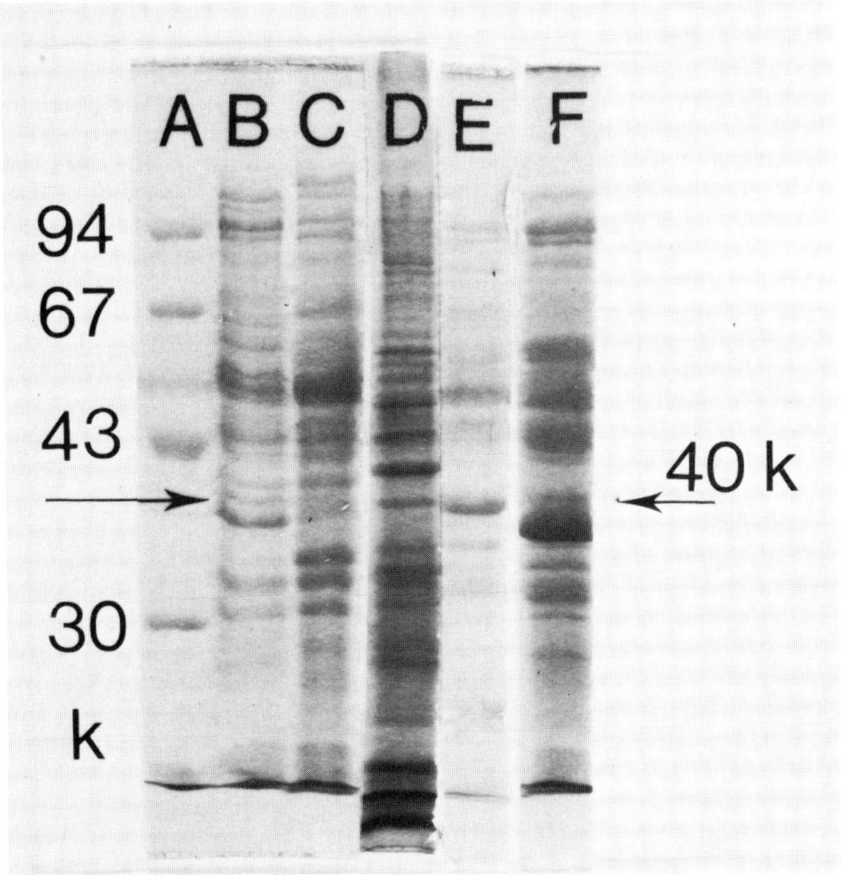

Figure 2. SDS-PAGE patterns of isolation steps of fimbriae from *B. fragilis* strain BE-1. Marker proteins (lane A); shear fraction of cells, grown at 37°C (lane B); shear fraction of cells, grown at 22°C (lane C); sucrose gradient fraction of the 37°C shear fraction (lane D); sucrose gradient fraction, subjected to gel filtration in the presence of 6 M urea (lane E); shear fraction of cells, cultured in excess haemin at 37°C (lane F).

observations suggested that the fimbriae from strain BE-1 were composed of subunits with a molecular weight of 40 000:

(1) the isolation procedure, in which the polypeptide is found in the void volume after gel filtration;
(2) the observation that the 40 000 D protein is not found in shear fractions of cells cultured at 22°C;
(3) electron microscopic observations of partially purified preparations.

Cells, grown in excess haemin produced more of the 40 000 D polypeptide, as determined by SDS-PAGE.

CONCLUSIONS

The molecular weight of the fimbriae subunit, partly purified from *B. fragilis* strain BE-1 is probably 40 000. The production seems to be dependent on the growth temperature and the concentration of haemin in the medium.

There seems to be no obvious homology between fimbriae of *B. nodosus*, and the *B. fragilis* strains, *B. gingivalis* strains and the species *B. ovatus* and *B. vulgatus* tested.

REFERENCES

1. Pruzzo, C., B. Dainelli and M. Ricchetti (1984). *Infect. Immun.* **43**, 189–194.
2. Elleman, T. C., P. A. Hoyne, D. L. Emery, D. J. Stewart and B. L. Clark (1984). *FEBS Lett.* **173**, 103–107.

Monoclonal Antibodies Raised Against Five Different P Fimbriae and Type 1A and 1C Fimbriae

J. M. de Ree[a], P. H. M. Savelkoul, P. Schwillens and J. F. van den Bosch

Department of Medical Microbiology, University of Limburg, Maastricht, The Netherlands

One of the most important virulence factors of uropathogenic *Escherichia coli* is the adhesive capacity of the bacteria to uroepithelial cells. Adhesion is mediated by fimbriae which are serologically heterogenous. Ørskov and Ørskov [1] could distinguish eight serologically different fimbriae from uropathogenic *E. coli* and Parry *et al.* [2] found seven serologically different fimbriae. Fimbriae from uropathogenic *E. coli* can also be distinguished by their receptor specificity:

(1) fimbriae that cause a mannose-sensitive haemagglutination of guinea pig erythrocytes (type 1 fimbriae);
(2) fimbriae that cause no haemagglutination of any erythrocytes (e.g. 1C);
(3) fimbriae that cause a mannose-resistant haemagglutination of human erythrocytes. This last group of fimbriae (3) can be subdivided on the basis of their receptor specifities into P, M, S and X fimbriae. The P fimbriae in particular have been associated with the pathogenesis of pyelonephritis.

Recently the genes encoding for the P fimbriae $F7_1$, $F7_2$, F9, F11 and pap (=F13) were cloned from uropathogenic *E. coli* strains. The 1C fimbriae were also cloned from a uropathogenic *E. coli* strain and the 1A fimbriae were cloned from an *E. coli* K12 strain. All these clones were transformed to an *E. coli* K12 strain which did not produce any fimbriae. Fimbriae were purified from these strains and purity was checked by SDS-PAGE and Western blots. Monoclonal antibodies (mabs) directed against these fimbriae were produced by the fusion

[a] *Telephone:* 31-43-888577

Protein-Carbohydrate Interactions in Biological Systems. ISBN 0 12 436665 1

of spleen cells from immunized BALB/c mice with SP2/0 myeloma cells. The resulting series of mabs were screened in an ELISA with eight different cloned and purified fimbriae. Four different $F7_1$ hybridomas produced mabs which recognized only epitopes on $F7_1$ fimbriae. Two $F7_2$ mabs recognized epitopes on $F7_2$ and F9 fimbriae, whereas another $F7_2$ mab recognized only an epitope on $F7_2$ fimbriae. Three mabs raised against F9 only reacted with epitopes on F9 fimbriae. Six mabs against F11 fimbriae could be divided into two groups: on the one hand 2 mabs recognizing F11, pap and $F7_2$ fimbriae, and on the other hand 4 mabs recognizing F11 and "Clegg" fimbriae. Five mabs against pap fimbriae only recognized epitopes on pap fimbriae, whereas another pap mab recognized an epitope on pap, $F7_2$ and F11 fimbriae. One mab against 1C fimbriae only reacted with an epitope on 1C fimbriae and the three mabs against 1A only reacted with 1A fimbriae.

In an ELISA with whole bacteria, the mabs could be used for determination of fimbriae on wild-type *E. coli* strains. With this simple and rapid method we are able to screen the various P fimbriae on clinical isolates of *E. coli*.

ACKNOWLEDGEMENT

This work was supported by the Dutch Kidney Foundation (Grant no. 81320).

REFERENCES

1. Ørskov, I. and F. Ørskov (1983). *In* "Host-Parasite Relationships in Gram-negative Infections" (Eds L. A. Hanson, P. Kallos and O. Westphal), Progress in Allergy, Vol. 33, pp. 80–105. Karger, Basel.
2. Parry, S. H., S. N. Abraham and M. Sussman (1982). *In* "Clinical, Bacteriological and Immunological Aspects of Urinary Tract Infections in Children" (Ed. H. Schulte-Wisseman), pp. 113–126. Thieme, Stuttgart.

Antigenic and Functional Properties of P- and X-Haemagglutinins of Extra-Intestinal *Escherichia coli*

Diana M. Rooke[a], Sarah Palmer and S. H. Parry*

Department of Microbiology, Medical School, University of Newcastle upon Tyne, UK, and
**Immunology Department, Unilever Research, Sharnbrook, Bedfordshire, UK*

Mannose-resistant (MR) adhesins are associated with human uroepathogenic *Escherichia coli* [1] in colonization of the urinary tract by adherence to uroepithelial cells [2] and the kidney. Such adhesins have been shown to display considerable antigenic heterogeneity [2] and more recently functional heterogeneity. P-fimbriae which recognize a digalactose receptor present in the P-blood group antigens show some antigenic variability. Non-P or X adhesins are functionally heterogeneous and, apart from M- and S-fimbriae, have undefined receptor specificities. The antigenic characteristics of X-specific haemagglutinins of extra-intestinal *E. coli* are poorly defined. Antigenic relationships are described in relation to haemagglutination receptor heterogeneity.

Strains were selected for the exclusive expression of P- or X-haemagglutinins by conventional methods and by haemagglutination inhibition of human red blood cells (RBC) by P_1 glycoprotein derived from hydatid cyst fluid [3]. P and X strains were grouped according to their HA reactions with RBC from different animals (Table 1). All strains with P adhesins had the same RBC agglutination profiles which correlated with the presence of Gal-Gal residues. In contrast, strains with X-haemagglutinins fell into six distinct groups on the basis of their HA profiles.

Serological relationships between adhesins were tested by examining antigenic cross-reactivity with polyclonal rabbit antisera raised against purified adhesin

[a]*Telephone:* 44-632-328511 ext. 3704

Protein-Carbohydrate Interactions in Biological Systems. ISBN 0 12 436665 1

Table 1. HA profiles of *E. coli* strains bearing P- or X-haemagglutinins

HA group			*X-specific antiserum from strain no. (HA group)*[a]				
(Total no. of strains)	*Strain*	*Source*	*SP60 (Ao)*	*SP780 (B)*	*SP386 (E)*	*SP577 (E)*	*SP793 (E)*
Ao (3)	SP60		+				
	SP898			+			
A (3)	SP411	Acute UTI		+			
	SP475	ABUP	+				
B (3)	SP780	Acute UTI		+			
D (11)	SP331	Acute UTI		+			
	SP833	Acute UTI		+			
	SP862	Acute UTI		+			
E (12)	SP386	Acute UTI			+		
	SP431	Acute UTI			+		
	SP474	Acute UTI			+		
	SP577	UTIP				+	+
	SP588	Acute UTI			+		
	SP625	Septicaemia			+		
	SP793	Acute UTI				+	+
	SP897	UTIP				+	+
	SP905	Faeces				+	+

[a] No cross-reactivity against type 1 fimbriae.
Acute UTI, acute lower urinary tract infection; ABUP, asymptomatic bacteriuria in pregnancy; UTIP, acute UTI in pregnancy.

Table 2. Serological cross-reactivity of X-haemagglutinins of *E. coli* determined by coagglutination with antibody-coated *Staphylococcus aureus*

		MRHA profile								
HA type	*HA group*	*Strains*	*Human*	*Pigeon*	*Chicken*	*Sheep*	*Bovine*	*Guinea pig*	*Horse*	*Mouse*
P		3	+	+	+/−	+	−	−	−	−
x	Ao	3	+	+	+	+	+	+	+	+
	A	3	+	+	+	+	+	+	+	−
	B	3	+	+	−/W	−	+	+	−/W	+
	C	1	+	W	−	−	+	−	−	−
	D	11	+	−	−	−/+	−	+	−	−
	E	12	+	−	−	−	−	−	−	−

+, positive HA; W, weak position HA; −, no HA.

preparations. Cross-reactions examined by slide coagglutination with two P-specific antisera indicated some heterogeneity of P-fimbriae, since all six P-bearing strains tested reacted with one or both antisera.

X-haemagglutinins showed great heterogeneity based on their cross-reactivity with 5 *X*-specific antisera (Table 2). This heterogeneity was further emphasized since 15/32 were not agglutinated by any antiserum. There was no apparent source-related pattern of serological specificity and in HA groups Ao, A, B and D there was no clear relationship of serological crossreaction with MRHA pattern. However, antisera raised against haemagglutinins of strains in group E, which agglutinated human RBC alone, reacted only with strains of the HA group. ELISA inhibition assays accorded with and confirmed the coagglutination data.

Some antigenic determinants seemed to be shared by P- and X-haemagglutinins of some strains, notably those of HA group D, despite the lack of evidence for functional P-haemagglutinins in the latter.

In conclusion, the X-haemagglutinins of extra-intestinal *E. coli* show marked functional and antigenic heterogeneity. The cross-reactivity seen between strains from different HA groups suggests the presence of multiple X-adhesins and the future identification of receptor specifications should allow the further elucidation of antigen–receptor relationships.

REFERENCES

1. Parry, S. H. *et al.* (1983). *Infection* **11**, 123–128.
2. Parry, S. H., S. N. Abraham and M. Sussman (1982). *In* "Clinical Bacteriological and Immunological Aspects of Urinary Tract Infections", pp. 113–126. Georg Thieme, Stuttgard and New York.
3. Parry, S. H., D. M. Rooke and M. Sussman (1984). *J. Microbiol. Methods* **2**, 323–331.

III. Role of Pili and Adhesins in Pathogenicity

Co-Chairpersons: David Lark
Catharina Svanborg-Edén

Genetic and *In Vivo* Studies With S-Fimbriae Antigens and Related Virulence Determinants of Extra-Intestinal *Escherichia coli* Strains

Jörg Hacker[a], T. Jarchau, S. Knapp, R. Marre*, G. Schmidt‡, T. Schmoll and W. Goebel

Institut für Genetik und Mikrobiologie, Würzburg,
**Institut für Medizinische Mikrobiologie, Lübeck, and*
‡Institut für Experimentelle Biologie und Medizin, Borstel, FRG

Bacterial adherence is the first step in most of the known infectious processes. This is also true for extra-intestinal *Escherichia coli* infections which include urinary tract infections (cystitis and pyelonephritis), neonatal sepsis and meningitis [1,2]. Adhesion of such extra-intestinal *E. coli* isolates is mediated by several adhesins, most of which consist of fimbriae and a binding part; the latter is also termed haemagglutinin or haemagglutination factor. These haemagglutination factors may interact with erythrocytes or other mammalian cells via specific eukaryotic receptor molecules.

Mannose-sensitive (MS) fimbriae which are produced by more than three-quarters of all extra-intestinal *E. coli* isolates adhere to mannose-containing receptor structures, and this adhesion can be blocked in the presence of 2% D-mannose. In contrast, mannose-resistant (MR) adhesion is independent of the presence of D-mannose. The P-fimbriae which belong to the MR-adhesion factors are present on more than 70% of the uropathogenic *E. coli* strains [1]. P-fimbriae recognize a receptor, the D-Gal-Gal-globoside which is part of the human P-blood group antigen. S-fimbriae have also been found among uropathogenic *E. coli* strains (especially among strains from cases of cystitis), but this adherence factor seems to be more frequently associated with neonatal sepsis and meningitis where

[a]*Telephone:* 49-931-31575

Protein-Carbohydrate Interactions in Biological Systems. ISBN 0 12 436665 1

it interacts with sialic acid-containing cell structures of various tissues including the brain. S-specific adherence is sensitive to treatment with neuraminidase and S-fimbriae are very often located on the surface of 018:K1 and 06:K^+ pathogenic *E. coli* strains [1,3].

CLONING OF THE DETERMINANT CODING FOR S-FIMBRIAE ANTIGEN (*sfa*)

The genetic determinant coding for S-specific, mannose-resistant haemagglutination (*mrh*) and fimbriae production (*fim*) has been cloned from the chromosome of the uropathogenic 06:K15:H31 *E. coli* strain 536 [3]. A cosmid gene bank from strain 536 was constructed and *E. coli* K-12 Mrh^+/Fim^+ cosmid clones were isolated. From one recombinant cosmid DNA (pANN801, see Table 1) several fragments carrying the whole *sfa*-determinant were ligated into the vector molecule pBR322 resulting in *mrh–fim* subclones. In addition, subclones have been isolated which only express the S-specific haemagglutination but not the fimbriae. These recombinant DNAs and the isolation of insertion mutants which exhibit either a Mrh^-/Fim^- or a Mrh^-/Fim^+ phenotype support the view that two different regions can be distinguished on the *sfa*-determinant, one coding for haemagglutination and the other for the fimbriae.

Table 1. Properties of wild-type strain 536, mutant strains and *E. coli* K-12 clones harbouring recombinant DNAs which were derived from a 536 gene bank

	S-fimbriae					
Strains	*Mrh*	*Fim*	*Type II fimbriae*	*MS fimbriae*	*Sre*[a]	*Hly*[b]
HB101	–	–	–	–	–	–
HB101 pANN801	+	+	–	–	–	–
HB101 pANN801-4	+	+	–	–	–	–
HB101 pANN801-13	+	+	–	–	–	–
HB101 pANN801-1	+	–	–	–	–	–
HB101 pANN801-13:Tn*5/38*	–	+	–	–	–	–
HB101 pANN801-13:Tn*5/7*	–	–	–	–	–	–
536 WT	+	+	+	+	+	+
536-21	–	–	–	–	–	–
536-111	–	–	–	–	–	–
536-21[a]	+	+	–	–	–	–
536-21[b]	+	+	–	–	–	–

[a]Resistance to 90% normal human serum.
[b]Haemolysin production.

GENETIC STRUCTURE OF THE DETERMINANT CODING FOR S-FIMBRIAE ANTIGEN

The recombinant plasmid pANN801-13 which codes for both S-specific functions, haemagglutination and fimbriae formation (Table 1), consists of a chromosomal *Bam*HI/*Eco*RI fragment inserted in the vector pBR322. A detailed restriction map of pANN801-13 is given in Fig. 1. One can distinguish between seven different *Pst*I fragments, designated as P4 to P10. The restriction pattern of the *sfa*-determinant seems to be rather different from those of the P-determinants, the physical maps of which show striking similarities to each other [4]. This similarity is also exhibited by the recently cloned P-fimbriae determinant of serotype F8 (Hacker *et al.*, submitted).

As demonstrated by the isolation of Tn*5* and Tn*1000* insertion mutants, the *sfa*-determinant spans a 6.5 kb stretch of DNA. The determinant starts 600 bp to the left of the first *Sma*I site and ends nearly 900 bp to the left of the *Pst*I fragment P9. The gene required for the generation of the 16.5 kD S-pilin protein is located between the second *Sma*I site and the *Cla*I site. A region which spans 200 bp from the right of the *Cla*I site to the *C*-terminal end of the determinant seems to be necessary for the Mrh$^+$ phenotype. Following complementation of Tn*1000* insertion mutants with different subclones, five complementation groups

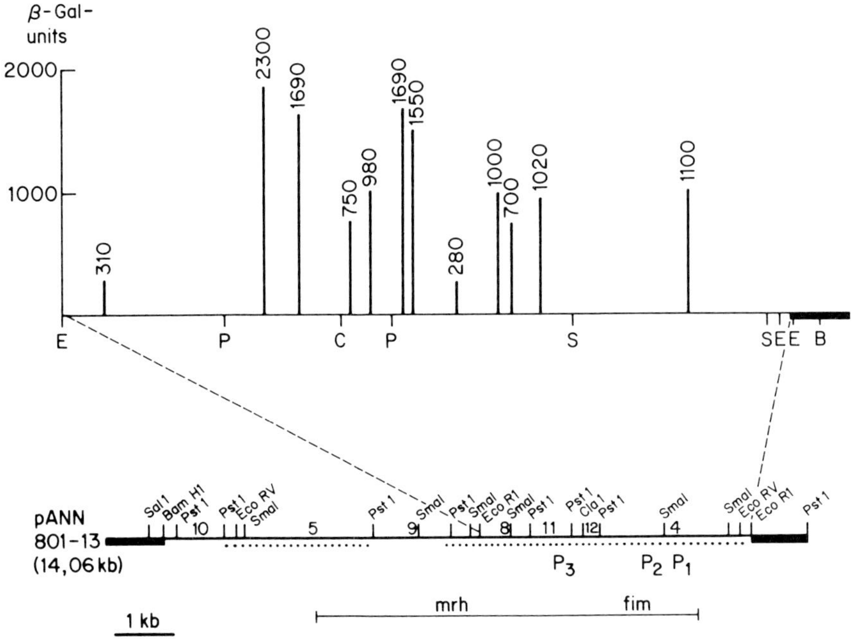

Figure 1. Restriction map of the *sfa*-determinant of strain 536. The underlined regions are analysed by means of λ*plac* Mu-insertions. The vertical lines in the upper part of the diagram represent the β-galactosidase activities of the fused determinants.

have been identified which could represent five different gene products (Hacker *et al.*, in prep.).

In order to identify the promoter regions from which the *sfa*-determinant is transcribed we inserted λp*lac*Mu-phages into suitable fragments of the *sfa*-determinant. Such phages contain the *lac*Z gene without its own promoter. Thus production of β-galactosidase must result from external transcription signals and the points of insertion should indicate the location of the promoter regions. Starting from the second *Sma*I site one promoter (P1) which is reading in the opposite direction to the determinant has been mapped (see Fig. 1). In analogy with a P-specific fimbriae determinant (*pap*), this signal could code for regulatory functions (Uhlin, pers. commun.). Two other promoters have been identified, one in front of the pilin gene (P2) and the other to the left of the *Cla*I site (P3). These start regions seem to be responsible for the transcription of the structural genes of the *sfa*-determinant.

SEROLOGICAL ANALYSIS OF FIMBRIAE ANTIGENS FROM STRAIN 536

The isolated subclones (PANN801-13 and others) exhibit neuraminidase-sensitive haemagglutination which is typical for S-fimbriae. Protein subunits of 16.5 kD in size could be detected following disintegration of fimbriae after SDS-polyacrylamide gel electrophoresis (PAGE). In order to obtain anti-S-fimbriae antiserum rabbits were injected with the pilin protein which had been purified by preparative PAGE. In Western blots protein extracts from an *E. coli* K-12 strain carrying the recombinant plasmid pANN801-13 and from the wild-type strain 536 react with the anti-S-fimbriae antibodies predominantly in the 16.5 kD S-pilin protein (Fig. 2). After growing the strains on solid medium or in a stationary liquid culture the S-pilin protein was detected.

As demonstrated in Table 1, in addition to the S-fimbriae, strain 536 exhibits two other fimbriae proteins: MS-fimbriae (also called common type I- or F1A-fimbriae) and a type II antigen which consists of 22 kD protein subunits and does not mediate agglutination with any of the different erythrocyte types tested. Neither the MS-fimbriae nor the "22 kD type II-fimbriae" of strain 536 were recognized by the S-specific antibodies, which is further evidence that:

(1) S-specific haemagglutination is associated only with the "16.5 kD-fimbriae"
(2) strain 536 produces three distinct adhesin factors, S, MS, and type II.

ISOLATION AND CHARACTERIZATION OF S-FIMBRIAE ANTIGEN (*Sfa*)-NEGATIVE MUTANTS

As shown earlier, the uropathogenic strain 536, which belongs to the serotype 06:K15:H31, produces the three adhesin factors S, MS and type II, is also resistant to human serum (Sre^+) and haemolytic (Hly^+). With relatively high frequencies (10^{-3}–10^{-4}), mutants were isolated which still retained their O- and capsule (K)-antigens but had lost the ability to produce adhesins [3] and became serum

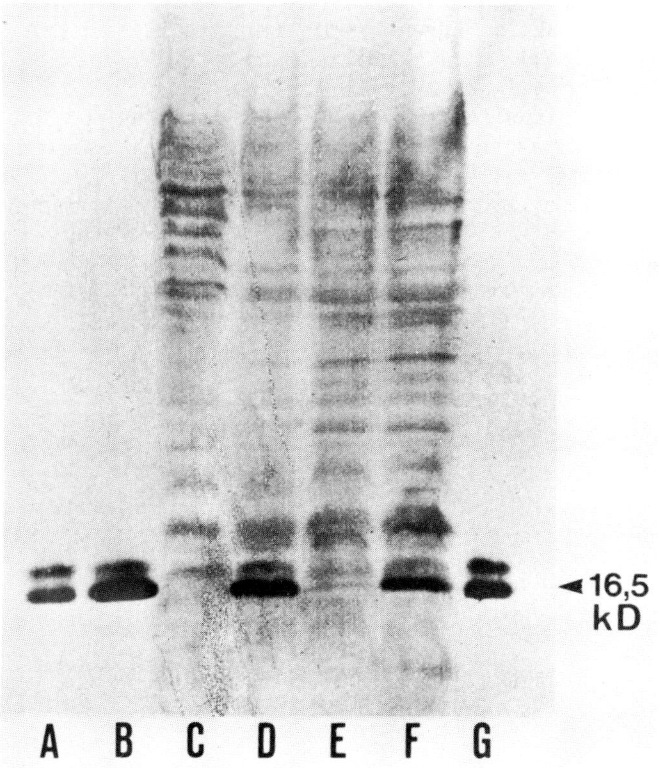

Figure 2. Western blots of whole-cell protein extracts from different strains probed against anti-S-fimbriae antiserum. Lanes A, G, purified S-fimbriae from HB101 pANN801-13 (control); lane B, HB 101 pANN801-13, grown on solid medium; lane C, 536-21, grown in stationary liquid culture; lane D, 536-WT, grown in stationary liquid culture; lane E, 536-21, grown on solid medium; lane F, 536-21 grown on solid medium.

sensitive and non-haemolytic (see also Table 1). Previous studies have shown [5,6] that the two haemolysin (*hly*)-determinants of this strain are deleted in these mutants together with two large regions of DNA 70–80 kb in size. When the *hly*-region II is deleted the mutants show a negative phenotype for the adhesins, as can be seen for the S-fimbriae in Fig. 2.

In contrast to the *hly* genes, the Sfa-coding region is not deleted in these mutants, rather it is still present on the chromosome. This has been demonstrated by:

(1) Southern hybridization studies (Knapp *et al.*, submitted);
(2) the isolation of Sfa^+ revertants from these Sfa^- mutants;
(3) the isolation of Sfa^+ cosmid clones from a gene bank which was derived from one of these mutants.

In order to find out whether the expression of the determinant is or is not regulated at the transcriptional level, total RNA was isolated from the *E. coli* K-12 strain carrying the cloned *sfa*-determinant, from the S-fimbriated wild-type strain 536 and its Sfa$^-$ counterpart 536-21, and hybridized against radioactively labelled DNA probes of the cloned S-fimbriae determinant. As shown in Fig. 3, RNA dot blot analysis indicates that the amount of RNA produced by the Sfa$^-$ mutant strain 536-21 is significantly decreased in comparison to that of the wild-type strain 536. Thus it seems that the lack of expression of the S-fimbriae in these mutants is due to the absence of transcription of the *sfa*-determinant.

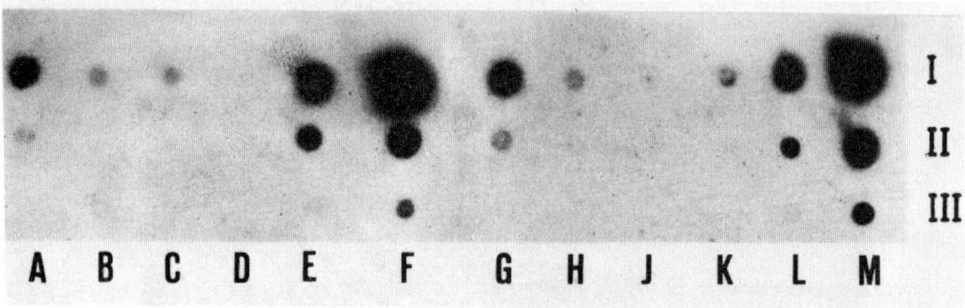

Figure 3. Dot blot analysis of total RNA isolated from different strains hybridized against ^{32}P-labelled DNA probes which consist of the 1.7 kb *Sma*I-*Cla*I fragment (lanes A-F) and the 1.6 kb *Cla*I-*Eco*RI fragment (lanes G-M) of the *sfa*-determinant (see Fig. 1). Lanes A, G, 536-WT, grown on solid medium; lanes B, H, 536-21, grown on solid medium; lanes C, J, 536-WT, grown in stationary liquid cultures; lanes D, K, 536-21, grown in stationary liquid cultures; lanes E, L, HB101 pANN801-13, grown on solid medium; lanes F, M, DNA-DNA control hybridization. In row I 4 μg RNA, in row II 0.4 μg RNA, in row III 0.04 μg RNA were spotted.

CONSTRUCTION OF STRAINS FOR *IN VIVO* TESTS

As already demonstrated, we were able to isolate mutants from the virulent 06:K15:H31 strain 536. These mutants had lost the presumptive virulence factors, S-fimbriae (Sfa$^-$), serum resistance (Sre$^-$) and haemolysin (Hly$^-$) and they became completely avirulent [7]. The re-introduction of cloned virulence genes by transformation into such mutants and the test of the transformants *in vivo* seems to be extremely suitable determining the contribution of these genes to virulence.

As indicated in Table 2 the genetic determinants for S- and for P(F8)-fimbriae were introduced into the Sfa$^-$, Sre$^-$, Hly$^-$ mutant 536-21. In addition the DNA coding for S-specific haemagglutination but not for fimbriae formation (*mrh*$^+$/*fim*$^-$, see above) and two recombinant plasmids carrying the determinants for haemolysin production from the 06 strain 536 and an 018 isolate were checked for the chromosomal and plasmid-encoded markers and only those colonies which still retained the recombinant plasmids were counted in the *in vivo* tests.

IN VIVO TESTS

The constructed strains representing different phenotypes were used in three *in vivo* test systems (Table 2). First, the adhesive capacity of the strains was measured. Bacteria were grown overnight and incubated with uroepithelial cells isolated from fresh morning urine as described [2]. After one hour the attached bacteria were counted. The introduction of both the S- and P-determinant influenced the attachment of the non-adhesive strain 536-21. While the P(F8)-determinant mediated adhesion to cells from the lower and upper urinary tracts, the *sfa*-determinant enhanced only attachment to cells from the bladder. This corresponds to the fact that S-fimbriated strains are connected more with cystitis than with pyelonephritis. The cloned haemolysin (*hly*)-determinants have no influence on the adhesion of the bacteria.

The toxicity of the strains was tested in a chicken embryo assay. Ten-day-old chicken embryos were infected with 1×10^6 bacteria. After two days the lethality was determined. As recently described for a mouse assay [7], the haemolysin determinants contribute to the toxicity of the strains but the re-introduction of the adhesins, S and P (F8), did not increase the lethality of the mutant strain 536-21.

Because each of the two test systems used, the adhesin assay and the chicken embryo test, only reflect one factor of the multifactorial *E. coli* virulence, we decided to use a rat pyelonephritis model to measure the renal bacterial counts seven days after intra-urethral injection of 7×10^7 bacteria. Injection of bacteria resulted in an acute pyelonephritis persisting for several weeks with relatively high renal counts (Marre *et al.*, submitted). While nearly 10^5 cells of the

Table 2. Virulence properties of wild-type strain 536 and mutant strains which carry recombinant DNAs coding for different virulent factors

Strains	*Phenotype*	*Adhesion*	*Toxicity*	*Nephro-pathogenicity*
536 WT	S-Mrh,S-Fim,Sre,Hly	6.8[a]	100[b]	91×10^{4}[c]
536-21		2.2	20	4×10^1
536-21 pANN801-4	S-Mrh,S-Fim	10.1	20	1×10^3
536-21 pANN801-1	S-Mrh	nt	15	5×10^1
536-21 pANN921	P-Mrh,P-Fim	13.5	nt	5×10^1
536-21 pANN5211-06	Hly (06)	1.8	85	4×10^2
536-21 pANN5311-018	Hly (018)	2.0	90	2×10^2

[a]Number of attached bacteria/uroepithelial cell.
[b]Percentage of killed chicken embryos after infection with 1×10^6 bacteria.
[c]Bacterial counts/g rat kidney.
nt, not tested.

wild-type strain 536 per gram of kidney could be isolated seven days after infection, the Sfa$^-$, Sre$^-$, Hly$^-$ mutant 536-21 led to dramatic reduction to only 40 cells per gram of kidney. The re-introduction of the S-fimbriae antigen determinant (*mrh/fim*) increases the virulence by a factor of 20 in comparison to that of the Mrh$^-$/Fim$^-$ mutant strain. No effect was observed following transformation of the *mrh*$^+$/*fim*$^-$ DNA. Thus both functions of the *sfa*-determinant—binding and fimbriae formation—are necessary for the production of *in vivo* virulence. Cloned P(F8)-fimbriae had no influence on the colonization of bacteria in the rat urinary tract because P-receptor structures are not present on rat tissue cells. In addition the *hly*-determinants increase the nephropathogenicity of strains by a factor of between five and ten.

SUMMARY

(1) The S-fimbriae antigen (*sfa*)-determinant has been cloned from the chromosome of the uropathogenic 06:K15:H31 *E. coli* strain 536. Two regions of the *sfa*-determinant can be distinguished, one coding for the 16.5 kD fimbrial structural protein (*fim*) and one which is responsible for mannose-resistant haemagglutination (*mrh*).

(2) The Sfa-specific Mrh$^+$/Fim$^+$ phenotype requires 6.5 kb of DNA. By complementation analysis five complementation groups have been identified. The *sfa*-determinant is transcribed from three different promoter regions.

(3) The strain 536 produces three different adhesin factors: S-fimbriae, MS-fimbriae and type II-fimbriae with 22 kD protein subunits.

(4) Sfa$^-$ mutants of strain 536 were obtained with frequencies of between 10^{-3} and 10^{-4}. These mutants had also lost the ability to produce haemolysin (Hly$^-$), Ms− and type II-fimbriae and they became serum sensitive (Sre$^-$). The absence of S-fimbriae is due to reduced transcriptional activity whereas the Hly$^-$ phenotype results from deletion events in the chromosome.

(5) A Sfa$^-$, Sre$^-$, Hly$^-$ mutant, 536-21, was used as a recipient strain for the introduction by transformation of the S- and P(F8)-adhesin determinants and of two haemolysin determinants.

(6) Both cloned fimbriae determinants, S and P(F8), contribute to the attachment of the non-adhesive strain 536-21 to uroepithelial cells. In contrast to the haemolysin (*hly*) genes, they have no influence on toxicity in a chicken embryo test. The wild-type strain 536 is virulent in a rat pyelonephritis model as indicated by high renal counts. The nephropathogenicity of the avirulent mutant 536-21 increases by a factor of 20 following re-introduction of the whole *sfa*-determinant (*mrh*$^+$, *fim*$^+$). The S-haemagglutination ability without fimbriae (*mrh*$^+$, *fim*$^-$) and P(F8) adhesin have no influence in this test.

ACKNOWLEDGEMENTS

The authors wish to thank H. Düwel and D. Teubel for excellent technical assistance and C. Kathariou for suggestions and critical reading of the manuscript.

The work was supported by the Deutsche Forschungsgemeinschaft (Go. 168-11-3 and Ma. 864-2-1).

REFERENCES

1. Korhonen, T. K., M. V. Valtonen, J. Parkkinen, V. Väisänen-Rhen, J. Finne, F. Ørskov, I. Ørskov, S. B. Svenson and H. P. Mäkelä (1985). Serotypes, hemolysin production, and receptor recognition of *Escherichia coli* strains associated with neonatal sepsis and meningitis. *Infect. Immun.* **48**, 486-491.
2. Svanborg Edén, C., L. Hagberg, L. A. Hanson, S. Hall, R. Hull, U. Jodal, H. Leffler, H. Lomberg and E. Straube (1983). Bacterial adherence—a pathogenetic mechanism in urinary tract infections caused by *Escherichia coli. Prog. Allergy* **33**, 175-188.
3. Hacker, J., G. Schmidt, C. Hughes, S. Knapp, M. Marget and W. Goebel (1985). Cloning and characterization of genes involved in production of mannose-resistant, neuraminidase-susceptible (X) fimbriae from a uropathogenic 06:K15:H31. *Infect. Immun.* **47**, 434-440.
4. Lund, B., F. P. Lindberg, M. Baga and S. Normark (1985). Globoside-specific adhesins of uropathogenic *Escherichia coli* are encoded by similar transcomplementable gene clusters. *J. Bacteriol.* **162**, 1293-1301.
5. Hacker, J., S. Knapp and W. Goebel (1983). Spontaneous deletions and flanking regions of the chromosomally inherited hemolysin determinant of an *Escherichia coli* 06 strain. *J. Bacteriol.* **154**, 1146-1152.
6. Knapp, S., J. Hacker, I. Then, D. Müller and W. Goebel (1984). Multiple copies of hemolysin genes and associated sequences in the chromosome of uropathogenic *Escherichia coli* strains. *J. Bacteriol.* **159**, 1027-1033.
7. Hacker, J., C. Hughes, H. Hof and W. Goebel (1983). Cloned hemolysin genes from *Escherichia coli* that cause urinary tract infection determine different levels of toxicity in mice. *Infect. Immun.* **42**, 57-63.

Bacterial Attachment to Glycoconjugate Receptors: Uropathogenic *Escherichia coli*

Catharina Svanborg-Edén[a], L. Hagberg, R. and S. Hull*, H. Leffler, H. Lomberg and B. Nilsson[‡]

*Departments of Clinical Immunology and Medical Biochemistry, University of Göteborg, Sweden, *Department of Microbiology, Baylor School of Medicine, Houston, Texas, USA, and ‡Swedish Sugar Company, Arlöv, Sweden*

The identification of glycoconjugates as receptors for attaching bacteria has opened new areas for research on protein-carbohydrate interaction. The demonstration that uropathogenic *E. coli* bind Galα1→4Galβ in the globo series of glycolipids has provided a useful system for detailed analysis [1]. Most wild-type bacteria, however, have the potential to recognize several receptor specificities. The binding to the functional target will thus be the sum of the adhesins expressed by the bacteria and the available receptors. The biological significance, and possibly usefulness of receptor analogues in diagnosis and therapy of infections, depends on the importance of the single binding specificity for colonization/infection. For uropathogenic *E. coli* recognizing Galα1→4Gal, considerable experience has been gathered by several groups. This brief review summarizes some of the background and future implications of this work.

Urinary tract infections, UTI, occur in ~2% of girls and increase in frequency in adult women. *E. coli* cause the majority of UTI. The bacteria become established in the urinary tract in spite of the mechanical rinsing by the urine flow. It was, thus, an apparent possibility that bacteria causing UTI need to attach to the mucosal lining to remain in the urinary tract [2].

The severity of UTI is determined by the balance between host defence and bacterial virulence. Bacteriuria is associated with several disease entities: acute pyelonephritis, py, involving the kidney; acute cystitis, cy, limited to the bladder;

[a]*Telephone:* 46-31-602082

Protein-Carbohydrate Interactions in Biological Systems. ISBN 0 12 436665 1

or asymptomatic bacteriuria, ABU, with no or few symptoms. The definition of virulence factors in uropathogenic *E. coli* is based on differences between py and ABU strains [3]. The py strains are selected out of the random *E. coli* flora as seen by a reduced genetic diversity [4,5] and a limited number of serotypes [4-7]. Strains of these serotypes coexpress adhesive properties, resistance to serum killing and haemolysin [3,8]. These characteristics are rare in ABU or normal faecal *E. coli* isolates, and are thus designated virulence factors. In different surveys, the frequency of strains attaching to human uroepithelial cells has been 75-90% in pyelonephritis [2,9-11] compared to <20% in ABU [2,10] isolates.

GLYCOCONJUGATE RECEPTORS

The search for uroepithelial receptors for attaching bacteria was guided by the apparent specificity for species, individual and tissue [12], and the fact that the composition of glycolipids varies in a similar way [1]. Uropathogenic *E. coli* attach to human uroepithelial cells but not to human small intestine, as tested with brush border preparations [13]. Consequently, the non-acid glycolipid fraction was extracted from both cell types, and used to inhibit adherence to uroepithelial cells. The urinary sediment epithelial glycolipid fraction completely blocked adherence; the small intestinal fraction did not [15]. We concluded that glycolipids can function as receptors for attaching bacteria, and proceeded to identify the active component by fractionation of the glycolipids and by testing of purified natural analogues to the ingoing components. Receptor activity was found in globotetraosylceramide, globotriaosylceramide and the Forssmann glycolipid hapten. Since the globotriaosylceramide was less active than globotetraosylceramide we suggested that Galα1→4Galβ was the minimal receptor, which was presented optimally in the tetrasaccharide chain of globotetraosylceramide. The relationship to the P blood group system was supported by the lack of inhibition by the glycolipid fraction from p̄ erythrocytes [14].

Källenius and coworkers arrived at a similar conclusion using a different approach [15]. The attaching bacteria will bind all cells containing the appropriate receptor [12]. Binding to erythrocytes results in haemagglutination. For uropathogenic *E. coli* mannose-resistant agglutination of human erythrocytes correlated with adherence [16]. Erythrocytes from individuals of the p̄ blood group, lacking the globo series of glycolipids were, however, not agglutinated [17].

Evidence for the receptor function of the globo series of glycolipids, and the specificity for Galα1→4Galβ has been collected by several groups, and may be summarized as follows:

(1) The receptors are found in the relevant tissue. In addition to the urinary sediment epithelial cells [14] the globo series glycolipids are major components of human kidney tissue and occur both in epithelial and non-epithelial components of the mucosae lining the urinary tract [18]. Individual variation in receptors may, indeed, influence the susceptibility to infection (Lomberg *et al.*, Lark *et al.*, this volume).

(2) The bacteria bind only to cells containing the receptors. This is shown by the lack of binding to epithelial cells from p̄ individuals [17]. Selective binding to different cell types within one individual also has interesting consequences. For example, polymorphonuclear leukocytes PMNL probably have a crucial role to clear the bacterial inoculum from, for example, infected kidneys. Human PMNL, however, lack the globo series of glycolipids. They contain small amounts of Galα1→4Gal-Cer [19]. Bacteria recognizing Galα1→4Gal will bind poorly to PMNL, and not induce an oxygen burst, uptake and killing [20]. The same binding specificity which promotes colonization of the mucosal surface will prevent contact with phagocytes, and thus enhance virulence by two mechanisms. Differential binding to subpopulations of lymphocytes might similarly influence the immune response to antigens on the attaching bacteria.

(3) The receptor activity of isolated glycolipids or synthetic oligosaccharides has been analysed by:

(a) *inhibition of attachment* to human uroepithelial cells or erythrocytes [17]. Globotetraosylceramide was more active than synthetic oligosaccharide derivatives of Galα1→4Galβ, including polyvalent neoglycoproteins, and neoglycolipids [1]. The effective inhibitory concentration required to reduce adherence to 50% of the saline control was around 0.2 mM to globotetraosylceramide, compared to ~ten-fold higher for Galα1→4Galβ-O-Etyl. The poor activity of Galα1→4Gal may be secondary to metabolization of the disaccharide and bacterial multiplication (data not shown).

(b) *coating.* Cells not containing the globo series glycolipids, e.g. erythrocytes, epithelial cells and PMNL, became reactive with bacteria after coating with, for example, globotetraosylceramide [14]. Thus, addition of the receptor was sufficient to restore the binding function to these cells.

(c) *binding to glycolipids on a solid phase.* The thin layer chromatogram binding assay was used to screen a wide range of natural glycolipids (see Karlsson *et al.*, this volume). The binding of *E. coli* 36692 was completely restricted to Galα1→4Galβ-containing glycolipids [21].

(d) *agglutination of latex beads.* Latex beads covalently linked to BSA were coupled to Galα1→4Gal or lactose or coated with globotetraosylceramide (see B. Nilsson *et al.*, this volume). Agglutination of latex beads associated with receptor has been described previously [22] but the ingoing components have not been described. The preliminary results of a survey of 450 clinical *E. coli* isolates are shown in Table 2.

Table 1. Inhibition of adhesion by Galα1→4Galβ-derivates

Inhibitor	EID_{50}* *(mM)*
GalNAcβ1→3Galα1→4Galβ1→4Glc	0.2
Galα1→4Galβ-O-Et	0.6
Galα1→4Galβ1→4Glcβ-O-Et	2
Galα1→4Gal	11

Table 2. Frequency of adhesive properties in *E. coli* isolates from the urine of patients with Py (acute pyelonephritis), Cy (acute cystitis), ABU (asymptomatic bacteriuria) or from the normal faecal flora, FN

	% Positive strains			
	Py *n=103*	*Cy* *n=119*	*ABU* *n=126*	*FN* *n=130*
Adherence >10 bact/cell	77	62	36	36
MR_{hum}[a]	77	35	18	16
Globoside-latex	73	22	10	10
Galα1→4Gal-latex	65	17	9	10

[a]Mannose-resistant agglutination of human AP_1 erythrocytes.

The globotetraosylceramide-coated beads are agglutinated by more strains and with greater intensity than the Galα1→4Gal-coupled beads. Extended surveys are continuing to determine the optimal composition of a diagnostic kit detecting attaching bacteria.

IN VIVO RELEVANCE OF ADHERENCE

The epidemiological studies provided the basis for the interest in adherence. Since adherence is only one of the clonally expressed properties in disease isolates, epidemiology does not permit conclusions to be drawn about the role of adherence as an isolated contributor to virulence [23]. That question may be approached in patients by administration of compounds selectively blocking adherence. As an alternative we have used experimental UTI with bacteria genetically manipulated to differ in adhesins [24,25]. Two sets of bacteria were used:

(a) Mutants of a wild-type pyelonephritis strain (075, K5, binding Galα1→4Gal and mannose) retaining either adhesin from the parent strain.

(b) Transformants. The *pap* operon from pRHU845 or the *pil* operon from SH2 was used to transform a non-adhering normal faecal isolate, *E. coli* 506. The transformants expressed either adhesins specific for Galα1→4Gal or for mannose [13].

The contribution of adherence in individual mice was measured by the relative persistence of the mutants differing in adhesins. Adhesins specific for the globo series of glycolipids increased the persistence of bacteria in the kidneys ~ten-fold. In the bladder, the combination of Galα1→4Gal and mannose-specific adhesins was optimal [25]. The difference in persistence between the mutants and transformants with the same adhesin was, however, ~100-fold. This emphasized the contribution of virulence factors in addition to adherence [25]. Recently, preliminary experiments were performed with mixed infections in primates. *E. coli* HU824 (Galα1→4Gal-specific) and HU742 (mannose-specific) were inoculated into the bladder of monkeys. The relative persistence of the mutants was analysed by daily urine samples. Initially, equal numbers were obtained. The Galα1→4Gal-binding mutant only had an advantage after 3-7 days (Table 3).

Table 3. Experimental bacteriuria in monkeys. The animals were infected with a mixture of two homogeneic mutants differing in adhesins

Receptor specificity of adhesin	*No. bact/ml urine*			
	Day 1	*Day 3*	*Day 7*	*Day 10*
Galα1→4Galβ	1.3×10^5	1×10^3	1.2×10^3	0
"Mannosides"	1.1×10^5	2×10^3	0	0

HOST DEFENCE IN RELATION TO BACTERIAL VIRULENCE

A number of inbred mouse strains with defects in antibacterial defence mechanisms are available for study. Screening of T lymphocyte (nude), B lymphocyte (xid), macrophage deficient (A/J) and complement deficient (AKR) mice did not reveal an increased susceptibility to experimental UTI. The Lps non-responder mouse, C3H/HeJ, however, was significantly more susceptible than the Lps responsive relatives C3H/HeN and C3HeB/FeJ [26,27].

The relative contribution of the defect in host defence and bacterial virulence was analysed by infection of resistant and susceptible mice with mutants and transformants differing in adhesins, O or K antigens. The relative persistence of the adhering and non-adhering mutants was similar in both mouse strains; the recovery from the kidneys of the susceptible mouse was ~2 logs higher than from the resistant mouse. Thus, in this system the host resistance factors contributed more than single bacterial traits, e.g. adherence.

Table 4. Contribution of adherence to the persistence of *E. coli* in the kidneys of C3H mice

Inoculum mixture (adhesins)	*Marker*	*Difference*	*Log bacterial recovery 24 h after injection*			
			C3H/HeN		*C3H/HeJ*	
			Total	*Ratio*	*Total*	*Ratio*
GS, MS	Str	Binding to	2.38	1.24	4.53	1.09
— MS		Galα1→4Gal	1.14		3.44	
GS, MS	Str	Binding to	2.62	1.81	4.29	1.63
GS —		"mannosides"	0.81		2.61	
GS —	Nal	Either	2.22	1.08	4.24	0.70
— MS		property	1.21		3.54	

GS = specific for the globo series of glycolipids; MS = specific for mannosides; — = neither specificity.

ACKNOWLEDGEMENTS

This work was supported by grants from the Swedish Medical Research Council (No. 215), The Medical Faculty, University of Göteborg, The Ellen, Walter and

Lennart Hesselmann Foundation for Scientific Research, The Swedish Sugar Company/Kabi Vitrum. The skilful typing of Ann-Charlotte Malmefeldt is greatly appreciated.

REFERENCES

1. Leffler, H. and C. Svanborg Edén (1985). *E. coli* lectins/adhesins binding to glycolipids. *In* "Bacterial Lectins" (Ed. D. Mirelman). Wiley, New York.
2. Svanborg Edén, C., L. Å. Hanson, U. Jodal, U. Lindberg and A. Sohl-Åkerlund (1976). Variable adhesion to normal urinary tract epithelial cells of *Escherichia coli* strains associated with various forms of urinary tract infection. *Lancet* **ii**, 490–492.
3. Svanborg Edén, C., L. Hagberg, L. Å. Hanson, T. K. Korhonen, H. Leffler and S. Olling (1981). Adhesion of *E. coli* in urinary tract infections. *In* "Adhesion and Microorganism Pathogenicity", Ciba Symposium 80, pp. 161–178. Pitman Medical, London.
4. Caugant, D. A., B. R. Levin, G. Lidin-Janson, T. S. Whittam, C. Svanborg Edén and R. K. Selander (1983). Genetic diversity and relationships among strains of *Escherichia coli* in the intestine and those causing urinary tract infections. *Prog. Allergy* **33**, 203–227.
5. Caugant, D. A., B. Levin, I. Ørskov, F. Ørskov, R. Selander and C. Svanborg Edén (1985). Genetic diversity within O, K and H serotypes of *Escherichia coli*. *Infect. Immun.*, in press.
6. Lidin-Janson, G., L. Å. Hanson, B. Kaijser, K. Lincoln, U. Lindberg, S. Olling and H. Wedel (1977). Comparison of *Escherichia coli* from bacteriuric patients with those from feces of healthy schoolchildren. *J. Infect. Dis.* **136**, 346–353.
7. Mabeck, C. E., F. Ørskov and I. Ørskov (1971). *Escherichia coli* serotypes and renal involvement in urinary tract infection. *Lancet* **i**, 1312.
8. Minshew, B. H., J. Jorgensen, M. Swanstrum, G. A. Grootes-Revvecamp and S. Falkow (1978). Some characteristics of *Escherichia coli* strains isolated from extraintestinal infections of humans. *J. Infect. Dis.* **137**, 648–654.
9. Källenius, G. and J. Winberg (1978). Bacterial adherence to periurethral epithelial cells in girls prone to urinary tract infections. *Lancet* **ii**, 540–543.
10. Svanborg Edén, C., B. Eriksson, L. Å. Hanson, U. Jodal, B. Kaijser, G. Lidin-Janson, U. Lindberg and S. Olling (1978). Adhesion to normal human uroepithelial cells of *Escherichia coli* from children with various forms of urinary tract infection. *J. Ped.* **93**, 398–403.
11. Väisänen, V., T. K. Korhonen, M. Jokinen, C. G. Gahmberg and C. Ehnholm (1982). Blood group M specific haemagglutination in pyelonephritogenic *Escherichia coli*. *Lancet* **i**, 1192.
12. Leffler, H., C. Svanborg Edén, G. Schoolnik and T. Wadström (1985). Glycosphingolipids as receptors for bacterial adhesion. *In* "Host Glycolipid Diversity and Other Selected Aspects" (Ed. E. C. Boedekker), Vol. II, pp. 177–187. CRC Press, Boca Raton.
13. Svanborg Edén, C., R. Hull, S. Hull, S. Falkow and H. Leffler (1983). Target cell Specificity of wild-type *E. coli* and mutants and clones with genetically defined adhesins. *Progr. Fd. Nutrition Sci. Res.* **7**, 75–89.
14. Svanborg Edén, C. and H. Leffler (1980). Glycosphingolipids of human urinary tract epithelial cells as possible receptors for adhering *Escherichia coli* bacteria. *Scand. J. Infect. Dis.* **S24**, 144–147.

15. Svanborg Edén, C. and H. Leffler (1980). Glycosphingolipids of human urinary tract epithelial cells as possible receptors for adhering *Escherichia coli* bacteria. *Scand. J. Infect. Dis.* **S24**, 144-147.
16. Källenius, G. and R. Möllby (1979). Adhesion of *Escherichia coli* to human periurethral cells correlated to mannose-resistant agglutination of human erythrocytes. *FEMS Microbiol. Lett.* **5**, 295-299.
17. Källenius, G., R. Möllby, S. B. Svensson, J. Winberg, A. Lundblad, S. Svensson and B. Cedergren (1980). The P^k antigen as receptor for the haemagglutinin of pyelonephritogenic *Escherichia coli*. *FEMS Microbiol. Lett.* **7**, 297.
18. Breimer, M. E., G. C. Hansson and H. Leffler (1985). A specific glycosphingolipid composition of human ureteral epithelial cells. *J. Urol.*, in press.
19. Macher, B. A. and J. C. Klock (1980). Isolation and chemical characterization of neutral glycosphingolipids of human neutrophils. *J. Biol. Chem.* **255**, 2092-2096.
20. Svanborg Edén, C., L. M. Bjursten, R. and S. Hull, Z. Moldovano and H. Leffler (1984). Influence of *E. coli* adhesins upon the interaction with human phagocytes. *Infect. Immun.* **44**, 672-680.
21. Bock, K., A. Brignole, M. E. Breimer, C. G. Hansson, K.-A. Karlsson, G. Larsson, H. Leffler, B. E. Samuelsson, N. Stromberg, C. Svanborg Edén and J. Thurin (1985). Glycolipids as receptors for adhesion of bacteria. *J. Biol. Chem.* **260**, 8545-8551.
22. Svensson, S. B., G. Källenius, R. Möllby, H. Hultberg and J. Winberg (1982). Rapid identification of P-fimbriated *Escherichia coli* by a receptor specific particle agglutination test. *Infection* **10**, 209.
23. Svenborg Edén, C., L. Hagberg, L. Å. Hanson, U. Jodal, H. Leffler, H. Lomberg and E. Straub (1983). Bacterial adherence—a pathogenetic mechanism in urinary tract infection. *Progr. Allergy* **33**, 175-189.
24. Hagberg, L., R. Freter, R. Hull, S. Hull and C. Svanborg Edén (1983). Ascending unobstructed primary tract infection in mice. A model for the study of bacterial attachment. *In* "Experimental, Bacterial and Parasitic Infections" (Eds J. Keusch and T. Wadström), pp. 119-123. Elsevier, Amsterdam.
25. Hagberg, L., R. Hull, S. Hull, S. Falkow, R. Freter and C. Svanborg Edén (1983). Contribution of adhesion to bacterial persistence in the mouse urinary tract. *Infect. Immun.* **40**, 265-272.
26. Hagberg, L., C. Svanborg Edén and D. Briles (1985). Evidence for separate genetic defects in C3H/HeJ and C3HeB/FeJ mice that affect the susceptibility to Gram-negative infections. *J. Immunol.*, in press.
27. Svanborg Edén, C., D. Briles, L. Hagberg, J. McGhee and S. Michalek (1984). Genetic factors in host resistance to urinary tract infection. *Infection* **12**, 132-137.

Fimbriae (Pili) Adhesins as Vaccines

M. Levine[a], J. G. Morris, G. Losonsky, E. Boedeker* and B. Rowe‡

*Center for Vaccine Development, Department of Medicine, University of Maryland, Baltimore, MD 21201, USA, *Walter Reed Army Institute of Research, Washington, DC 20307, USA, and ‡the Central Public Health Laboratory, London, UK*

Certain fimbriae (pili) found only on enterotoxigenic *Escherichia coli* (ETEC), such as colonization factor antigens I and II (CFA/I, II) and E8775 fimbriae serve as virulence properties by allowing attachment of ETEC to enterocytes in the proximal small intestine [1]. Veterinary studies have shown that antibody directed against analogous fimbriae in animal ETEC can suppress bacterial attachment and prevent diarrhoea [2-5]. We have studied ways to stimulate intestinal secretory IgA anti-CFA/II in animals and man using purified fimbriae or a live attenuated, non-enterotoxinogenic *E. coli* strain as oral vaccines [1,6,7]. Eight 2.0 mg doses of purified CFA/II (CS1 and CS3 antigens) applied to the mucosa of an exteriorized loop of rabbit intestine stimulated a brisk SIgA anti-CFA/II response. Thirteen rabbits immunized orally with eight 0.8 mg doses of purified CFA/II fimbriae vaccine and 13 control rabbits were challenged with 10^{10} enterotoxigenic *E. coli* (RITARD method) bearing CFA/II. Diarrhoea occurred in 8 of 13 of each group.

Ten volunteers were immunized with 2.0 mg of purified CFA/II orally twice weekly for four weeks (following suppression of gastric acid with cimetidine and $NaHCO_3$). Only 2 of 10 developed significant rises in intestinal SIgA or serum IgG anti-CFA/II. This cohort of volunteers was not significantly protected, in comparison with unimmunized controls, when challenged with pathogenic ETEC; diarrhoea occurred in 6 of 9 controls and 3 of 8 vaccinees with no difference in disease severity. In view of the excellent immunogenicity of this fimbriae vaccine in stimulating specific SIgA antibody when applied to the mucosa of exteriorized loops of rabbit intestine, we are distressed at the poor immune response following ingestion of multiple oral doses in man. We suspected that

[a] *Telephone:* 1-301-528-7588

Protein-Carbohydrate Interactions in Biological Systems. ISBN 0 12 436665 1

despite cimetidine and $NaHCO_3$ treatment to neutralize gastric acid, it is possible that the fimbrial protein was being adversely affected during its passage through the stomach. This seemed more probable when Schmidt *et al.* [8] showed that gastric acid, even at neutral pH, adversely affects fimbrial protein. Accordingly, we elected to administer three 5 mg doses of the CFA/II fimbrial vaccine directly into the duodenum via intestinal tube, thereby bypassing the stomach. Vaccine was given on days 0, 14, and 28 and intestinal fluid was collected to measure specific anti-fimbrial antibody on days 0, 14, 28 and 35. Given by this route, the CFA/II fimbrial vaccine stimulated significant rises in SIgA anti-CFA/II antibody in four of five vaccines (Table 1). These data emphasize the importance of protecting fimbrial vaccine proteins as they transit the human stomach. Research will have to be undertaken to identify practical and effective delivery systems.

Table 1. Significant rises in intestinal fluid SIgA antibody to CFA/II (CS1, CS3) antigens in volunteers given three 5 mg doses of CFA/II fimbrial vaccine via intestinal tube on days 0, 14 and 28

	SIgA anti-CFA/II Titre[a]	
Volunteer	*Pre-*	*Peak*
1	4	4
3	<4	16
5	<4	1024
6	16	256
7	4	16

[a] Measured by IgA-ELISA.

Nineteen volunteers were fed a single 10^9, 10^{10} or 6×10^{10} organism dose of a non-toxigenic, CFA/II-positive 06:H16 strain (E1392-75-2A) as an oral vaccine [6,7]. Two developed mild diarrhoea, all had positive coprocultures and most had positive duodenal fluid cultures. Paired pre- and post-vaccination intestinal fluids were available from 14 of the 19 volunteers to measure local SIgA antibody. Significant rises in intestinal SIgA antibody to CFA/II antigens were detected in 11 of 14 including 6 of 6 who ingested a dose of 6×10^{10} vaccine organisms. Twelve volunteers who ingested a single 5×10^{10} organism dose of E1392-75-2A live oral vaccine and six unimmunized control volunteers were challenged with 5×10^8 enterotoxigenic *E. coli* of a distinct O:H serotype (0139:H28, LT+/ST+) but bearing the identical CFA/II antigens (CS1 and CS3) as the vaccine strain. Diarrhoea occurred in 6 of 6 controls but in only 3 of 12 vaccinees ($p < 0.005$). Although the challenge strain was excreted in the same concentration in stool cultures of the two groups, a significant difference was noted in duodenal fluid cultures. The enterotoxigenic challenge strain was recovered from 5 of 6 controls (mean 7×10^3 *E. coli*/ml but from only 1 of 12 vaccinees (10^1 *E. coli*/ml) ($p < 0.004$). These studies point to the superiority of live oral vaccines in stimulating anti-fimbrial antibody.

ACKNOWLEDGEMENTS

These studies were supported by research contracts NO1AI42553 from the National Institute of Allergy and Infectious Diseases and DAMD 17-78-C-8011 from the US Army Medical Research and Development Command.

REFERENCES

1 Levine, M. M., J. B. Kaper, R. E. Black and M. L. Clements (1983). New knowledge on pathogenesis of bacterial enteric infections as applied to vaccine development. *Microbiol. Rev.* **47**, 510-550.
2. Acres, S. D., R. E. Isaacson, L. A. Babiuk and R. A. Kapitany (1979). Immunization of calves against enterotoxigenic colibacillosis by vaccinating dams with purified K99 antigen and whole cell bacterins. *Infect. Immun.* **25**, 121-126.
3. Morgan, R. L., R. E. Isaacson, H. W. Moon, C. C. Brinton and C. C. To (1978). Immunization of suckling pigs against enterotoxigenic *Escherichia coli*-induced diarrheal disease by vaccination dams with purified 987 or K99 pili: protection correlates with pilus homology of vaccine and challenge. *Infect. Immun.* **22**, 771-777.
4. Nagy, B., H. W. Moon, R. E. Isaacson, C. C. To and C. C. Brinton (1978). Immunization of suckling pigs against enteric enterotoxigenic *Escherichia coli* infection by vaccinating dams with purified pili. *Infect. Immun.* **21**, 269-274.
5. Rutter, J. M. and G. W. Jones (1973). Protection against enteric disease caused by *Escherichia coli* — a model for vaccination with a virulence determinant. *Nature (Lond.)* **242**, 531-532.
6. Levine, M. M. (1983). Travellers' diarrhoea: prospects for successful immunoprophylaxis. *Scand. J. Gastroenterol.* **18** (Suppl. 84), 121-134.
7. Levine, M. M., R. E. Black, M. L. Clements, C. R. Young, C. P. Cheney, P. Schad, H. Collins and E. C. Boedeker (1984). Prevention of enterotoxigenic *Escherichia coli* diarrheal infection in many by vaccines that stimulate antiadhesion (anti-pili) immunity. *In* "Attachment of Organisms to the Gut Mucosa", (Ed. E. C. Boedeker), pp. 223-244. CRC Press, Boca Raton.
8. Schmidt, M., E. P. Kelley, L. Y. Tseng and E. C. Boedeker (1985). Towards an oral *E. coli* pilus vaccine for travelers diarrhea: susceptibility to proteolytic digestion. *Gastroenterology* **1985**, 1575.

Development of Enteric Vaccines Based on Synergism Between Antitoxin and Anti-Colonization Immunity

A.-M. Svennerholm[a], C. Åhrén, Y. Lopez-Vidal,
N. Lycke and J. Holmgren

Department of Medical Microbiology, University of Göteborg, Göteborg, Sweden

The major pathogenic events in cholera and other enterotoxin-induced diarrhoeal diseases include colonization and multiplication of the bacteria in the small intestine with penetration of the mucus layer, adherence to the surface of the mucosal cells, and elaboration of one or more heat-labile (LT) and/or heat-stable (ST) enterotoxins that are responsible for the increased water and electrolyte secretion in the intestinal lumen. Development of effective vaccines against these diseases will be based on the use of immunogens which give rise to an immune response interfering efficiently with one or more of these pathogenic events. A prerequisite of such vaccines is that they can stimulate the local immune system of the gut in an efficient manner. Studies in experimental animals indicate that locally produced secretory IgA (SIgA) antibodies and immunologic memory for production of this class of antibodies are of importance for protection against enterotoxin-induced diarrhoeal disease. Such antibodies can be effectively stimulated by oral antigen administration.

EVIDENCE FOR SYNERGISTIC PROTECTIVE IMMUNITY AGAINST CHOLERA

In previous studies we have shown that immunization with a combination of *Vibrio cholerae* enterotoxin (cholera toxin) and lipopolysaccharide (LPS) affords protection against experimental cholera in rabbit ileal loops. The magnitude of

[a] *Telephone:* 46-31-603738

Protein-Carbohydrate Interactions in Biological Systems. ISBN 0 12 436665 1

this protection markedly exceeds the sum of the protective effects induced by each antigen alone, providing evidence of true synergism [1]. A similar synergistic effect has been observed after immunization with whole cell cholera vaccine and the non-toxic B subunit portion of the cholera toxin molecule [2]. This synergistic protection is due to the ability of the intestinal immunity to interfere with different pathogenic steps. An adjuvant action from either of the immunogens on the immune response to the other vaccine component that could explain the synergy is less likely, since neither the antitoxic nor the antibacterial antibody responses in intestine or serum were higher after immunization with a combination of somatic and toxin antigens than after administration of each component alone [1]. Furthermore, the combined vaccine was not more effective than cholera toxin alone in inducing protection against toxin challenge, and the synergistic cooperation could be reproduced in rabbit ileal loop systems using antibody preparations rather than active immunization (unpubl. data).

PROTECTIVE IMMUNE MECHANISMS IN EXPERIMENTAL CHOLERA

Antitoxic immunity in cholera is predominantly mediated by antibodies directed against the B subunit portion of the toxin molecule. Recovery from clinical cholera, or immunization with cholera toxin consistently gives rise to higher titres against subunit B than against subunit A. Also, mole for mole anti-B antibody is considerably more effective than anti-A antibodies in neutralizing cholera toxin [2]. Anti-B antibodies protect by preventing binding of the toxin molecule to its specific receptor, the ganglioside GM1.

The nature of antibacterial immunity against cholera is not as well understood. Local antibacterial antibodies may protect against disease by interfering with vibrio colonization in the small intestine. Such protection can not be fully evaluated in the ligated loops, since the mechanical defence against bacterial colonization obtained by, for example peristalsis, is not operating in this system. Therefore, we have evaluated the extent and nature of antibacterial cholera immunity also in a non-ligated intestine experimental model called the RITARD (Reversible Intestinal Tie Adult Rabbit Diarrhoea) model [3]. In this model immunocompetent rabbits can be made susceptible to intestinal colonization with *V. cholerae* or enterotoxin-producing *Escherichia coli* with a relatively minimal alteration of small intestinal functions. By infecting rabbits with live *V. cholerae* in this model, an immune response can be evoked that may be evaluated for its protective effect against subsequent challenge with fully virulent *V. cholerae* bacteria [3].

We found that an initial infection with *V. cholerae* bacteria in the RITARD model effectively prevented vibrio colonization of the small intestine after re-infection with both homologous and heterologous *V. cholerae* strains. Whereas the *V. cholerae* strains studied adhered in great numbers to the intestinal mucosa 3 days after the initial infection, no vibrios were found in any part of the small intestine at that time after re-infection. Furthermore, faecal excretion of the challenge strains was only seen in a few rabbits (Table 1) and then only during

Table 1. Protection against re-infection in the RITARD model induced by a prior infection with *V. cholerae*

Initial infection[a]	*Challenge infection*[b]	
	T19479	*T19766*
	[Diarrhoea attack rate (faecal excretion)]	
T19479 (El Tor, Inaba)	0/5 (0/5)[c]	0/6 (1/6)
T19766 (Classical, Ogawa)	0/8 (0/8)	0/5 (2/5)

[a]With a dose of *V. cholerae* that causes diarrhoea in >90% (ED_{90}) of the animals challenged.
[b]With 100-1000×$ED_{>90}$ of the respective strain 10 days after the initial infection.
[c]Number of animals responding with diarrhoea (vibrio excretion in faeces) in relation to total number of animals challenged.

the first day after infection; such excretion occurred for 2-4 days after the initial infection.

The immunity evoked by the initial cholera infection in the RITARD model was highly protective. Thus, infection with a dose of *V. cholerae* that induced diarrhoea in >90% of previously non-infected animals ($ED_{>90}$) gave rise to solid protection against development of disease by a 1000-fold higher dose of the same strain given 10 days later (Table 1). The initial *V. cholerae* infection also gave rise to complete protection against a *V. cholerae* strain representing the heterologous serotype as well as the heterologous biotype given in a dose of 100-1000×$ED_{>90}$ (Table 1).

In previous ligated-loop experiments we have shown that the LPS is responsible for most of the protective immunity induced by killed vibrios, and that this protection is mainly serotype-specific, i.e. directed against the unshared determinants on Ogawa and Inaba LPS, respectively [2]. The findings in the RITARD model of a very strong cross-protection also against the heterologous serotype could suggest that other antigens in the outer membrane of live cholera vibrios might induce protective immunity, too. Possible candidates of such structures might be the cell-bound haemagglutinins, since these proteins are presumed to function as bacterial adhesins. However, the strong cross-protection in the RITARD studies against vibrios, also of the heterologous biotype, makes it less likely that these haemagglutinins would be the only such structures involved in inducing broad cross-protective immunity against bacterial colonization, since these antigens seem to be strongly biotype-associated, at least *in vitro*, with a fucose-sensitive haemagglutinin (HA) being associated with classical vibrios and a mannose-sensitive HA associated with El Tor vibrios.

We also evaluated whether symptomatic cholera in the RITARD model may induce protective antitoxic immunity. This was tested by comparing the fluid response to cholera toxin challenge in infected and concurrently operated but non-infected rabbits. It was found that the infected rabbits were significantly protected against fluid secretion induced by graded doses of purified cholera toxin given in ileal loops 14 days after the "immunizing" infection, the difference between pre- and post-infection being more than ten-fold (unpublished data).

DEVELOPMENT OF A COMBINED TOXOID-SOMATIC ANTIGEN CHOLERA VACCINE

The clarification of the subunit structure–function relationship for cholera toxin, the observation of synergy between antibacterial and antitoxic immunity for protection against cholera, and the recent knowledge of the best route for stimulating the gut mucosal immune system preferentially were the three main reasons for our development of an oral cholera vaccine consisting of purified cholera B subunit, and both formalin- and heat-inactivated cholera vibrios. The B subunit component is prepared from the culture filtrate of fermentor-grown *V. cholerae* by affinity chromatography on a specific toxin-binding GM1 ganglioside-Spherosil® column, followed by gel filtration in acid buffer and dialysis. By this procedure, about 30–50 g of highly purified B subunit can be obtained from each 1000-litre culture batch.

The B subunit-whole cell cholera vaccine (B+WCV) has proved to be completely safe when tested in Swedish, Bangladeshi and American volunteers. Studies in Bangladeshi volunteers showed that oral administration of B+WCV gave rise to local IgA antitoxic and antibacterial immune responses in intestine and also to a local gut mucosal immunologic memory comparable to those induced by cholera disease itself [4]; the disease is known to result in very effective long-lasting immunization against cholera. Furthermore, in American volunteers, oral B+WCV induced significant protection against subsequent challenge with an ID_{100} dose of live cholera vibrios [5].

The promising results with B+WCV in a number of small-scale trials in cholera-endemic and non-endemic areas [4–6] have led to the initiation of a field trial of the vaccine in Bangladesh by the International Center for Diarrhoeal Disease Research. In that study, the protective efficacy of B+WCV will be compared with that of the WCV component alone and of an oral *E. coli* K12 placebo preparation (Table 2, [7]). Hitherto, approximately 90 000 persons have received 3 doses of one of the three preparations in a randomized, double-blinded fashion. The efficacy of the two vaccines to prevent cholera and enterotoxin-induced diarrhoeas caused by, e.g. *E. coli* and NAG vibrios, will be evaluated by comparing the number of persons developing these diseases in the vaccine groups with that in the placebo group.

Table 2. Immunization groups in 1985 ICDDR,B field trial of oral cholera vaccines

	Immunization[a]	
Group[b]	*Immunogen*	*Dose*
1	B+WCV	1 mg B sub+1×10^{11} *V. cholerae*
2	WCV	1×10^{11} *V. cholerae*
3	Placebo	1×10^{11} *E. coli* K12

[a]Three identical oral immunization given with 6 (3–8) week intervals.
[b]Recruited among non-pregnant women >14 years and children 2–14 years; ~30 000 persons in each group.

SYNERGISTIC PROTECTIVE IMMUNITY AGAINST ETEC-INDUCED DIARRHOEA

As observed in experimental cholera, we have also found that antibodies against LT and the colonization factor antigens CFA/I and CFA/II may afford synergistic protection against diarrhoea induced by enterotoxin-producing *Escherichia coli* (ETEC) carrying the homologous CFA [8]. Furthermore, using the RITARD model we have shown that *E. coli* strains carrying CFA and producing LT were more effective than CFA-carrying, LT-negative ETEC strains or CFA-negative, LT-producing strains in conferring protection against diarrhoea by subsequent challenge with LT-producing strains carrying the homologous CFA (unpublished data). Thus, previous infection with CFA/I-carrying, ST/LT-producing *E. coli* protected all animals re-infected with an otherwise highly diarrhoeogenic dose of the same strain as well as against challenge with a CFA/I-carrying strain with different O, K and H antigens (Table 3). Faecal excretion of the bacteria used for re-challenge was also significantly reduced in the CFA immunized animals, although not complete. When only one of the two antigens CFA/I and LT was shared by the immunizing and re-challenge strains, only partial protection was evident, consistent with independent antibacterial (anti-CFA) and antitoxic (anti-LT) immune mechanisms (Table 3).

Table 3. Protection against challenge with an *E. coli* CFA/I, ST/LT strain by initial infection with serotype heterologous strains carrying the homologous CFA and/or enterotoxin

Immunizing infection[a] E. coli *strain*	*Challenge infection*[b]	
	Diarrhoea attack rate	*Protection*[c]
258909-3 (CFA/I, ST/LT)	0/14	$p<0.001$
1392-75 (CFA/II, ST/LT)	3/7	$p<0.01$
304688-2 (CFA/I, ST)	1/7	$p<0.001$

[a]Infection was performed in the RITARD model with a dose of bacteria that caused diarrhoea in ≥85% of the animals challenged.
[b]With strain 258909-3 10–14 days after the initial infection.
[c]In comparison with the diarrhoea attack rate (41/44) after infection with a similar dose of the challenge strain in noninfected rabbits; Fisher's exact test.

POTENTIAL FOR A COMBINED VACCINE AGAINST ETEC

In many geographic areas ETEC strains causing disease via the non-immunogenic ST are common and many ETEC strains lack CFA/I or CFA/II. By coupling ST to the B subunit portion of LT or cholera toxin non-toxic conjugates, high antibody titres may be achieved against both ST and LT. Such toxin conjugates may be more effective than an LT toxoid alone in conferring protection against strains producing both LT and ST, and may afford protection against ST-only strains. Indeed, Klipstein *et al.* [9] have reported that synthetic vaccines such ST-LT B conjugates are highly effective in conferring protection against

ST/LT-induced *E. coli* fluid secretion in the small intestine. The protective effects of such ST-LT toxoids would be especially pronounced if they could be designed to cooperate synergistically with somatic antigens on ETEC bacteria.

The ideal somatic antigen in an ETEC vaccine is one that affords cross-protection against ETEC representing different CFAs and many different O, K and H serotypes. Recently, we have isolated a spontaneous CFA/I-deficient mutant from a CFA/I-carrying ETEC strain that may induce such cross-protective, antibacterial immunity. This mutant colonized well in the intestine of RITARD rabbits. The excretion pattern of this mutant in the stool did not differ significantly from that of the CFA/I-carrying original strain but the mean excretion time was significantly ($p < 0.05$) longer than that of a CFA-deficient, non-enterotoxinogenic *E. coli* control strain. Infection with the CFA-deficient mutant in the RITARD model resulted in protection against disease caused by subsequent challenge with the original CFA/I-carrying ST/LT strain but was also effective against disease by challenge with serotype heterologous, CFA/II-positive ETEC bacteria (Table 4). In addition to protecting against symptomatic disease, the initial infection with the mutant strain resulted in reduced excretion of both the CFA/I- and the CFA/II-carrying ETEC strains used for re-challenge. These results suggest that the suppression of the defined adhesin may have resulted in expression or increase of a membrane structure with the potential of inducing cross-protective antibacterial immunity breaking serotype as well as CFA-type barriers. The nature of this "cross-protective antigen" is presently studied.

Table 4. Protection in the RITARD model against challenge with a CFA/I- or CFA/II-carrying ETEC strain by prior infection with a CFA-deficient ETEC mutant

	Challenge infection[b]	
Immunizing infection[a]*: Strain*	*CFA/I, LT/ST* (*Diarrhoea attack rate*)	*CFA/II, LT/ST*
258909-3 M (CFA^-, LT)	1/7**	0/**
258909-3 (CFA/I, LT/ST)	0/9***	N.T.
E. coli control (CFA^-, tox^-)	3/5	N.T.

[a]Infection with 10^{11} bacteria of the respective strain.
[b]With 10^{11} bacteria of strain 258909-3 (CFA/I) or 1392-75 (CFA/II) 14 days after the initial infection.

SUMMARY

Immunization with a combination of *V. cholerae* enterotoxin-derived and somatic antigens affords protection against experimental cholera which markedly exceeds the sum of the effects induced by each type of antigen alone. Similarly, antibodies against heat-labile enterotoxin (LT) and colonization factor antigens (CFA) gave synergistic protection against diarrhoea by enterotoxinogenic *E. coli* (ETEC) carrying the homologous type of CFA. These synergistic effects seem to be due to interference of the mucosal immunity with bacterial colonization and toxin action rather than to an adjuvant action of either the toxin or the somatic antigens.

The synergistic principle has been exploited for the development of an oral cholera vaccine consisting of B subunit and whole cell vaccine. Based on promising results in several human volunteer studies, this vaccine is presently being evaluated in a large field trial in Bangladesh for protective efficacy and duration of protection against endemic cholera.

A spontaneous CFA-deficient mutant has been isolated that colonizes well in the small intestine and that induces effective protection against subsequent challenge with the original CFA/I-carrying ST/LT strain but also against challenge with serotype-heterologous CFA/II-positive bacteria. The potential of using such a strain in combination with an ST/LT toxoid as a vaccine against ETEC disease is discussed.

ACKNOWLEDGEMENTS

Financial support for the studies described herein was obtained from the Swedish Medical Research Council (grant no 16X-3382) and the Swedish Agency for Research Cooperation with Developing Countries (SAREC).

REFERENCES

1. Svennerholm, A.-M. and J. Holmgren (1976). Synergistic protective effect in rabbits of immunization with *Vibrio cholerae* lipopolysaccharide and toxin/toxoid. *Infect. Immun.* **13**, 735–740.
2. Svennerholm, A.-M. (1980). The nature of protective immunity in cholera. *In* "Cholera and Related Diarrheas" (Eds Ö. Ouchterlony and J. Holmgren), pp. 171–184. S. Karger, Basel.
3. Spira, W. M., R. B. Sack and J. L. Froehlich (1981). Simple adult rabbit model for *Vibrio cholerae* and enterotoxinogenic *Escherichia coli* diarrhea. *Infect. Immun.* **32**, 739–747.
4. Svennerholm, A.-M., M. Jertborn, L. Gothefors, AMMM, Karim, D. A. Sack and J. Holmgren (1984). Comparison of mucosal antitoxic and antibacterial immune responses after clinical cholera and after oral immunization with a combined B subunit-whole cell vaccine. *J. Infect. Dis.* **149**, 884–893.
5. Black, R. E., M. M. Levine, M.-L. Clements, C. R. Young, A.-M. Svennerholm, J. Holmgren and R. Germanier (1983). Oral immunization with killed whole vibrio and B subunit or procholeragenoid combination cholera vaccines: immune response and protection from *V. cholerae* challenge. *In* 19th Joint US–Japan Cholera Meeting, Washington, October 1983, in press.
6. Jertborn, M., A.-M. Svennerholm and J. Holmgren (1984). Gut mucosal, salivary and serum antitoxic and antibacterial antibody responses in Swedes after oral immunization with B subunit-whole cell cholera vaccine. *Int. Arch. Allgergy Appl. Immunol.* **75**, 38–43.
7. Clemens, J. D. (1984). Field trial of oral B subunit/whole cell and whole-cell cholera vaccines. ICDDR,B Protocol, International Centre for Diarrhoeal Disease Research, Bangladesh.

8. Åhrén, C. and A.-M. Svennerholm (1982). Synergistic protective effect of antibodies against *Escherichia coli* enterotoxin and colonization factor antigens. *Infect. Immun.* **38**, 75–79.
9. Klipstein, F., R. F. Engert and R. A. Houghton (1983). Protection in rabbits immunized with a vaccine of *Escherichia coli* heat-stable toxin cross-linked to the heat-labile toxin B subunit. *Infect. Immun.* **40**, 888–893.

Pilus-Mediated Interactions of the *Escherichia coli* Strain RDEC-1 With Intestinal Mucus

P. M. Sherman[a], E. P. Kelly and E. C. Boedeker

Department of Gastroenterology, Division of Medicine, Walter Reed Army Institute of Research, Washington DC 20307, USA

Escherichia coli strain RDEC-1 is an effacing adherent rabbit pathogen which provides an animal model for human enteropathogenic *E. coli*. This strain adheres to enterocytes *in vivo* and to microvillus membranes (MVMs) *in vitro* [1]. *In vitro* interactions are mediated by both non-mannose-sensitive (AF/R1) and type 1 pili expressed on the surface of RDEC-1 [2]. To examine whether similar interactions occur with intestinal mucus, we first fed RDEC-1 to rabbits and examined the luminal mucus for colonization by RDEC-1. Following *in vivo* RDEC-1 infection, RDEC-1 organisms were seen in the lumen enmeshed in PAS and Alcian blue-positive material. Subsequently, preparations of rabbit luminal mucus were tested for their ability to agglutinate RDEC-1 *in vitro*. These *in vitro* RDEC-1/mucin interactions were compared to known interactions of RDEC-1 with MVMs and with a glycoprotein fraction removed from MVMs by papain digestion. *In vitro* interactions of RDEC-1 with rabbit crude Luminal Mucus and more purified Luminal Mucus Glycoprotein fractions, were measured by direct observation, in microtitre plates and in an aggregometer. Non-piliated RDEC-1 did not interact with any rabbit mucus fraction, nor with MVMs or papain digest. RDEC-1 expressing AF/R1 or type 1 pili were agglutinated by both rabbit mucus preparations (and by MVMs and papain digest), but only type 1 mediated interactions were inhibited by the presence of D-mannose. These studies demonstrate that RDEC-1 are associated with the luminal mucus during *in vivo* infection, and that piliated, but not non-piliated, RDEC-1 organisms can interact with rabbit mucus glycoproteins *in vitro*. The mucus

[a] *Telephone:* 1-202-576-1493

Protein-Carbohydrate Interactions in Biological Systems. ISBN 0 12 436665 1

layer could serve as a site for bacterial replication and colonization prior to enteroadherence.

REFERENCES

1. Cheney, C. P., E. C. Boedeker and S. B. Formal (1979). Quantitation of adherence of an enteropathogenic *Escherichia coli* to isolated rabbit intestinal epithelial cells. *Infect. Immun.* **26**, 736-743.
2. Sherman, P. M., W. L. Houston and E. C. Boedeker (1985). Functional heterogeneity of type 1 somatic pili (fimbriae) expressed by intestinal *Escherichia coli* strains: assessment of bacterial adherence to intestinal membranes and surface hydrophobicity. *Infect. Immun.* **49**, 797-804.

The Role of Fimbriae of Uropathogenic *Escherichia coli* as Carriers of the Adhesin Involved in Mannose-Resistant Haemagglutination

Irma van Die, Elly Zuidweg, Wiel Hoekstra and Hans Bergmans

Department of Molecular Cell Biology, State University of Utrecht, The Netherlands

INTRODUCTION

The DNA encoding expression of a number of P fimbriae has been cloned by several groups. Analysis of this DNA has demonstrated that a number of genes are involved in expression of MRHA and fimbriae:

(1) a gene encoding expression of the fimbrial subunit; and
(2) four to seven additional genes, which probably are involved in transport, or assembly of the subunits into fimbriae.

It has been shown for the Pap-fimbrial gene cluster that one of the additional genes represents the actual adhesin. *Escherichia coli* cells harbouring a gene cluster with a mutated fimbrillin gene still showed MRHA and adherence although no fimbriae were formed [1]. Here we show that possession of $F7_1$, $F7_2$, F9 or F11 fimbriae is not strictly necessary for adherence of *E. coli* K12 cells. However, some wild-type *E. coli* strains apparently do need fimbriae for adhesion, probably as a carrier for the adhesion protein.

RESULTS

A number of plasmids were constructed without the gene encoding the fimbrial subunit protein (Fig. 1). The MRHA of *E. coli* K12-cells, harbouring these

[a] *Telephone:* 31-30-533184

Protein-Carbohydrate Interactions in Biological Systems. ISBN 0 12 436665 1

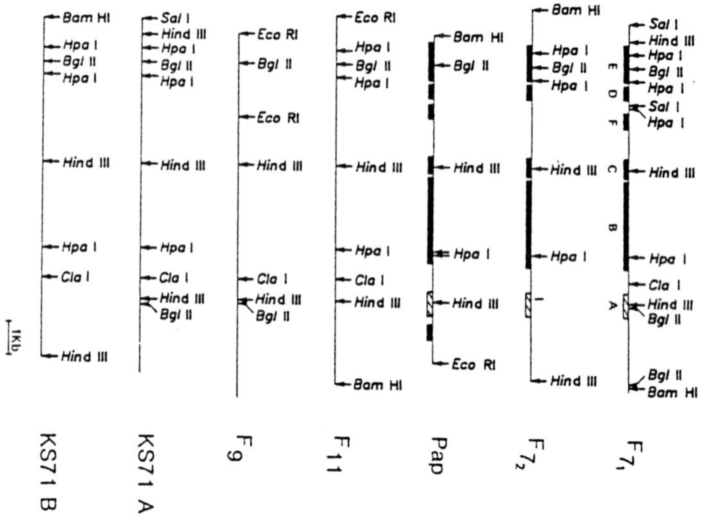

Figure 1. Recombinant plasmids that lack the gene encoding the fimbrial subunit but harbour the genes involved in expression of MRHA. The recombinant plasmids pPIL110-373, pPIL110-701, pPIL288-124 and pPIL291-152 were derived from pPIL110-37 [ref. 2], pPIL110-70 [ref. 3], pPIL288-12 [ref. 4] and pPIL291-15 [ref. 5], respectively. The thin black lines represent cloned DNA fragments.

Table 1. MRHA titres of *E. coli* K12, some wild-type *E. coli* strains (U20; F3; F8) and LPS mutants derived from these wild-type strains (U20-3; F3-2; F8-5) harbouring different recombinant plasmids

Bacterial strains	*MHRA titre of bacteria harbouring different gene clusters*						
	$F7_1$	*$F7_1$ -subunit*	*F_9*	*F_9 -subunit*	*F11*	*F11 -subunit*	*pBR322*
E. coli K_{12}	1/128	1/16	1/64	1/16	1/128	1/16	0
U20	1/128	1/2	1/64	0	1/64	1/2	0
U20-3	1/128	1/8	1/64	1/8	1/64	1/8	0
F3	1/64	0	1/54	0	1/64	1/4	0
F3-2	1/64	1/8	1/64	1/8	1/64	1/8	0
F8	1/64	1/2	1/128	1/2	1/64	1/4	1/2
F8-5	1/64	1/16	1/64	1/32	1/64	1/32	1/2

plasmids was compared with the MRHA capacity of *E. coli* K12-cells harbouring plasmids with the complete gene cluster (Table 1). The results show that the presence of $F7_1$, F9 or F11 fimbriae is not a prerequisite for adhesion in *E. coli* K12; this was also observed for $F7_2$ fimbriae (results not shown).

It has been suggested [1] that the adhesin of the Pap-fimbrial gene cluster is normally associated with the fimbriae, possibly at the tip of individual fimbriae.

Since in *E. coli* K12, adhesion can occur without the presence of fimbriae, the question can be raised as to what function fimbriae do have.

A possible explanation is that fimbriae are a carrier for the adhesion protein, to overcome the LPS barrier which could shield the adhesion protein if it were located in the outer membrane. To study this, we have chosen three wild-type *E. coli* strains with different O-serotypes, isolated form faeces (F) or urinary tract (U) that do not show MRHA F3 (041), F8 (0156) and U20 (015), and LPS mutants of these strains that lack most or all O-antigen. These strains were tested for MRHA after introduction of plasmids harbouring complete fimbrial gene clusters, or the gene clusters that miss the subunit gene (Table 1). The results show that the MRHA activity of these wild-type *E. coli* strains with a complete O-antigen, harbouring the gene clusters that miss the fimbrial subunit gene, is clearly reduced compared with the LPS mutant strains and *E. coli* K12 harbouring the same plasmids.

These results strongly suggest that for some wild-type *E. coli* strains fimbriae are required as carriers for the adhesion protein to overcome the LPS-O-antigen barrier that shields the outer membrane.

ACKNOWLEDGEMENTS

We thank Peter van der Ley for the gift of wild-type *E. coli* strains and their LPS-mutant derivatives, Hans van Veggel for assistance in part of the experimental work and Han de Ree for the gift of antisera.

REFERENCES

1. Lindberg, F. P., B. Lund and S. Normark (1984). *EMBO J.* **3**, 1167-1173.
2. Van Die, I., I. Van Megen, W. Hoekstra and H. Bergmans (1984). *Mol. Gen. Genet.* **194**, 528-533.
3. Van Die, I., G. Spierings, I. Van Megen, E. Zuidweg, W. Hoekstra and H. Bergmans (1985). *FEMS Microbiol. Lett.*, in press.
4. De Ree, J. M., P. Schwillens, L. Promes, I. Van Die, H. Bergmans and J. F. Van den Bosch (1985). *FEMS Microbiol. Lett.* **26**, 163-169.
5. De Ree, J. M., P. Schwillens and J. F. Van den Bosch (1985). *FEMS Microbiol. Lett.* **29**, 91-97.

Carriage of the Mannose-Resistant Haemagglutination Gene by an R-Plasmid which Persists Through a Recurrent Urinary Tract Infection

B. A. Hales[a] and S. G. B. Amyes

Department of Bacteriology, University Medical School, Edinburgh, UK

A study was carried out over a period of 25 months on a female patient suffering from recurrent urinary tract infections. Urine specimens were collected regularly and the bacterial pathogen isolated and identified. Initially the infection was treated with co-trimoxazole, but the infecting *Escherichia coli* strain developed resistance to trimethoprim (Tp). The Tp R-plasmid-containing strain persisted for 11 months even in the absence of Tp therapy. Treatment with cephradine cured the infection [1], but strains possessing the Tp R-plasmid subsequently reappeared in a further infection. These results indicated that some factor was playing a role in the persistence of the R-plasmid in the absence of selective pressure.

Ninety-five of the strains isolated over this period transferred trimethoprim resistance to *E. coli* K12 strain J62-2 and, in each case the resistance determinants for ampicillin, streptomycin, spectinomycin and sulphamethoxazole were co-transferred. The Tp-resistant transconjugants were assayed for the ability to cause mannose-resistant haemagglutination (MRHA) of human group A erythrocytes and 24 (25.3%) were shown to be positive. These MRHA-positive transconjugants were more prevalent from isolates of later infections—9 out of 65 transconjugants in the first 19 months, and 15 out of 30 in the last 6 months of the study. The presence of the MRHA gene in *E. coli* K12 suggested that the gene was plasmid encoded and its efficiency of transfer increasing with the number of re-infections.

[a] *Telephone:* 44-31-667-1011 ext. 2211

Protein-Carbohydrate Interactions in Biological Systems. ISBN 0 12 436665 1

The plasmid DNA from the MRHA-positive transconjugants was purified and examined by agarose gel electrophoresis. All strains possessed the same 75 kbase (kb) plasmid. The DNA from a selection of MRHA-negative, Tp-resistant transconjugants was also examined for comparison. These strains also contained a single plasmid, but it was 4.5 kb smaller. Inability to transfer the MRHA gene obviously did not correlate with the loss of a plasmid, and the difference in molecular size of the plasmids suggested that the MRHA-positive transconjugants possessed an additional piece of DNA. We had previously found the MRHA gene to be carried on a transposable element in certain uropathogenic strains [2] thus, in order to determine if this MRHA gene was carried on a transposon, the R-plasmid Sa was introduced into 8 of the MRHA-positive transconjugants to allow its transposition into a known plasmid vector, which was then transferred to *E. coli* K12 strain J53. The adhesive nature of the Sa-containing J53 transconjugants was then determined with the haemagglutination assay.

Plasmid DNA from the resultant MRHA-positive strains was purified and examined as before. A molecular size of 35.2 kb was obtained for the standard Sa compared to 44.4 kb calculated for the plasmids in the MRHA-positive J53 strains. This gives a more accurate estimate of the size of the insertion as 9.2 kb.

ACKNOWLEDGEMENT

We wish to thank the Scottish Home and Health Department for the grant which funded this research.

REFERENCES

1. Amyes, S. G. B., C. J. McMillan and J. L. Drysdale (1981). *In* "New Trends in Antibiotics: Research and Therapy" (Eds G. G. Grassi and L. D. Sabath), pp. 325-327. Elsevier, Amsterdam.
2. Hales, B. A. and S. G. B. Amyes (1983). The genetic carrier of bacterial attachment. *J. Pharm. Pharmacol.* **35** Suppl. 50P.

IV. Carbohydrate Receptors: Identification, Analysis and their Biological Role

Co-Chairmen: Karl-Anders Karlsson
Klaus Bock

Monoclonal Antibodies to Bacterial O-Antigens in Combination with Synthetic Glycoconjugates for Mapping the Combined Site

D. R. Bundle[a]

Division of Biological Sciences,
National Research Council of Canada, Ottawa, Ontario, Canada

The nature of the physical forces and atomic interactions which create the exquisite specificity of the carbohydrate binding sites of antibodies and lectins have been suggested to have their origins in short-range multipoint van der Waals forces of attraction between complementary and largely non-polar surfaces in combination with a crucial polar interaction involving a tight cluster of two to three hydroxyl groups. This concept has been termed the hydrated polar-group gate effect [1]. Since oligosaccharides may be expected to be extensively hydrogen bonded to solvent water, the simple exchange to an array within a combining site, itself extensively hydrated, would not be expected to contribute significantly to the energy of association. Although this idea of complementary and essentially non-polar surfaces has received strong support from recent site mapping studies of blood-group specific lectins and monoclonal antibodies [1], X-ray data for other carbohydrate binding proteins, *Escherichia coli*, L-arabinose binding protein [2] and phospharylase [3] have indicated extensive networks of hydrogen bonds between the protein and saccharide. The published X-ray data on antibody combining sites has been acquired on myeloma proteins for which the antigenic stimulus is unknown and whose combining sites were by no means filled by the best-fitting ligand [4]. It is desirable, therefore, to study in detail several antibodies with well defined binding characteristics, and for which the complementary antigenic determinant is both known and available in quantity.

[a]*Telephone:* 1-613-990 0838

Protein-Carbohydrate Interactions
in Biological Systems. ISBN 0 12 436665 1

Chemical synthesis provides both ligands and glycoconjugates whose three-dimensional shape in solution may be well appreciated [5]. The hybrid-myeloma methodology is the method of choice for obtaining large quantities of carbohydrate binding protein with predetermined specificity, provided the binding characteristics and heavy chain isotope of the Ig molecule may be easily screened. In conjunction with synthetic antigens these objectives may be satisfied.

Mouse monoclonal antibodies have been produced to two non-charged, linear O-antigens, the *Shigella flexneri* Y-LPS, which possesses a tetrasaccharide repeating unit [5] and the second, a simple homopolymeric O-antigen of *Brucella abortus* and *Yersinia enterocolitica* 0:9 [6]. The *Salmonella typhimurium* serogroup B LPS provides an example of a branched tetrasaccharide repeating sequence, which differs from the *Salmonella* O-antigens of serogroups A and D_1 by chiral inversion at single sites within the 3,6-dideoxyhexose moiety [7]. Since univalent Fab fragments are preferred for binding studies and X-ray diffraction analysis, IgG producing clones with the desired binding profile were sought. In the case of *Yersinia enterocolitica* several of these antibodies were obtained, and for *Salmonella typhimurium* one $IgG_1\lambda$, that binds the branched abequose tetrasaccharide has yielded a crystalline Fab fragment, which is largely sequenced in both the Fd and λ chains.

The solution conformations of the *S. flexneri* (Fig. 1) [5] and *S. typhimurium* determinants [7] have been determined by n.m.r. and HSEA calculations and both are conformationally well defined. In this connection it is assumed as a simplifying hypothesis that the saccharide is bound in the combining site in a conformation close to the HSEA calculated minimum energy conformation supported by n.m.r. data [1,5]. The linear sequence of three rhamnose and one *N*-acetyl glucosamine residues (Fig. 1) has been shown to correspond to the biological repeating unit [8] of the O-chain and one possible complementary surface was speculatively identified based on the prominent disposition of rhamnose *C*-methyl groups and an intervening acetamido group along one edge of the polymer chain [5] (Fig. 1a). In order to obtain antibodies directed toward determinants located along the polysaccharide rather than at its tip, culture supernatants from cell-fusion experiments were screened against synthetic glycoconjugates with two linear sequences. The biological repeating unit,

---

Figure 1 (opposite). Two potential antigenic sites of the *S. flexneri* Y O-antigen presented as shaded CPK plots of an octasaccharide (two repeating units of sequence abcdabcd) in its HSEA minimum energy conformation. From left to right the sequence Rha→Rha→Rha→GlcNAc (abcd) represents the biological repeating unit of the O-chains. (a) A view displaying the juxtaposition of the acetamido group of unit-d with adjacent *C*-methyl groups of rhamnose provided by unit-b of the first repeating sequence and rhamnose a from the second sequence. (b) Alternate sequence of the octasaccharide showing the exposure of rhamnose c with the neighbouring hydrophobic α-face of the 2-acetamido-2-deoxy-β-D-glucose unit d. The absence of the amide hydrogen from this α-face is due to its exclusion from HSEA calculations, but clearly its presence completes an extensive array of five to six carbon bonded H atoms around one face of the GlcNAc unit.

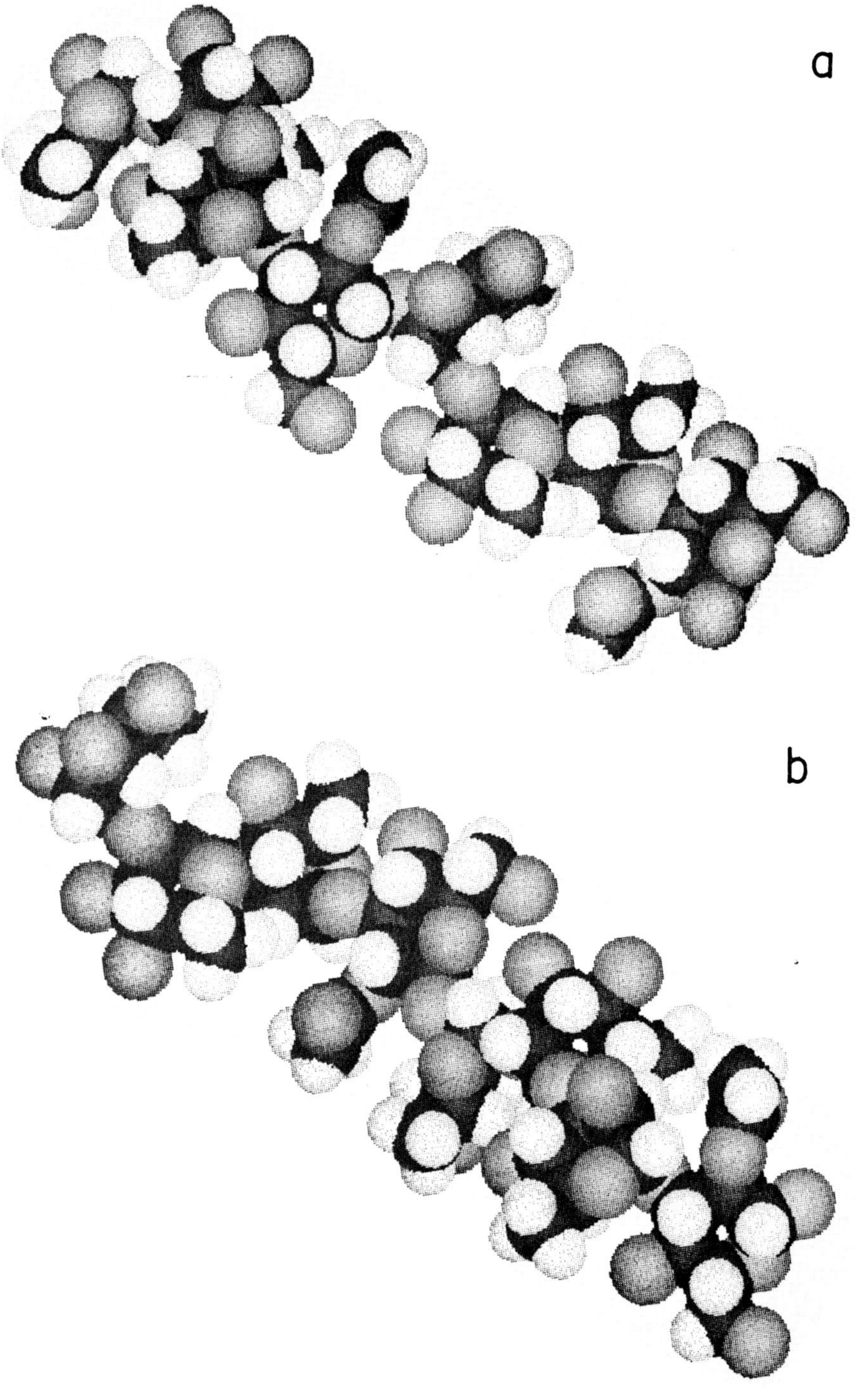
a
b

α−L−Rha(1→2)−α−L−Rha(1→3)−α−L−Rha(1→3)−β−D−GlcNac 1→OR

(sequence abcd) and its single frame shifted counterpart

α−L−Rha(1→3)−α−L−Rha(1→3)−β−D−GlcNAc(1→2)−α−L−Rha 1→OR

(sequence bcda) each coupled to BSA [9] were used in initial ELISA screening assays (Table 1). Clones producing antibody toward the oligosacchrides were tested for precipitation with O-polysaccharide and alkali treated LPS (Table 1). Those precipitating antibodies SfYGc-4 and -6 were judged to recognize internal-chain determinants (Fig. 2) [6]. Isotyping experiments established that the

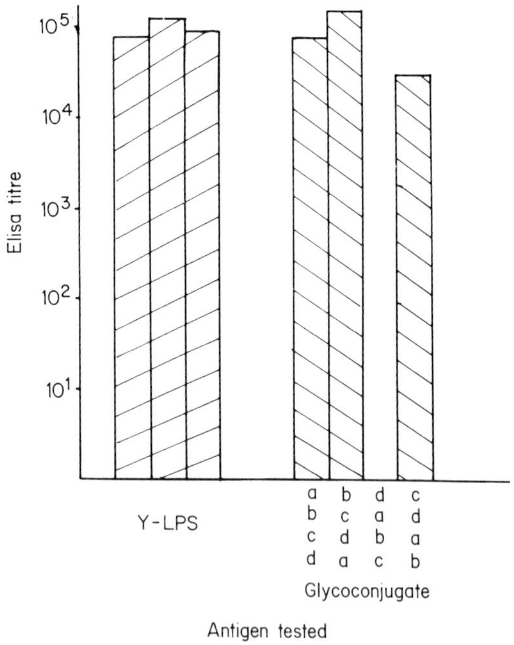

Figure 2. Reciprocal ELISA end-point titres of ascites containing monoclonal antibody SfYGc-4 with polysaccharide and synthetic antigens. The cell line producing this antibody was selected from a fusion experiment on the basis of high titres exhibited toward the synthetic antigens Rha→Rha→Rha→GlcNAc and Rha→Rha→GlcNAc→Rha→BSA. The profile of binding characteristics appears to indicate that the sequence Rha→GlcNac→Rha is crucial for binding.

Table 1. ELISA characteristics and immunoprecipitation of *S. flexneri* monoclonal antibodies with synthetic antigens and Y-polysaccharide

		ELISA Titre[a] of ascites fluid				*Immunoprecipitation*			
Monoclonal antibody	*Ig heavy chain*	*Glycoconjugates[b]* *a b c d*	*b c d a*	*d a b c*	*Y-LPS*	*Alkali-treated LPS*	*Polysaccharide[c] O-chain*	*Glycoconjugates[b]* *a b c d*	*b c d a*
SfYGc-1	μ	1×10^2	1×10^2	n.d.	1×10^3	−ve	−ve	−ve	−ve
SfYGc-2	μ	1×10^2	5×10^3	n.d.	1×10^4	−ve	−ve	−/+	+ve
SfYGc-3	μ	1×10^2	1×10^2	n.d.	1×10^3	−ve	−ve	−ve	−ve
SfYGc-4	μ	5×10^4	1×10^5	−ve	1×10^5	+ve	+ve	+ve	+ve
SfYGc-5	μ	1×10^2	1×10^2	n.d.	5×10^3	−ve	−ve	−	−
SfYGc-6	μ	5×10^3	1×10^4	−ve	1×10^4	+ve	+ve	+ve	+ve
SfYGc-7	μ	5×10	1×10^2	n.d.	5×10^2	−ve	−ve	−ve	−ve
SfYGc-8	μ	1×10^3	1×10^3	n.d.	5×10^3	−ve	−ve	−ve	−ve

[a] Titre is the reciprocal of the dilution giving an OD of 0.1 above background after 60 min incubation of plates with enzyme-substrate.
[b] Synthetic glycoconjugates prepared by coupling of synthetic ligands to BSA(9) and used for coating ELISA plates at 10 μg/ml.
[c] Polysaccharide O-chain liberated from Y-LPS by mild acid hydrolysis (1% acetic acid for 60 min at 100°C).

glycoconjugate selected, clones were exclusively IgM, despite intensive immunization protocols using whole bacterial cells (5×10^8/injection). Although numerous attempts over several fusion experiments were made to secure IgG producing clones specific for a tetrasaccharide-sized determinant, none have been forthcoming. Consequently, attempts are in progress to select switched heavy chain variants by the sib selection technique [10] from either the SfYGc-4 or -6 cell lines. In order to map the combining sites of the *S. flexneri* antibodies various modified tetrasacchrides have been synthesized in order to test the hypothesized binding surface (Fig. 1).

These include:

α-L-Man(1→3)-α-L-Rha(1→3)-β-D-GlcNAc(1→2)-α-L-Rha 1→OCH_3

α-L-Rha(1→3)-α-L-Man(1→3)-β-D-GlcNAc(1→2)-α-L-Rha 1→OCh_3

α-L-lyx(1→3)-α-L-Rha(1→3)-β-D-GlcNAc(1→2) α-L-Rha 1→OCH_3

each of which is designed to test the degree to which the *C*-methyl groups of each rhamnose unit contributes to binding (cf. Figs 1a dn 1b). Trivial changes of the *N*-acyl group will also permit accessments of the involvement of the acetamido group. Extension of these mapping strategies based on the work of Lemieux [1] involve deoxygenation at sites suspected of involvement in polar contacts. In such cases deoxygenation is expected to inactivate the inhibitor, whereas when this replacement involves OH groups located in a hydrophobic area of the oligosaccharide, enhanced activity is anticipated.

Glycoconjugates and chemically synthesized haptens specifically modified to test hypotheses of protein-carbohydrate interaction are well suited to the generation, purification and mapping of hybrid-myeloma antibody combining sites. The three antigen-antibody systems described here are intended to provide detailed information on the features of the complementary combining sites of neutral oligosaccharides and eventually a set of guiding principles for such specific binding.

REFERENCES

1. Lemieux, R. U. (1984). The hydrated polar-group "gate" effect on the specificity and strength of the binding of oligosaccharides by protein receptor sites. *In* "Proceedings of the 8th International Symposium on Medicinal Chemistry", Aug. 1984, Uppsala, Sweden. Swedish Pharmaceutical Society, in press.
2. Quiocho, F. A. and N. K. Vyas (1984). Novel stereospecificity of the L-arabinose-binding protein. *Nature* **310**, 381-386.
3. Goldsmith, E. and R. J. Fletterick (1983). Oligosaccharide conformation and protein saccharide interactions in solution. *Pure Appl. Chem.* **55**, 577-588.
4. Davies, D. R. and H. Metzger (1983). Structural basis of antibody function. *Ann. Rev. Immunol.* **1**, 87-117.

5. Bundle, D. R., M. A. J. Gidney, S. Josephson and H.-P. Wessel (1983). Synthesis of *Shigella flexneri* O-antigenic repeating units; conformational probes and aids to monoclonal antibody production. *Am. Chem. Soc. Symp. Ser* **231**, 49-63.
6. Bundle, D. R., M. A. J. Gidney, M. B. Perry, J. R. Duncan and J. W. Cherwonogrodzky (1984). Serological confirmation of *Brucella abortus* and *Yersinia enterocolitica* 0:9 O-antigens by monoclonal antibodies. *Infect. Immun.* **46**, 389-393.
7. Bock, K., M. Meldal, D. R. Bundle, T. Iversen, P.-J. Garegg, T. Norberg, A. A. Lindberg and S. B. Svenson (1984). The conformation of *Salmonella* O-antigenic polysaccharide chains of serogroup A, B and D_1 predicted by semi-emperical, hard-sphere (HSEA) calculations. *Carbohydr. Res.* **130**, 23-34.
8. Carlin, N. I. A., A. A. Lindberg, K. Bock and D. R. Bundle (1984). The *Shigella flexneri* O-antigenic polysaccharide chain: nature of the biological repeating unit. *Eur. J. Biochem.* **139**, 189-194.
9. Pinto, B. M. and D. R. Bundle (1983). Preparation of glycoconjugates for use as artificial antigens: a simplified procedure. *Carbohydr. Res.* **124**, 313-318.
10. Spira, G., A. Bargellesi, J.-L. Teillund and M. D. Scharff (1984). The identification of monoclonal class switch variants by sib selection and ELISA assay. *J. Immunol. Methods* **74**, 307-315.

Protein–Carbohydrate Interactions: the Substrate Specificity of Amyloglycosidase (EC 3.2.1.3)

Klaus Bock[a] and Henrik Pedersen

Department of Organic Chemistry,
The Technical University of Denmark, Lyngby, Denmark

INTRODUCTION

The present communication describes an investigation of the carbohydrate–protein interaction using an enzyme, amyloglycosidase (AMG) (EC 3.2.1.3) which cleaves α-glycosidic linkages, e.g. in maltose, to two molecules of D-glucose.

The enzyme AMG is produced by the microorganism *A. Niger* and is commercially available in large amounts due to its technical use in starch processing. The enzyme is a glycoprotein with the composition indicated in

Table 1. Glucoamylase (*Aspergillus niger*) (1→4-α-D-glucan glucohydrolase, EC 3.2.1.3)

Multiple forms	G1	G2
Digestion of raw starch	+	−
Molecular weight	82 500	72 500
Amino acid residues	616	512
Neutral sugar residues	99	99
O-Glycosylated positions	47	50
N-Glycosylated positions	2	2

The amino acid sequence of the protein, the nucleotide sequence of the cloned cDNA, and the gene structure are known.

[a] *Telephone:* 45-2-882566

Protein–Carbohydrate Interactions in Biological Systems. ISBN 0 12 436665 1

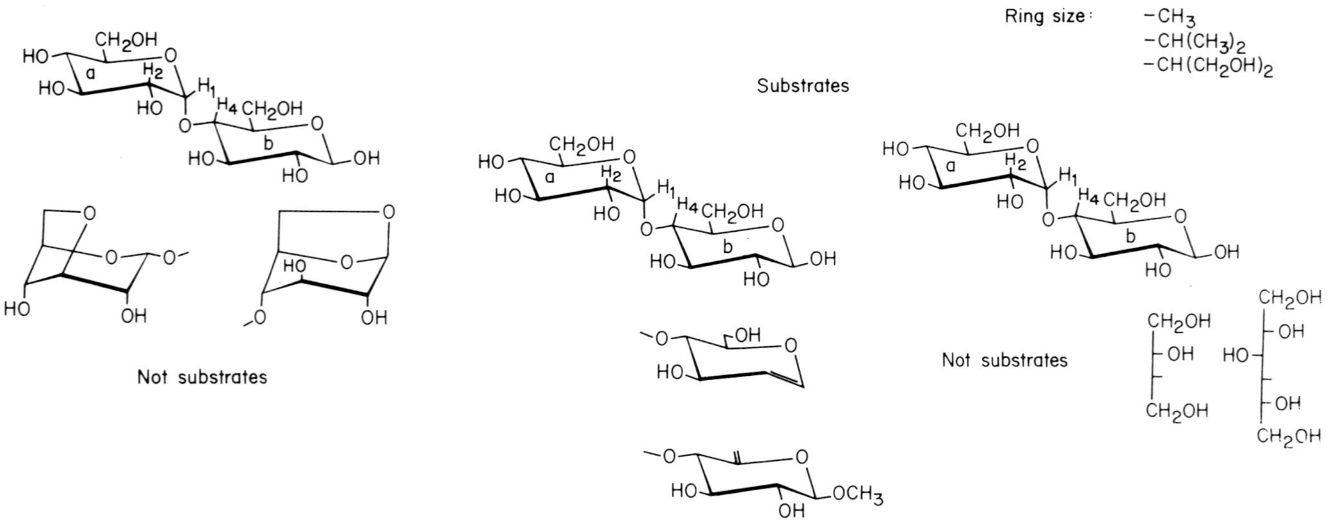

Figure 1. Substrate specificity of AMG: ring conformation.

Table 1. The amino acid sequence has been determined recently by Svensson *et al.* [1] but it has not been possible to crystallize the enzyme, in order to obtain a picture of the three-dimensional structure. It was, therefore, decided to study the reaction mechanism of this enzyme and the forces involved in the recognition of substrates in the active site through chemical synthesis of modified substrates. The reaction kinetics of these substrates has furthermore been determined using high field proton N.M.R. spectroscopy and the results correlated with the preferred three-dimensional structure of the substrates in aqueous solutions.

SUBSTRATE SPECIFICITY

The behaviour of chemically modified maltose derivatives towards AMG was first investigated. The results in Fig. 1 show that the active site of the enzyme require that both D-glucopyranose units in maltose should be present and that the non-reducing unit exists in a $^4C_1$-conformation. The conformational requirements for the reducing glucopyranose-unit are less stringent because it is seen that unsaturation between C-1 and C-2 or C-5 and C-6 does not eliminate the enzymatic reaction, but complete inversion of the pyranose-ring to a $^1C_4$-conformation is not incompatible with the enzymatic reaction.

The results from introduction of deoxy-functions (i.e. selective removal of hydroxy groups) are shown in Fig. 2. From these it is seen that the hydroxy groups, OH-3 of the reducing unit, OH-4 and OH-6 of the non-reducing unit, all are essential for the enzyme in order to cleave the substrates. On the other hand removal of the other hydroxy groups does not prohibit the enzyme from reacting with these substrates and the compounds with deoxy groups in the 6-position of the reducing unit and in the 3-position of the non-reducing unit are even better substrates for the enzyme than the corresponding hydroxylated (natural) substrates. Epimerization of hydroxy groups at different centres have also been investigated and the results are shown in Fig. 3. From these results it again follows that the hydroxy group at position 3 of the reducing end is essential for enzymatic function.

It has furthermore been shown that this group has to be a hydroxy group because the 3-*O*-methyl or 3-bromo-3-deoxy derivatives are also not substrates for AMG. Epimerization of the OH-2 of the reducing unit is acceptable whereas epimerization of the OH-2 of the non-reducing unit gives a compound which does not react with the enzyme. On the other hand extensive chemical modification of the hydroxy group in the 6-position of the reducing unit is tolerated by the enzyme, as shown in Fig. 4. The derivatives with the neutral substituents F−, Cl−, Br−, I−, N_3−, NHAc and COOMe are all substrates for the enzyme (although with different affinities and rate constants, see below) and even the branch point trisaccharide of amylopectin (i.e. an α-D-glucopyranosyl unit in the 6-position) is a substrate for AMG. However, if the substituent is changed to a charged group like NH_3 or COOH, then the enzyme is not capable of using the compounds as substrates.

The affinity of some of the above-mentioned compounds towards AMG has been studied using high field 500 MHz ^1H−N.M.R. spectroscopy which allows

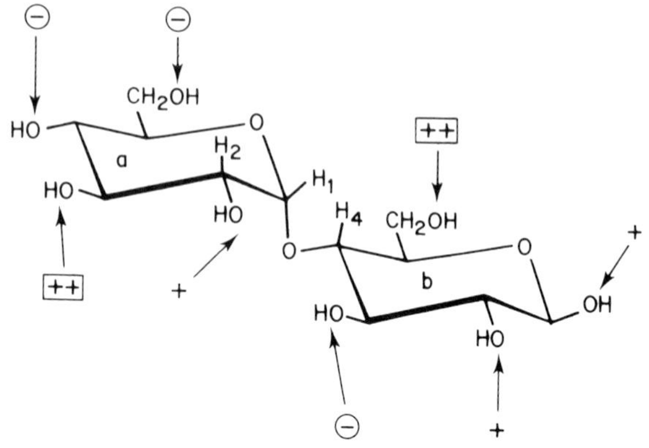

Figure 2. Substrate specificity of AMG: deoxy derivatives.

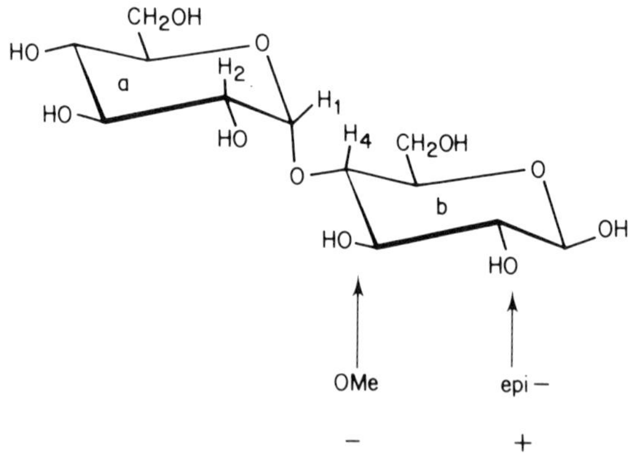

Figure 3. Substrate specificity of AMG: other functionalities.

one to determine the rate constants for the enzymatic reaction rates of pure compounds, which shows (Fig. 5A) that the enzymatic reaction proceeds under inversion of the configuration at the anomeric centre of the cleaved glycosidic linkage. This technique is, however, much more powerful because it is possible to determine the rate constants of the individual compounds from mixtures of substrates or inhibitors as shown in Fig. 5B. Results from this type of investigation are also shown in Table 2.

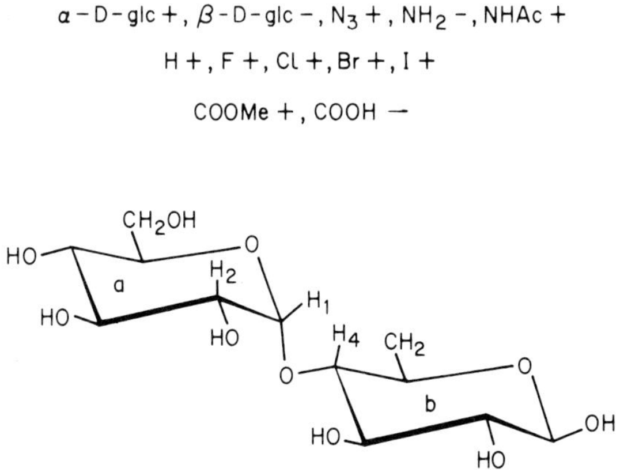

Figure 4. Substrate specificity of AMG: other functionalities. No charged groups in the hydrophobic surface.

Table 2. Reaction rates relative to the rate of cleavage of methyl β-maltoside

Calculated from completion experiments		*Calculated from the rate of release of glucose*	
Methyl 6-deoxy-β-maltoside	1.64	Maltose	1.49
Maltose	1.46	Methyl 6-deoxy-β-maltoside	1.27
Methyl 6-azido-β-maltoside	1.22	Methyl β-maltoside	1.00
Methyl β-maltoside	1.00	Methyl 6-acetamido-β-maltoside	0.68
Methyl 6-chloro-β-maltoside	0.84	Methyl 6-azido-β-maltoside	0.25
Methyl 6-iodo-β-maltoside	0.70	Methyl 6-iodo-β-maltoside	0.17
Methyl 4-*O*-(α-D-glucopyranosyl)-α-D-xylopyranoside	0.62	Methyl 6-chloro-β-maltoside	0.11
Methyl 6-acetamido-β-maltoside	0.31		
Methyl 6,6′-dideoxy-α-maltoside	0.25		

CONFORMATIONAL ANALYSIS

Methods to determine the preferred conformation of oligosaccharides in aqueous solutions based on simple hard sphere *exo*-anomeric effect (HSEA) calculations combined with experimental verification using high field proton and carbon-13 N.M.R. spectroscopy have been developed during the last five years [2-11]. The sum of interactions between the H−, C−, and O− atoms in the two

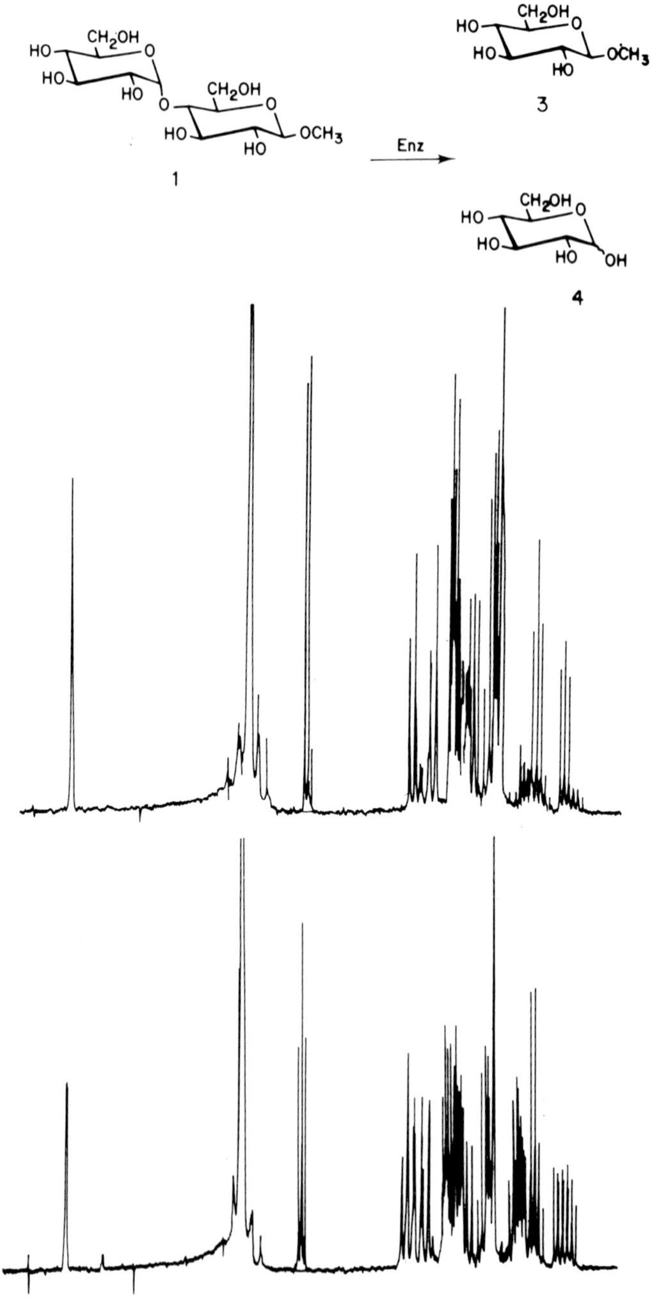

Figure 5A.

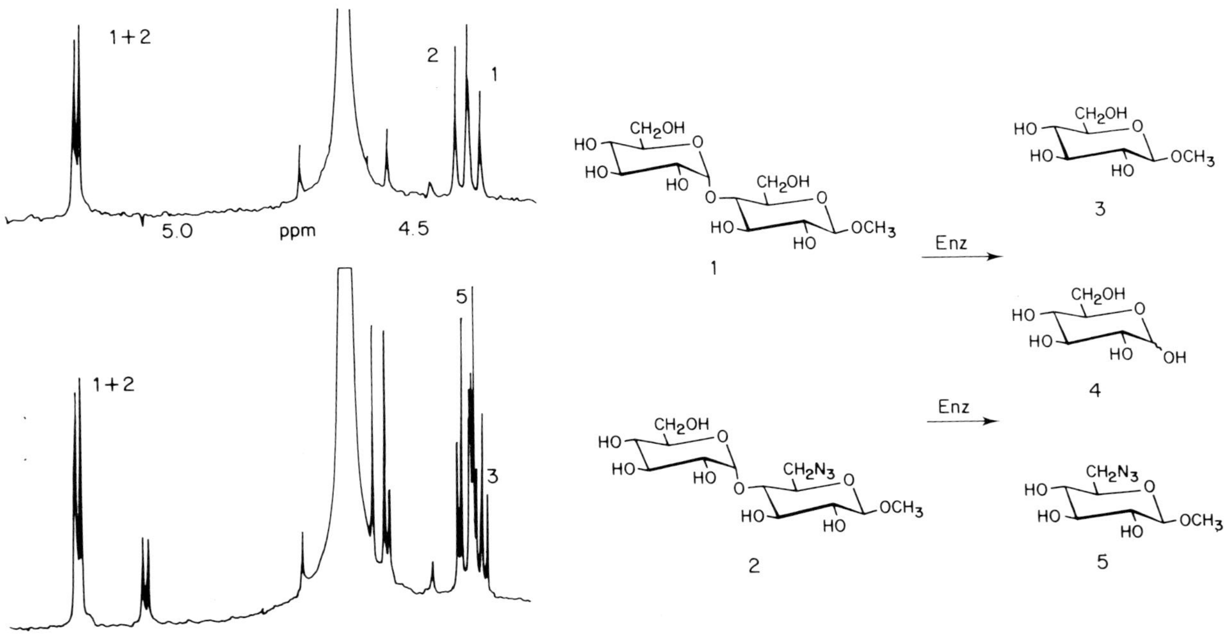

Figure 5B.

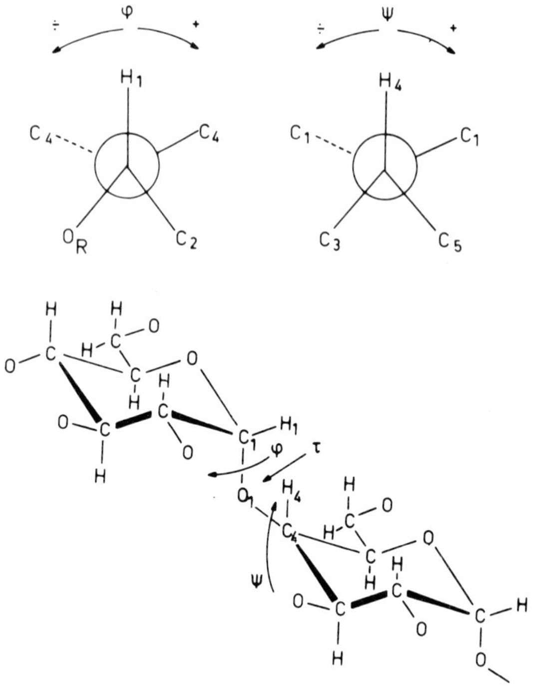

Figure 6. The definition of the torsion angles ϕ and ψ which define the conformation of the glycosidic linkage.

monosaccharide units linked, e.g. α 1-4 as shown in Fig. 6 for maltose, are calculated for different pairs of torsion angles according to the potential functions given by Kitaygorodsky [12]. This results in a set of energies as a function of the rotation about the ϕ/ψ angles and the energy contribution from the *exo*-anomeric effect [4] is added and the minimum energy conformation(s) of the molecule can be determined and finally inspected using different graphical representations as shown in Fig. 7. The conformations thus predicted using this approach can be experimentally supported, but it is beyond the scope of this contribution to discuss this and the reader is referred to reviews about this subject [2,13].

The above-mentioned method has been used to determine the preferred conformation of a strong inhibitor of AMG the pseudotetrasaccharide Acarbose [14] (see Fig. 7). From these results it is clearly seen that the molecule in its minimum energy conformation has a three-dimensional structure which is very similar to that of maltose, but when an acid functionality from the enzyme is close to the glycosidic oxygen, it will form a salt with acarbose, which thus acts as "a suicide" inhibitor.

A

B

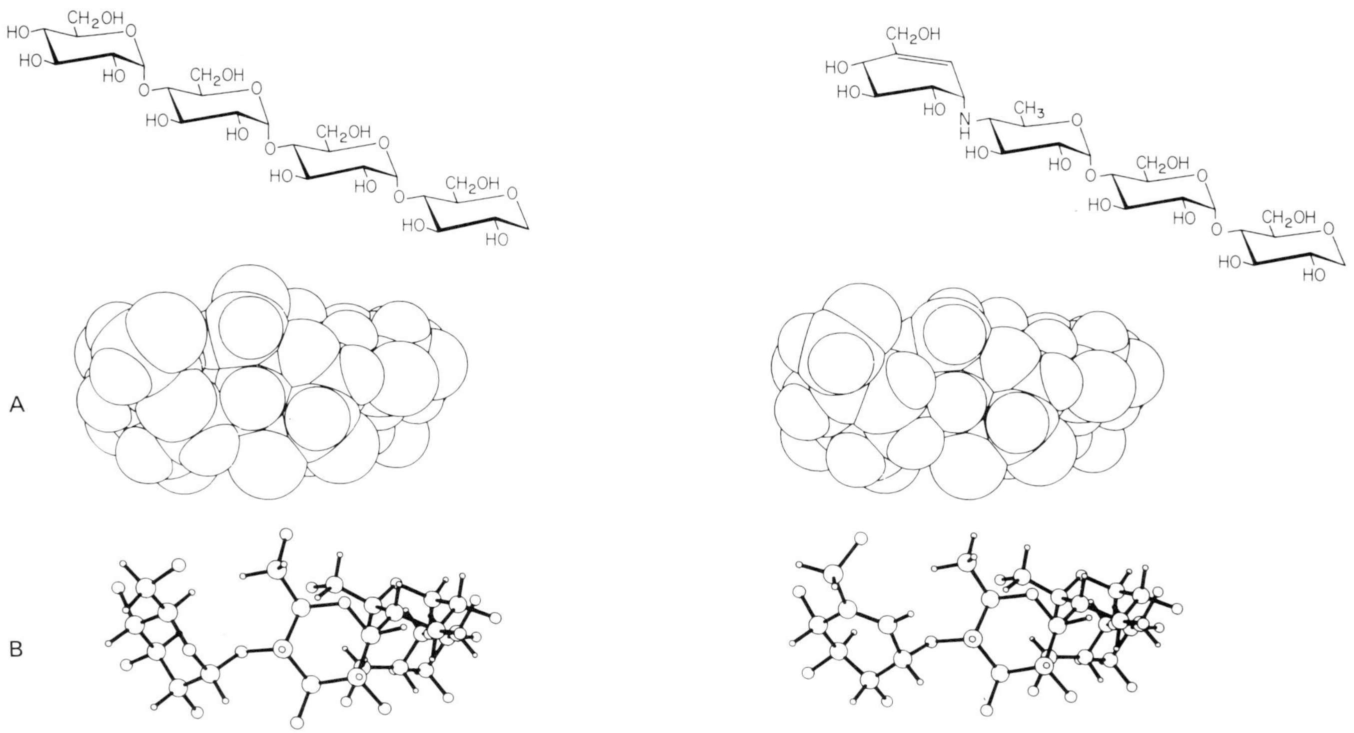

Figure 7. Minimum energy conformation of maltotetraose and acarbose (1) calculated using the HSEA method: (A) CPK models; (B) stick and ball models. The ϕ/ψ values used for the maltose glycosidic bonds are $-25°/-20°$, and the ϕ/ψ angles for the pseudoglycosidic bond between the valienamine unit and the 4-amino-4,6-dideoxy-α-D-glucopyranose unit are $-39°/-19°$.

SUMMARY

From the above-mentioned results it can be concluded that:

(1) The substrate specificity of maltose derivatives towards AMG can be correlated with their ground state conformations.

(2) The hydroxy group in position 3 of the reducing unit of maltose is essential for the enzymatic function.

(3) Several major changes in the hydrophobic region of the substrate can be tolerated and binding enhanced for more hydrophobic molecules but charged groups are not allowed in the hydrophobic region as these compounds do not bind to the enzyme.

(4) The HSEA calculations provide a simple method to evaluated the preferred solution conformation of oligosaccharides and it appears that the most important non-covalent forces for biological associations in aqueous solutions are electrostatic, van der Waals' and hydrophobic. Electrostatic interactions are probably more important in providing specificity than in contributing to the overall thermodynamic driving force for association. It is likely that van der Waals' interactions (complementarity) are important and hydrophobic terms essential in most protein-carbohydrate interactions.

ACKNOWLEDGEMENTS

The authors wish to thank Novo A/S and The Danish Technical Research Council for support. The 500 MHz spectrometer was provided by The Danish Natural Science Research Council and The Carlsberg Foundation.

REFERENCES

1. Svensson, B., K. Larsen and I. Svendsen (1983). *Carlsberg Res. Commun.* **48**, 517-527.
2. Bock, K. (1983). *Pure Appl. Chem.* **55**, 605-622.
3. Lemieux, R. U., K. Bock, L. T. Delbaere, S. Koto and V. S. Rao (1980). *Can J. Chem.* **58**, 631-653.
4. Thøgersen, H., R. U. Lemieux, K. Bock and B. Meyer (1982). *Can. J. Chem.* **60**, 44-57.
5. Bock, K., D. Bundle and S. Josephsen (1982). *J. Chem. Soc. Perkin* **II**, 59-70.
6. Lemieux, R. U. and K. Bock (1983). *Arch. Biochem. Biophys.* **221**, 125-134.
7. Paulsen, H., T. Peters, W. Sinnwell, R. Lebuhn and B. Meyer (1985). *Liebig's Ann. Chem.* 489-509.
8. Bock, K., J. Arnarp and J. Lönngren (1982). *Eur. J. Biochem.* **129**, 171-178.
9. Bock, K., M. Meldal, D. R. Bundle, T. Iversen, P. J. Garegg, T. Norberg, A. A. Lindberg and S. B. Svenson (1984). *Carbohydr. Res.* **130**, 23-34.
10. Khare D. P., Hindsgaul O., and Lemieux, R. U. (1985). *Carbohydr. Res.* **136**, 285-308.
11. Lemieux, R. U. (1984). VIII International Symposium on Medicinal Chemistry, Aug. 1984, Uppsala, Sweden. Swedish Pharmaceutical Society, in press.
12. Kitaygorodsky (1978). Chem. Soc. Rev. 7, 133-162.
13. Bock, K. and H. Thøgersen (1982). *Ann. Rep. N.M.R. Spectrosc.* **13**, 1-57.
14. Bock, K. and H. Pedersen (1984). *Carbohydr. Res.* **132**, 142-149.

0.17 nm X-Ray Structure of an L-Arabinose Binding Protein-Ligand Complex: Detailed New Understanding of Protein-Sugar Interactions

Florante A. Quiocho[a] and Nand K. Vyas

Department of Biochemistry, Rice University Houston, TX 77251, USA

INTRODUCTION

In order to elucidate the molecular details of protein-ligand complex, it is necessary to obtain an accurate and unbiased tertiary structure of the complex by crystallographic refinement methods at better than 0.2 nm resolution. We have extensively refined the structure of the liganded form of the L-arabinose binding protein (ABP) at 0.17 nm resolution to an *R*-factor of 13.7% [1]. This analysis has provided new and fundamental understanding of sugar-protein interaction.

ABP is a member of a large class of proteins which serves as initial components of osmotic shock-sensitive active transport systems for a large variety of carbohydrates, amino acids and ions in Gram-negative bacteria. Several of the sugar-binding proteins also act as initial receptors for chemotaxis. Binding proteins exhibit monomeric molecular weights in the range 25 000–45 000 and contain one tight ligand binding site with dissociation constants of about 10^{-7} M. Protein components embedded in the cytoplasmic membrane, distinct for either chemotaxis or transport, are further required for both processes. In this report we summarize the essential features of ABP-sugar complex and further amplify basic concepts of protein-carbohydrate interactions.

[a] *Telephone:* 1-713-527-4872

Protein-Carbohydrate Interactions in Biological Systems. ISBN 0 12 436665 1

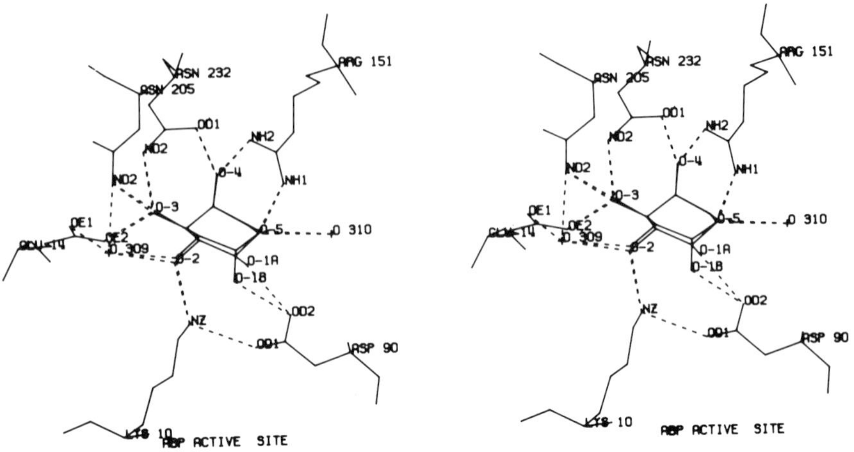

Figure 1. Stereo view of the interaction between arabinose-binding protein and α or β anomer of L-arabinose as determined by crystallographic structure refinement at 0.17 nm resolution. The α-anomeric hydroxyl is labelled 0-1A and the β-hydroxyl 0-1B. Note that refined positions of both sugars are only partially coincident.

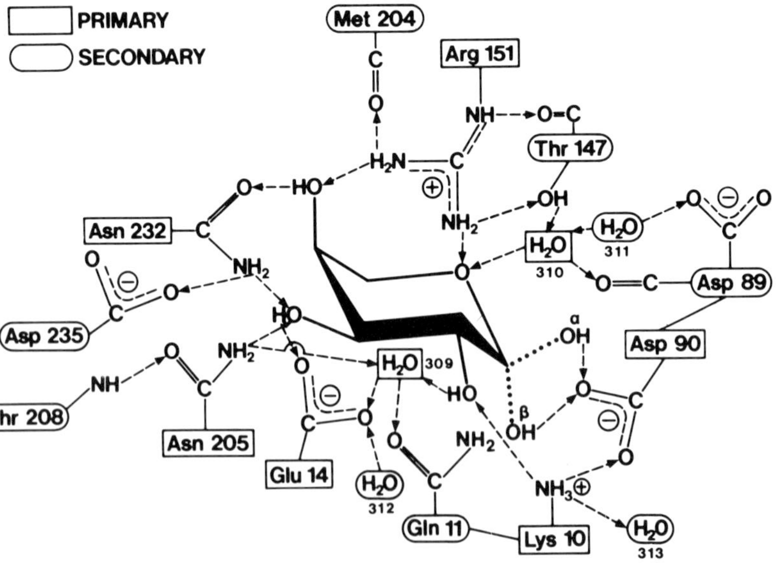

Figure 2. Schematic diagram of the intricate networks of hydrogen bonds formed in the ABP:arabinose complex. Reproduced with permission from *Nature* **310**, 381-386 (1984) [1]. Note that Gln 11 was incorrectly identified as Glu 11 in the original figure published in *Nature*.

Table 1. Interactions between arabinose-binding protein and L-arabinose

(A) **Hydrogen bonds** [data from *Nature* **310**, 381–386 (1984)]

		α-L-arabinose		β-L-arabinose	
Donors (X)	*Acceptors (Y)*	*X...Y* (nm)	*XĤ...Y* (°)	*X...Y* (nm)	*X–Ĥ...Y* (°)
O-1	Asp 90 OD2	0.277	159	0.274	166
O-2	Wat 309 O	0.261	155	0.250	156
O-3	Glu 14 OE2	0.277	176	0.256	173
O-4	Asn 232 OD1	0.262	178	0.262	176
Lys 10 NZ	O-2	0.273	146	0.286	151
Asn 205 ND2	O-3	0.303	169	0.309	167
Asn 232 ND2	O-3	0.297	162	0.298	159
Arg 151 NH2	O-4	0.282	167	0.281	168
Arg 151 NH1	O-5	0.305	161	0.299	160
Wat 310 0	O-5	0.280	—	0.274	—
Overall mean		0.2 (0.015)	164 (10)	0.281 (0.016)	164 (8)

(B) **van der Waals' interactions** (maximum distance = 0.45 nm)

Sugar atoms	*No. of contacts with protein (non-hydrogen atoms)*		*Mean distance (nm)*
C-1	9		0.39
C-2	8		0.40
C-3	10		0.42
C-4	13		0.40
C-5	14		0.40
O-1 (α)	9		0.37
O-1 (β)	9		0.37
O-2	7		0.40
O-3	8		0.38
O-4	6		0.37
O-5	7		0.40
Total	100	Overall mean	0.39

THE ABP-ARABINOSE INTERACTION

The details of the mode of binding of L-arabinose to ABP are shown in Figs 1 and 2 and Table 1. Several unusual features of the ABP-arabinose complex have emerged. The most notable is the binding site geometry which is designed to accommodate either the α or β anomeric form of the L-arabinose substrate. Both bound sugar anomers, which refined to the normal *Cl* pyranose full chair conformation, are only partially coincident; of the atoms common to both sugars, the positions of C-4 and adjacent atoms are superimposable and the positions of C-1 atoms are separated by 0.031 nm. Nevertheless, both anomers are held in place by the same hydrogen bonds. Of critical importance to this interaction is the precise alignment of atom OD2 of residue Asp 90 which enables it to accept a hydrogen bond from either the α- (equatorial) or the β- (axial) anomeric hydroxyl. Moreover, as the locations of the anomeric hydroxyls are related by a local mirror plane formed by Lys 10 NZ, Asp 90 OD1, CG and OD2, and a point midway between the C-1 atoms of both sugars, each anomeric hydroxyl interacts equivalently with atoms in the plane, especially with Asp 90 OD2 and Lys 10. The rest of the hydroxyls and ring oxygen of both anomers are all involved in identical extensive networks of hydrogen bonds (see Figs 1 and 2). The following are noteworthy features of these networks. First, whereas the anomeric hydroxyls participate solely as hydrogen bond donors, the rest of the sugar hydroxyl groups simultaneously donate and accept hydrogen bonds. This is consistent with the observation that whereas the anomeric hydroxyl is a poor hydrogen bond acceptor, the rest of the sugar hydroxyls can effectively serve simultaneously as both hydrogen bond donors and acceptors [2]. Second, Lys 10 is critically positioned to donate a hydrogen bond to O-2, to stabilize the alignment of Asp 90 via a salt-link, and to make van der Waals' contact with either anomeric hydroxyls. Third, by donating one and accepting two hydrogen bonds, the O-3 of the sugars is fully coordinated, including the -C-O bond. The coordination is approximately tetrahedral. Fourth, all the functional groups of the branched, multidentate essential residues are used to bind the sugar or to form hydrogen bond networks with other residues or with isolated water molecules. Asn 232 and Arg 151 are prime examples (see Fig. 2). Fifth, the sugar ring oxygen (O-5) is almost tetrahedrally coordinated, hence each of the two ring oxygen *p* orbitals is directed at a hydrogen bond donor. Sixth, all of the potential hydrogen bond donor groups from the sugar and from side chains of the active site residues Lys 10, Arg 151, Asn 205, and Asn 232 are fully utilized. Seventh, two sequestered water molecules (309 and 310) play dual roles; they mediate through hydrogen bonds the interactions between sugars and ligand site residues and provide additional non-covalent linkages between the two domains at the opening of the cleft. Water 309 is central to a network of hydrogen bonds linking Glu 14 and Gln 11 of the *N*-terminal domain to Asp 205 of the *C*-terminal domain. On the other hand, water 310 is part of a chain of interaction consisting of Asp 90 and Asp 89 of the *N*-terminal domain and Thr 147, Arg 151 and Met 204 of the *C*-terminal domain. These linkages, in addition to those formed via the bound sugars, stabilize the closed or liganded form of ABP.

As summarized in Table 1, there are numerous van der Waals' contacts to the bound sugar anomers which provide additional stability to the complex. One of these, which is of interest as it has been shown that the binding of ligand causes protein fluorescence change [3,4], is between a hydrophobic patch of the arabinose (consisting of C-3, C-4, and C-5) and Trp 16 and Phe 17. It is important to note that though the numbers of van der Waals' contacts to both anomeric hydroxyls are the same, the types of contacts are not entirely identical owing to the different positions of the C-1 hydroxyls and to protein asymmetry.

That the sugars bound in the cleft and most of the essential residues in the binding site are inaccessible to the solvent provides clear evidence for a ligand-induced protein conformational change [1]. A relative twisting motion between the two domains could easily modulate the closing and opening of the cleft and further juxtapose upon ligand binding the essential residues poised in both domains. The importance of ligand-induced conformational changes in transport and chemotaxis has been presented [1].

ANOMERIC EFFECT

Due to the differences in the electronic structure of bonds to the anomeric carbon atom C-1 [5], sugar molecules with equatorial anomeric substituent are less stable than those with axial substituent; this is referred to as the *anomeric effect*, discovered by Edwards [6] and Lemieux [7]. Thus, the α anomeric form of L-arabinose (with equatorial anomeric hydroxyl) is less stable than the β anomer (with axial hydroxyl). Moreover, as the anomeric effect is enhanced by a low dielectric constant environment, the binding site of ABP would further destabilize the α anomer or would favour the β sugar substrate. Consequently, our discovery that almost equal amounts of both anomers are bound to ABP [1] is unexpected. How is the anomeric effect minimized or neutralized? We originally proposed that this is accomplished by the formation of two hydrogen bonds between the sugar ring oxygen and the guanidinium group of Arg 151 and the water molecule 310 (see Figs 1 and 2). Though both hydrogen bonds are formed with both anomers, the ring oxygen of the equatorial anomer should delocalize electron density toward these hydrogen bonds more readily than does the axial anomer.

The structure analysis of ABP provides the first direct and detailed picture of a binding site exactly complementary to both α and β anomeric forms of a sugar substrate. This novel stereospecificity is consistent with previous rapid kinetic measurements in solution indicating that ABP binds both sugar anomers with the same affinity and at equal rates [4]. The lack of anomeric specificity correlates with the transport function of the ABP. As the open-chain, aldehyde form of L-arabinose is utilized in the biosynthesis of pentose phosphates [8], translocation of both sugar anomers by the binding protein-dependent transport system is an obvious advantage.

CORRELATION WITH OTHER STUDIES

The detailed knowledge of the molecular interaction betwen L-arabinose and ABP provides an excellent basis for interpreting previous findings from sugar binding studies in solution. This includes the fluorescence change indicated above. Furthermore, the binding of both anomers at equal rates [4] may now be readily explained and the tight binding [4] can be seen as the net result of the displacement of several water molecules from the cleft upon sugar binding (unpubl. data), dehydration of solvated sugar, the formation of extensive networks of hydrogen bonds, and the sequestering of the ligands within the cleft. The binding of D-galactose to ABP with affinity and rate constants comparable to those observed for L-arabinose [4] can now also be understood. The binding site cleft contains a cavity where the C-6 hydroxymethyl substituent can be easily accommodated with retention of the same binding mode determined for the

L-arabinose for the remainder of the D-galactose molecule (unpubl. data). In addition, the known exclusion of epimers of L-arabinose and D-galactose and the requirement for all of the free sugar hydroxyls for binding are entirely predictable from the structure of the complex. The presence of Glu 14, Asp 90, Lys 10 and Arg 151 as the only ionizing groups hydrogen-bonded to the sugar is consistent with a broad pH range of maximal binding activity, extending from pH 5 to 9.

HYDROGEN BONDS IN CARBOHYDRATE-PROTEIN COMPLEXES

Hydrogen bonding is a predominant type of interaction encountered in binding of carbohydrates to proteins (unpubl. survey). This is amply and unambiguously demonstrated in the case of the ABP-arabinose complex.

With the exception of the anomeric hydroxyl (vide infra), the sugar hydroxyl groups can simultaneously serve as both hydrogen bond donors and acceptors. These dual cooperative properties of hydroxyl groups create stronger hydrogen bonds than those where the hydroxyl group is a hydrogen bond donor only as in the case of the anomeric hydroxyls [2]. They also lead to the formation of extensive hydrogen bond networks with binding site residues.

As hydrogen bonds are highly directional, they are mainly responsible for conferring geometrical specificity on the binding site and ensuring correctness of fit for the substrates. It is for these reasons that hydrogen bonds are more important than hydrophobic interactions. Moreover, especially in the case of the transport binding proteins, hydrogen bonds are stable enough to provide significant ligand binding but are of sufficiently low strength to allow rapid ligand dissociation. Tight ligand affinity and the kinetics of ligand binding (on and off rate constants) are fundamentally related to the functions of binding proteins in transport and chemotaxis [1,4].

The finding that the thermodynamic parameters $\Delta G°$, $\Delta H°$, $\Delta S°$ and $\Delta C_p°$ for the binding of sugars to ABP all have negative signs [10] accords with the conclusion that the stability of the ABP-sugar complex is dependent mainly upon electrostatic interactions of hydrogen bonds and van der Waals' contacts. Other protein-carbohydrate complexes also show similar results.

In our recent survey of several structures of protein-carbohydrate complex, we observed that the residues utilized most in hydrogen bonding are the type with multifunctional groups (e.g. Asn, Asp, Glu, Gln, Arg) (details will be published in a review article). This observation is clearly manifested in the ABP-sugar complex; of the six residues directly hydrogen-bonded to the sugars, only Lys 10 does not belong to this type of residue, although it still forms several interactions via hydrogen bonds and salt-linkage (see Figs 1 and 2). Often some of the charged residues are not neutralized by salt-links. Residues Glu 14 and Arg 151 in the binding site of ABP are prime examples [1]; interestingly, both side chains are also buried and inaccessible to the solvent. As best exemplified in the ABP-arabinose complex, the preponderant involvement of this type of residue is the result of the formation of bidentate hydrogen bonds with the sugar (e.g. Asn 232 and Arg 151) and hydrogen bond networks. These intricate networks fix the orientation of the active site residues which are required for the specificity of the site.

Therefore, replacements of these active site residues by site specific mutagenesis offers very little by way of understanding sugar-protein interactions. On the other hand, binding and X-ray studies of the interaction of modified sugar substrates (by systematic removal of hydroxyl groups) should yield important results.

We have summarized the most salient features of the ABP-arabinose interaction as revealed by structure refinement at atomic resolution. These features, which are a reflection of the stereospecificity of ABP, are vital to the understanding of the function of the protein. Furthermore, the studies reported here constitute an excellent basis for understanding other carbohydrate-protein complexes.

ACKNOWLEDGEMENT

This work was supported by grants from NIH (GM-21371) and the Welch Foundation (C-581).

REFERENCES

1. Quiocho, F. A. and N. K. Vyas (1984). *Nature* **310**, 381-386.
2. Jeffrey, G. A. and L. Lewis. (1978). *Carbohyd. Res.* **60**, 179-182.
3. Parsons, R. G. and R. W. Hogg (1974). *J. Biol. Chem.* **249**, 3602-3607.
4. Miller, D. M., J. S. Olson, J. W. Pflugrath and F. A. Quiocho (1983). *J. Biol. Chem.* **258**, 13 665-13 672.
5. Jeffrey, G. A., J. A. Pople, J. S. Binkley and S. Vishveshwara (1978). *J. Am. Chem. Soc.* **100**, 373-379.
6. Edward, J. T. (1955). *Chem. Ind.* 1120-1104.
7. Lemieux, R. M. (1963). *In* "Molecular Rearrangements" (Ed. P. de Mayo), pp. 713-769. Wiley-Interscience, New York.
8. Englesberg, E. (1963). *In* "Metabolic Pathways: Metabolic Regulation" (Ed. H. J. Vogel), Vol. 1, p. 257. Academic Press, New York.
9. Newcomer, M. E., B. A. Lewis and F. A. Quiocho (1981). *J. Biol. Chem.* **256**, 13 218-13 222.
10. Fukada, H., J. M. Sturtevant and F. A. Quiocho (1983). *J. Biol. Chem.* **258**, 13 193-13 198.

Protein-Bond Carbohydrate Involvement in Plasma Membrane Assembly: The Retinal Rod Photoreceptor Cell as a Model

S. J. Fliesler[a], M. E. Rayborn* and J. G. Hollyfield*

*Departments of Ophthalmology and Biochemistry, University of Miami School of Medicine, Miami, FL 33101, USA, and *Cullen Eye Institute, Baylor College of Medicine, Houston, TX 77030, USA*

The retinal rod photoreceptor is a neuronal cell which displays a remarkable degree of structural and functional compartmentalization (see Fig. 1). The rod outer segment (ROS) consists of a highly ordered stack of several hundred flattened membrane saccules (*disks*) bordered by the plasma membrane of the cell. The membrane constituents of the ROS are renewed throughout the life of the cell: new disks are added at the base of the ROS and packets of older disks are detached from the distal tip of the ROS, thus maintaining a relatively constant ROS length [1]. In the process of ROS renewal, new membrane constituents are transported from the rod inner segment (RIS) to a region near the base of the ROS, where they are added to the plasma membrane. The plasma membrane then undergoes mechanical deformation, evaginating outward from the ciliary stalk toward the opposite cellular border, thereby forming a so-called *open disk*. This open disk then fuses circumferentially along its rim with the overlying ROS plasma membrane; the fused membranes subsequently detach from the plasma membrane proper to become a *closed disk* which is sequestered within the ROS. This process by which the ROS plasma membrane undergoes structural differentiation to form disks is known as *disk morphogenesis* [2,3].

During disk morphogenesis, extensive planar domains of plasma membrane are brought into close proximity. These membranes exhibit a highly negative

[a]*Telephone:* 1-305-326-6329

Protein-Carbohydrate Interactions in Biological Systems. ISBN 0 12 436665 1

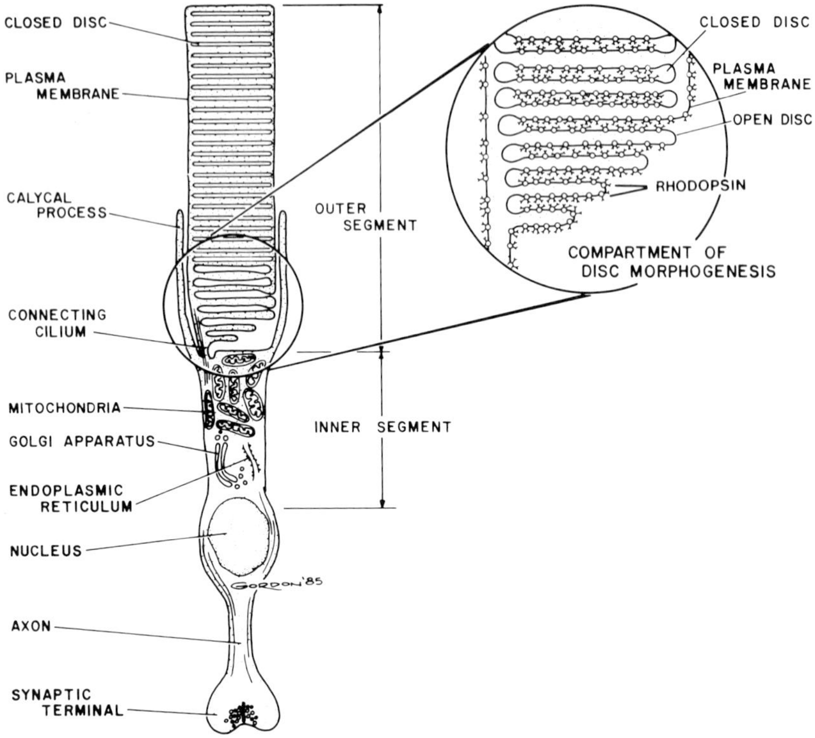

Figure 1. Schematic diagram of a rod photoreceptor cell, illustrating the cellular compartments and organelles. Insert: Magnified view of the basal region of the rod outer segment, depicting the structural relationship between the ROS plasma membrane and the open and closed disks, and the localization of rhodopsin to these structures.

surface charge (due largely to their phospholipid content) and also contain cell surface glycoconjugates (primarily glycoproteins), two features which might be expected to prevent direct membrane apposition. In addition, rather than forming blebs or exocytotic vesicles, the evaginating ciliary plasma membrane maintains a characteristic disk-like morphology while expanding outward over a relatively large distance (e.g 5–7 μm in amphibians, 1–2 μm in mammals). The molecular details of the mechanism which both permits and promotes such membrane dynamics are not well understood.

We have been employing known inhibitors of glycoprotein synthesis and processing to determine the importance of oligosaccharide moieties in directing the intracellular transport of opsin and the process of disk morphogenesis. Opsin, the apoglycoprotein component of the rod visual pigment rhodopsin and the major ROS protein constituent, consists of a single polypeptide chain ($M_r \sim 37$ kD) to

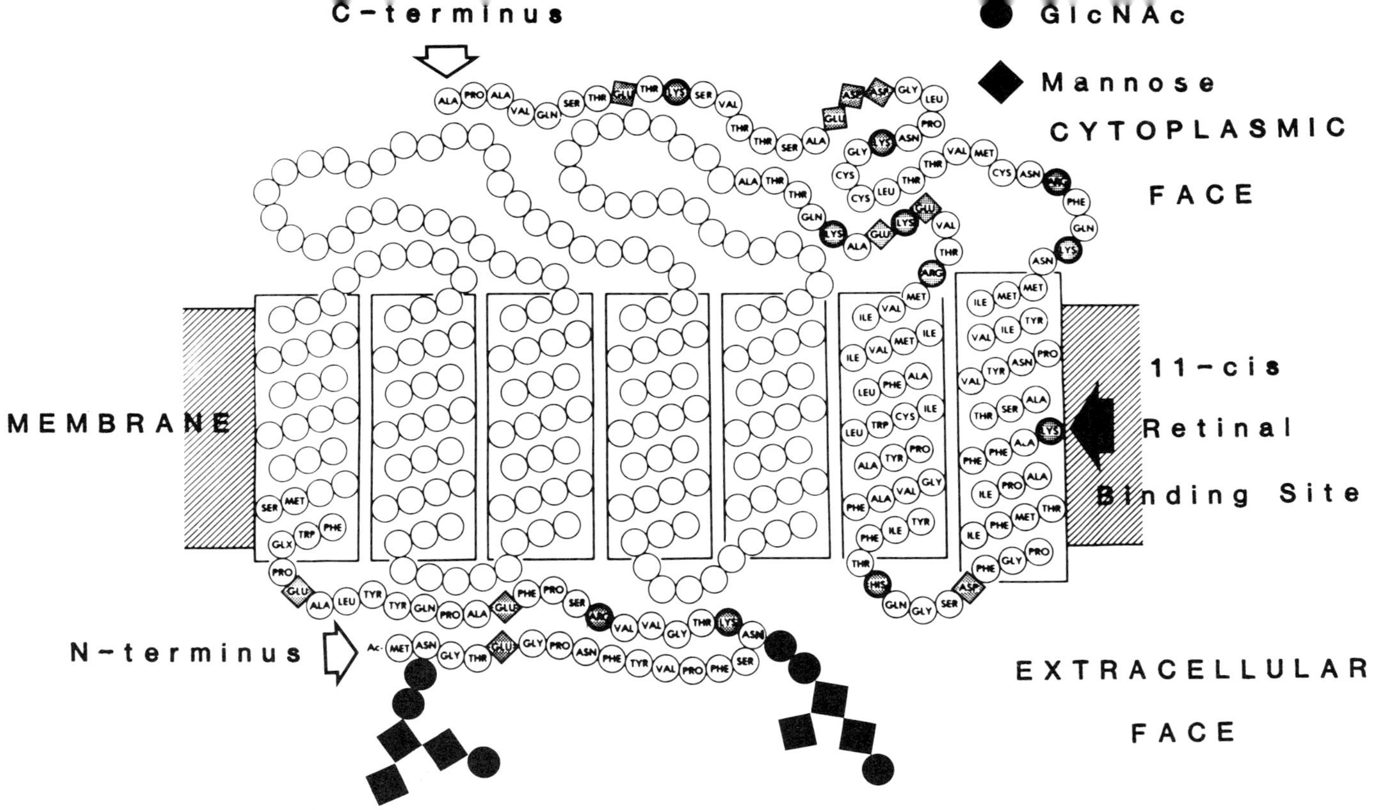

Figure 2. Hypothetical model for the arrangement of opsin's polypeptide chain in the disk membrane, depicting seven transmembrane barrel helices in "exploded view" and the two N-linked oligosaccharide chains attached to the *N*-terminal regions of the polypeptide. (Modified after Hargrave [4].)

which two short, neutral, N-linked oligosaccharide chains are attached [4] (Fig. 2). The visual pigment is an integral membrane protein localized to the ROS plasma membrane and disks; most of its mass is deeply embedded within the hydrophobic interior of the membrane, with its *N*-terminus exposed at the extracellular (or intradisk) face and its *C*-terminus exposed at the cytoplasmic (or interdisk) face. The oligosaccharide composition and structures are known only for bovine opsin: each chain has the predominent composition $Man_3GlcNAc_3$, with lesser amounts of $Man_4GlcNAc_3$ and $Man_5GlcNAc_3$ [5-7]. Herein we review experimental data and propose an hypothesis which implicate the participation of N-linked oligosaccharide chains (particularly those of opsin) in the mechanism of disk morphogenesis.

EFFECT OF TUNICAMYCIN ON OPSIN GLYCOSYLATION AND DISK MORPHOGENESIS

Tunicamycin (TM) is an antibiotic which blocks the formation of N-linked oligosaccharides utilized for the glycosylation of proteins via the *lipid intermediate pathway* [8,9]. TM has been shown to prevent the glycosylation of opsin *in vitro* in bovine [10], amphibian [11,12], and human [13] retinas. Opsins from various animal species are very similar both in the size and amino acid composition of their polypeptide moiety [14,15]. However, while TM produces a marked shift to lower M_r in bovine [10] and frog [11] opsin, there is no appreciable effect of TM on the M_r of *Xenopus* [12] or human [13] opsin, suggesting species-specific differences in the number and/or structure of oligosaccharide chains attached to opsin.

Using biochemical methods as well as light and electron microscopic autoradiography, TM has been shown to disrupt disk morphogenesis in *Xenopus* retinas *in vitro* [12]. When control retinas are incubated in the presence of a radiolabelled glycoprotein precursor (e.g. ^{3}H-leucine, ^{3}H-mannose, etc.), the resulting electron microscopic autoradiograms exhibit a discrete concentration of silver grains in a *band* over the basal ROS disks, indicative of the assembly of newly synthesized proteins (glycoproteins) into disk membranes (see Fig. 3a). However, when retinas are incubated with such labelled precursors in the presence of TM, several distinct morphological and autoradiographic differences are observed (Fig. 3b and 3c). Instead of the normally close apposition between the plasma membrane surfaces of the ROS and RIS, the extracellular space between the ROS and RIS (the *intersegment space*) is grossly dilated and filled with heterogeneous, vesicle-like membranes. The results of a recent scanning electron microscopy study [16] indicate that this membrane material consists of a complex network of interconnected tubular cisternae as well as individual vesicles of varying sizes, closely associated with the basal ROS surface. Open disks were not observed in the rods of TM-treated retinas, although they were observed with relatively high frequency in control-incubated retinas.

The membrane material in the intersegment space exhibits incorporation of ^{3}H-leucine (Fig. 3b) but not ^{3}H-mannose (Fig. 3c). Biochemical analyses (e.g. specific radioactivity of TCA-precipitable vs. TCA-soluble material, and

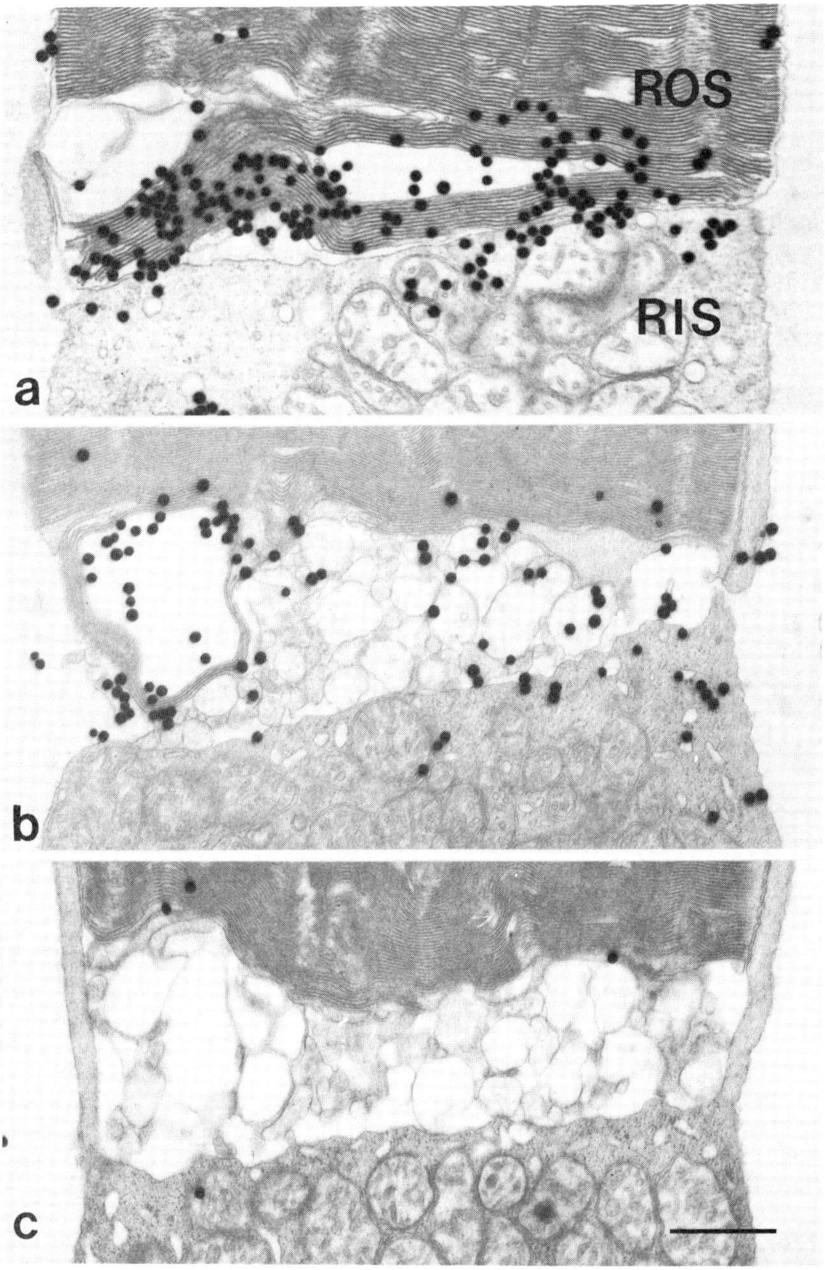

Figure 3. Electron microscopic autoradiograms of the inner segment (RIS)/outer segment (ROS) junctional region of rod cells from *Xenopus* retinas incubated in the presence (b, c) or absence (a) of tunicamycin, with either ^{3}H-mannose (a, c) or ^{3}H-leucine (b) as radiolabelled substrate. Note accumulation of membrane vesicles in the dilated space between the ROS and RIS in the tunicamycin-treated samples. Scale bar = 1 μm.

SDS-PAGE) indicated that TM did not perturb either the cellular uptake of radiolabelled precursors by the retina or the incorporation of ^{3}H-leucine into retinal proteins (particularly opsin), whereas incorporation of ^{3}H-mannose into retinal proteins was virtually abolished [12]. These results indicate that the membrane material in the intersegment space of TM-treated retinas contains newly synthesize polypeptides which lack N-linked oligosaccharide chains, and demonstrate that this membrane material does not arise by degradation of pre-existing ROS disks.

From these data it is apparent that the lack of N-linked oligosaccharides on the newly synthesized membrane proteins does not prevent their routeing through the RIS to the site of membrane assembly. Rather, the process of disk morphogenesis is specifically perturbed, resulting in the formation of structurally aberrant ROS membranes. Freeze-fracture analysis of TM-treated *Xenopus* retinas [17] has revealed that both the size and density of intramembrane particles (IMPs) in these aberrant membranes are virtually identical to those of normal closed disks. Thus, the assembly of ROS membrane proteins (predominantly opsin) into a spatial arrangement characteristic of the disk membrane is independent of protein glycosylation. However, it is apparent that such lattice formation *precedes* disk morphogenesis, and that the "normality" of this aspect of membrane molecular architecture does not necessarily guarantee normal disk membrane morphogenesis.

The ability of TM to inhibit disk morphogenesis has also been demonstrated *in vivo* by intravitreal injection in the frog eye [18]. Addition of new disks to the base of the ROS ceases while shedding of disks from the distal ROS tip continues (perhaps in even greater than normal amounts), resulting in a progressive shortening of the ROS and eventually complete degeneration of the photoreceptors.

EFFECT OF CASTANOSPERMINE ON OPSIN GLYCOSYLATION AND DISK MORPHOGENESIS

Castanospermine (CAS) is an indolizine alkaloid which has been shown to be a potent α-glucosidase inhibitor [19,20]. In the course of glycoprotein maturation (i.e. oligosaccharide "trimming" and other post-translational modifications), the conversion of the precursor N-linked oligosaccharide $Glc_3Man_9GlcNAc_2$ to fully processed oligosaccharide structures lacking terminal glucose residues is blocked by CAS [21,22]. This results in the synthesis of glycoproteins with elevated M_rs, due to the presence of larger than normal, mannose-enriched oligosaccharides.

Frog retinas incubated with CAS in a dual-label protocol exhibit a selective enhancement in the incorporation of ^{3}H-mannose into TCA-precipitable material in both retinas (*c.* 1.5-fold) as well as isolated ROS membranes (*c.* 2.4-fold) without a significant effect on ^{14}C-leucine incorporation, relative to controls (Table 1). The incorporation of these labelled substrates into both whole retina and ROS proteins was further examined by SDS-PAGE/fluorography (Fig. 4). Under the given fluorographic conditions, only the incorporation of the ^{14}C-leucine can be detected; hence, the fluorogram depicts newly synthesized

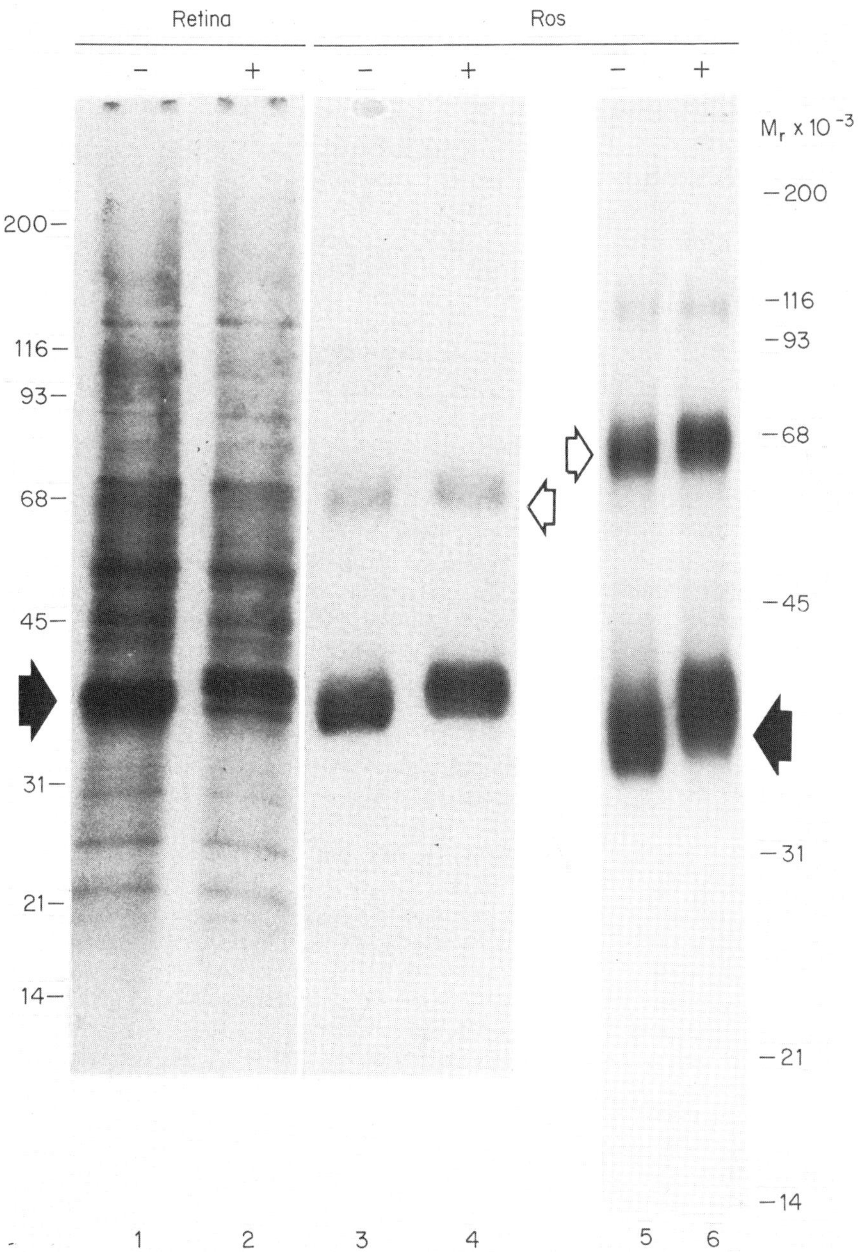

Figure 4. SDS-PAGE fluorograms of radiolabelled proteins from frog retinas (lanes 1 and 2) and isolated ROS membranes (lanes 3-6) obtained after incubation retinas with ^{14}C-leucine and ^{3}H-mannose in the presence (+) or absence (−) of castanospermine. Solid and open arrows denote the monomeric ($M_r \sim 36$ kD) and dimeric ($M_r \sim 68$ kD) forms of opsin, respectively. Slab gels containing either a 7.5-15% acrylamide gradient (lanes 1-4) or 10% acrylamide (lanes 5 and 6) were employed.

Table 1. Effect of castanospermine (CAS) on incorporation of ^{3}H-mannose and ^{14}C-leucine into TCA-precipitable material in frog retinas and ROS membranes

	^{3}H-mannose		*^{14}C-leucine*	
	dpm/μg protein	*% control*	*dpm/μg protein*	*% control*
Retinas				
Control	201±48 (3)	100	1071±90 (3)	100
CAS	307±21 (3)	153[a]	1021±153 (3)	95[b]
Ros membranes				
Control	299 (2)	100	907 (2)	100
CAS	706 (2)	236	1032 (2)	114

Values are the mean ±SD of triplicate specific radioactivity determinations, with the number of independent samples given in parentheses.
[a] $p > 0.02$.
[b] Not significant.

polypeptides. In the fluorograms obtained from whole retinas (Fig. 4, lanes 1 and 2), both the number and relative intensities of detectable components were similar when comparing the control and CAS-treated samples, indicating that CAS did not significantly perturb protein synthesis. In both instances opsin is the major labelled component, and it is also the only labelled component whose migration on the gel is appreciably altered by CAS. In the presence of CAS the M_r of opsin is shifted higher by ~2-3 kD relative to its normal M_r of ~36 kD. This is clearly observed in the fluorograms of purified ROS membranes (lanes 3-6), where the monomeric and dimeric forms of opsin are virtually the only components visible.

The 10% SDS-PAGE gel containing the ROS proteins was cut into individual lanes and each lane was cut into 2 mm slices, solubilized, and assayed for ^{3}H and ^{14}C. The results are shown in Fig. 5. The gel of the control ROS sample (Fig. 5b) exhibited two major dual-labelled peaks of radioactivity, corresponding to the dimeric (peak fraction 21) and monomeric (peak fraction 37) forms of opsin. The corresponding gel from the CAS sample (Fig. 5a) appeared quite similar, except that the peaks representing the opsin dimer and monomer (peak fractions 20 and 36, respectively) were shifted to slightly higher M_r. Furthermore, the ^{3}H/^{14}C ratios in the gel slices corresponding to the opsin components in the ROS sample from the CAS-treated retinas were about two-fold higher than those of the control ROS sample. These data indicate that CAS caused a selective enrichment in the ^{3}H-mannose content of newly synthesized opsin as well as increasing its M_r by ~2-3 kD, consistant with hyperglycosylation due to inhibition of oligosaccharide processing.

Using the same *in vitro* protocol, similar results have been obtained with *Xenopus* retinas [23]. In direct analogy to the studies described above concerning the effects of TM on disk morphogenesis, we employed electron microscopic autoradiography to further examine the effects of CAS on disk morphogenesis in *Xenopus* retinas, using ^{3}H-mannose as a radiolabelled substrate. Under

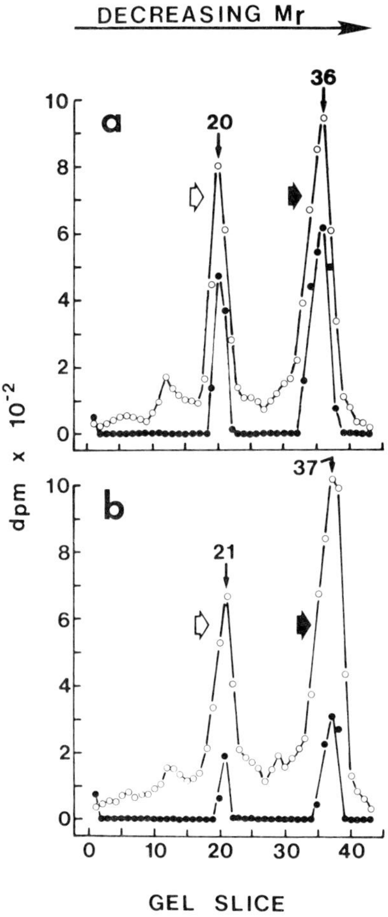

SLICE	$^3H/^{14}C$
19	0.30
20	0.58
21	0.61
34	0.65
35	0.64
36	0.65
37	0.82

CAS

SLICE	$^3H/^{14}C$
20	0.11
21	0.29
35	0.07
36	0.27
37	0.30
38	0.28

Control

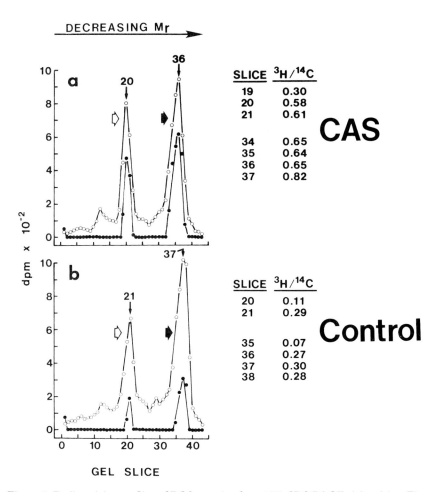

Figure 5. Radioactivity profiles of ROS proteins from 10% SDS-PAGE slab gel (see Fig. 4, lanes 5 and 6), following incubation of retinas with ^{14}C-leucine (●---●) and ^3H-mannose (○---○) in the presence or absence of castanospermine (CAS). The two major peaks in each electrophoretogram correspond to the monomeric (solid arrow) and dimeric (open arrow) forms of opsin.

conditions where CAS produced about a two-fold stimulation of ^3H-mannose incorporation into opsin, both the number and morphology of newly assembled ROS disks were *not* significantly different when compared with control retinas from companion incubations. The number of newly assembled disks in rods from CAS-treated retinas was 9.3 ± 2.6 (mean ± SD, $N = 19$), compared with 8.4 ± 3.0 ($N = 22$) in control retinas. Furthermore, quantitation of silver grains over identical areas of newly formed disks revealed that the grain density was about 2.6-fold greater in the ROS membranes of CAS-treated retinas relative to controls, in reasonable agreement with the biochemical results.

Therefore, under *in vitro* conditions which lead to *hyperglycosylation* of opsin, disk morphogenesis does not appear to be significantly perturbed. Clearly, these results are in striking contrast to those obtained with TM, where the *absence* of newly synthesized N-linked oligosaccharides resulted in marked disruption of disk morphogenesis. It is not clearly, however, that the structural stability of the ROS will persist if the retina is exposed to CAS over a duration sufficient to allow the accumulation of a large number of disks containing hyperglycosylated opsin (e.g. following intra-vitreal injection of CAS).

HYPOTHESIS FOR THE PARTICIPATION OF N-LINKED OLIGOSACCHARIDES IN DISK MORPHOGENESIS

From the results of the tunicamycin experiments described above, it can be concluded that N-linked oligosaccharides (particularly those of opsin) are involved in some critical manner in the process of disk morphogenesis. However, the results of the castanospermine experiments indicate that the structural requirements which the oligosaccharides must meet in order to fulfill their role in this biological process may not be specified stringently. That is, the mechanism of disk morphogenesis appears to be able to accommodate opsin oligosaccharide structures which vary between $Man_{3-5}GlcNAc_3$ (the presumed composition of amphibian opsin carbohydrate chains, in analogy to that of bovine opsin) and the unprocessed precursor structure $Glc_3Man_{7-9}GlcNAc_2$.

We propose that, in addition to its physiological role in photon capture, rhodopsin (opsin) performs a structural role as an *adhesion* molecule during disk morphogenesis. The participation of glycoproteins in eukaryotic cell-cell and

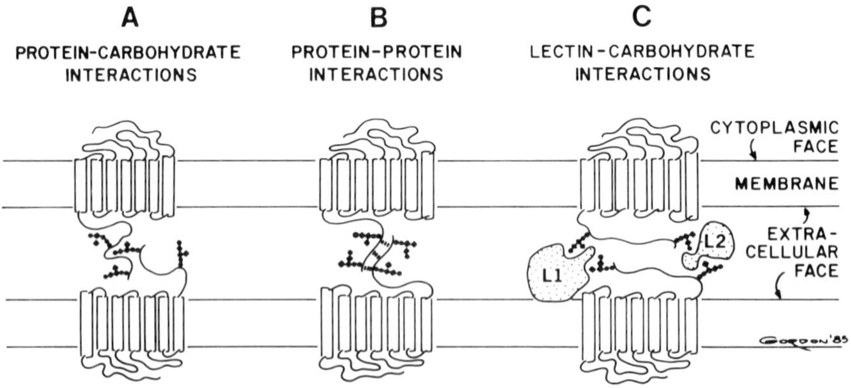

Figure 6. Hypothetical models for the molecular basis of adhesion between membrane surfaces during disk morphogenesis. Opsin molecules may adhere to one another (homophilic bonding) via (a) protein-carbohydrate or (b) protein–protein interactions. Alternatively, (c) adhesion may involve interactions between opsin and other molecules (heterophilic bonding), such as membrane-associated (L1) or soluble (L2) lectins.

cell-substrate adhesion events has been described [24-26]. The basic physical principles of such events appear to be applicable to microbe-host cell attachment during the initial phase of pathogenesis (see elsewhere, this volume). Although the chemical and molecular nature of such adhesion events are not well understood, various mechanisms might be envisioned which would promote and stabilize close membrane apposition during outward expansion of the plasma membrane from the ciliary stalk (see Fig. 6). For example, the *N*-terminal domains of opsin molecules on one membrane face might interact with those on the opposing membrane face; this type of interaction between like molecules is termed *homophilic* bonding [25]. Such interactions are likely to involve primarily weak bonds (e.g. van der Waals' forces and hydrogen bonds) (see discussions by Bock, Bundle and Quiocho, this volume). Three major types of interactions are possible: protein-carbohydrate (Fig. 6a), protein-protein (Fig. 6b), and carbohydrate-carbohydrate. Of these, the third type is the least likely to occur, primarily due to steric constraints. Since rhodopsin can undergo both rotation about an axis perpendicular to the plane of the membrane [27,28] as well as translation within the plane of the membrane [29,30] wth great facility, such bonds between opsin molecules would have to be made and broken rapidly. The apparent lateral diffusion constant of rhodopsin in disk membranes is about $3.5\text{-}5.5 \times 10^{-9}$ cm^2/sec [29,30]; for comparison, the value obtained for N-CAM (the neuron-specific cell adhesion molecule) is approximately 6×10^{-10} cm^2/sec [25], indicating that N-CAM also exhibits a high degree of lateral mobility.

The overall efficacy of such molecular interactions would depend upon several factors, including the number of bonds, their energies and lifetimes, and the relative magnitude of opposing forces (e.g. thermal energy, kinetic energy of translation and rotation, electrostatic and other repulsions). There are on the order of 10^6 rhodopsin molecules per disk membrane; in theory, therefore, the summation of weak bonding interactions among even a small percentage of the rhodopsin molecules on apposing membrane surfaces could approach covalent bonding energies. The critical requirement for such mechanism is that, at any given time during the course of membrane evagination, a sufficient population of rhodopsin molecules would need to be engaged in effective bonding in order to stabilize membrane-membrane associations. One might envisage this process as a "molecular zipper", where the "teeth" engage and disengage while maintaining the zipper intact overall. Molecular modelling of homophilic bonding interactions between opsin molecules is hampered by the fact that an accurate description of the three-dimensional structure of opsin is not yet available [4,6], especially with regard to conformation of the *N*-terminal peptide and the spatial disposition of the oligosaccharide chains. Thus, one cannot accurately predict how the presence or absence of a given oligosaccharide structure (such as $Man_{3\text{-}5}GlcNAc_2$ or $Glc_3Man_{7\text{-}9}GlcNAc_2$) might affect intermolecular bonding of opsin domains, or what affect these structures might have on disk morphogenesis.

Alternatively, there is the possibility that one or more lectins might be involved in bridging the gap between opposing plasma membrane surfaces via the carbohydrate chains of opsin (Fig. 6c). However, since lectins have not been identified as endogenous constituents of ROS membranes, such molecules would have to be either transiently associated with the plasma membrane (i.e. loosely bound) or else be soluble (i.e. components of the interphotoreceptor matrix). Endogenous lectins have been described in retin and other tissues [31,32], although the localization of lectins to the compartments relevant to disk morphogenesis has not been demonstrated. Furthermore, such a lectin would need to be at least divalent to serve as a cross-bridge in binding oligosaccharides from opposing membrane surfaces, and would also need to have a binding specificity which would account for recognition of opsin-like oligosaccharide structures as well as alternate, related structures (to be consistent with the castanospermine results). Although these multiple caveats render the lectin-mediated mechanism less attractive, such a heterophilic mechanism would specifically implicate the oligosaccharide chains as ligands, and would therefore be consistent with the lack of normal disk morphogenesis in the presence of tunicamycin.

In addition, hydrophobic interactions (i.e. ordering of water molecules near the apposed membrane surfaces) could contribute to lowering the free energy of the system, thus making such membrane-membrane associations more energetically favourable. Presumably, the kinds of molecular interactions described above could be involved in maintaining the structure of mature (i.e. closed) disks, where the opposing lumenal surfaces are separated by a distance of only about 2 nm.

The particiption of protein-bound carbohydrates may be involved in other biological systems where membrane surfaces are brought into close apposition in order to form morphologically and functionally specialized membrane structures. A case in point is the formation of myelin in the peripheral nervous system (PNS). PNS myelin is a multilamellar structure which arises by the repeated, spiral envelopment of an axon by a neighbouring Schwann cell. Two different regional forms of myelin can be distinguished, based upon the spacing between the opposing extracellular leaflets of the Schwann cell plasmalemma which constitute the "minor dense lines": *semi-compact* myelin lies closest to the axonal plasmalemma and has a 12–14 nm gap between the external leaflets, whereas the more distal *compact* myelin has an interleaflet gap of only 2 nm. The major glycoprotein of PNS myelin, P_0, is a small protein ($M_r \sim 30$ kD) which represents about 50–70% of the total myelin protein [33,34]. Thus P_0 constitutes a major structural element of PNS myelin. Its carbohydrate moiety consists of a single, N-linked nonasaccharide having the composition NANA(Gal)$Man_3GlcNAc_3$Fuc, which represents about 5% of the M_r of P_0 [33,35]. P_0 is exclusively localized to the compact PNS myelin, wherein it is uniformly distributed [36–38]. Furthermore, experimentally induced Wallerian degeneration (where the myelin lamellae become unravelled and disorganized) has been correlated with a marked reduction in P_0 content, and the rate of this reduction is greater than for any other myelin-associated protein [39,40]. Therefore, the PNS myelin glycoprotein P_0 may serve as an adhesion molecule to both establish and preserve the characteristic 2 nm spacing between the external membrane leaflets which constitute compact myelin. Interestingly, this spacing is identical to that of the lumen which separates the internal membrane leaflets in mature ROS disks.

ACKNOWLEDGEMENTS

This work was supported, in part, by grants EY04738, EY06045 (SJF), and EY02363 (JGH) from the US Department of Health and Human Services (NIH/NEI, Bethesda, MD), by a Center Grant from the National Retinitis Pigmentosa Foundation (Baltimore, MD), and by grants from Research to Prevent Blindness, Inc. (New York, NY) and the Retina Research Foundation (Houston, TX). We thank Ms Sandra Wisely-Carr and Ms Ann Osterfeld for technical assistance, Mr Gary Rutheford and Ms Barbara French for photographic services, and Dr William C. Gordon for technical illustrations.

REFERENCES

1. Young, R. W. (1976). Visual cells and the concept of renewal. *Invest. Ophthalmol.* **15**, 700-725.
2. Papermaster, D. S. and B. G. Schneider (1982). Biosynthesis and morphogenesis of outer segment membranes in vertebrate photoreceptor cells. *In* "Cell Biology of the Eye" (Ed. D. McDevitt), pp. 475-531. Academic Press, New York and London.
3. Steinberg, R. H., S. K. Fischer and D. H. Anderson (1980). Disc morphogenesis in vertebrate photoreceptors. *J. Comp. Neurol.* **190**, 501-518.
4. Hargrave, P. A. (1982). Rhodopsin chemistry, structure and topography. *In* "Progress in Retinal Research" (Eds N. Osborne and G. Chader), Vol. 1, pp. 1-51. Pergamon Press, New York.
5. Fukuda, M. N., D. S. Papermaster and P. A. Hargrave (1979). Rhodopsin carbohydrate: structure of small oligosaccharides attached at two sites near the NH_2-terminus. *J. Biol. Chem.* **254**, 8201-8207.
6. Hargrave, P. A., J. H. McDowell, R. J. Feldman, P. H. Atkinson, J. K. M. Rao and P. Argos (1984). Rhodopsin's protein and carbohydrate structure: selected aspects. *Vision Res.* **24**, 1487-1499.
7. Liang, C. J., K. Yamashita, H. Schichi, C. G. Muellenberg and A. Kobata (1979). Structure of the carbohydrate moiety of bovine rhodopsin. *J. Biol. Chem.* **254**, 6414-6418.
8. Elbein, A. (1984). Inhibition of the biosynthesis and processing of N-linked oligosaccharides. *CRC Crit. Rev. Biochem.* **161**, 21-49.
9. Schwartz, R. T. and R. Datema (1982). The lipid intermediate pathway of protein glycosylation and its inhibitors: the biological significance of protein-bound carbohydrate. *Adv. Carbohydr. Chem. Biochem.* **40**, 287-379.
10. Plantner, J. J., L. Poncz and E. L. Kean (1980). Effect of tunicamycin on glycosylation of rhodopsin. *Arch. Biochem. Biophys.* **201**, 527-532.
11. Fliesler, S. J. and S. F. Basinger (1985). Tunicamycin blocks the incorporation of opsin into retinal rod outer segment membranes. *Proc. Natl. Acad. Sci. USA* **82**, 1116-1120.
12. Fliesler, S. J., M. E. Rayborn and J. G. Hollyfield (1985). Membrane morphogenesis in retinal rod outer segments: inhibition by tunicamycin. *J. Cell. Biol.* **100**, 574-587.
13. Fliesler, S. J., G. A. Tabor and J. G. Hollyfield (1984). Glycoprotein synthesis in the human retina: localization of the lipid intermediate pathway. *Exp. Eye Res.* **39**, 153-173.

14. Abrahamson, E. W., R.S. Fager and W. T. Mason (1974). Comparative properties of vertebrate and invertebrate photoreceptors. *Exp. Eye Res.* **18**, 51-67.
15. Papermaster, D. S. and W. J. Dreyer (1974). Rhodopsin content of the outer segment membranes of bovine and frog retinal rods. *Biochemistry* **13**, 2438-2444.
16. Ulshafer, R. J., C. B. Allen and S. J. Fliesler (1985). Tunicamycin-induced dysgenesis of retinal rod outer segment membranes. I. A scanning electron microscopy study. (Submitted.)
17. Defoe, D. M., J. C. Besharse and S. J. Fliesler (1985). Tunicamycin-induced dysgenesis of retinal rod outer segment membranes. II. Quantitative freeze-fracture analysis. (Submitted).
18. Fliesler, S. J., L. M. Rapp and J. G. Hollyfield (1984). Photoreceptor-specific degeneration caused by tunicamycin. *Nature (Lond.)* **311**, 575-577.
19. Saul, R., J. P. Chambers, R. J. Molyneux and A. D. Elbein (1983). Castanospermine, a tetrahydroxylated alkaloid that inhibits β-glucosidase and β-glucocerebrosidase. *Arch. Biochem. Biophys.* **221**, 593-597.
20. Saul, R., R. J. Molyneux and A. D. Elbein (1984). Studies on the mechanism of castanospermine inhibition of α- and β-glucosidases. *Arch. Biochem. Biophys.* **230**, 668-675.
21. Hori, H., Y. T. Pan, R. J. Molyneux and A. D. Elbein (1984). Inhibition of processing of plant N-linked oligosaccharides by castanospermine. *Arch. Biochem. Biophys.* **228**, 525-533.
22. Pan, Y. T., H. Hori, R. Saul, B. A. Sanford, R. J. Molyneux and A. D. Elbein (1983). Castanospermine inhibits the processing of the oligosaccharide portion of the influenza viral hemagglutinin. *Biochemistry* **22**, 3975-3984.
23. Fliesler, S. J., M. E. Rayborn and J. G. Hollyfield (1985). Castanospermine effects on opsin glycosylation and rod outer segment disc morphogenesis. *J. Cell Biol. 100 (Abstr.)*, in press.
24. Damsky, C. H., K. A. Knudsen and C. A. Buck (1984). Integral membrane glycoproteins in cell-cell and cell-substrate adhesion. *In* "The Biology of Glycoproteins" (Ed. R. J. Ivatt), pp. 1-64. Plenum Press, New York.
25. Edelman, G. M. (1985). Cell adhesion and the molecular process of morphogenesis. *Ann. Rev. Biochem.* **54**, 135-169.
26. Rauvala, H. (1983). Cell surface carbohydrates and cell adhesion. *Trends Biochem. Sci.* **8**, 323-325.
27. Brown, P. K. (1972). Rhodopsin rotates in the visual receptor membrane. *Nature (Lond.), New Biol.* **236**, 35-38.
28. Cone, R. A. (1972). Rotational diffusion of rhodopsin in the visual receptor membrane. *Nature (Lond.), New biol.* **236**, 39-43.
29. Liebman, P. A. and G. Entine (1974). Lateral diffusion of visual pigment in photoreceptor disk membranes. *Science* **185**, 457-459.
30. Poo, M. M. and R. A. Cone (1974). Lateral diffusion of rhodopsin in the photoreceptor membrane. *Nature (Lond.)* **247**, 438-441.
31. Ashwell, G. and J. Harford (1982). Carbohydrate-specific receptors of the liver. *Ann. Rev. Biochem.* **51**, 531-554.
32. Barondes, S. H. (1984). Soluble lectins: a new class of extracellular proteins. *Science* **233**, 1259-1264.
33. Ishaque, A., M. W. Roomi, I. Szymanska, S. Kowalski and E. H. Eylar (1980). The P_0 glycoprotein of peripheral nerve myelin. *Can. J. Biochem.* **58**, 913-921.
34. Roomi, M. W., A. Ishaque, N. R. Khan and E. H. Eylar (1978). The P_0 protein: the major glycoprotein of peripheral nerve myelin. *Biochim. Biophys. Acta* **536**, 112-121.

35. Roomi, M. W. and E. H. Eylar (1978). Isolation of a product from the trypsin-digested glycoprotein of sciatic nerve myelin. *Biochim. Biophys. Act* **536**, 122-133.
36. Peterson, R. G. and R. W. Gruener (1978). Morphological localization of PNS myelin proteins. *Brain Res.* **152**, 17-29.
37. Trapp, B. D. and R. H. Quarles (1982). Presence of the myelin-associated glycoprotein correlates with alterations in the periodicity of peripheral myelin. *J. Cell Biol.* **92**, 877-882.
38. Trapp, B. D., Y. Itoyama, N. H. Sternberger, R. H. Quarles and H. F. Webster (1981). Immunocytochemical localization of P_0 protein in Golgi complex membranes and myelin of developing rat Schwann cells. *J. Cell. Biol.* **90**, 1-6.
39. Luttges, M. W., P. T. Kelly and R. H. Gerren (1976). Degenerative changes in mouse sciatic nerves: electrophoretic and electrophysiological characterizations. *Ex. Neurol.* **50**, 706-733.
40. Wood, J. G. and R. M. C. Dawson (1974). Lipid and protein changes in sciatic nerve during Wallerian degeneration. *J. Neurochem.* **22**, 631-635.

Fine Dissection of Binding Epitopes on Carbohydrate Receptors for Microbiological Ligands

Karl-Anders Karlsson[a], Klaus Bock*,
Nicklas Strömberg and Susann Teneberg

*Department of Medical Biochemistry, University of Göteborg, Göteborg, Sweden, and *Department of Organic Chemistry, The Technical University of Denmark, Lyngby, Denmark*

The richness in carbohydrate at the animal cell surface may in part explain why bacteria [1], bacterial toxins [2] and viruses [3] appear to have selected primarily carbohydrate receptors for adhesion or penetration. The number of known carbohydrate sequences is rapidly increasing. There is also an improving documentation on the specificity in the expression of surface carbohydrate related to animal species, individuals, tissues and cells which is in line with a diversity and tropism at the receptor level. Our knowledge of the structure of specific receptors on animal cell surfaces has however progressed rather slowly, in part explained by the chemical complexity of carbohydrate compared to protein, but also due to the physical properties of membrane-bound substances, being amphipathic and difficult to resolve or handle in classical inhibition experiments. The present report will briefly summarize a novel approach based on recent technical improvements which dramatically facilitate the accessibility to detailed receptor information.

SOLID-PHASE BINDING ASSAY BASED ON A THIN-LAYER CHROMATOGRAM WITH SEPARATED GLYCOLIPIDS BEING RECEPTOR CANDIDATES

The problems referred to above have in part been overcome by an overlay assay, adapted for viruses [4] and bacteria [5]. After separation in several lanes of

[a] *Telephone:* 46-31-853490

Protein-Carbohydrate Interactions in Biological Systems. ISBN 0 12 436665 1

pure or mixtures of glycolipids on a layer of silica gel coated on a sheet of alumina, the chromatogram is treated by short dipping in a diluted solution of polyisobutylmethacrylate in diethylether, followed by coating of excess hydrophobic surface with bovine serum albumin. The plate is then overlayered with ligand suspension. After careful washings bound ligand may be detected by autoradiography, either of externally or metabolically labelled ligand or after binding with specific antibody and labelled anti-antibody. Alternatively an ELISA technique may be used. One advantage of the method is the possible analysis in the same single assay of many receptor candidates with built-in controls. This allows a screening of both a large number of ligands and glycolipids. The multivalent presentation affords a binding also in case of low-affinity binding sites. The plastic treatment appears to be critical and probably induces a presentation of the amphipathic glycolipids with similarity to the natural cell membrane. The final coating with albumin avoids unspecific hydrophobic interaction, which may be one of the major drawbacks in traditional inhibition assays using solubilized substance in various micellar forms. We have added various improvements of this assay including avidity estimation from curves based on binding in microtiter wells [6]. An example of overlay assay is shown in Fig. 1 for *Escherichia coli* and recognition of Galα1→4Gal-containing glycolipids.

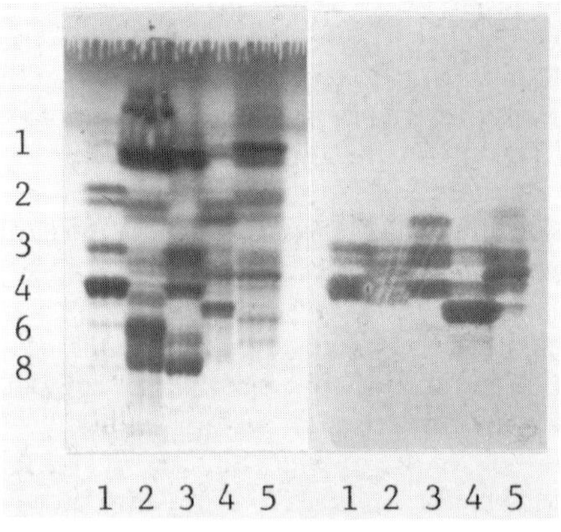

Figure 1. Thin layer chromatogram detected with anisaldehyde (left) and autoradiogram (right) after binding of ^{35}S-labelled uropathogenic *E. coli* (typical for strain J 96) to non-acid glycolipids of the following sources: human erythrocytes (1); human meconium (2); intestine from *Macaca cynomolgus* (3); dog small intestine (4); and rabbit small intestine (5). Numbers to the left indicate the approximate number of sugars. Autoradiography for 24 h.

ISOLATION AND STRUCTURAL ANALYSIS OF THE GLYCOLIPID RECEPTOR AND CONFORMATION ANALYSIS OF THE OLIGOSACCHARIDE

The overlay assay may be used to monitor the isolation to homogeneity of a particular receptor glycolipid, using various techniques for separation and structural analysis (see [7]) including high-technology mass spectrometry and N.M.R. spectroscopy (see [8]). Recent developments allow the assignment of preferred solution conformations of oligosaccharides by simple computer calculations (HSEA calculations), which are in excellent agreement with direct N.M.R. analysis (see [9]). An illustration of this will be given below, comparing two closely related interaction systems.

MICROBIOLOGICAL LIGANDS RECOGNIZE INTERNAL SEQUENCES OF OLIGOSACCHARIDE CHAINS

Our novel set of methods for receptor analysis has accumulated evidence that bacteria, bacterial toxins and viruses appear to have selected the common property of binding to internal parts of oligosaccharide chains. This is not the role for antibody-carbohydrate interactions, where immunodominant groups are most often terminal sequences which differ between individuals. The selection by microbiological ligands of internal core sequences may be to avoid such differences. Whether the character of the binding sites of the two types of actual proteins differ will be interesting to find out. Examples of internal bindings are shown in Table 1. Uropathogenic *E. coli* binds to Galα1→4Gal with about the same avidity regardless of the nature of the saccharide extension [10]. We have several other examples of this, including lactose-recognizing bacteria [11], Sendai virus [12], and Shiga toxin (Table 1). Also other toxins appear to bind internal sequences like tetanus toxin [2] and cholera toxin which are at present being analyzed in detail by us. Thus cholera toxin tolerates some but not other saccharide extensions of the best binder, the ganglioside GM1 [2].

Table 1. Results from binding of *E. coli* and the toxin from *Shigella dysenteriae* to various glycolipids

No.		E. coli	*Shiga toxin*
1	*Galα1→4Galβ*Cer	+	+
2	*Galα1→4Galβ*1→4GlcβCer	+	+
3	GalNAcβ1→3*Galα1→4Galβ*1→4GlcβCer	+	(+)
4	GalNAcβ1→3GalNAcβ1→3*Galα1→4Galβ*1→4GlcβCer	+	−
5	GalNAcα1→3GalNAcβ1→3*Galα1→4Galβ*1→4GlcβCer	+	−
6	Galβ1→3GalNAcβ1→3*Galα1→4Galβ*1→4GlcβCer	+	−
7	Fucα1→2Galβ1→3GalNAcβ1→3*Galα1→4Galβ*1→4GlcβCer	+	−
8	GalNAcα1→3(Fucα1→2)Galβ1→3GalNAcβ1→3*Galα1→* *4Galβ*1→4GlcβCer	+	−
9	NeuAcα2→3Galβ1→3GalNAcβ1→3*Galα1→4Galβ*1→4GlcβCer	+	−
10	NeuAcα2→6(NeuAcα2→3)Galβ1→3GalNAcβ1→3*Galα1→* *4Galβ*1→4GlcβCer	+	−
11	Galaα1→3*Galα1→4Galβ*1→4GlcβCer	+	+
12	$(\text{Gala}\alpha1\to3)_{2\text{-}5}$*Galα1→4Galβ*1→4GlcβCer	+	−
13	GalNAcβ1→3$(\text{Gala}\alpha1\to3)_{1\text{-}5}$*Galα1→4Galβ*1→4GlcβCer	+	−
14	Gal*α1→4Galβ*1→4GlcβCer	−	−
15	*Galα1→Galβ1*→4GlcNAcβ1→3Galβ1→4GlcβCer	+	+

THERE ARE VARIANTS WITH CLOSELY RELATED BINDING SPECIFICITY

The two examples of Table 1 will be discussed in greater detail. Apparently, uropathogenic *E. coli* and the Shiga toxin recognize the same disaccharide but with a slightly different epitope, causing some extensions on Galα1→4Gal to completely block toxin binding in contrast to *E. coli*. Similar variants have been shown in case of lactose and *Propionibacterium* [11], and for gangliosides and Sendai virus [12]. This is analogous to the recently shown variants of influenza virus [13], but in this case the position of binding of the receptor disaccharide differed, being NeuAcα2→3Gal and NeuAcα2→6Gal, respectively, corresponding to only one amino acid difference in the binding site of the haemagglutinin, Gln and Leu, respectively. Our variants should also be very similar in the binding sites of the lectin-like proteins.

ASSIGNMENT OF A PROBABLE DIFFERENCE IN BINDING EPITOPE FOR *E. COLI* AND THE SHIGA TOXIN, BOTH RECOGNIZING Galα1→4Gal

We have worked out the detailed specificity for the binding of *E. coli* [10] and the Shiga toxin [14] to natural glycolipids using the overlay technique (see also A. A. Lindberg, this volume). The glycolipids were isolated from a wide range of sources and several of them represent new oligosaccharide structures characterized during the course of this investigation. Both the bacterium and the toxin apparently have absolute specificity for Galα1→4Gal (Table 1), since no other naturally occurring disaccharide shows any activity, including Galα1→3Gal. As shown in the Table the bacterium binds to any glycolipid containing this disaccharide in terminal or internal position regardless of whether or not the extension contains a bulky disialo grouping as in glycolipid no. 10, or a blood group A determinant as in no. 8, and the binding in these cases is of about the same strength. The toxin on the other hand is completely blocked in its binding by most extensions. Galα1→3 is well tolerated (no. 11) but GalNAcβ1→3 (no. 3) partially interferes. Further additions to these two monosaccharides abolish the binding (nos 4–10, 12, 13). One conclusion from this difference between the two ligands may be that the toxin binding is shifted more to one end of the disaccharide compared to the bacterium, and that some substitutions of the disaccharide at Galα by steric hindrance may block toxin but not bacterium access to the respective binding epitope. The preferred conformations of these glycolipids as obtained by HSEA calculations support this reasoning [10]. The molecular model of Fig. 2 for glycolipid no. 5 of Table 1, able to bind the bacterium but not the toxin, will help to illustrate our findings.

The fact that there is a 90 degrees bend or knee in the oligosaccharide at the Galα1→4Gal, and that all glycolipids analysed bind the bacterium, together make likely that the binding epitope should reside on the accessible convex side of the bend. This exposes the lipophilic α side of Galβ (H-1, H-3, H-4 and H-5, marked in the figure) and H-1 and H-2 of Galα (also marked), providing a continuous

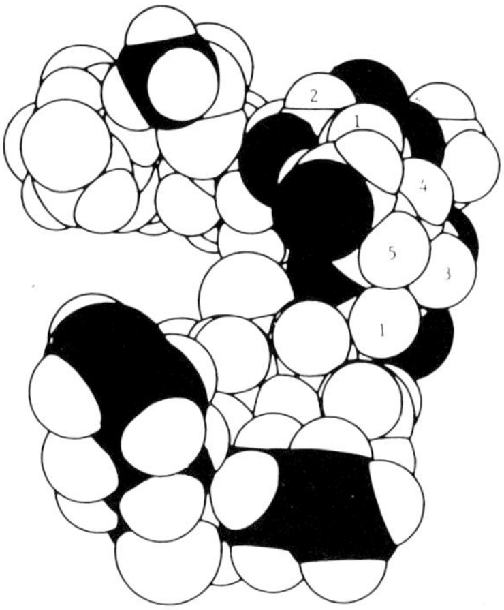

Figure 2. Molecular model of the HSEA-calculated conformation of Forssman glycolipid (no. 5 of Table 1). The model is projected to visualize the probable binding epitope for *E. coli* and Shiga toxin on the convex side of Galα1→4Gal (marked with numbered ring hydrogens). In the bottom the cut chains (black) of the fatty acid and base are seen projecting towards the viewer. For more details, see text.

hydrophobic surface of the disaccharide, of possible importance for the strength of binding to peptide (see [9,15]). At the borders of this hydrophobic surface there are several oxygens (shaded larger atoms) which may provide specificity in the binding through the hydrated polar-group gate according to Lemieux [16]. Concerning the difference between the bacterium and the toxin, the toxin binding is blocked in the molecule of Fig. 2 and binds only weakly to no. 3 of Table 1. This supports the assumption that the toxin binding epitope is shifted more to the top of Fig. 2, towards the Galα part, compared to the bacterium binding epitope. The acetamido substituent of GalNAcβ (partly shaded) may thus induce a steric hindrance in case of the toxin but not for the bacterium. Our results from quantitation of binding in microtitre wells support this, showing a 5-10 times weaker binding of the toxin than the bacterium to the disaccharide placed in terminal position (no. 15 of Table 1). This is expected if only part of the hydrophobic α side of Galβ is involved in toxin binding.

OUTLINE OF A GENERAL APPROACH FOR THE ASSIGNMENT OF DETAILED BINDING EPITOPES ON CARBOHYDRATE RECEPTORS

Our experience from more than three years of mainly methodological work on carbohydrate receptors for microbiological ligands, has afforded the following route for assignment of binding epitopes, of theoretical importance for revealing precise protein-carbohydrate interactions, and of practical importance when applying the knowledge in medicine.

(1) *Identification of receptors.* This is efficiently done by the overlay assay [4–6, 10–12, 14] supplemented by avidity estimation in microtiter wells [6,12]. This is at present practically limited to glycolipids, but these represent most known cell-bound oligosaccharides outside proteoglycans and correspond to peptide-bound oligosaccharide in 0-glycosidic linkage including also secreted mucins. We are at present extending the approach to free oligosaccharides and glycoproteins.

(2) *Determining the primary structure.* This is done by modern high-resolution isolation and analysis methods including mass spectrometry and N.M.R. spectroscopy.

(3) *Preliminary assignment of binding epitope within the receptor sequence.* As illustrated in the present paper the knowledge of binding preferences in case of natural glycolipids with internally placed binding sequences, in combination with knowledge of preferred conformations, makes it possible to encircle a reasonable epitope of the receptor saccharide. This would not have been possible with only terminally placed sequences, and this is a considerable advantage in relation to anti-carbohydrate antibodies, which differ in mostly recognizing terminal parts (see also Lark *et al.*, this volume). Although it may be hard work to prepare glycolipid isoreceptors from diverse sources [10], the importance of gaining this knowledge lies in that these natural sequences are part of the selection system for the ligands to optimize attachment.

(4) *Define assignment of binding epitope.* This requires organic synthesis of a large number of chemical variants of the receptor sequences. The starting alternatives may, however, be drastically reduced by the knowledge gained under (3). In the end, detailed direct analysis has to be done on peptide–oligosaccharide complexes, providing that the binding constants allow univalent complex formation [17].

ACKNOWLEDGEMENTS

We are indebted to Dr B.-E. Uhlin for providing the bacteria for Fig. 1. The work was supported by a grant from the Swedish Medical Research Council (no. 3967).

REFERENCES

1. Jones, G. W. and R. E. Isaacson (1983). *CRC Crit. Rev. Microbiol.* **10**, 229.
2. Eidels, L., R. L. Proia and D. A. Hart (1983). *Microbiol. Rev.* **47**, 596.
3. Dimmock, N. J. (1982). *J. Gen. Virol.* **59**, 1.
4. Hansson, G. C., K.-A. Karlsson, G. Larson, N. Strömberg, J. Thurin, C. Örvell and E. Norrby (1984). *FEBS Lett.* **170**, 15.
5. Hansson, G. C., K.-A. Karlsson, G. Larson, N. Strömberg and J. Thurin (1985). *Anal. Biochem.* **146**, 158.
6. Karlsson, K.-A. and N. Strömberg (1986). *Methods Enzymol.*, in press.
7. Hakomori, S.-I. (1983). *In* "Sphingolipid Biochemistry" (Eds J. Kanfer, S.-I. Hakomori), Handbook of Lipid Research, Vol. 3, p. 1. Plenum Press, New York.
8. Breimer, M. E., K.-E. Falk, G. C. Hansson and K.-A. Karlsson (1982). *J. Biol. Chem.* **257**, 50.
9. Sabesan, S., K. Bock and R. U. Lemieux (1984). *Can. J. Chem.* **62**, 1034.

10. Bock, K., M. E. Breimer, A. Brignole, G. C. Hansson, K.-A. Karlsson, G. Larson, H. Leffler, B. E. Samuelsson, N. Strömberg, C. Svanborg Edén and J. Thurin (1985). *J. Biol. Chem.* **260**, 8545.
11. Hansson, G. C., K.-A. Karlsson, G. Larson, A. A. Lindberg, N. Strömberg and J. Thurin (1983). *In* 'Glycoconjugates" (Eds A. Chester, D. Heinegård, A. Lundblad and S. Svensson), Proc. 7th Int. Symp. Glycoconj., p. 631. Rahms i Lund, Lund, Sweden.
12. Holgersson, J., K.-A. Karlsson, P. Karlsson, E. Norrby, C. Örvell and N. Strömberg (1985). *In* "World's Debt to Pasteur" (Eds H. Koprowski and S. Plotkin), p. 273. Alan Liss, New York.
13. Rogers, G. N., J. C. Paulson, R. S. Daniels, J. J. Skehel, I. A. Wilson and D. C. Wiley (1983). *Nature* **304**, 76.
14. J. E. Brown, K.-A. Karlsson, A. A. Lindberg, N. Strömberg and J. Thurin (1983). *In* "Glycoconjugates" (Eds A. Chester, D. Heinegård, A. Lundblad and S. Svensson), Proc. 7th Int. Symp. Glycoconj., p. 678. Rahms i. Lund, Lund, Sweden.
15. Lemieux, R. U., T. C. Wong, J. Liao and E. A. Kabat (1984). *Mol. Immunol* **21**, 751.
16. Lemieux, R. U. (1984). *In* "Proceedings of the 8th International Symposium on Medicinal Chemistry" Aug. 1984, Uppsala, Sweden. Swedish Pharmaceutical Society, in press.
17. Quiocho, F. A. and N. K. Vyas (1984). *Nature* **310**, 381.

Synthesis of Neo-Glycoconjugates

Göran Magnusson[a]

Organic Chemistry 2, Chemical Center,
The Lund Institute of Technology, Lund, Sweden

A growing appreciation of the role of glycoconjugates in biological receptor interactions (Fig. 1) has recently directed much interest into the chemistry and molecular biology of these compounds. A sound knowledge of the molecular mechanisms underlying these phenomena will undoubtedly lead to the development in the near future of new principles of diagnosis and therapy.

CARBOHYDRATE RECEPTORS

"Carbohydrate"	"Lectin"
Erythrocytes	Antibodies
Mammalian epithelial cells	Bacteria, viruses, toxins
Tumour antigens	NK-cells
Egg	Sperm
Sugar; initiating defense-response	Plant-cell; close to injured site

Figure 1.

Natural cell-surface glycoconjugates are divided into the groups glycolipids [1] and glycoproteins [2], each of which can be subdivided into different classes of compounds. Isolation of these natural products from biological sources is normally rather cumbersome, consequently the preparation of large amounts of

[a] *Telephone:* 46-46-108215

Protein-Carbohydrate Interactions in Biological Systems. ISBN 0 12 436665 1

pure substances in this fashion is impractical. Similarly, chemical synthesis of most of these complex molecules on a large scale is not feasible due to the inadequacy of present methodology. Total synthesis of complete glycoprotein molecules, where suitably protected amino acid glycosides should be used as starting materials, has not yet been realized. However, syntheses of large glycoprotein-related oligosaccharides have been reported recently [3].

For the total synthesis of glycolipids, other problems arise, mainly connected with the difficulty of preparing enantiomerically pure sphingosine aglycons (Fig. 2) and to use these as starting materials in glycoside syntheses. For example, the use of racemic sphingosine derivatives would lead to diastereomeric products, and only in rare cases would it be possible to separate these on a large scale.

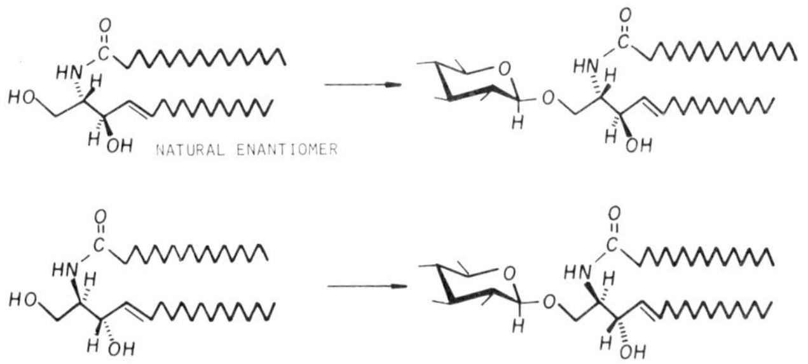

Figure 2.

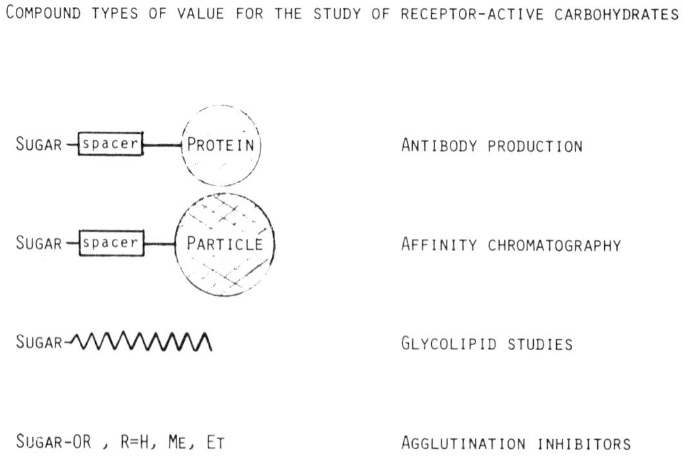

Figure 3.

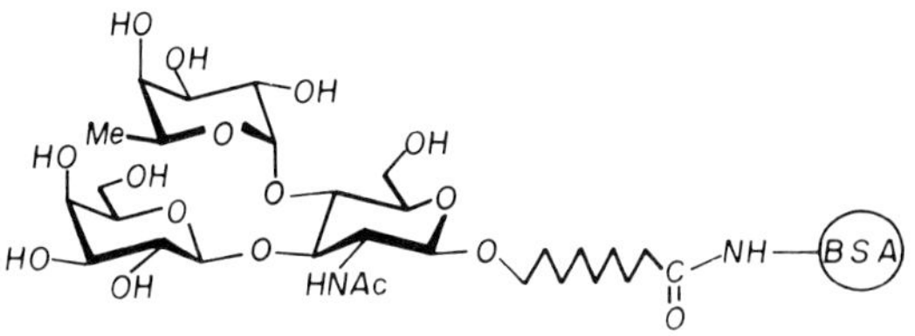

Figure 4

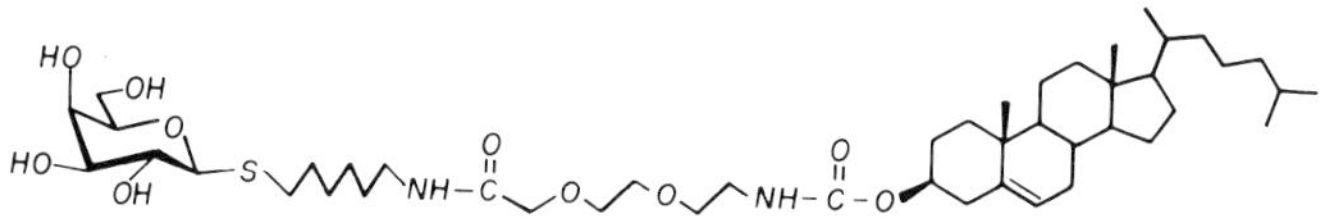

Figure 5

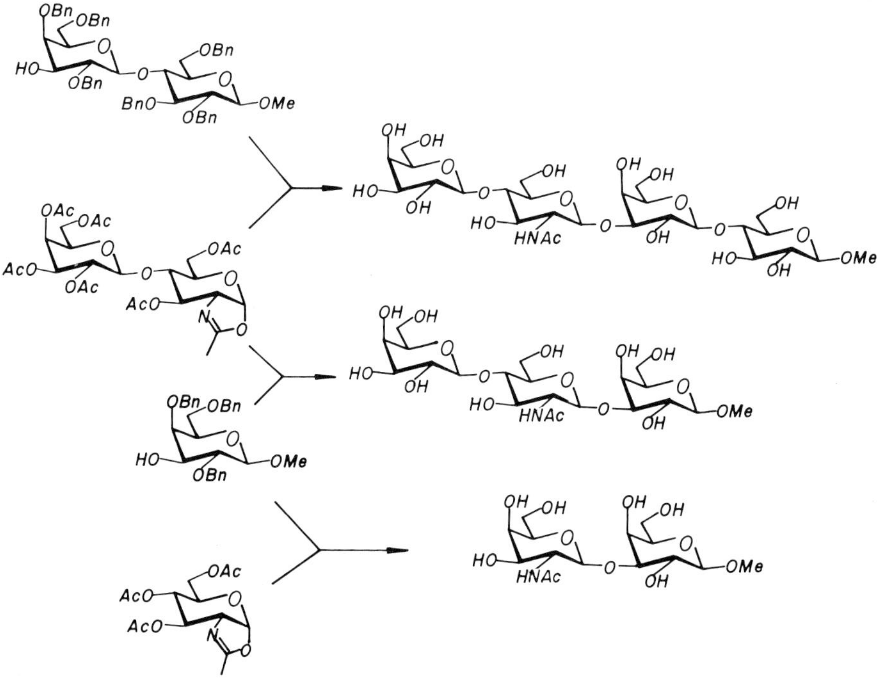

Figure 6

All this means that the only biologically interesting compounds that are potentially available in reasonably large quantities (for use in drug development for example) are the so-called neo-glycoconjugates (Fig. 3), compounds that can be constructed from relevant carbohydrates and model (fatty) alcohols, proteins and particles (such as latex and glass beads). Examples of neo-glycoproteins and -glycolipids are shown in Figs 4 and 5. (The former compounds were used as antigens to raise antibodies [4] and the latter were incorporated into liposomes, thus making these agglutinizable by lectins [5].) It should also be noted that glycolipids that carry glycoprotein-derived oligosaccharides and vice versa, can only be obtained in neo-glycoconjugate form. Furthermore, simple glycosides (such as methyl and ethyl) can be used as agglutination inhibitors. This is exemplified in Fig. 6 with three oligosaccharides [6] that were used to determine a receptor [7] for *Streptococcus pneumoniae*.

Recently, Lemieux [8], Kabat [9] and their coworkers have investigated optimized carbohydrate receptors for lectins and antibodies. These receptors had to be synthesised and the carbohydrate units had to be modified in many ways (for example specific removal of hydroxyl groups, exchange of these for fluorine, elongation of the carbon skeleton of the sugars) in order to pinpoint the binding epitope of the oligosaccharide. This means that synthetic carbohydrate chemistry has expanded into areas such as immunology and cell-surface biochemistry.

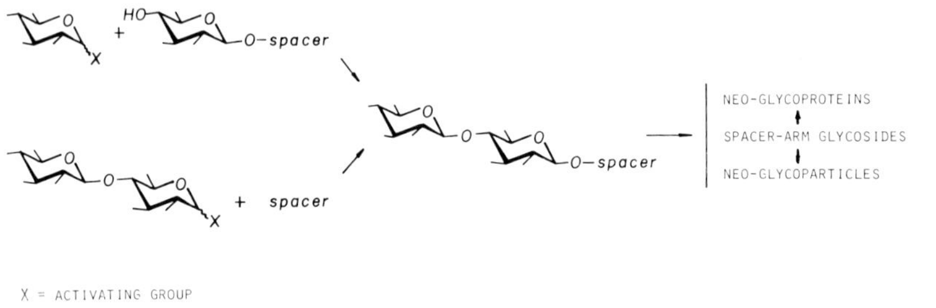

Figure 7.

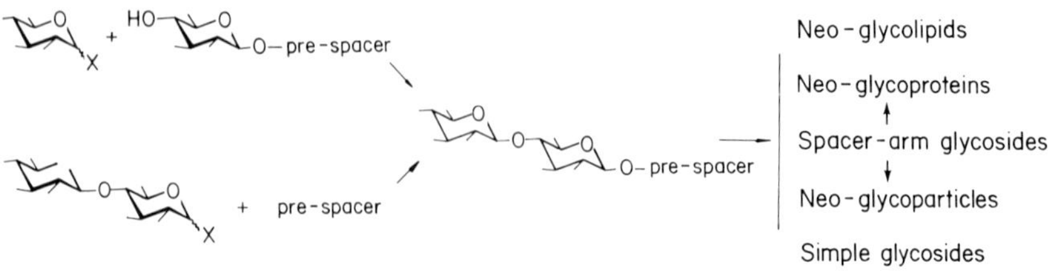

Figure 8.

The formation of glycosidic bonds by organic synthetic methods gives in most cases mixtures of α and β glycosides that have to be separated. When spacer-arm oligosaccharides are formed from spacer-arm *mono*saccharides and activated sugars, the inter-sugar glycosidic bond is responsible for the α/β mixture, whereas reaction of a spacer-arm alcohol with an activated *oligo*saccharide leads to the α/β mixture (Fig. 7; only one anomer is shown). In rare cases, rather complex oligosaccharides with anomerically pure inter-saccharide linkages can be obtained by controlled enzymatic hydrolysis of structurally regular, inexpensive polysaccharides. These oligosaccharides can then be transformed into well-defined neo-glycoconjugates (see for example Fig. 22).

It is of substantial interest to be able to synthesise glycoconjugates of different kinds where the carbohydrate portion is the same in all compounds. Since the separation of different stereoisomers is a very time-consuming part of most synthetic work, it would be of great value if a common type of intermediate could be found that would expedite the further preparation of all the desired glycoconjugate types. In other words, some pre-spacer alcohol should be used that is compatible with chemical methods of carbohydrate synthesis. It should be possible to use this over-all synthetic strategy so that a wide selection of final products could be obtained (Fig. 8).

A few years ago, we started to investigate this problem. We reasoned that pre-spacer glycosides having an aglycon with alkylating properties should be suitable provided that the pre-spacer would survive the various reaction conditions of

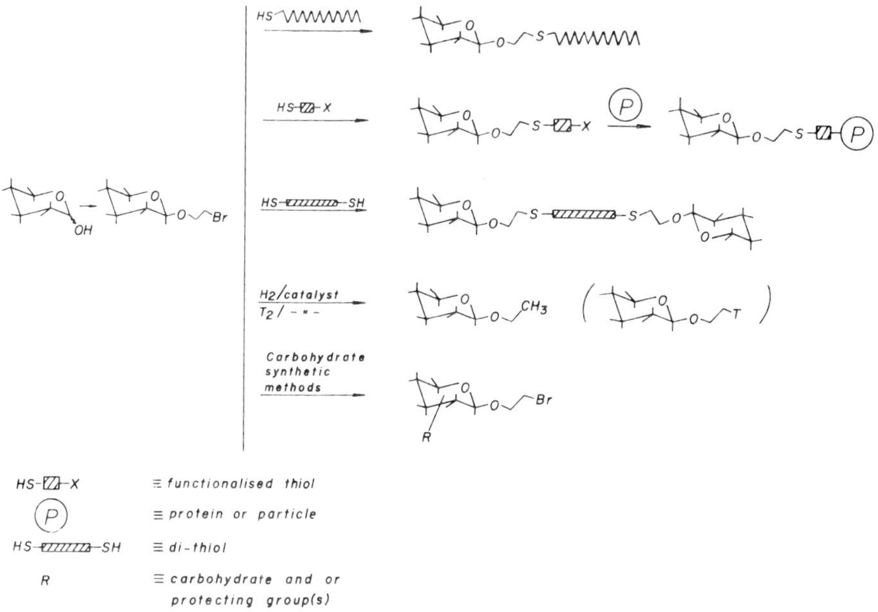

Figure 9.

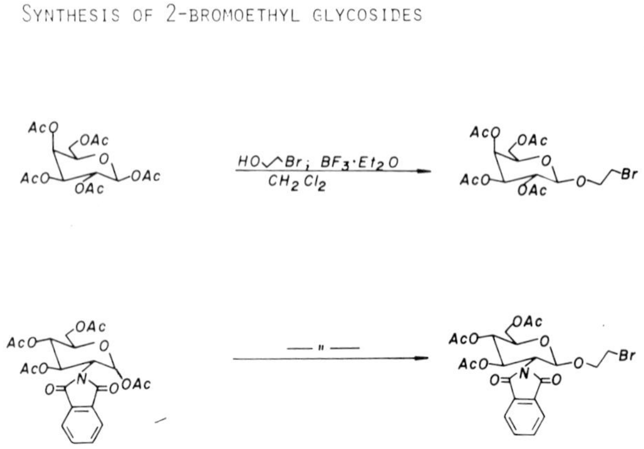

Figure 10.

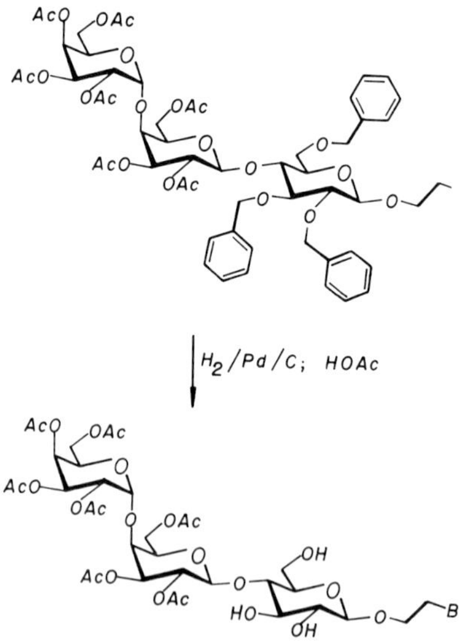

Figure 11.

typical carbohydrate synthesis. The pre-spacer glycoside had to permit the further preparation of simple glycosides, spacer-arm glycosides and neo-glycolipids and the aglycon of the simple glycosides (for use as inhibitors) should be small enough so as not to convey lipidic character to the final product.

We have found that 2-bromoethyl glycosides [10] (Fig. 9) fulfill all these criteria in a highly acceptable way. They can be synthesized directly from sugar acetates by boron trifluoride-induced glycosidation [11,12] as shown in Fig. 10, as well as by other methods. They are stable under many standard organic reaction conditions, easily reduced by hydrogenation to yield ethyl glycosides, and they can be transformed into spacer-arm and lipid glycosides by alkylation of different functionalized thiols. As a bonus, the sulphur atom of the spacer-arm is well suited for a combustion-analytical determination of the number of sugar haptens in the neo-glycoproteins [10]. The different uses of 2-bromoethyl glycosides is summarized schematically in Fig. 9.

Removal of benzyl protecting groups by hydrogenolysis under acidic conditions can be performed without the concomitant removal of the bromine atom of the 2-bromoethyl glycoside [13] (Fig. 11). Furthermore, formation of ethyl glycosides (suitable as agglutination inhibitors [14]) by hydrogenolysis under basic conditions can be done without removal of benzyl groups [15] (Fig. 12). This is important since hydrogenation reactions are very common in synthetic carbohydrate chemistry.

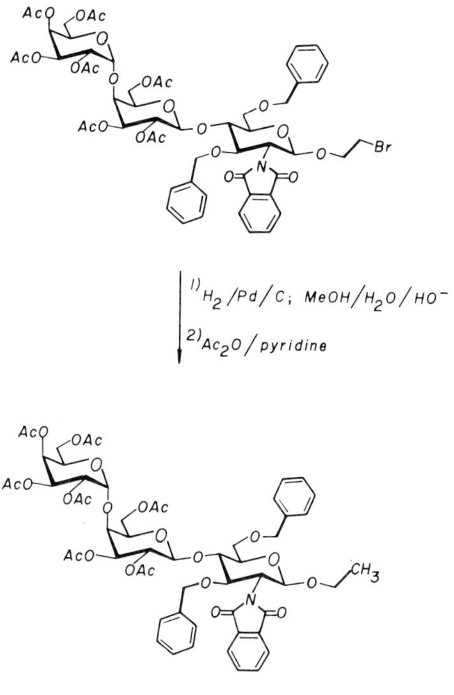

Figure 12.

The main practical draw-back of the 2-bromoethyl glycosides is that they are not fully stable under strongly basic reaction conditions. Intramolecular alkylation to give 1,4-dioxane derivatives [13] (Fig. 13) or β-elimination to give (unstable) vinyl glycosides are the important side reactions. Strongly basic conditions are encountered mainly in *O*-alkylation reactions (such as benzylation); here, phase-transfer methods can be of some use [13].

The problems above can of course be avoided if the formation of the 2-bromoethyl glycoside is performed after the different protection/de-protection and glycoside synthesis steps. However, this calls for some other glycoside synthesis steps. However, this calls for some other aglycon (to protect the reducing end during synthesis) that should be stable to all normally used reaction conditions, and then be selectively removed. Recently, Lipshutz *et al.* [16] reported the preparation of β-trimethylsilylethyl glycosides and selective removal of the trimethylsilylethyl group by lithium tetrafluoroborate. We have recently found [17] that this reaction can be catalized by trifluoroacetic anhydride and that addition of acetic anhydride permits the *selective* preparation of 1,2-*trans* acetates (*trans:cis* > 3:1) in a "one-pot" procedure. It is well known that 1,2-*trans* sugar acetates are much more efficient in lewis acid-induced glycoside synthesis that then corresponding 1,2-*cis* compounds. The reaction sequence where a trimethylsilylethyl glycoside is transformed into the corresponding 2-bromoethyl glycoside is shown in Fig. 14.

Thus armed with efficient methods for the preparation of novel neo-glycoconjugates, we embarked on a programme for the synthesis of several biologically important compounds, where each one was to become available as neo-glycolipid, ethyl glycoside, spacer-arm glycoside, and neo-glycoprotein. All these compounds (except the ethyl glycosides) were prepared via routes that all started with a nucleophilic substitution of bromine by a thiol, as exemplified in Fig. 15. It should be noted that the thiols attack only the 2-bromoethyl group despite the presence of numerous acetate groups. Figure 16 shows a few representative compounds [18] that were prepared from 2-bromoethyl glycosides.

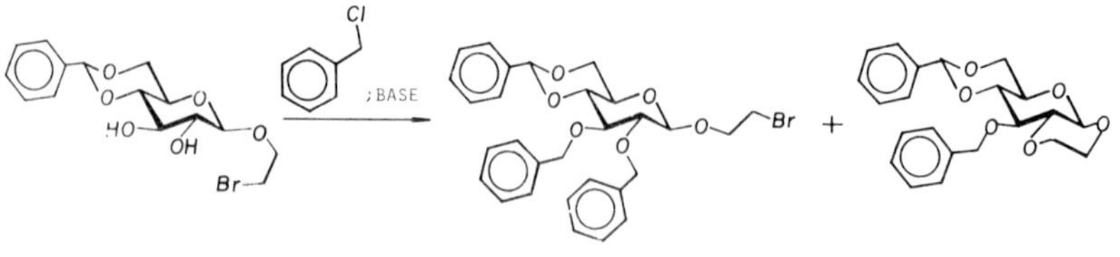

Figure 13.

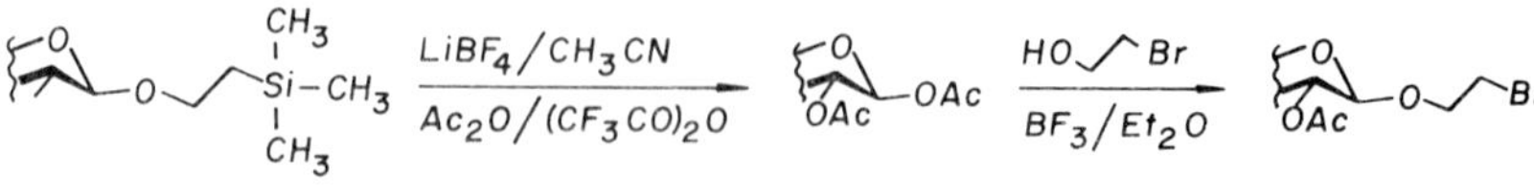

Figure 14.

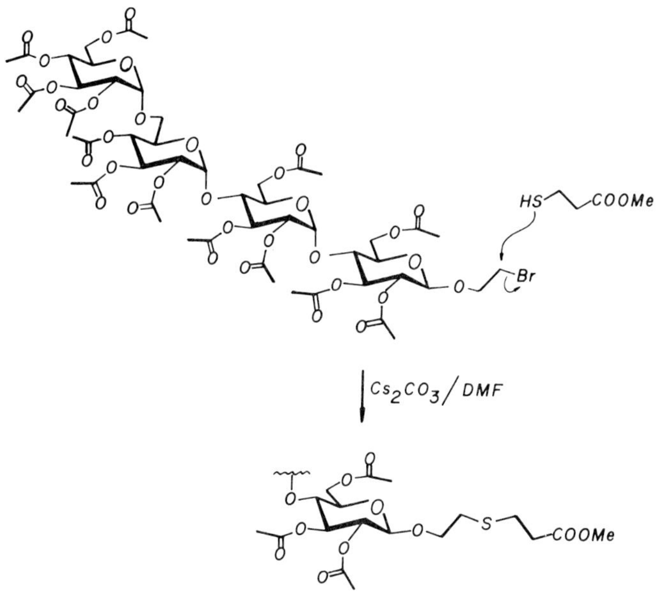

Figure 15.

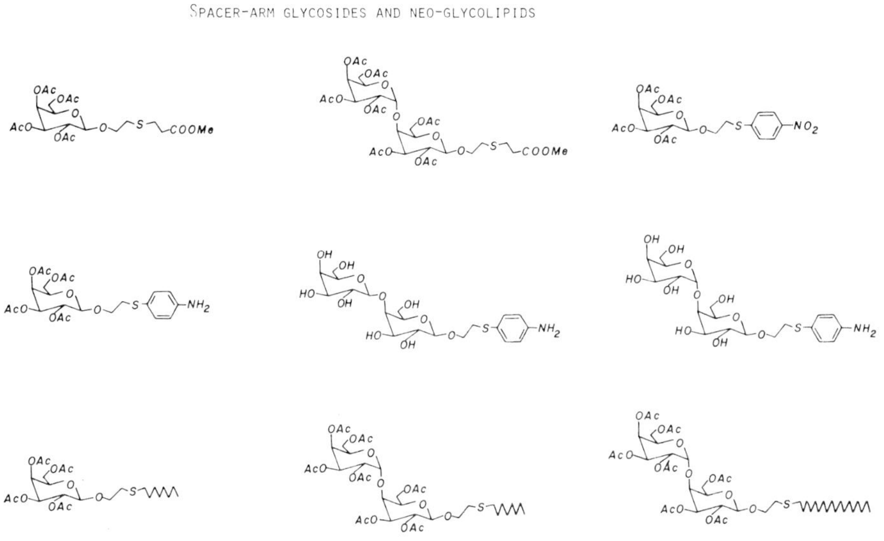

Figure 16.

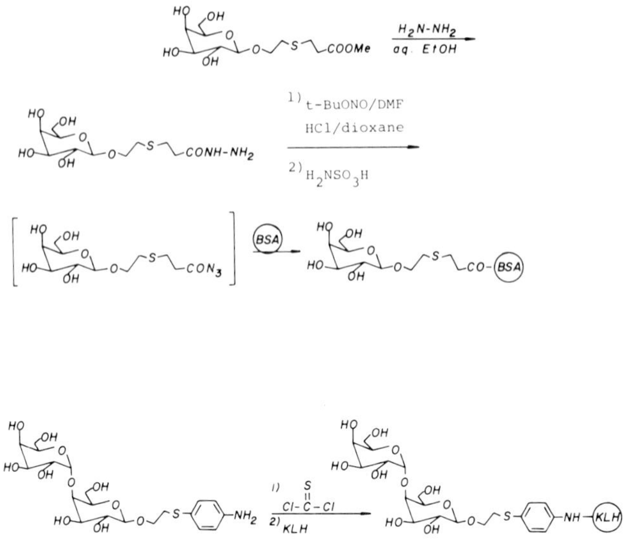

Figure 17.

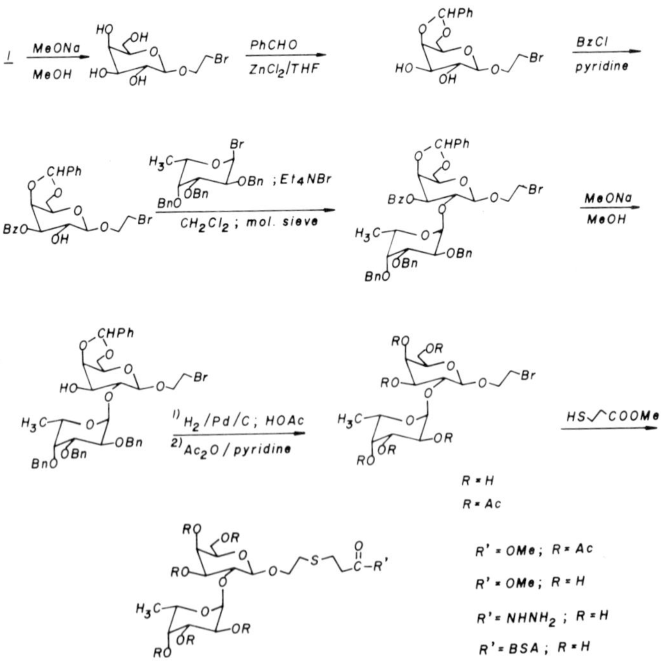

Figure 18.

The spacer-arm glycosides that carry ester or amine functionalities were then coupled to bovine serum albumin (BSA) and key-hole limpet haemocyanine (KLH) [19] (Fig. 17).

BIOCHEMICAL APPLICATIONS OF 2-BROMOETHYL GLYCOSIDES

The blood-group H disaccharide (Fucα1-2Gal) and *N*-acetyllactosamine [19] (an important constituent of natural glycoproteins) were synthesized in the form of 2-bromoethyl glycosides and transformed into neo-glycoproteins (Fig. 18).

Uropathogenic *Escherichia coli* bacteria adhere with Galα1-4Gal specificity to glycolipids of the human urinary tract [20,21]. We have prepared Galα1-4Gal (galabiose) on a fairly large scale by controlled enzymatic hydrolysis of polygalacturonic acid (easily prepared from pectin) followed by chemical transformation of digalacturonic acid [22] (Fig. 19). Several Galα1-4Gal derivatives (Figs 16 and 17) were then prepared via the 2-bromoethyl glycoside [18]. Acetobromogalabiose and the tetradeuterated analogue were used for the preparation [13,15] of the P^k- and P_1-antigens in neo-glycoconjugate form (Figs 20 and 21). Here, preformed and suitably protected 2-bromoethyl glycosides were used as aglycons in silver triflate promoted glycoside syntheses. The final products have been used to raise antibodies and as receptor analogues for the study of *E. coli* binding.

The sugar part of the glucose tetrasaccharide derivative shown in Fig. 22 is normally excreted in human urine [23]. Increased excretion is observed in patients having glycogenosis, in Duchenne muscular dystrophy and during normal

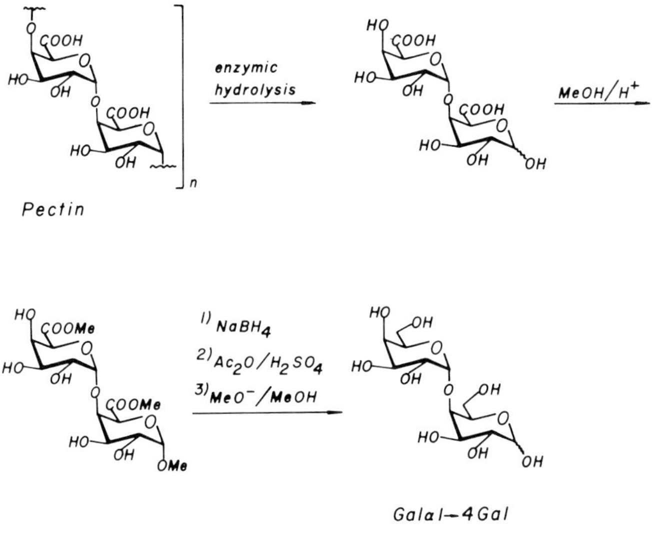

Figure 19.

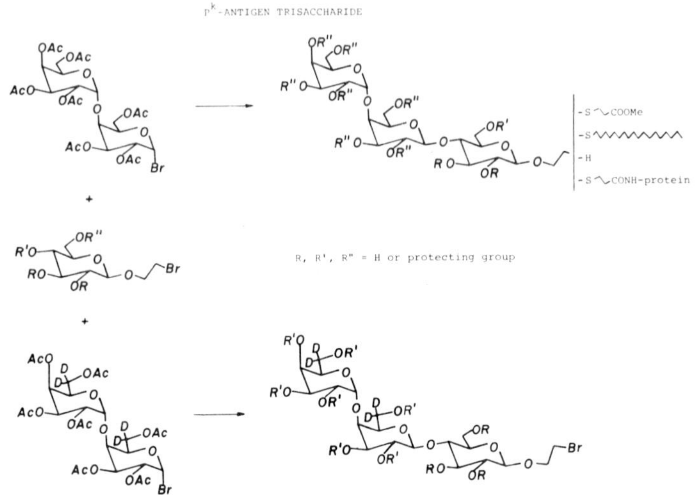

Figure 20.

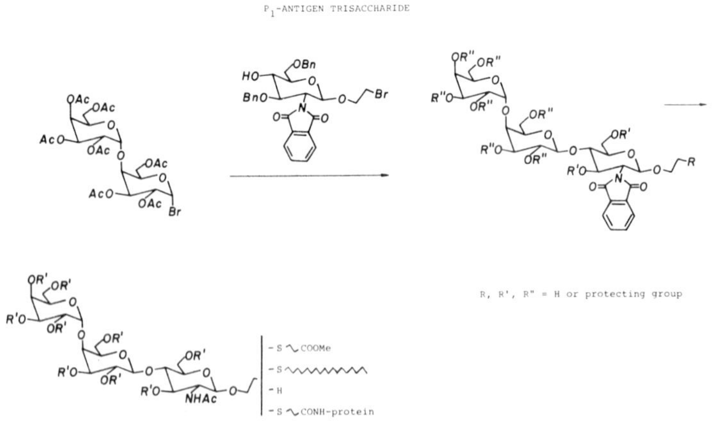

Figure 21.

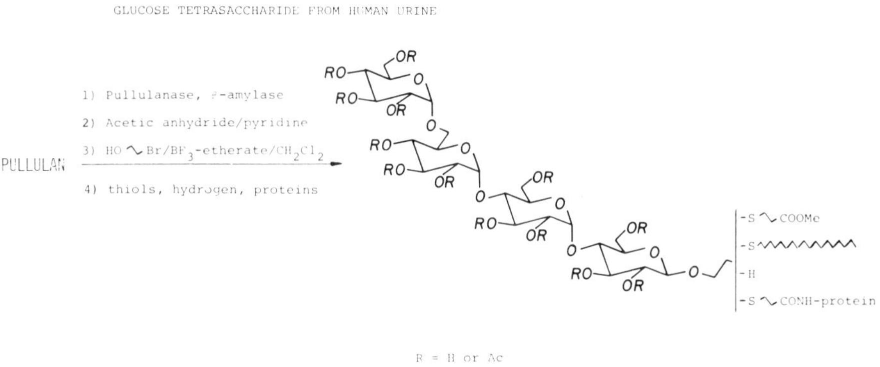

Figure 22.

pregnancy [24]. We have prepared the tetrasaccharide by controlled enzymatic hydrolysis of pullulan, followed by acetylation of the crude reaction mixture and isolation of the fully acetylated material [25]. Boron trifluoride-induced glycosidation with 2-bromoethanol gave the 2-bromoethyl glycoside, which was then treated in the usual way to give the different compounds shown in Fig. 22. The neo-glycoprotein was used to raise antibodies against the tetrasaccharide. Such antibodies are valuable for rapid detection of the compound in urine samples [26].

In summary, 2-bromoethyl glycosides have been shown to be very versatile intermediates for the simple preparation of biologically interesting neo-glycoconjugates. However, the aglycon part does not have much structural resemblance with that of natural compounds, which can be a draw-back in certain biological applications. The preparation of neo-glycoconjugates that are mimics of natural products on the molecular level is currently being investigated in our laboratory.

ACKNOWLEDGEMENTS

I am grateful to my former coworkers at the Research Laboratory of the Swedish Sugar Company for their devoted work. Part of the recent work presented here was supported by the Swedish Natural Science Research Council and the Swedish Board for Technical Development.

REFERENCES

1. Hakomori, S. (1981). *Ann. Rev. Biochem.* **50**, 733.
2. Montreuil, J. (1980). *Adv. Carbohydr. Chem. Biochem.* **37**, 157.
3. Lönn, H. and J. Lönngren (1983). *Carbohydr. Res.* **120**, 17.
4. Lemieux, R. U., D. R. Bundle and D. A. Baker (1975). *J. Am. Chem. Soc.* **97**, 4076.
5. Slama, J. S. and R. R. Rando (1980). *Biochemistry* **19**, 4595.

6. Dahmen, J., G. Gnosspelius, A.-C. Larsson, T. Lave, G. Noori, K. Palsson, T. Frejd and G. Magnusson (1985). *Carbohydr. Res.* **138**, 17.
7. Andersson, B., J. Dahmen, T. Frejd, H. Leffler, G. Magnusson, G. Norri and C. Svanborg Edén (1983). *J. Exp. Med.* **158**, 559.
8. Lemieux, R. U. (1981). Front. Chem. Plenary Keynote Lecture, 28th IUPAC Congress (Ed. K. J. Laidler), Vol. 3. Pergamon Press, Oxford.
9. Kabat, E. (1980). *Methods Enzymol.* **70**, 3.
10. Dahmen, J., T. Frejd, G. Magnusson and G. Noori (1982). *Carbohydr. Res.* **111**, c1.
11. Dahmen, J., T. Frejd, G. Grönberg, T. Lave, G. Magnusson and G. Noori (1983). *Carbohydr. Res.* **116**, 303.
12. Dahmen, J., T. Frejd, G. Magnusson and G. Noori (1983). *Carbohydr. Res.* **114**, 328.
13. Dahmen, J., T. Frejd, G. Magnusson, G. Noori and A.-S. Carlström (1984). *Carbohydr. Res.* **127**, 15.
14. Bock, K., M. E. Breimer, A. Brignole, G. C. Hansson, K.-A. Karlsson, G. Larsson, H. Leffler, B. E. Samuelsson, N. Strömberg, C. Svanborg Edèn and J. Thurin (1985). *J. Biol. Chem.* **260**, 8545.
15. Dahmen, J., T. Frejd, G. Magnusson, G. Noori and A.-S. Carlström (1984). *Carbohydr. Res.* **129**, 63.
16. Lipshutz, B. H., J. J. Pegram and M. C. Morey (1981). *Tetrhedr. Lett.* **22**, 4603.
17. Jansson, K. and G. Magnusson, unpublished results.
18. Dahmen, J., T. Frejd, G. Grönberg, T. Lave, G. Magnusson and G. Noori (1983). *Carbohydr. Res.* **118**, 292.
19. Dahmen, J., T. Frejd, G. Magnusson, G. Noori and A.-S. Carlström (1984). *Carbohydr. Res.* **125**, 237.
20. Källenius, G., R. Möllby, S. B. Svenson, J. Winberg, A. Lundblad, S. Svensson and B. Cedergren (1980). *FEMS Lett.* **7**, 297.
21. Leffler, H. and C. Svanborg Edén (1980). *FEMS Lett.* **8**, 127.
22. Dahmen, J., T. Frejd, T. Lave, F. Lindh, G. Magnusson, G. Noori and K. Palsson (1983). *Carbohydr. Res.* **113**, 219.
23. Hallgren, P., G. Hansson, K. G. Henriksson, A. Häger, A. Lundblad and S. Svensson (1974). *Eur. J. Clin. Invest.* **4**, 429.
24. Lundblad, A., S. Svensson, I. Yamashina and M. Ohta (1979). *FEBS Lett.* **97**, 249; and references cited therein.
25. Dahmen, J., T. Frejd, G. Magnusson, G. Noori and A.-S. Carlström (1984). *Carbohydr. Res.* **127**, 27.
26. Zopf, D. A., R. E. Levinsson and A. Lundblad (1982). *J. Immunol. Methods* **48**, 109.

Poster Session

Host-Parasite Interactions Underlying Non-Secretion of Blood Group Antigens and Susceptibility to Recurrent Urinary Tract Infections

C. C. Blackwell[a], S. J. May,
R. P. Brettle*, C. J. MacCallum and D. M. Weir

*Department of Bacteriology, The Medical School, University of Edinburgh, Edinburgh, Scotland, and *Pyelonephritis Clinic and Division of Infectious Diseases, City Hospital, Edinburgh, Scotland*

In 1982 we reported an increased incidence of patients who are blood group B and/or non-secretors of blood group antigens among a group of women under investigation for urinary tract infection (UTI). The present study examined two populations of women. Group I consisted of 671 women referred for specialist investigation of recurrent UTI, and group II consisted of 135 women from whom a positive urine culture had been obtained but who did not have recurrent infections. There was a significant increase in the incidence of non-secretors among the women with recurrent infections (33.2%, $\chi^2 = 4.21$, $p < 0.05$), but this was not found for group II. From their records, 138 women in group I who had been monitored for 20 years were diagnosed as having pyelonephritis or cystitis. Pyelonephritis was defined as radiological evidence of kidney scarring and cystitis as no evidence of scarring but recurrent symptoms and positive urine cultures. Those who were judged to have improved had fewer infections during the second decade and those who had not improved had the same number or more infections during the second decade.

Hypothesis I. In vitro the MR adhesins CFA I and CFA I associated with enteropathogenic *Escherichia coli* can be inhibited by oligosaccharides in human

[a]*Telephone:* 41-31-667-1011 ext. 2330

Protein-Carbohydrate Interactions in Biological Systems. ISBN 0 12 436665 1

milk [1]. If a similar set of interactions occurs *in vivo*, the secreted blood group antigens might inhibit attachment of the MR adhesins associated with uropathogenic strains of *E. coli.* We predicted that there would be a higher proportion of strains with MR adhesins isolated from the urine of non-secretors. Of the 229 strains examined, 107 were isolated from women in group I and 122 from group II. There was no difference in the proportion of strains with MR adhesins obtained from secretors or non-secretors of group I and there was a lower proportion of those with MR adhesins found among those obtained from group II. In both groups the proportion of strains with no detectable haemagglutinins (MS or MR) was higher for non-secretors. These results suggest that the MS and MR adhesins might be of value to the bacteria for colonization of the gut if the host is a secretor.

Hypothesis II. If IgA plays a role in protection of the urinary tract from infection, the lower levels of serum and secretory IgA reported for non-secretors might be significant [2]. In contrast to findings reported by Grundbacher [2], the geometric means for IgA were higher for non-secretors than for secretors, as were their IgG levels. The immunoglobulin levels for the two diagnostic categories were surprising. The IgA levels were the same for patients with pyelonephritis who had improved and those who had not; the IgG levels for those who had not improved were very low. There was no difference in the geometric means of the immunoglobulin levels for patients with cystitis, regardless of improvement.

These findings suggest that both innate and specific host defences are involved in the response to urinary pathogens. Most scarring occurs before the age of 4 when Iga levels are about 32% that of the adult. There is a significantly higher incidence of non-secretors among our patients in whom symptoms began at an early age (14–24, $p=0.0009$). A self-immunization process has been suggested to explain the observation that most patients improve regardless of age or therapy. The higher immunoglobulin levels of non-secretors may reflect an increased dependence on the specific immune response.

REFERENCES

1. Holmgren, J., A.-M. Svennerholm and C. Ahren (1981). Non-immunoglobulin fraction of human milk inhibits bacterial adhesion (haemagglutination) and enterotoxin binding of *Escherichia coli* and *Vibrio cholerae. Infect. Immun.* **33**, 136–141.
2. Grundbacher, F. J. (1972). Immunoglobulins, secretor status and the incidence of rheumatic fever and rheumatic heart disease. *Hum. Hered.* **25**, 399–404.

Host-Parasite Interactions Underlying Non-Secretion of Blood Group Antigens and Susceptibility to Infections by *Candida albicans*

C. C. Blackwell[a], S. M. Thom, D. M. Weir, D. F. Kinane* and F. D. Johnstone‡

Department of Bacteriology and ‡Department of Obstretrics and Gynaecology, The Medical School, University of Edinburgh, Scotland and
**Department of Peridontology, The Dental School, University of Dundee, Scotland*

There was evidence that suggested non-secretion of blood group antigens might be a predisposing factor for infections due to *Candida albicans.* Salivary glycoproteins with blood group activity can inhibit or even reverse binding of *Streptococcus salivarius* to buccal epithelial cells [1]. The binding of *C. albicans* to human epithelial cells can be inhibited by fucose [2], the immunodominant sugar of blood group O which is found in the body fluids of all secretors. Our hypothesis was that if there are adhesins on the yeast that recognize carbohydrates with blood group activity on epithelial cells, interactions similar to those observed for *S. salivarius* might occur on the surfaces of secretors but not on those of non-secretors. We predicted that there would be a higher incidence of non-secretors among patients with candida infection and that the boiled saliva of secretors, but not that of non-secretors, would inhibit binding of the yeast to buccal cells *in vitro.*

The local incidence of non-secretors is 26.6%, but we found a significant increase of this figure among pregnant women with candida vaginitis (82.3%) and elderly dental patients of both sexes with oral candida infections (62.9%). As predicted, the saliva samples of secretors of blood groups A, B and O were able to inhibit binding of candida strains associated with oral infections serotypes A (3091) and

[a]*Telephone:* 44-31-667-1011 ext. 2330

Protein-Carbohydrate Interactions in Biological Systems. ISBN 0 12 436665 1

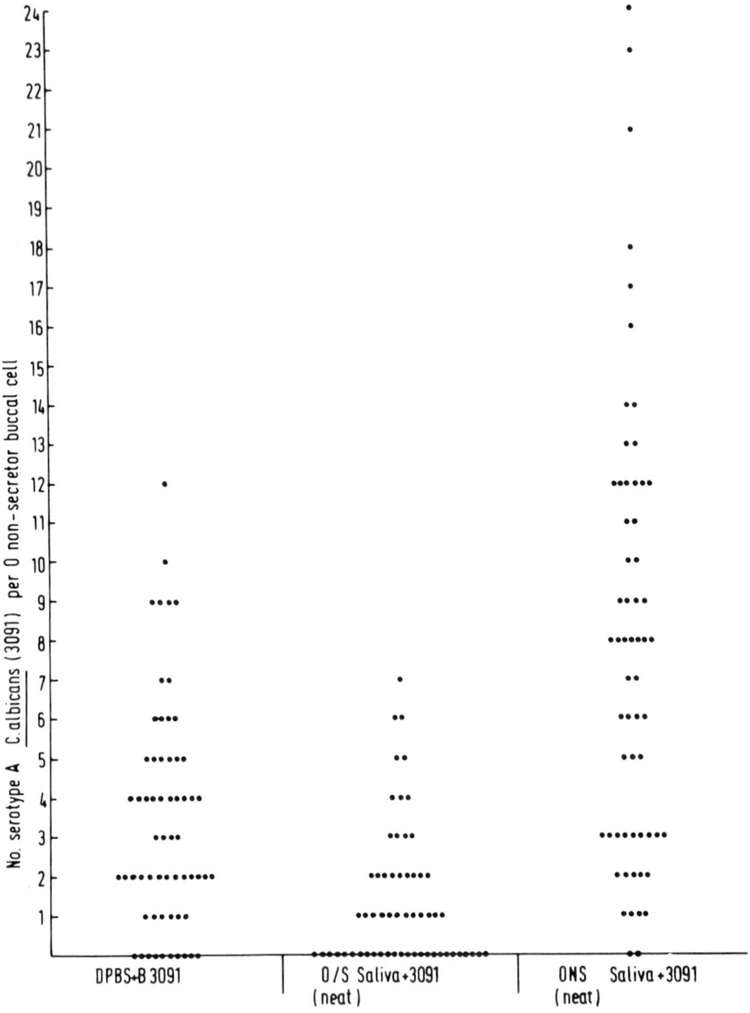

Figure 1. Effect of pre-incubation of serotype A *Candida albicans* with Dulbecco's phosphate buffered saline supplemented with Ca^{2+} and Mg^{2+} (DPBS+B), saliva from a blood group O secretor (O/S) anu ı blood group O non-secretor (ONS) on bindng of the yeast to buccal cells from an O non-secretor.

B (3118C) to buccal cells. The saliva of non-secretors did not inhibit binding but often enhanced it. At dilutions of secretor saliva in which there was no detectable blood group antigen, the inhibitory effect was not present.

These results suggest that secreted blood group antigens play a role in the innate defences of the mucosal surfaces, an effect seen particularly in patients who are immunocompromised—pregnancy or old age. We would therefore predict an increased incidence of non-secretors among patients undergoing immunosuppression who develop superficial candida infections. The findings also suggest an approach to isolation of microbial antigens involved in colonization. If epidemiological data can be confirmed by inhibition of binding, as in these studies, the carbohydrates involved might be utilized to isolate the adhesin by affinity chromotography.

REFERENCES

1. Williams, R. C. and R. J. Gibbons (1975). Inhibition of streptococcal attachment to receptors on human buccal epithelial cells by antigenically similar salivary glycoproteins. *Infect. Immun.* **11**, 711-718.
2. Sobel, J. D., P. G. Myers, D. Kaye and M. E. Lawson (1982). Adherence of *Candida albicans* to human vaginal and buccal epithelial cells. *J. Infect. Dis.* **143**, 76-82.

Influence of Secretor Status on the Availability of Receptors for Attaching *Escherichia coli* on Human Uroepithelial Cells

H. Lomberg[a], H. Leffler and Catharina Svanborg-Edén

Department of Clinical Immunology, University of Göteborg, Göteborg, Sweden

INTRODUCTION

Uroepithelial cells express a variety of complex glycoconjugates. Some of these, e.g. the globo series of glycolipids, containing the disaccharide Galα1→4Galβ, have been shown to act as receptors for the majority of attaching *Escherichia coli*. Individual variation in the composition and amount of receptor-active glycoconjugates may determine the extent of bacterial attachment, and thus affect the susceptibility to infection. For example, individuals of blood group P_1 synthesize the P_1, P and p^k antigens; P_2 individuals lack the P_1 antigen and $\bar{p}$ individuals do not produce detectable amounts of these antigens. The P_1 blood group is overrepresented in patients with recurrent pyelonephritis without reflux [1].

In this study bacteria with known specificity for the globo series glycolipids or other receptors were used to characterize the distribution of receptors at different levels of the urinary tract, variation in receptors related to P blood group, and variation in receptors related to other blood group determinants.

MATERIALS AND METHODS

To determine *E. coli* adhesins specificity and adherence the following tests were performed:

[a]*Telephone:* 46-31-602082

Protein-Carbohydrate Interactions in Biological Systems. ISBN 0 12 436665 1

Haemagglutination. A crude classification of the bacterial adhesins was obtained by haemagglutination in the presence or absence of 0.1 mM α-methyl-mannoside. Human erythrocyte of blood group P_1, P_2, $\bar{p}$ and guinea pig erythrocytes were used.

Receptor-coated erythrocytes. Guinea pig erythrocytes were coated with globotetraosylceramide, neolactotetraosylceramide or sialylneolactotetraosylceramide.

Latex beads. The receptor specificity of the strains was defined by agglutination of synthetic Galα1→4Galβ covalently linked via a spacer arm to BSA-latex beads. Synthetic Galα1→4Glc covalently linked to the latex beads were used as controls.

Adherence. Bacterial binding to uroepithelial cells from donors of varying blood group. Adherence to 20 squamous and 20 transitional epithelial cells was registered separately.

RESULTS

Characterization of the adhesins

The receptor specificity of the adhesins was defined as shown in Table 1. MR:GS = adhesins specific for Galα1→4Galβ. MR: non-GS = strains with MR adhesins not recognize Galα1→4Gal ~ BSA latex and agglutinating human erythrocytes independent of P blood group. Strains with both specificities were designated MR:GS + MR:non-GS. Non-MR strains were used as controls. The receptor specificity of the MR:non-GS strains was analysed further using glycolipid-coated erythrocytes. No positive reactions were obtained using globotetraosylceramide, neolacto- and sialylneolactotetraosylceramide. The receptor specificities of these strains remain undefined.

Target cell specificity

The MR:GS strains attached avidly both to squamous and transitional epithelial cells. In contrast, the MR:non-GS strains preferentially bound the squamous epithelial cells.

Table 1. Definition of the receptor specificity *E. coli* adhesins

		Specificity testing			
		Synthetic oligosaccharides		*Human erythrocytes*	
Designation	*Receptor specificity*	*Galα1→4Galβ ~ Latex*	*Galα1→4Glc ~ Latex*	P_1	$\bar{p}$
MR:GS	Galα1→ 4Galβ	+ + +	−	+ + +	−
MR:non-GS	several?	−	−	+ + +	+ +
MR:GS + MR:non-GS		+ + +	−	+ + +	+ +
non-MR	−	−	−	−	

Adhesion in relation to P blood group of cell donor

The requirement of P blood group determinants for the adherence of MR:GS strains was shown by the lack of attachment to cells from $\bar{p}$ individuals (Table 2). Contrary to what might have been expected, no difference in receptor density could be registered between cells from P_1 and P_2 individuals (Table 3). The binding of MR:non-GS strains was independent of P blood group. The strain with both MR:GS and MR:non-GS adhesins bound like an MR:non-GS strain to $\bar{p}$ cells and in an additive manner to the P_1 cells (Table 2).

ABH blood group

Differences in the A, B and H blood group determinants did not consistently affect adhesion. Variation between cell donors thus appeared independent of ABH and P_1-P_2 blood group (Table 3).

Table 2. Influence of P blood group on the attachment to human uroepithelial cells

Adhesion designation	No. of strains[a]	*Bacterial adhesion (bacteria/cell)* P_1 Squamous	P_1 Transitional	$\bar{p}$ Squamous	$\bar{p}$ Transitional
MR:GS	5	76	87	2	0
MR: non-GS	5	73	11	77	15
MR:GS + MR non-GS	1	144	105	60	13
non-MR	5	0	0	0	0

[a] Mean of ≥2 experiments per strain and cell donor.

Table 3. Pairwise comparison of *E. coli* attachment to uroepithelial cells from donors at different blood groups

Blood group of cell donor	No. of paired tests	*Mean pairwise differences in adherence (bacteria/cell)* MR:GS Squamous	MR:GS Transitional	MR:non-GS Squamous	MR:non-GS Transitional
P_1-P_2	29	12	4	−15	−14
A_2-B	19	8	7	−21	−5
Secretor-non-secretor	23	−54[a]	8	−37[a]	1

[a] $P<0.001$.

Secretor status

The receptivity for attaching bacteria was significantly higher of cells from non-secretors compared to secretors ($p<0.001$) (Table 3).

CONCLUSIONS

(1) Bacteria with adhesins of defined receptor specificity can be used as probes to analyse the distribution of receptors in the tissues.

(2) The distribution of the globo series glycolipids receptors was; similar on squamous and transitional epithelial cells; dependent on P blood group. Individuals of blood group $\bar{p}$ had no receptor activity; independent of P_1 or P_2 phenotype.

(3) The distribution of receptors for MR:non-GS adhesins differed from that of the globo series of glycolipids. They were found mainly on squamous epithelial cells; they appeared independent and additive to the globo series glycolipid receptors.

(4) The secretor status influenced the availability of receptors.

(5) Individual variation in receptors may affect the susceptibility to infection in the urinary tract.

ACKNOWLEDGEMENTS

This study was supported by grants from the Swedish Medical Research Council (No. 215 and 3967), the Medical Faculty, University of Göteborg, Sweden, the Ellen, Walter and Lennart Hesselman Foundation for Scientific Research and the BACH project Division, SSA, Kabi Sweden.

REFERENCE

1. Lomberg, H., L. Å. Hanson, B. Jacobsson, U. Jodal, H. Leffler and C. Svanborg Edén (1983). Correlation of P blood group vesicoureteral reflux and bacterial attachment in patients with recurrent pyelonephritis. *New Engl. J. Med.* **308**, 1189–1192.

Diagnostic Kits for Typing p-Fimbriated *Escherichia coli*

Bertil Nilsson, Anne-Sofie Carlström and Catharina Svanborg-Edén*

*Swedish Sugar Company, R&D, Box 6, S-232 00 Arlöv, Sweden, and *Department of Clinical Immunology, University of Göteborg, S-413 46 Göteborg, Sweden*

Various diagnostic kits, based on particle agglutination, have been developed and tested for typing p-fimbriated *Escherichia coli*. It is known that p-fimbriated *E. coli* interacts *in vivo* with the urinary epithelial glycolipid globoside[1]. The host binding site for the bacteria has been identified to the disaccharide Galα1→4Galβ[2]. Latex bead kits composed of this structural unit (and homologues) have been synthesized according to the scheme below. Different parameters such as oligosaccharide concentration, latex bead diameter, spacer arm length, pH and ion strength have been varied.

The sensitivity of the kits for different strains (manipulated strains or strains isolated from patients with pyelonephritis) has been tested. The results of this optimization show that covalently bonded Galα1→4Galβ-O-BSA-latex was comparable in detection response to the coated BSA-latex kit. The specificity for both these kits were excellent when tested on strains with known binding specificity.

REFERENCES

1. Leffler, H. and C. Svanborg Edén (1980). *FEMS Microbiol. Lett.* **8**, 127.
2. Källenius, G., S. B. Svensson, R. Möllby, B. Cedergren, H. Hultberg and J. Winberg (1981). *Lancet* **ii**, 604.

[a]*Telephone:* 46-40-431530

Protein-Carbohydrate Interactions in Biological Systems. ISBN 0 12 436665 1

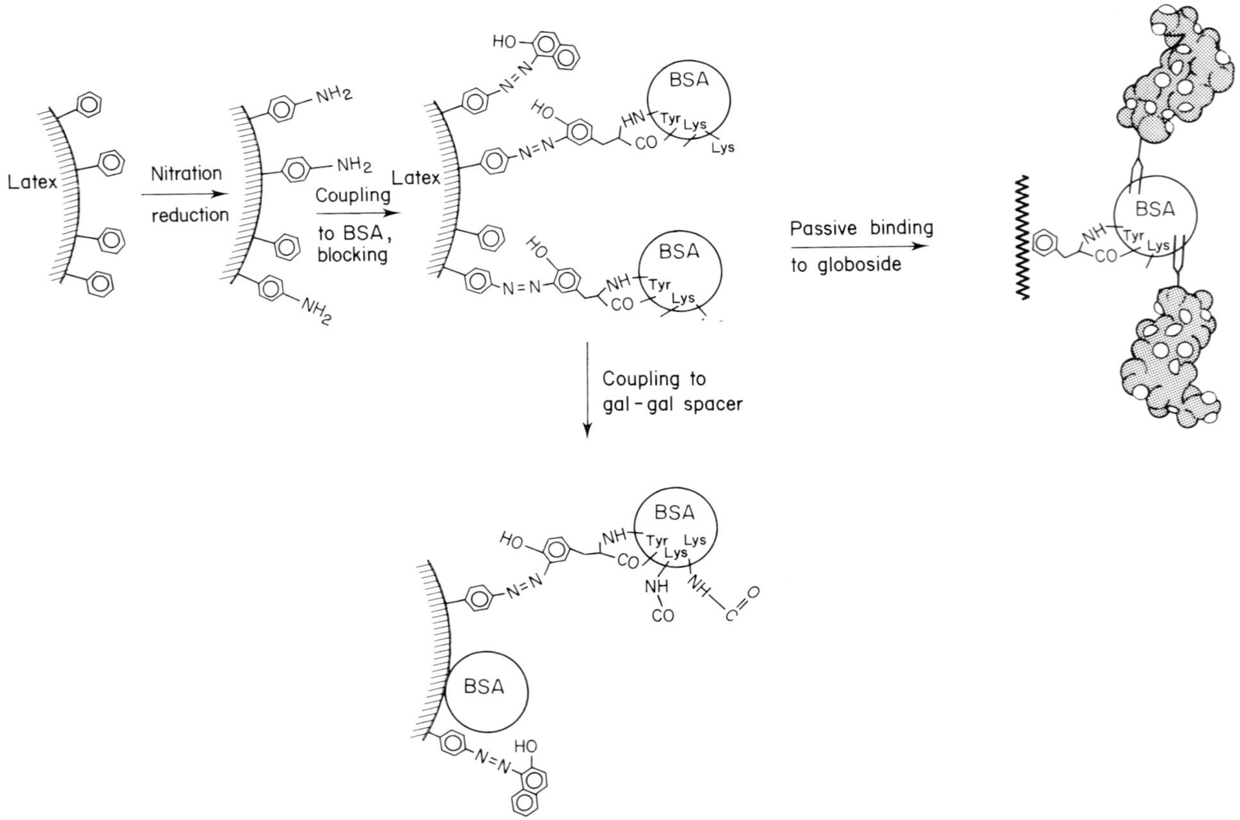

Latex
Nitration
reduction
NH2
Coupling
to BSA,
blocking
Latex
HO
N=N
BSA
HN
Tyr
Lys
CO
NH
Passive binding
to globoside
Coupling to
gal - gal spacer

Figure 1.

Pertussis Toxin: Identification of the Carbohydrate Receptor

Ronald D. Sekura[a] and Marie-Jose Quentin-Millet*

*Laboratory of Developmental and Molecular Immunity, National Institute of Child Health and Human Development, National Institutes of Health, Bethesda, MD 20205, USA, and *Institute Merieux, Marcy l'Etoile, Charbonnieres/Bains Cedex, France*

Intoxication of mammalian cells by pertussis toxin (PT) involves a sequence of events which include: (1) binding of PT to the cell surface; (2) entry of the appropriate PT subunit into the cell; and (3) PT-catalysed ADP-ribosylation of the adenylate cyclase regulatory component N_i. Since, PT is known to interact with a series of glycoproteins such as haptoglobin, ceruloplasmin and fetuin [1] studies were initiated to examine the nature of this interaction and to establish whether PT binding to cells utilizes the same mechanism. Since fetuin is a readily available glycoprotein with known carbohydrate structure ([2] Fig. 1), it was used as a model to examine the mechanism of toxin binding.

The interaction of PT with fetuin was monitored by a solid phase binding assay in which binding of ^{125}I-fetuin to immobilized PT was followed. Fetuin binding was found to be concentration dependent ($k_b = 4$–7×10^{-8}mol) and saturable; in addition, specific binding could be inhibited by adding excess quantities of unlabelled fetuin. This feature was used to explore the structural elements of

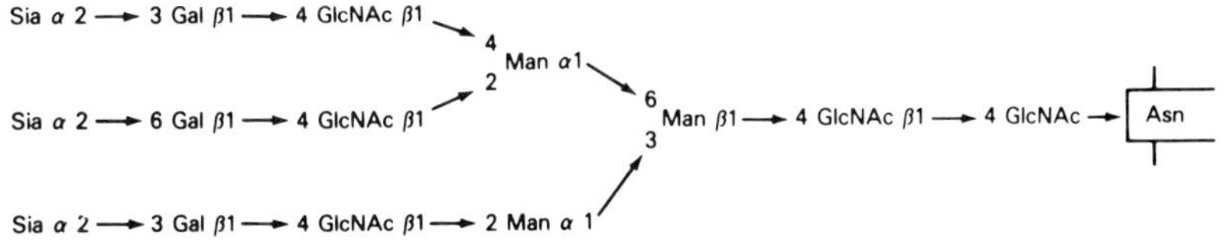

Figure 1. The asparagine-linked carbohydrate moiety of fetuin [2].

[a]*Telephone:* 1-301-496-1185

Protein-Carbohydrate Interactions in Biological Systems. ISBN 0 12 436665 1

fetuin which are involved in PT binding. Fetuin possesses both asparagine-linked and O-linked carbohydrate moieties. Use of aglycofetuin, fetuin with O-linked carbohydrate removed, and the corresponding O-linked and asparagine-linked oligosaccharides (or glycopeptide) as inhibitors of ^{125}I-fetuin binding established that the PT-fetuin interaction is carbohydrate dependent and involves the asparagine-linked carbohydrate moiety.

The nature of the specific carbohydrate residues involved in PT binding was examined by sequential removal of terminal carbohydrate residues by chemical and enzymatic treatment. Use of these derivatives as inhibitors of ^{125}I-fetuin binding demonstrated that terminal sialic acid and galactose residues do not affect inhibition of binding. In contrast, when mannose linked *N*-acetlyglucosamine was removed a substantial reduction in potency as an inhibitor was seen. This observation points to the importance of *N*-acetylglucosamine substituted mannose core in directing the interaction of PT with fetuin. ^{125}I-fetuin binding is also inhibited by the various classes of immunoglobulin with IgE being the most potent inhibitor. This observation is especially significant since the immunoglobulin associated carbohydrate differs from that of fetuin in that it is biantennary, lacking the β-(1–4) linked *N*-acetylglucosamine branch. On the basis of the observation and in conjunction with other data presented here, it is suggested that the minimal structure required for PT binding is the biantennary mannose core substituted with β-(1–2) linked *N*-acetylglucosamine.

The role of fetuin associated carbohydrate on the interaction of PT with cells was examined by monitoring the potency of fetuin and fetuin derivatives as inhibitors of PT-mediated agglutination of goose erythrocytes. These studies yielded results which were qualitatively similar to those observed for ^{125}I-fetuin binding, i.e. aglycofetuin derivatives were not potent inhibitors, and sequential removal of carbohydrate residues did not result in substantial reduction in inhibitor potency until *N*-acelylglucosamine was removed. These data confirm the importance of carbohydrate structure for the interaction of PT with cells and suggest that receptors with similar carbohydrate structure may exist on mammalian cells and play an important role in PT binding and intoxication.

REFERENCES

1. Sekura, R. D., F. Fish, B. Meade, C. R. Manclark and Y. Zheng (1983). Pertussis toxin: affinity purification of a new ADP-ribosyltransferase. *J. Biol. Chem.* **258**, 14 647–14 651.
2. Baenzinger, J. U. and D. Fiete (1979). Structure of the complex oligosaccharides of fetuin. *J. Biol. Chem.* **254**, 789–795.

V. Adhesins: Their Receptors and Interactions

Co-Chairmen: Alf Lindberg
Timo Korhonen

Fimbrial Phase Variation in *Escherichia coli*: A Mechanism of Bacterial Virulence?

Timo K. Korhonen[a], Bogdan Nowicki, Mikael Rhen, Vuokko Väisänen-Rhen, Auli Pere, Ritva Virkola, Jukka Tenhunen, Kirski Saukkonen* and Jaana Vuopio-Varkila*

*Department of General Microbiology, University of Helsinki, and *National Public Health Institute, SF-00280, Helsinki, Finland*

Our analysis of *Escherichia coli* strains isolated from children with urinary tract infections [1-3] or from cases of neonatal sepsis and meningitis [3,4] have shown that most of the pathogenic strains express many fimbrial antigens. These antigens can be distinguished either by their binding specificities or by serological properties. The pathogenic role and biological significance of some of the adhesins, e.g. the P fimbria, are currently understood, whereas the function and the clinical significance of the other adhesins are not yet clear.

Table 1 summarizes the occurrence of various adhesins on *E. coli* strains isolated from extraintestinal infections in Finland. The type-1 fimbria are common on all strains, whereas the P fimbria correlate with pyelonephritis and the S fimbria with septicemia. It should be noted that very few of the pathogenic strains are devoid of detectable adhesins and that most pathogenic strains have many fimbrial types.

The strains have also been analysed for other properties, and it became evident that the different fimbrial types were associated with certain *E. coli* serotypes [2-4]. In addition to fimbriae, many of the strains with certain O:K:H types were found to possess other common factors, such as the outer membrane protein pattern, possession or lack of haemolytic activity, and plasmid content [2,4]. This allowed identification of certain groups of relatively homogeneous bacteria,

[a]*Telephone:* 358-0-47351

Protein-Carbohydrate Interactions In Biological Systems ISBN 0 12 436665 1

Table 1. Occurrence of adhesins on *E. coli* strains of different pathogenic origin. The data is based on [2-4]

Adhesin	*Percentage of strains having the adhesin*				
	PN[b] (n=*67)*	*CYS* (n=*60)*	*ABU* (n=*60)*	*F* n=*50)*	*SEP* (n=*63)*
Type-1 fimbria	91	83	82	76	81
Type-1C fimbria	24	13	10	9	13
P fimbria	76	23	18	16	38
S fimbria	7	2	2	6	29
X adhesins[a]	13	8	7	4	19
No adhesins	0	12	15	18	9

[a] Refers to a group of heterogeneous adhesins.
[b] PN, pyelonephritis; CYS, cystitis; ABU, asymptomatic bacteriuria; F, faecal; SEP, sepsis.

which were named clonal groups after Achtman *et al.* [5]. It is probable that the strains within a clonal group have a common evolutionary origin and share properties which make them highly capable of causing extraintestinal diseases. In pyelonephritis, these bacterial factors include P fimbria, haemolytic activity, aerobactin production and acidic capsules [2]. Many of the pyelonephritis-associated strains possess all these factors, which probably act in concert to bring about the disease.

CELL POPULATIONS OF FIMBRIATE STRAINS ARE HETEROGENEOUS

When testing our model strains, *E. coli* KS71 (serotype 04:K12) and 3040 (018ac:K1:H7), for agglutination with specific anti-fimbriae sera, we observed that only a fraction of the cells were precipitated, i.e. carried the fimbrial antigen tested [6]. It was also observed that bacterial cells could be separated into distinct subpopulations by precipitation with fimbria-specific antisera [6] or by adsorption onto erythrocytes or yeast cells [7].

A quantitative estimation of the composition of bacterial cultures was made by immunofluorescence assays (IFA) using fimbria-specific FITC- or TRITC-labelled antibodies (Fig. 1; [7,8]). These assays showed that the different fimbrial types of a strain, at any time, mostly occur on separate cells (Fig. 1; Table 2). Only a small proportion (3–5%) of KS71 and 3040 cells carried more than one fimbrial type, and many non-fimbriate cells were observed (Fig. 1; Table 2).

Agar-grown subcultures of fractionated subpopulations of KS71 and 3040, consisting of cells carrying only one fimbrial type or lacking fimbriae [6,7], gave colonies that expressed the same fimbrial antigens as did the original non-fractionated cultures [6,7]. This indicated that the phase shifts from one fimbrial phase to another, or from a non-fimbriate to a fimbriate phase and vice versa, were reversible. Thus the phenomenon resembled phase variation.

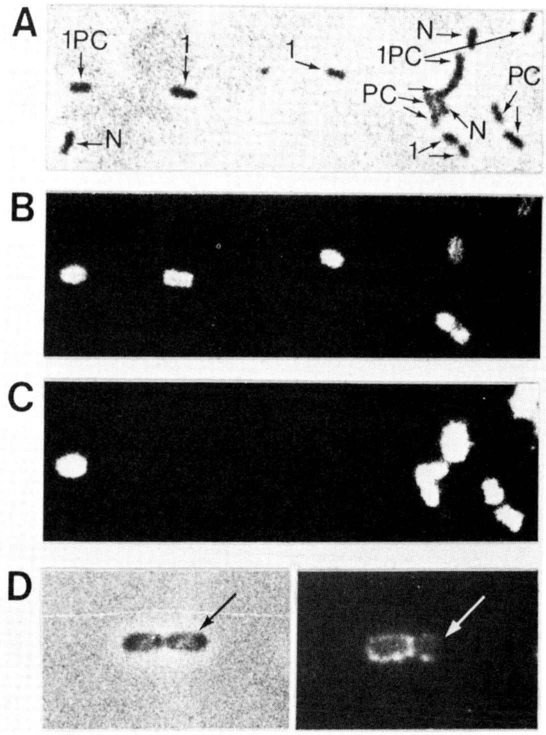

Figure 1. Examples of staining of bacteria with fimbria-specific antibodies. (A) is a phase contrast micrograph of KS71 cells that are shown for anti-type-1-FITC staining in (B) and for anti-type-1C- and anti-P-fimbria-TRITC staining in (C). Staining reaction of each cell is marked in (A); note that only two cells [marked by 1PC in (A)] show reaction in both (B) and (C) and that some cells are not stained at all [marked by N in (A)]. (D) shows a dividing S-fimbriate cell of *E. coli* 3040 that shows polar staining with anti-S-FITC conjugate and probably is under a shift to a non-fimbriate phase. (D) was reproduced from [10] with the permission of the Federation of European Microbiological Societies.

Table 2. Composition of bacterial cultures[a]. The data is from [7] and [8]

E. coli *3040*		E. coli *KS71*	
S-fimbriate	33%	P-fimbriate[b]	17%
Type-1-fimbriate	52%	Type-1-fimbriate	20%
With both fimbria	3%	Type-1C-fimbriate	16%
Non-fimbriate	12%	With more than one fimbria	5%
		Non-fimbriate	42%

[a] 35-h cultures in static Luria broth.
[b] The strain has two P fimbriae termed KS71A and KS71B.

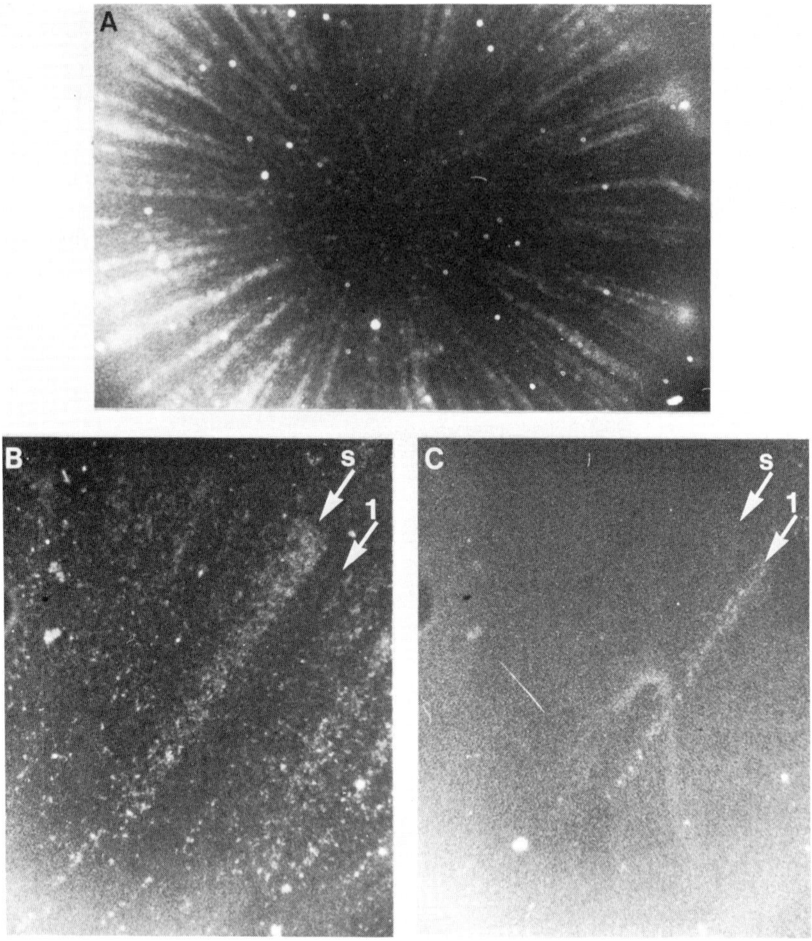

Figure 2. (A) shows a colony obtained from S-fimbriate subpopulation of *E. coli* 3040 and stained with anti-S-FITC conjugate. (B) and (C) show a close-up of a similar colony stained simultaneously with anti-S-FITC (B) and anti-type-1-TRITC (C) conjugates. Note sectorial staining in each micrograph; cells stained with anti-S-FITC or with anti-type-1-TRITC in (B) and (C) are marked by s and 1, respectively.

An intriguing feature of fimbrial phase variation in *E. coli* is its rapidity: single colonies are heterogeneous with respect to fimbriation. This could be visualized by immunofluorescence staining of colonies obtained from fractionated subpopulations of *E. coli* 3040 [9]. The method consisted of cultivating the fractionated bacteria as colonies on a nitrocellulose filter mounted on an agar plate and staining the colonies with fluorochrome-conjugated anti-type-1- or anti-S-fimbria antibodies. Figure 2 shows a typical result of such a staining: the colony is stained as distinct sectors. The length and width of the sectors depend on the fimbrial phase of the progeny cell [9]. Thus there is organized heterogeneity in colonies of fimbriate bacteria, and the sectors probably result from rapid phase variation of fimbrial antigens.

We then tried to determine the rates of phase shifts by preparing highly pure subpopulations of 3040 and by following the expression of heterologous fimbrial antigens by immunofluorescence [10]. We were able to determine the rates of shifts from the S- and type-1-phases to the non-fimbriate phase, and in both cases it was about 1×10^{-2} per cell generation. This is high enough to explain the heterogeneity of individual colonies. It was also apparent that the phase shifts were not totally random; the shifts from the S- or the type-1-phase to the non-fimbriate phase took place more frequently than did direct shifts to the opposite fimbriate phase [10], although both types of shifts are possible.

IFA assays of individual cells showed dividing cells that were polarly stained with the anti-fimbriae antibodies ([10]; Fig. 1D). These were frequently observed during the logarithmic growth phase of the fimbriate subpopulations, which suggested that the shift from a fimbriate phase to a non-fimbriate one results from sectorial growth of cell wall devoid of fimbria. In addition, we observed 3040 cells that were evenly stained with both anti-fimbriae sera [10], which indicates that there is another type of mechanism for the direct shift.

GENETICS OF FIMBRIAL PHASE VARIATION

It is obvious that the genetic mechanisms responsible for phase variation are complex. We cloned separately the genes encoding the four fimbrial antigens of strain KS71 [11]. Each of the fimbrial antigens is coded by a separate chromosomal DNA segment. The gene clusters for the two P fimbriae of KS71 (KS71A and KS71B) are homogeneous and their transport and assembly genes can complement each other [12], whereas the type-1C (KS71C) fimbrial gene cluster is unrelated to the KS71A and KS71B gene clusters and is not able to complement them. This indicates that the machinery needed in constructing the P fimbriae of KS71 is different from that of the KS71C fimbriae. The molecular mechanisms of phase variation are now being studied; so far we have been able to demonstrate regulatory interactions between the cloned fimbrial gene clusters of KS71 [12].

We have constructed recombinant strains that are either fimbriate and non-haemagglutinating or non-fimbriate and haemagglutinating, which is in accordance with the observation of Normark *et al.* [13] showing that fimbriation and haemagglutination are genetically separable properties. We have found that antibodies raised against highly purified fimbriae also inhibit haemagglutination by the non-fimbriate mutant strain, which suggests that the P-specific lectin is physically associated with the fimbriae. This is in accordance with the fractionation experiments described above, where the binding properties (or the lectins) were always expressed together with the fimbrial filament [6–8]. A tight association of a lectin with certain fimbrial structural proteins is further supported by the fact that we have never detected a fimbrial filament in the wild-type KS71 that would be "a mixed protein", e.g. serologically KS71C but functionally P.

FIMBRIAL PHASE VARIATION *IN VIVO*

The experiments described above have demonstrated a rapid fimbrial phase variation under laboratory growth conditions. It was of interest to us to see whether a similar variation takes place also *in vivo*, e.g. in patient samples or in animals challenged with fimbriate bacteria. We have analysed by IFA about 30 urine samples from patients with acute pyelonephritis and found that, in each case, only a fraction of *E. coli* cells in the urine (a patient is normally infected by one strain only) are stained with anti-P-fimbriae antibodies (B. Nowicki, A. Pere, J. Elo and T. K. Korhonen, unpubl. res.). This finding is to be expected on the basis of phase variation.

Another approach used in assaying possible *in vivo* fimbrial phase variation was to challenge newborn mice and 5-day-old rats with S-, type-1- and non-fimbriate phases of 3040 and to follow by IFA the expression of the fimbriae in the test animals (K. Saukkonen, J. Vuopio-Varkila, B. Nowicki, M. Leinonen, T. K. Korhonen and P. H. Mäkelä, unpubl. res.). The results demonstrated rapid shifts to a predominantly S-phase population in the peritoneal fluid, blood (and in the rats, also cerebrospinal fluid) in the intraperitoneally infected animals. Moreover, it seems that the S-phase cells are more virulent than the type-1-phase cells.

CONCLUSIONS

We have demonstrated in pathogenic *E. coli* strains a rapid fimbrial phase variation that takes place both *in vitro* and *in vivo*. There are a number of ways how the phase variation might increase the pathogenic potential of a strain. Different adhesins are probably functioning at different stages of the pathogenic process, and it might be advantageous to the bacteria to vary their expression. The type-1-fimbriae are known to associate the bacteria with phagocytic cells, and therefore their expression in the tissues could be harmful to the invading bacteria. Phase variation might also help the invading bacteria to avoid specific immune defences of the host.

ACKNOWLEDGEMENTS

This work has been supported by the Academy of Finland, the Yrjö Jahnsson Foundation and the Sigrid Juselius Foundation.

REFERENCES

1. Väisänen, V., J. Elo, L. G. Tallgren, A. Sittonen, P. H. Mäkelä, C. Svanborg Edén, G. Källenius, S. B. Svenson, H. Hultberg and T. K. Korhonen (1981).

Mannose-resistant haemagglutination and P-antigen-recognition are characteristic of *Escherichia coli* causing primary pyelonephritis. *Lancet* **ii**, 1366–1369.
2. Väisänen-Rhen, V., J. Elo, E. Väisänen, A. Siitonen, I. Ørskov, F. Ørskov, S. B. Svenson, P. H. Mäkelä and T. K. Korhonen (1984). P-fimbriated clones among uropathogenic *Escherichia coli*. *Infect. Immun.* **43**, 149–155.
3. Pere, A., M. Leinonen, V. Väisänen-Rhen, M. Rhen, and T. K. Korhonen (1985). Occurrence of type-1C fimbriae on *Escherichia coli* strains isolated from human extraintestinal infections. *J. Gen. Microbiol.* **131**, 1705–1711.
4. Korhonen, T. K., M. V. Valtonen, J. Parkkinen, V. Väisänen-Rhen, J. Finne, F. Ørskov, I. Ørskov, S. B. Sevenson and P. H. Mäkelä (1985). Serotypes, hemolysin production, and receptor recognition of *Escherichia coli* strains associated with neonatal sepsis and meningitis. *Infect. Immun.* **48**, 486-491.
5. Achtman, M., A. Mercer, B. Kusecek, A. Pohl, M. Heuzenroeder, W. Aaronson, A. Sutton and R. P. Silver (1983). Six wide-spread bacterial clones among *Escherichia coli* K1 isolates. *Infect. Immun.* **39**, 315-335.
6. Rhen, M., P. H. Mäkelä and T. K. Korhonen (1983). P fimbriae of *Escherichia coli* are subject to phase variation. *FEMS Microbiol. Lett.* **19**, 267–271.
7. Nowicki, B., M. Rhen, V. Väisänen-Rhen, A. Pere and T. K. Korhonen (1985). Fractionation of a bacterial cell population by adsorption to erythrocytes and yeast cells. *FEMS Microbiol. Lett.* **26**, 35-40.
8. Nowicki, B., M. Rhen, V. Väisänen-Rhen, A. Pere and T. K. Korhonen (1984). Immunofluorescence study of fimbrial phase variation in *Escherichia coli* KS71. *J. Bacteriol.* **160**, 691-695.
9. Nowicki, B., M. Rhen, V. Väisänen-Rhen, A. Pere and T. K. Korhonen (1985). Organization of fimbriate cells in colonies of *Escherichia coli* strain 3040. *J. Gen. Microbiol.* **131**, 1263-1266.
10. Nowicki, B., M. Rhen, V. Väisänen-Rhen, A. Pere and T. K. Korhonen (1985). Kinetics of phase variation between S and type-1 fimbriae of *Escherichia coli*. *FEMS Microbiol. Lett.* **28**, 237–242.
11. Rhen, M., J. Knowles, M. Penttilä, M. Sarvas and T. K. Korhonen (1983). P fimbriae of *Escherichia coli*: molecular cloning of DNA fragments containing the structural genes. *FEMS Microbiol. Lett.* **19**, 119–123.
12. Rhen, M., V. Väisänen-Rhen, A. Pere and T. K. Korhonen (1985). Complementation and regulatory interaction between two cloned fimbrial gene clusters of *Escherichia coli* strain KS71. *Mol. Gen. Genet.*, **200**, 60-64.
13. Normark, S., D. Lark, R. Hull, M. Norgren, M. Båga, P. O'Hanley, G. Schoolnik and S. Falkow (1983). Genetics of a digalectoside binding adhesin from a uropathogenic *Escherichia coli* strain. *Infect. Immun.* **41**, 942-948.

Characterization and Receptor Binding Specificities of the X-Binding UTI *Escherichia coli* Adhesin AFA-I

M. Alexander Schmidt*, Waltraud Walz* and Gary K. Schoolnik[a]

Department of Medical Microbiology, School of Medicine, Stanford University, Stanford, CA 94305, USA

As a prerequisite for infection, adherence of pathogenic bacteria to host epithelial cells is mediated by adhesins associated with the bacterial surface either as filamentous appendages, termed pili or fimbriae [1], or as afimbrial adhesins [2]. These molecules recognize specific structures on epithelial cells and thus enable bacteria to bind to and to colonize mucosal surfaces. In *Escherichia coli*, three functional adhesin classes have been described which are distinguished by their receptor specificity. Type I, or common fimbriae, are mannose-sensitive (MS); they recognize α-D-mannose residues on mammalian cells. P-specific adhesins have been shown to bind the globo series glycolipids containing the structural element α-Gal-(1-4)-β-Gal. They agglutinate human erythrocytes containing P blood group antigens and attach to uroepithelia. "X"-adhesins exhibit mannose-resistant agglutination activity that is distinct from the receptor-specificity of P-pili. "X"-adhesins are functionally heterogenous. Since most *E. coli* pyelonephritis isolates express either p- or "X"-adhesin activity, these adhesin classes are important virulence determinants.

AFA-I, an afimbrial "X"-adhesin derived from a human pyelonephritis strain has recently been cloned and expressed in a neutral *E. coli* background [2]. Here we describe the isolation, purification and characterization of the AFA-I protein from a recombinant strain. It is found to be an afimbrial "X"-binding adhesin

[a] *Telephone:* 1-415-497-3083

*Present address: Zentrum für Molekulare Biologie der Universität Heidelberg, Im Neuenheimer Feld 282, D-6900 Heidelberg, FRG.

Protein-Carbohydrate Interactions in Biological Systems. ISBN 0 12 436665 1

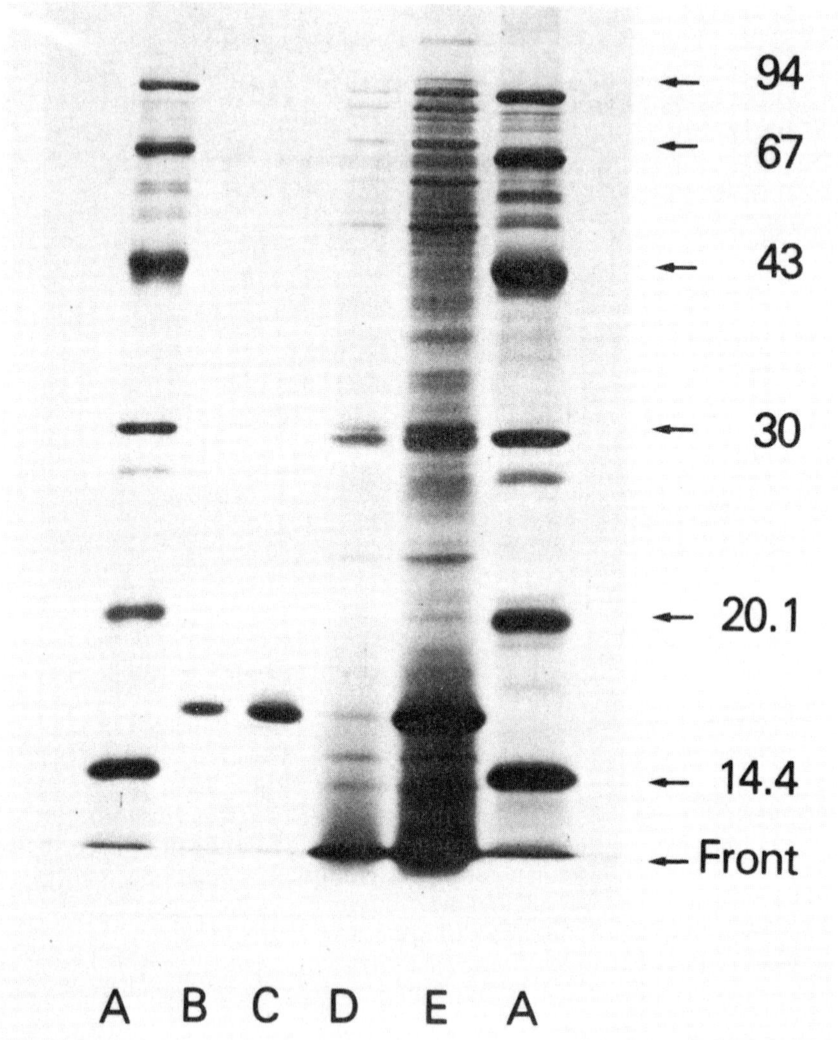

Figure 1. Monitoring of the AFA-I protein purification by SDS-PAGE. A, molecular weight standards; B, AFA-I after Biogel A 1.5 m (0.1% SDS/PBS); C, AFA-I after Biogel in PBS; D, ammonium sulphate precipitate; E, single colony of HB 101 (pIL 14).

with distinctive chemical, functional and serological properties, possibly exemplary of an adhesin class commonly associated with uropathogenic *E. coli*.

PURIFICATION OF AFA-I

E. coli strains KS 52, a human pyelonephritis isolate, and HB 101 (pIL 14), which harbours a recombinant plasmid containing a 6.7 kb KS 52 chromosomal fragment, produces a mannose-resistant, P-independent, non-fimbrial adhesin (AFA-I) that agglutinates human erythrocytes and binds uroepithelial cells. HB 101 (pIL 14) was grown in L-broth containing 100 μg/ml ampicillin and AFA-I was purified from the cell-free supernatant of spent culture medium [3]. In the late logarithmic phase of growth, the bacteria were removed by centrifugation and discarded. AFA-I was precipitated from the supernatant with ammonium sulphate and the precipitate was collected by centrifugation. After chromatography of the PBS soluble material on a Biogel A 1.5 m column the fractions were monitored for absorbance at 230 nm and for haemagglutination activity. Only fractions eluting with the void volume agglutinated human erythrocytes. These were combined and analysed by SDS-PAGE (Fig. 1). After staining with Coomassie blue AFA-I was detected as a single band of 16 000 D apparent molecular weight.

Minor low molecular weight contaminants detected by silver staining were removed by chromatography of AFA-I in PBS containing 0.1% SDS. The resulting profile in SDS-PAGE showed AFA-I as a single band upon silver staining. This preparation was considered to be homogeneous and was used in subsequent experiments. The yield of purified AFA-I was estimated to be 500 μg/l of culture supernatant using the Biorad assay with BSA as the standard protein.

AFA-I AMINO ACID COMPOSITION

The amino acid compositions of AFA-I and of the afimbrial haemagglutinins prepared by Sheladia [4] from strain GV-12 (01:H$^-$) and by Williams *et al.* [5] from *E. coli* strains 444-3 (0?:H4) and 468-3 (021:H$^-$) are presented in Table 1. Also depicted are the amino acid compositions of the P and MS fimbrial proteins from recombinant strains HU 849 and SH 48 [6]. The most remarkable difference between the afimbrial adhesins and pili is the low content of non-polar hydrophobic amino acids (Ala, Val, Ile, Leu and Phe) in AFA-I and the afimbrial haemagglutinins from strains GV-12, 444-3 and 469-3 (22–27%) and the relatively high content of hydrophobic residues in P pili (34%) and MS fimbriae (48%). The identification of 2.5 cystein residues per subunit is in agreement with the DNA sequence of the AFA-I structural gene [7]. No interchain disulphide bridges were detected.

AFA-I *N*-TERMINAL AMINO ACID SEQUENCE

The AFA-I *N*-terminal amino acid sequence was determined through residue 24 by automated Edman degradation (Table 2). Except for residues 8–11 of the AFA-I

Table 1. Amino acid compositions of afimbrial adhesins in relation to P- and MS-fimbriae

	Number of residues per subunit					
	Afimbrial adhesins				*Fimbriae*	
Amino acid	*AFA-I*	*GV-12*	*444-3*	*469-3*	*HU 849*	*SH 48*
Asx	15	17	17	15	19	18
Thr	15	13	10	8	12	20
Ser	11	8	25	21	11	9
Glx	10	9	20	17	13	16
Pro	2	Tr	2	4	5	2
Gly	19	28	24	20	21	21
Ala	6	6	10	12	17	35
Cys	2.5[a]	n.d.	n.d.	n.d.	2[a]	2[a]
Val	7	7	6	7	17	14
Met	3[a]	1	2	2	1[a]	n.d.
Ile	4	4	4	5	6	5
Leu	8	8	7	8	9	14
Tyr	n.d.	3	4	2	2	2
Phe	2	4	5	5	7	8
Lys	4	6	3	4	10	4
His	3	4	4	3	2	2
Arg	6	13	0	2	n.d.	2

[a] Cysteine was analysed as cysteic acid and methionine as methionine sulphone.
n.d. No amino acid was detected.

Table 2. *N*-terminal amino acid sequence of AFA-I in comparison with *E. coli* fimbrial proteins

	Residue number																								
Protein	*1*	*2*	*3*	*4*	*5*	*6*	*7*	*8*	*9*	*10*	*11*	*12*	*13*	*14*	*15*	*16*	*17*	*18*	*19*	*20*	*21*	*22*	*23*	*24*	*25*
AFA-I	N	F	T	S	S	G	T	N	G	K	V	D	L	T	I	T	E	E	C	R	V	T	V	E	
J 96	A	P	T	I	P	Q	G	Q	G	K	V	T	F	N	G	T	V	V	D	A	P	C	S	I	S
F 12	A	P	T	I	P	E	G	Q	–	K	V	T	P	N	G	T	V	V							
F 7	A	A	T	I	P	Q	G	Q	G	E	V	A	P	K	G	T	V	V	N/D	A	P				
KS 71A	A	A	T	I	P	Q	G	Q	G	E	V	T	F	K	G	T	V	V	N	A					
F 9	E	T	T	P	T	T	V	N	G	G	T	V	H	F	K	G	E	V	V	N	A	A			
Type 1A			A	A	T	T	V	N	G	G	T	V	H	F	K	G	E	V	V	N	A	A			
ER2B2				V	T	T	V	N	G	G	T	V	V	F	K	G									
CFA I		V	E	K	N	I	T	V	T	A	Ş	V	D	P	V	I	D	L	LQ	A	D	G	N	A	L
K 88ab	W	M	T	G	D	F	N	G	S	V	D	I	G	G	S	I	T	A	D	D	Y	R	Q	K	W

sequence (-Asn-Gly-Lys-Val-) and the corresponding residues in HU 849 P pili (-Gln-Gly-Lys-Val-) no homology was evident in the *N*-terminal sequence of other *E. coli* adhesins.

SEROLOGICAL PROPERTIES OF THE AFA-I PROTEIN

In a solid phase binding assay employing purified HU 849 (P) pili and SH 48 (MS) pili as solid phase, AFA-I antiserum did not bind P- or MS-pili, indicating lack of shared antigenicity.

Other "X"-binding *E. coli* urine isolates were tested for the presence of AFA-I protein by Western blotting. Pyelonephritis or cystitis strains that agglutinated human erythrocytes in the presence of D-mannose and synthetic α-Gal-(1-4)-β-Gal were removed as single colonies from L-agar, subjected to SDS-PAGE and subsequently blotted on nitrocellulose paper. Four out of sixteen "X"-binding *E. coli* isolates exhibited an AFA-I positive, 16 000 D protein; an 18 000 D protein was also detected in some strains.

SURFACE LOCALIZATION OF AFA-I BY IMMUNO-GOLD LABELLING

AFA-I was localized on the bacterial surface employing the immuno-gold labelling technique. Bacteria grown overnight on agar plates were adsorbed to Formvar-coated copper grids. The grids were blocked with BSA and treated with anti-AFA-I antibodies in 0.1% BSA/PBS. Bound antibody was visualized with protein A-coated colloidal gold (Fig. 2, A-D) and the bacteria were subsequently negatively stained with 1% phosphotungstic acid. The afimbrial character of the AFA-I protein on the cell surface is clearly demonstrated. Higher magnification (Fig. 2D) indicates the presence of an almost "capsular" material bound by anti-AFA-I antibodies on the surface of the cells.

AFA-I AGGLUTINATION AND BINDING PROPERTIES

E. coli strains KS 52 and the AFA-I clone HB 101 (pIL 14) agglutinate human erythrocytes and adhere to voided uroepithelial cells [2]. The haemagglutination specificity was examined with human and animal erythrocytes (Table 3). For AFA-I on the surface of bacteria the receptor sites seem to be much more accessible on red cells of man and the anthropoid apes. The purified AFA-I protein, however, also agglutinated erythrocytes of several other species.

AFA-I BINDING SITES ON TISSUE CULTURE CELLS

The receptivity of HeLa tissue culture cells for the AFA-I clone HB 101 (pIL 14) and the purified AFA-I protein was assessed by indirect immunofluorescent microscopy in order to determine the topography of the receptor on the epithelial cell surface. The cell-bound AFA-I protein was detected with anti-AFA-I rabbit antiserum and FITC-labelled goat anti-rabbit IgG as second antibody. While

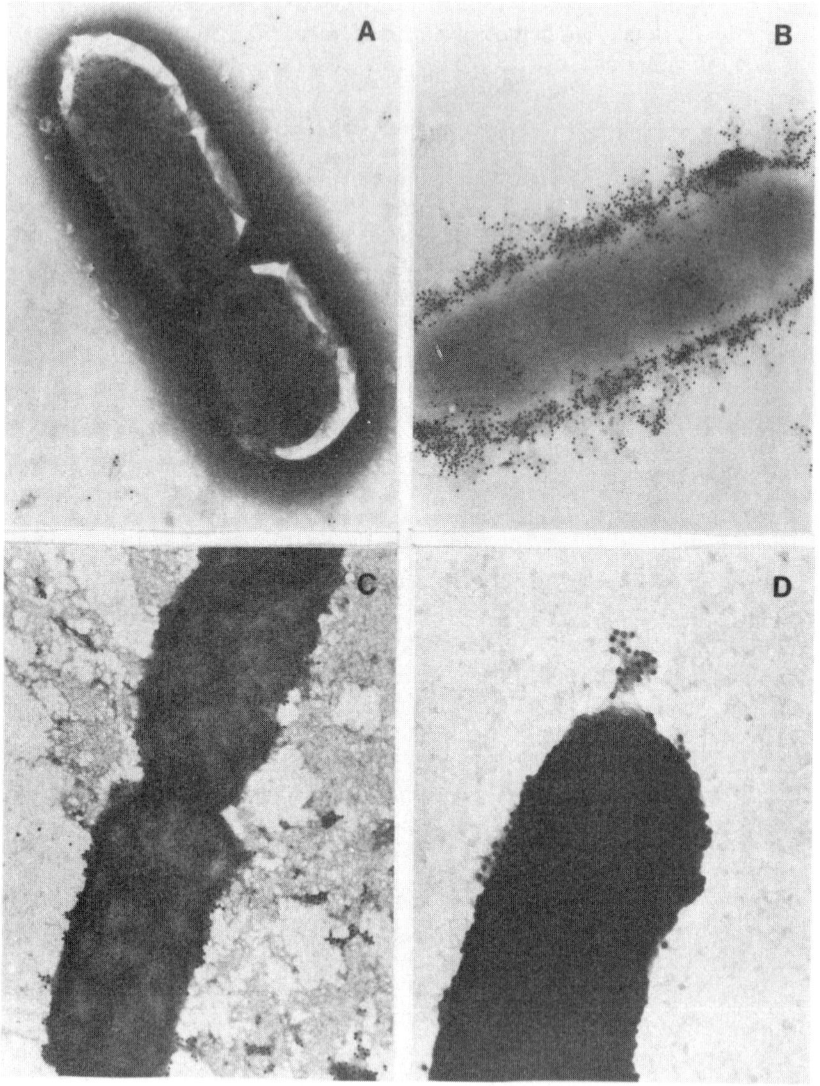

Figure 2. Immunogold-labelling of AFA-I on the bacterial cell surface. (A) HB 101 negative control; (B) KS 52 wild-type strain (×30 000); (C) HB 101 (pIL 14) (×30 000); (D) HB 101 (pil 14) (×70 000).

HB101 (pIL 14) and KS 52 bacteria bound to HeLa cells more localized and in relatively low numbers, the AFA-I protein bound diffusely but with a particular high density at the periphery of single cells. These studies indicate that the AFA-I receptor is rather evenly distributed over the epithelial cell surface.

Table 3. Agglutination of human and animal erythrocytes by AFA-I

	Agglutinin	
Erythrocyte	*HB 101 (pIL 14)*	*AFA-I*
Human[a]	+	+
Gorilla	+	+
Rhesus	−	+
Squirrel	−	−
Guinea pig	−	+
Rabbit	−	+/−
Rat	−	+/−
Mouse	−	−
Horse	−	+/−
Goat	−	−
Turkey	−	−
Cow	−	−
Dog	−	−
Sheep	−	−
Cat	−	−
Pig	−	−

[a] HB 101 (pIL 14) and AFA-I agglutinated the erythrocytes of all human blood groups available at the Stanford Medical Center Blood Bank.

The difference in binding pattern is thought to be due to a difference in accessibility for AFA-I on the surface of whole bacterial cells and isolated AFA-I protein.

IDENTITY OF THE AFA-I RECEPTOR ON HUMAN ERYTHROCYTES

The identity of the AFA-I receptor was sought by attempting to find a haemagglutinin-resistant human blood group. All human blood groups available from the Stanford Blood Bank (e.g. O, A, AB, B, MM, MN, Rh^+, Rh^-, Le^a, Le^b, En(a^-) and neuraminidase-treated cells) were agglutinated by AFA-I. However, AFA-I failed to agglutinate latex beads coated with α-Gal-(1-4)-β-Gal (P), lactose or synthetic oligosaccharides specifying the A, B, H, Le^a and Le^b human blood group antigens [2]. Likewise AFA-I did not bind purified glycophorin A and haemagglutination could not be inhibited by α-NeuNAc-(2-3/6)-lactose. AFA-I is therefore distinct from *E. coli* "X"-adhesins with "M"- and "S"-specificity [8,9].

Total lipid extracts from human erythrocytes and voided uroepithelial proved negative for specific binding of AFA-I using the HPTLC-overlayer binding technique [10].

Western blotting of erythrocyte ghosts revealed specific binding to a doublet of peripheral (glyco) proteins with an approximate molecular weight between 96 000 and 98 000 (Fig. 3). The isolation and characterization of these proteins is currently being pursued in our laboratory.

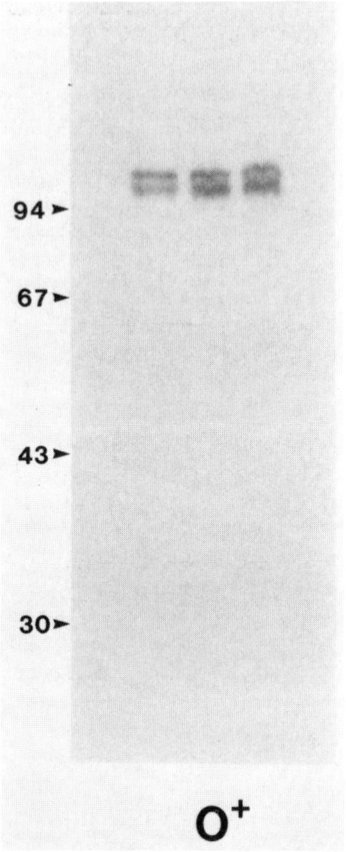

Figure 3. Binding of AFA-I to transblotted proteins from 0^+ erythrocyte ghosts. After SDS-PAGE the proteins of 0^+-ghosts were transblotted on nitrocelluse, blocked with BSA and incubated with a solution of AFA-I in 0.1% BSA/PBS. Bound AFA-I protein was detected with anti-AFA-I antibodies followed by ^{123}I-labelled protein A.

SUMMARY

(1) AFA-I, a mannose-resistant, P-independent, "X"-binding afimbrial *E. coli* adhesin was purified from a recombinant strain and chemically, functionally and serologically characterized. AFA-I exists on the bacterial surface and free as a macromolecular aggregate in the supernatant of spent culture medium. It is composed of a single (repeating) 16 000 D polypeptide subunit. The AFA-I protein amino acid composition is remarkable for the presence of 2.5–3.0 cysteines per subunit and for a marked decrease in hydrophobic amino acids as compared to

subunits of *E. coli* pili. Since AFA-I travels as a monomer in SDS-PAGE under non-reducing conditions, no disulphide bonds exist between subunits and at least one free sulphydryl per subunit is available. The AFA-I *N*-terminal amino acid sequence through residue 24 was unrelated to any known *E. coli* fimbrial sequence. Immuno-gold labelling demonstrated the afimbrial nature of the AFA-I protein on the bacterial cell surface.

(2) Anti-AFA-I sera bound AFA-I in Western blots of four of sixteen "X"-binding *E. coli* urine isolates. They did not bind MS or P pili.

(3) HB 101 (pIL 14), the AFA-I recombinant strain, agglutinated only human or gorilla erythrocytes, indicating a preference for receptor molecules on the red cells of man and the anthropoid apes. AFA-I did not bind glycophorin A or sialyl glycosides and is therefore distinct from the *E. coli* "X"-binding adhesins with "M" and "S" specificity. The AFA-I receptor was found to be abundant and diffusely distributed on HeLa tissue culture monolayer cells surfaces by indirect fluorescent microscopy.

(4) Total lipid extracts of human erythrocytes and voided uroepithelial cells proved negative for specific binding of AFA-I. The AFA-I protein was shown to bind to a doublet of probably peripheral (glyco)proteins from human erythrocyte ghosts of approximately 96 000–98 000 D molecular weight. The characteristics of these (glyco)proteins are currently under investigation.

REFERENCES

1. Duguid, J. P., S. Clegg and M. I. Wilson (1979). The fimbrial and non-fimbrial hemagglutinins of *Escherichia coli*. *J. Med. Microbiol.* **12**, 213–227.
2. Labigne-Roussel, A. F., D. Lark, G. K. Schoolnik and S. Falkow (1984). Cloning and expression of an afimbrial adhesin (AFA-I) responsible for P-blood-group-independent mannose resistant hemagglutination from a pyelonephritic *Escherichia coli* strain. *Infect. Immun.* **46**, 251–259.
3. Walz, W., M. A. Schmidt, A. F. Labigne-Roussel, S. Falkow and G. K. Schoolnik (1985). AFA-I—a cloned afimbrial X-type adhesin from a human pyelonephritic *E. coli* strain. *Eur. J. Biochem.* **152**, 315–321.
4. Sheladia, V. L., J. P. Chambers, J. Guevara Jr and D. J. Evans (1982). Isolation, purification and partial characterization of type V-A hemagglutinin from *E. coli* GV-12 (01:H^-). *J. Bacteriol.* **152**, 757–761.
5. Williams, P. H., S. Knutton, M. G. M. Brown, D. C. A. Candy and A. S. NcNeish (1984). Characterization of nonfimbrial mannose-resistant protein hemagglutinins of two *Escherichia coli* strains isolated from infants with enteritis. *Infect. Immun.* **44**, 592–598.
6. O'Hanley, P., D. Lark, S. Normark, S. Falkow and G. K. Schoolnik (1983). Mannose-sensitive and Gal–Gal binding *Escherichia coli* pili from recombinant strains. *J. Exp. Med.* **158**, 1713–1719.
7. Labigne-Roussel, A. F., M. A. Schmidt, W. Walz and S. Falkow (1985). Genetic organization of the afimbrial adhesin operon and nucleotide sequence from an uropathogenic *Escherichia coli* gene encoding an afimbrial adhesin. *J. Bacteriol.* **162**, 1285–1292.

8. Väisänen, V., T. K. Korhonen, M. Jokinen, C. G. G. Ahmberg and C. Ehnholm (1982). Blood group M specific hemagglutinin in pyelonephritogenic *Escherichia coli. Lancet* **i**, 1192.
9. Parkkinen, J., J. Finne, M. Achtman, V. Väisänen and T. K. Korhonen (1983). *Escherichia coli* strains binding neuraminyl-a-α-(2-3)-galactoside. *Biochem. Biophys. Res. Commun.* **111**, 456–461.
10. Magnani, J. L., D. F. Smith and V. Ginsburg (1980). Detection of gangliosides that bind Cholera toxin: Direct binding of ^{125}I-labeled toxin to thin-layer chromatograms. *Anal. Biochem.* **109**, 399–402.

Characterization of a Fibronectin Binding Protein of *Staphylococcus aureus*

G. Fröman[a], L. Switalski*, B. Guss‡, M. Lindberg‡, M. Höök* and T. Wadström‡

*Department of Bacteriology and Epizootology, Swedish University of Agricultural Sciences, Uppsala, Sweden, *Connective Tissue Laboratory, Diabetes Hospital, University of Alabama in Birmingham, Birmingham Al 35294, USA, and ‡Department of Microbiology, University of Uppsala, Uppsala, Sweden*

Staphylococcus aureus is the most common cause of post-traumatic infection in skin and soft tissue lesions and a common cause of infections after surgery. Despite the observation more than 70 years ago that *S. aureus* cells clump in the presence of fresh human plasma, it was only several decades later that fibrinogen (fb) was identified as the protein inducing clumping by binding to a specific staphylococcal cell surface component called clumping factor (CF). However, some strains were also found to clump in fibrinogen-depleted plasma. Some decades later Kuusela [1] discovered that heat- or formalin-killed *S. aureus* bound another major serum protein, i.e. fibronectin (fn) and thus the cell clumping of some strains in fb-depleted plasma could be explained.

Specific fibronectin binding has been demonstrated among both Gram-positive and Gram-negative bacteria [1-3]. Staphylococci and Streptococci interact with at least two sites in the fibronectin molecule, one site located in the amino-terminal domain and another in the carboxy-terminal region [4]. Espersen and Clemmensen [5] isolated a fibronectin binding protein from sonicated *S. aureus* by affinity chromatography on fibronectin-Sepharose. The molecular weight, as estimated by SDS-polyacrylamide gel electrophoresis, was 197 kD.

[a] *Telephone:* 46-18-174097

Protein-Carbohydrate Interactions in Biological Systems. ISBN 0 12 436665 1

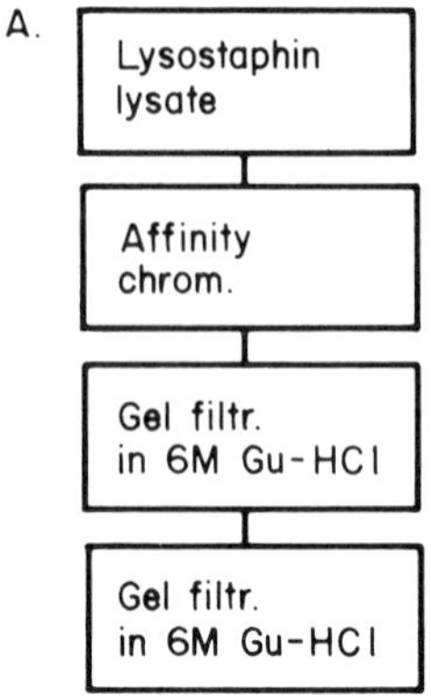

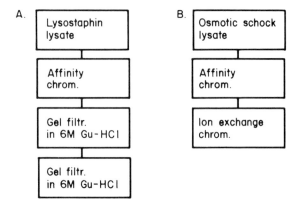

Figure 1. Purification schemes for the isolation of fn-binding protein from *S. aureus* strain Newman (A) and from an *E. coli* clone (B) carrying the *S. aureus* gene for a fn-binding protein.

CLONING OF FN-BINDING PROTEIN IN *E. COLI*

A gene bank of *S. aureus* was constructed in *Escherichia coli* as previously described [6]. Five hundred clones were screened for the expression of fn-binding protein and one clone which released the protein upon cold osmotic shock was isolated. The insert in the pBR 322 vector was 6.5 kilobases.

PURIFICATION OF FN-BINDING PROTEIN(S)

Fn-binding proteins of *S. aureus* have been purified from strain Newman (Fröhman *et al.*, submitted) and from *E. coli* with a cloned staphylococcal gene for the fn-binding protein using the procedures outlined in Fig. 1. The purified proteins had a molecular weight of 210 kD (from *S. aureus*) or 165 kD and 87 kD (from *E. coli* clone). The two high molecular weight proteins showed a similar and very high specific inhibitory activity of fn-binding to *S. aureus*. The protein pattern during the different purification steps are shown in Fig. 2.

The amino acid composition of the purified proteins are shown in Table 1. The 87 kD protein showed a high similarity as compared to the 165 kD and 210 kD proteins. This might imply that the fn-binding protein from *S. aureus* is composed of repeating domains with similar amino acid composition, each of which is capable of binding to fibronectin and that the 87 kD protein is a proteolytic degradation product of the 165 kD protein. The yield of protein was about 1 mg per 4 l culture of bacteria for the 210 kD protein and for the 165 kD protein about 1 mg per 25 l culture.

Antibodies raised against the purified 210 kD protein were immobilized on Sepharose CL-4B and affinity purified fn-binding protein was passed through a column substituted with either immune- or preimmune IgG antibodies (see Table 2). The immune IgG-Sepharose adsorbed about 60% of the inhibitory

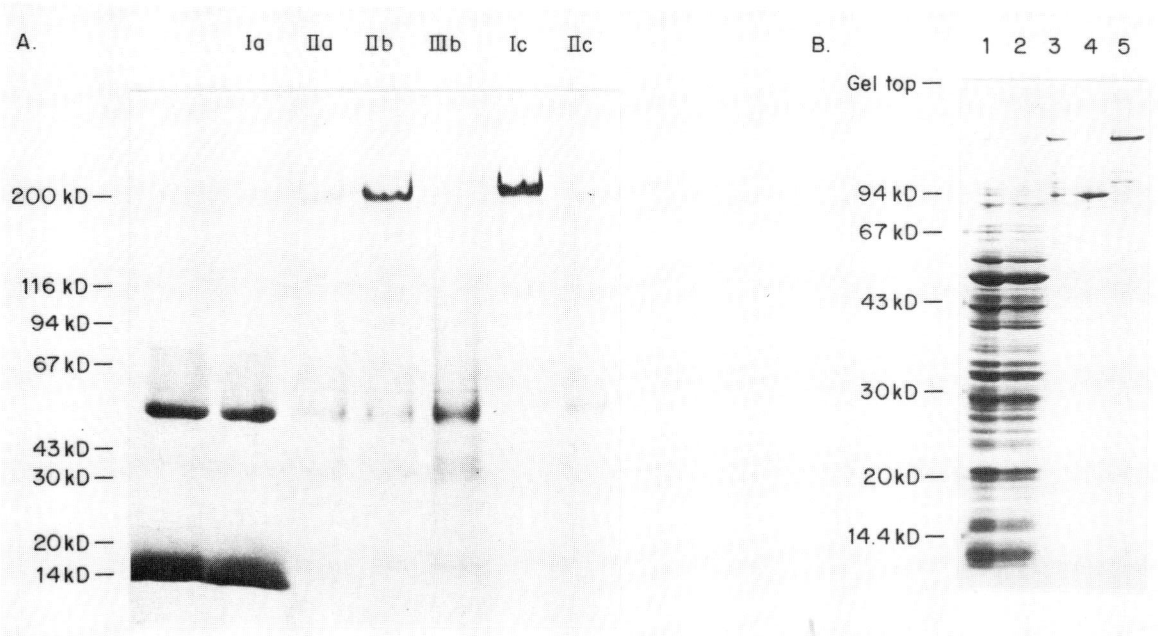

Figure 2. Gel electrophoresis of materials in the different purification steps. (A) Purification from lysostaphin lysate. Affinity chromatography on fn-fragment (amino-terminal 29 kD) Sepharose. First lane, material applied to the column, unabsorbed (lane Ia) and adsorbed (lane IIa) material. Gel filtration in Sephacryl S-400 in 6 M guanidinium hydrochloride: first active pool (lane IIb); second active pool (lane IIIb). Rechromatography (of pool IIb) on Sephacryl S-400: first active pool (lane Ic); second active pool (lane IIc).

(B) Purification from osmotic shock lysate. Material applied (lane 1) to the affinity column (fn-Sepharose), unadsorbed (lane 2) and adsorbed (lane 3) material. Ion exchange chromatography of affinity purified material on a Mono Q FPLC column (Pharmacia AB, Sweden): first active peak (lane 4); second active peak (lane 5).

Table 1.

Amino acid	*Composition (mol %)* 210 kD	165 kD	87 kD
Aspartic acid	15.4	14.6	13.4
Threonine	9.7	10.7	10.3
Serine	7.4	6.5	8.0
Glutamic acid	17.9	17.1	15.1
Proline	6.3	6.2	5.8
Glycine	8.9	7.9	8.4
Alanine	5.3	4.6	4.7
Half-cystine	0.1	0.2[a]	n.d.
Valine	7.2	7.8	8.6
Methionine	0.6	0.6[a]	n.d.
Isoleucine	4.3	4.7	3.8
Leucine	4.1	4.0	4.6
Tyrosine	2.1	2.3	4.1
Phenylalanine	2.4	2.0	3.6
Histidine	1.0	3.2	3.0
Lysine	5.3	6.3	6.6
Tryptophan	n.d.	n.d.	n.d.
Arginine	1.9	1.2	—

[a] Determined after performic acid oxidation.
n.d. Not determined.

Table 2. Immunosorbent chromatography of fibronectin binding components of *Staphylococcus aureus* strain Newman

Fractions from the columns	*Sepharose CL-4B substituted with* *Immune IgG*	*Preimmune IgG*
Unadsorbed material	29%	95%
Adsorbed material	61%	Trace

Data are given in per cent of total inhibitory activity applied to the columns.

activity applied to the column while the preimmune IgG-Sepharose adsorbed none. This shows that the immune serum contains antibodies directed against the fn-binding protein.

During the purification procedures we noticed that the high molecular weight compounds (210 kD and 165 kD) were very rapidly degraded to smaller peptides which seemed to retain their fn-binding capacity. To test this hypothesis, the 210 kD protein was ^{125}I-labelled and digested with staphylococcus V8 protease. Fractionation before and after protease digestion on a column of Sepharose substituted with the 29 kD fibronectin fragment was carried out and fractions from the column were analysed by SDS-acrylamide gel electrophoresis (Fig. 3). The protease treatment of the 210 kD fn-binding protein generated a large number

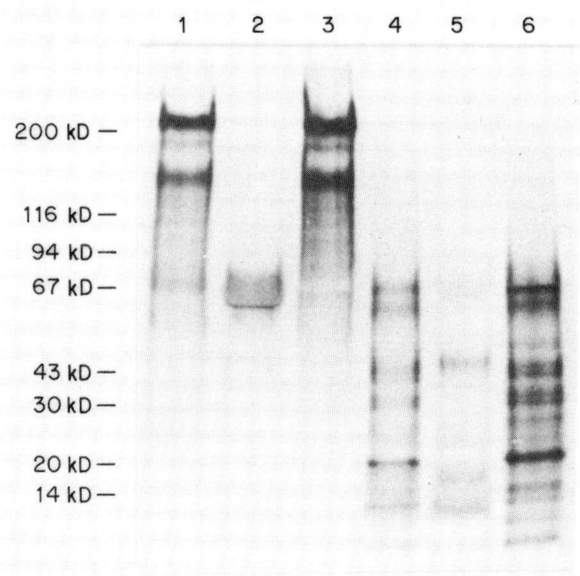

Figure 3. Gel electrophoresis of ^{125}I-labelled fn-binding proteins. Affinity chromatography of undigested (lanes 1–3) and digested (lanes 4–6) protein. Material applied to the column (lanes 1 and 4). Unadsorbed (lanes 2 and 5) and adsorbed (lanes 3 and 6) material.

of peptide fragments that retained their ability to adsorb to the 29 kD fn-fragment-Sepharose.

Binding of ^{125}fn to both live and heat-killed staphylococcal cells is a high affinity binding process which probably enables the pathogens to colonize "irreversibly" in open wound tissues and blood clots. The modulation of cell surface fn-binding proteins by both staphylococcal as well as tissue proteases released into injured tissues with microthrombs and blood-clots may then be an important mechanism for the pathogen to allow efficient spreading of daughter cells into the environment. This promotes the infection process by supplying nutrients released upon tissue degradation by alphatoxin and a number of tissue degrading staphylococcal enzymes [7] such as proteases, lipases, nucleases, and hyaluronate lyase.

ACKNOWLEDGEMENT

We gratefully acknowledge Birgitta Evremar's help in preparing the manuscript. This work was supported by grants from the Swedish Medical Research Council (16X-07168 and 16X-04723) and by Public Health Service Grant AM 7807 from the National Institute of Health. M.H. is a recipient of an Established Investigator Award from the American Heart Association.

REFERENCES

1. Kuusela, P. (1978). *Nature* **276**, 718-720.
2. Fröman, G., L. M., Switalski, A. Faris, T. Wadström and M. Höök (1984). *J. Biol. Chem.* **259**, 14899-14905.
3. Baloda, S., A. Faris, G. Fröman and T. Wadström (1985). *FEMS Microbiol. Lett.* **28**, 1-5.
4. Kuusela, P., T. Vartio, M. Vuento and E. B. Myhre (1984). *Infect. Immun.* **45**, 433-436.
5. Esperson, F. and I. Clemmensen (1982). *Infect. Immun.* **37**, 526-531.
6. Löfdahl, S., B. Guss, M. Uhlén, L. Philipson and M. Lindberg (1983). *Proc. Natl. Acad. Sci. USA* **80**, 697-701.
7. Wadström, T. (1983). In "Staphylococci and Staphylococcal Infection" (Eds C. S. F. Easmon and C. Adlam), pp. 671-704. Academic Press, London and New York.

Biophysical Properties of Adhesins and Other Surface Antigens

Karl-Eric Magnusson[a], Erik Kihlström , Gizela Maluszynska,
Lena Öhman and Asa Walan

Department of Medical Microbiology, University of Linköping, Linköping, Sweden

CHEMICAL COMPOSITION AND BIOPHYSICAL PROPERTIES OF THE ENVELOPE OF GRAM-NEGATIVE BACTERIA

From a chemical point of view, the surface of bacteria displays a great variety of distinct molecules, which to the immune system act as individual antigens. From a cellular physiological aspect, some of these surface structures provide a protective layer, e.g. the lipopolysaccharide and capsular substance in Gram-negative bacteria. Others are involved in various transport functions across the outer membrane, or communication between the surrounding and the cytoplasmic space, such as the adhesion zones (Bayer's patches). Certain fimbrial surface appendages have an affinity for different carbohydrate structures, i.e. lectin-like activity [1]. Since they promote attachment to other cells and surfaces which display the appropriate sugar, they are also named adhesins. In *Escherichia coli* there are adhesins with mannose-sensitive (MS) as well as mannose-resistant binding (MR; [1-3]). The globotetraosylceramide-sensitive (GS) adhesins have been shown to be particularly important for the development of ascending urinary tract infections in women [4]. In other compartments of man, or in other species, other specific recognition systems are operative. Some bacterial "lectins" are not surface-associated, e.g. the cholera toxin. The function of cholera toxin was shown very early on to be critically dependent on the binding of the toxin to the GM_1-ganglioside receptor.

The major carbohydrate components of the envelope of Gram-negative bacteria are found in the lipopolysaccharide (LPS; O-antigen) and the capsular antigen (K-antigen). The composition of the repeating oligosaccharide unit in *E. coli* or

[a]*Telephone:* 46-13-192053

Protein–Carbohydrate Interactions in Biological Systems. ISBN 0 12 436665 1

Salmonella bacteria is highly specific. Thus, it is recognized by bacteriophages, but probably also by mammalian lectins present on, for example, circulating or tissue-associated macrophages. In the liver, these cells recognize, among others, mannose galactose and fucose residues on circulating antigens or immune complexes. Thus, there is a clear reciprocal presentation of carbohydrates and lectin activity in bacteria and mammalian cells.

From a biophysical point of view the complexity of the membrane structure is reduced to carbohydrate affinity and exposure, and physico-chemical surface properties, namely the tendency to hydrophobic interaction and the surface charge. To infer the specific effect of a particular surface component, defined alterations of the amount and quality of that component, e.g. by genetic engineering, has proven to be essential. We have for instance studied the surface properties of mutants of *Salmonella typhimurium* and *E. coli* as well as of clinical isolates of *Neisseriae gonorrhoeae*, *Yersinia enterocolitica* and *E. coli* [5,6].

METHODS USED FOR THE ASSESSMENT OF BIOPHYSICAL SURFACE PROPERTIES

The envelope of the bacteria was studied with respect to surface charge and hydrophobicity. These properties were assessed with aqueous two-phase partitioning (TPP) in dextran(Dx)/polyethylene glycol (PEG) systems with positively or negatively charged or hydrophobic ligands attached to PEG, with ion exchange chromatography (IEC) on DEAE-Sepharose and hydrophobic interaction chromatography (HIC) on Octyl-Sepharose [6]. The carbohydrate affinity was measured by the binding to appropriate carbohydrates exposed on erythrocytes or glycolipid-containing liposomes, respectively. Details on the methodology have been presented elsewhere [3,6,7]. The rationale for using two different methods to assess hydrophobic interaction is that the carrier of the ligand (fatty acid) in TPP is water-soluble PEG 4000 or 6000, and in HIC, a Sepharose matrix. This may affect ligand accessibility as well as side-effects induced by the polymers (PEG and Dx) or by the solid support. Other methods available for biophysical surface analysis have recently been reviewed [6].

BIOPHYSICAL SURFACE PROPERTIES

Effect of lipopolysaccharide mutations

By the use of a series of mutants derived from *Salmonella typhimurium* 395 MS and *S. minnesota* S99, we have found that S-R mutation is accompanied by increased negative surface charge and increased hydrophobic properties. We have also observed that "leaky mutants", i.e. R-type bacteria with some synthesis of complete LPS (S-type), have irregular partition in the two-phase system, and unexpected affinity for positively charged or hydrophobic gels. When similar LPS mutations were studied in a laboratory strain of *E. coli* K12, the effects on the surface properties were reduced. This was probably due to the presence of an acidic capsular (K−) antigen [5]. Öhman *et al.* further found that heat treatment

(70°C) which destroys the K-antigen affected the aqueous two-phase partitioning, unravelling the underlying S-R character of the LPS (O. Stendahl and L. Öhman, this volume). In a very recent series of experiments together with Dr C. Svanborg Edén and Dr L. Hagberg in Göteborg, we have studied a number of *E. coli* mutants derived from *E. coli* GR 12 (strain Nos. Hu 972, 973, 997, 998, 824 and 734) with various differences in the synthesis of O−(075+/−) antigen (unpubl. obs.). The results confirmed previous findings with *Salmonella* and *E. coli* bacteria.

To summarize, S-type O-antigen gives a hydrophilic surface with little negative charge, R-type O-antigen makes the bacteria liable to hydrophobic interaction and negatively charged, and K-antigen increases the hydrophilic properties and the negative surface charge. Furthermore, in relation to the interaction with various mammalian cells it appears that hydrophilic character of the envelope counteracts association with animal cells [5,6]. A negative surface charge *per se* does not seem to prevent attachment. The effect depends on whether the surface can be characterized as hydrophilic or hydrophobic.

Effect of the adhesins (fimbriae, pili)

It is well established that fimbriae promote attachment to animal cells and colonization of biological and other types of surfaces. This is of course particularly important when the biophysical properties of the envelope counteract attachment. We have studied the physico-chemical properties of two types of *E. coli* fimbriae, namely the MS and GS adhesins [3,4,7]. The bacteria studied have either been derived from patients with urinary tract infections, e.g. strains PN7 and ABU2 [3], or from faecal *E. coli* 506 bacteria redesigned with fimbriae-encoding plasmids from different sources [4]. For MS-adhesins we found a positive correlation between mannose-sensitive haemagglutination and the affinity for hydrophobic Octyl-Sepharose [3]. The bacteria also displayed mannose-specific binding to lipid vesicles containing a mannose-analogue (maltobioamide). The hydrophobic nature of different *E. coli* fimbriae has been graduated by Wadström and coworkers, who found that CFA/I > CFA/II > K88 > > K99 > Type 1 with respect to tendency to hydrophobic interaction [2]. To test the hydrophobic interaction of, for instance *E. coli* with type 1 fimbriae, the bacteria can be suspended in media with gradually reduced surface tension and then passed through Octyl-Sepharose. We thus found that cultivation in glucose-enriched broth reduced the tendency to hydrophobic interaction of the three *E. coli* strains isolated from patients with urinary tract infections (PN7, ABU2, and PN12), and that increasing concentrations of ethylene glycol (EG; 0–40% v/v) prevented a binding to the hydrophobic gel [7]. Furthermore, when 10% EG in phosphate-buffered saline, pH 7.3 (PBS) was made 0.1 M with respect to D-mannose, the affinity was even more reduced. This indicates that D-mannose reduced the hydrophobic character of the fimbriae, perhaps by blocking the hydrophobic combining site or by inducing a conformational change that denatured the adhesin. This effect was also seen in the *E. coli* 506 (MS+) strain [4]. Only D-mannose, and not D-glucose, D-galactose nor L-fucose, had this capacity [3]. This indicates that in the fimbriae–ligand interaction there was also a cooperation between the carbohydrate-specific part and a non-specific hydrophobic factor. By contrast, when a very hydrophobic bacterium like *E. coli* 506 without adhesins (MS −, GS −), was compared with the transformants with MS or GS adhesins, the latter had a hydrophilic effect. The percentage bacteria bound to Octyl-Sepharose, when suspended in PBS, was 90, 80 and 70% respectively [4]. With respect to the negative surface charge the consequences of a loss of mannose-sensitive binding depended on other surface structures as well. In the transformants of *E. coli* 506 it appeared that MS-adhesins decreased the negative charge as evidenced by the association to and elution with 0.5–1.0 M NaCl from DEAE-Sepharose. In contrast, GS-adhesins increased the negative charge [4]. In an otherwise hydrophilic bacterium, *E. coli* strain Hu 734 as compared to Hu 824, the effect was the opposite (unpubl. obs.).

Effect of environmental growth conditions

We have already described two environmental factors that strongly influence the surface properties of *E. coli* bacteria, namely the concentration of glucose in the growth medium and the culture period [7]. More recently we have found that anaerobic conditions with reduced redox potential increased the hydrophobicity of some *Salmonella* and *E. coli* bacteria, particularly of those with a complete O-antigen [8]. This suggests that the amount of O-antigen and/or K-antigen may be reduced under these circumstances. We find this observation challenging since in the intestine there is a largely anaerobic milieu, whereas aerobic conditions prevail in blood and other tissues with maintained blood flow. Therefore hydrophobic properties might promote the attachment to non-professional phagocytes of the intestinal wall, but the bacteria could then increase their resistance to phagocyosis by becoming more hydrophilic when they encounter professional phagocytes and other host defence measures in oxygen-rich compartments. For other bacteria like *Yersinia* the growth temperature is essential. We found for several *Yersinia enterocolitica* strains and *Yersinia enterocolitica*-like bacteria that the change of the temperature from 37°C to 22°C greatly influenced both the surface charge and hydrophobicity, and the expression of different virulence attributes. Recent experiments have shown that anaerobic growth conditions also affect the surface properties of the *Yersinia* bacteria. During an infection with *Gonococci* in the genital tract of women, the concentration of iron and the pH are two important variables [5]. *Gonococci* are in general much more hydrophobic than wild-type *Salmonella* and *E. coli* bacteria. They also very avidly bind to phagocytic cells and HeLa cells [5]. A low pH and a high concentration of iron increased the hydrophilicity of the gonococcal strains, thereby possibly also the virulence [6]. In this case, the surface properties could be linked to the presence of pili and colony-opacity associated proteins. Incidentally, a fresh clinical isolate appeared more hydrophilic than a laboratory-maintained strain (unpubl. obs.).

Role of antibodies and complement

It is well known that specific antibodies IgG promote contact and phagocytosis by polymorphonuclear leukocytes (PMNL) by increased interaction with the

Fc-receptor on the PMNL membrane. Furthermore, opsonization with complement enhances the interaction with the phagocytes through receptors for complement, primarily C3b, on the PMNL. Using two-phase partition we found that increasing concentrations of specific anti-*Salmonella typhimurium* 395MS IgG gradually transferred the bacteria from the PEG-rich top to the Dx-rich bottom phase. A hydrophobic effect was confirmed by the affinity for Octyl-Sepharose [6,9]. In either assay, opsonization occurred with such a small amount of IgG that the hydrophobic properties were barely affected; subsequent complement opsonization increased the tendency to hydrophobic interaction [6]. In contrast to IgG, secretory IgA (SIgA) had a hydrophilic effect, and reduced the association with PMNL [9]. When IgG and SIgA were reacted with the same bacterial strain, *S. typhimurium* 395MR10, SIgA modulated the biophysical effects of IgG, as well as the interaction with PMNL. Although SIgA reduced the association with PMNL, it augmented the binding to a mucus gel [9]. The affinity for mucus was established early, when SIgA was purified from mucosal secretions.

Both IgG and SIgA are glycoproteins, which may introduce new carbohydrate groups on the target cells, or by blocking characteristic bacterial sugars on the O- or K-antigen. Using a fluorescent lectin binding assay on IgG-opsonized *S. typhimurium* 395MS and MR10, and SIgA-sensitized *S. typhimurium 395MR10* or *E. coli* 086, we found that IgG introduced primarily galactose residues, whereas SIgA increased the binding of lectins specific for galactose, mannose and *N*-acetylglucosamine [10]. These findings may be of significance for the interaction with different phagocytic cells, e.g. in the liver. The clearance of particles by the liver depends on non-specific factors, as well as on specific lectin-like interactions with affinity for mannose and galactose.

SUMMARY

In order to describe general principles for the adhesion between micro-organism and eukaryotic cells, Gram-negative bacterial surface properties have been investigated with respect to surface charge and hydrophobic interaction, and lectin activity and lectin binding. How sensitization with antibodies SIgA and/or IgG or complement modify these properties was also studied. It was found that both S-type LPS and K-antigen give a hydrophilic character to the cell envelope, whereas R-LPS and type 1 pili increase the hydrophobicity. Furthermore, SIgA has a hydrophilic effect, while IgG and complement have a hydrophobic effect. Thus, there is a correlation between adhesion and hydrophobic interaction tendency, and hydrophilic characteristics and virulence/antiphagocytosis.

ACKNOWLEDGEMENTS

The studies reported here have been supported by the Swedish Medical Research Council (Project Nos. 5968 and 6251), King Gustaf Vth 80 year Fund and the Medical Research Council of the Swedish Life Insurance Companies.

REFERENCES

1. Firon, N., I. Ofek and N. Sharon (1983). Carbohydrate specificity of the surface lectins of *Escherichia coli, Klebsiella pneumoniae* and *Salmonella typhimurium. Carbohydr. Res.* **120**, 235-249.
2. Wadström, T., A. Faris, J. Freer, D. Habte, D. Hallberg and A. Ljung (1980). Hydrophobic surface properties of enterotoxigenic *E. coli* (ETEC) with different colonization factors (CFA/I, CFA/II, K88 and K99) and attachment to intestinal epithelial cells. *Scand. J. Infect. Dis. Suppl.* **24**, 148-153.
3. Öhman, L., K.-E. Magnusson and O. Stendahl (1982). The mannose-specific lectin activity of *Escherichia coli* type 1 fimbriae assayed by agglutination of glycolipid-containing liposomes, erythrocytes, and yeast cells and hydrophobic interaction chromatography. *FEMS Microbiol. Lett.* **14**, 149-153.
4. Svanborg-Eden, C., L.-M. Bjursten, R. Hull, S. Hull, K.-E. Magnusson, Z. Moldovano and H. Leffler (1984). Influence of adhesins on the interaction of *Escherichia coli* with human phagocytes. *Infect. Immun.* **44**, 672-680.
5. Magnusson, K.-E., J. Davies, T. Grundström, E. Kihlström and S. Normark (1980). Surface charge and hydrophobicity of *Salmonella, E. coli* and *Gonococci* in relation to their tendency to associate with animal cells. *Scand. J. Infect. Dis. Suppl.* **24**, 135-140.
6. Magnusson, K.-E., C. Dahlgren, G. Malusynska, E. Kihlström, T. Skogh, O. Stendahl, G. Söderlund, L. Öhman and A. Walan (1985). Non-specific and specific recognition mechanisms of bacterial and mammalian cell membranes. *J. Dispersion Sci. Technol.* **6**, 69-89.
7. Öhman, L., K.-E. Magnusson and O. Stendahl (1985). Effect of monosaccharides and ethyleneglycol on the interaction between *Escherichia coli* bacteria and Octyl-Sepharose. *Acta Path. Microbiol. Immunol. Scand. Sect. B* **93**, 133-138.
8. Maluszynska, G., O. Stendahl and K.-E. Magnusson (1985). Interaction between human polymorphonuclear leukocytes (PMNL) and bacteria cultivated in aerobic and anerobic conditions. *Acta Path. Microbiol. Immunol. Scand. Sect. B* **93**, 139-143.
9. Magnusson, K.-E. and I. Stjernström (1982). Mucosal barrier mechanisms. Interplay between secretory IgA (SIgA), IgG and mucins on the surface properties and association of salmonellae with intestine and granulocytes. *Immunology* **45**, 239-248.
10. Magnusson, K.-E. and L. Edebo (1984). Carbohydrate exposure on *Salmonella* and *E. coli* bacteria after reaction with antibody IgG and secretory IgA (SIgA) assessed with fluorescent lectins. *Immunol. Commun.* **13**, 151-160.

Interaction of Bacterial Adhesins with Inflammatory Cells

Olle Stendahl[a] and Lena Öhman

Department of Medical Microbiology, University of Linköping, Linköping, Sweden

The interplay between invading bacteria and the host defence is governed by specific ligand-receptor interaction in conjunction with more general physico-chemical surface properties [1]. How these entities interact is vital for the outcome of cell-bacteria interaction. In the presence of serum, the "immunological" ligands or opsonins are primarily immunoglobulin G (IgG) and activated complement factors C3b/bi. These molecules will interact with specific Fc and C3b/bi receptors on the granulocyte or macrophage. However, the cellular response elicited by these two ligand-receptor interactions are different. The activation of the Fc receptor leads to metabolic activation, ingestion and degranulation, whereas the activation of the C3b/bi receptors leads to a weak cellular response and slow phagocytosis [2].

It is well established that phagocytosis also occurs in the absence of serum. In this "non-immunological" milieu both non-specific physico-chemical properties and specific ligands, such as bacterial adhesins interacting with sugar residues on the eukaryotic cell, are operative. When evaluating the host-parasite interaction, both these mechanisms must be taken into account.

TYPE 1 FIMBRIAE MEDIATED INTERACTION WITH PHAGOCYTES

Type 1 fimbriae are hydrophobic mannose-specific bacterial lectin present on *Escherichia coli*, *Klebsiella* and *Salmonella* bacteria. Their expression on the bacterial surface is affected by various environmental factors, such as culture time and conditions [3]. The presence of type 1 fimbriae is associated with increased bacterial adherence both to granulocytes [3,4] and macrophages [5]. There is a good correlation between the extent of fimbriae expression and adherence [3],

[a]*Telephone:* 46-13-192050

Protein-Carbohydrate Interactions in Biological Systems. ISBN 0 12 436665 1

Table 1. Metabolic activation of granulocytes mediated by type 1 fimbriae interaction

	−D-mannose		+D-mannose	
Bacteria	*Iodination*[a]	*CL*[b]	*Iodination*	*CL*
PN 7	0.11	5.6	0.03	0.6
ABU2	0.10	6.2	0.02	0

[a] Expressed as nmol of I^- precipitated per 10^6 cells, 60 min.
[b] Chemiluminescence is expressed as peak value within 45 min (mV).

whereas other general physico-chemical surface properties, such as charge and hydrophobicity, have little influence on this lectin-mediated attachment. The subsequent bacterial interplay with the granulocyte is, however, affected by the general physico-chemical surface properties.

The lectin-mediated interaction leads to metabolic activation of the granulocyte with release of oxygen radicals (O_2^-, H_2O_2, $OH^·$) and myeloperoxidase-mediated protein iodination (Table 1). However, when phagocytic cells encounter invading bacteria, the ultimate goal for the host defence is irradication of the microbe. This is primarily accomplished after ingestion and intracellular killing.

To evaluate whether type 1 mediated adherence leads to ingestion, different smooth (PN7) and rough (ABU2) *E. coli* strains expressing type 1 fimbriae were tested. Whereas the smooth and rough bacteria adhered to the granulocytes to similar extent, the ingestion varied greatly. Seventy-five per cent of the rough ABU2 strain was ingested whereas less than 25% of the smooth strain was ingested (Table 2). When a bactericidal assay was used instead of the fluorescent quenching technique, no clear difference was observed between the two strains. This suggests that the activated granulocytes may kill associated bacteria either intracellularly through different mechanisms or extracellularly through generation and release of oxygen-derived reactive metabolites (O_2^-, H_2O_2, $OH^·$). These extra- and intracellular microbicidal mechanisms were recently proposed by Elsbach *et al.* (unpubl. obs.). These data show that the subsequent triggering of ingestion by adherent bacteria is not only dependent on the lectin-sugar interaction, but that the underlying physicochemical properties play a vital role—a rough, hydrophobic surface will facilitate phagocytosis, whereas a smooth hydrophilic surface will counteract ingestion.

When correlating the extent of type 1 fimbriae expression (assayed as mannose-sensitive liposome agglutination) with bacteria-granulocyte interaction, we found

Table 2. The mannose-sensitive association and ingestion of smooth (PN7) and rough (ABU2) *E. coli* bacteria

Bacteria	*Association*[a]		*Ingestion (%)*[b]
	−*mannose*	+ *mannose*	
PN7	365	10	25
ABU	375	10	75

[a] Number of bacteria/100 cells.
[b] Per cent of associated bacteria that were ingested.

Table 3. Mannose-sensitive association of *E. coli* strains to and ingestion by human granulocytes. Effect of different culture time

	Association/Ingestion[a]		
Bacteria	*16 h*[b]	*36 h*[b]	*48+48 h*[b]
PN7	180/10	200/15	360/10
ABU2	175/88	330/220	340/230

[a]Number of bacteria associated/ingested per 100 cells.
[b]Culture time.

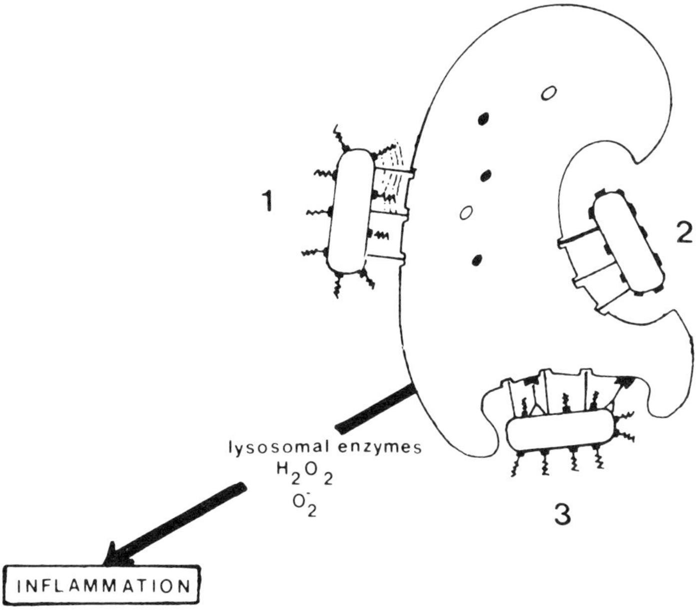

Figure 1. A proposed model of interaction between different type-1 fimbriae bearing *Escherichia coli* strains and human polymorphonuclear leukocytes. (1) A strain with hydrophilic properties (S-LPS and K-antigen) resists ingestion, but will stimulate the PMNL metabolically with release of potentially toxic oxidative metabolism. (2) A strain with hydrophobic properties (R-LPS) is avidly ingested and killed. (3) An opsonized, hydrophilic strain is avidly ingested and killed.

that association to and metabolic activation of granulocytes with increasing expression of type 1 fimbriae. Although the overall ingestion also increased, the percentage of associated bacteria that were ingested did not increase (Table 3). This is in contrast to Silverblatt *et al.* [6], who found a positive correlation between degree of piliation and susceptibility to phagocytosis. It has however also been pointed out that not only the extent, but also the organization, of the mannose-binding lectin on the bacterial surface is most important in promoting ingestion of bacteria attached to macrophages [5]. The role of type 1 fimbriae in affecting phagocytosis resembles that of C3b and fibronectin. Both these ligands will enhance association, but will only enhance ingestion when the particles express some other phagocytosis-promoting structures.

These data show that both granulocytes and macrophages expose mannose residues that are able to interact with type 1 fimbriae. However, other fimbriae, such as those binding to gal-gal chains of the globo series of glycolipids, are important adhesins promoting associations to eukaryotic cells [7]. This adhesin will however only bind to human granulocytes after the cells have been coated with the appropriate receptor glycolipid [4]. This suggests that this important virulence factor has little effect on the subsequent interaction with inflammatory cells.

EFFECT OF ANTIBODIES ON LECTIN-MEDIATED INTERACTION WITH PHAGOCYTES

When type 1 fimbriated bacteria are exposed to serum, the ingestion of the attached bacteria is greatly enhanced (>95% of the attached bacteria are ingested). However, D-mannose still partly inhibited the interaction, indicating that complement factor C3b and the lectin cooperate in this interaction [8]. It is, however, difficult to evaluate the specific role of complement in relation to mannose-specific lectins, since there appears to be a direct lectin–complement interaction (Öhman *et al.*, unpubl. obs.). To investigate how specific antibodies affect the lectin-mediated interaction, we used monoclonal antibodies to type 1 fimbriae and capsular (K13) antigen.

Addition of type 1 fimbriae antibodies increased the association and ingestion of the bacteria and metabolic activation of the granulocytes. This antibody-mediated stimulation was, however, inhibited by D-mannose [9]. One possible explanation for the enhancement of the bacterial interaction by type 1 fimbriae antibodies could be aggregation of the fimbriae. It would then be expected that mannose should inhibit the interaction also at high antibody concentration. The fact that the inhibitory effect of D-mannose decreased with increasing antibody concentration rather implies that the antibody interacts with the granulocyte through its Fc-part and cooperates with the mannose-specific lectin interaction. In this interplay the lectin promotes attachment and the Fc-moiety triggers ingestion. Also anti-capsular antibodies enhanced attachment and ingestion of type 1 fimbriated bacteria and interestingly, this interaction was completely inhibited by D-mannose [9]. This effect of anti-capsular antibodies on type 1 fimbriated bacteria also supports the suggested cooperation between Fc− and lectin-mediated interaction with granulocytes.

THE EFFECT OF BACTERIAL LECTINS ON INFLAMMATION

It is well known that soluble lectins such as Concanavalin A may activate granulocytes and macrophages. Several experiments also show that the mannose-specific bacterial lectin (type 1 fimbriae) triggers the generation and release of inflammatory mediators, such as oxygen-derived metabolites (O_2^-, H_2O_2, $OH^\cdot$) and granulae enzymes. This process is particularly augmented when the lectin-associated bacteria are located extracellularly [8] on the surface of granulocytes and macrophages. We can thus postulate that type 1 fimbriae may enhance uptake and inactivation of rough, non-virulent bacteria, whereas the same lectin on smooth and virulent bacteria that resist phagocytosis, may rather augment the inflammatory response. Antibodies to the lectins and other surface components promote ingestion in cooperation with the lectin. It is thus important to take into account the different virulence-promoting surface properties on *E. coli*, lipopolysaccharide, and capsular antigen when evaluating the role of mannose-specific type 1 fimbriae and other adhesins in bacterial invasion and pathogenicity.

ACKNOWLEDGEMENTS

Part of the work referred to in this paper was supported by grants from the Swedish Medical Research Council (5968, 6251), King Gustaf V:s 80-year Foundation and Östergötlands läns landsting.

REFERENCES

1. Stendahl, O. (1983). The physicochemical basis of surface interaction between bacteria and phagocytic cells. In "Medical Microbiology", Vol. 3, pp. 137-152. Academic Press, London and New York.
2. Hed, J. and O. Stendahl (1982). Differences in the ingestion mechanisms of IgG and C3b particles in the phagocytosis by neutrophils. *Immunol.* **45**, 727-736.
3. Öhman, L., K-E. Magnusson and O. Stendahl (1985). Mannose-specific and hydrophobic interaction between *Escherichia coli* and polymorphonuclear leukocytes—influence of bacterial culture periods. *Acta Path. Microbiol. Scand. Sect. B* **93**, 125-133.
4. Svanborg Edén, C., L-M. Bjursten, R. Hull, S. Hull, K-E. Magnusson, Moldovano and H. Leffler (1984). Influence of adhesions in the interaction of *Escherichia coli* with human phagocytes. *Infect. Immun.* **44**, 672-680.
5. Bar-Shavit, Z., R. Goldman, I. Ofek, M. Sharon and D. Mirelman (1980). Mannose binding activity of *Escherichia coli*: a determinant of attachment and ingestion of the bacteria by macrophages. *Infect Immun.* **29**, 417-424.
6. Silverblatt, F. J., J. S. Dreyer and S. Schauer (1979). Effect of pili on susceptibility of *Escherichia coli* to phagocytosis. *Infect. Immun.* **24**, 218-223.
7. Leffler, H. and C. Svanborg Edén (1980). Chemical identification of a glycosphingolipid receptor for *Escherichia coli* attaching to human urinary tract epithelial cells and agglutinating human erythrocytes. *FEMS Microbiol. Lett.* **8**, 127-134.
8. Öhman, L., J. Hed and O. Stendahl (1982). Interaction between human polymorphonuclear leukocytes and two different strains of type 1 fimbriae-bearing *E. coli* *J. Infect. Dis.* **146**, 751-757.
9 Söderström, T. and L. Öhman (1984). The effect of monoclonal antibodies against *E. coli* type 1 pili and capsular polysaccharides on the interaction between bacteria and human granulocytes. *Scand. J. Immunol.* **20**, 299-305.

Poster Session

Breast Milk Inhibition of Adhesion of *S. pneumoniae* and *H. influenzae*

B. Andersson[a], O. Porras, L. Å. Hanson, H. Leffler* and C. Svanborg-Edén

*Department of Clinical Immunology and *Department of Medical Biochemistry, University of Göteborg, Göteborg, Sweden*

Microbial attachment to host mucosal surfaces is one of the first events to take place in host-parasite interactions. *Streptococcus pneumoniae* and *Haemophilus influenzae* are the major causes of upper respiratory tract infections and otitis media. Both species show the ability to adhere to human upper respiratory tract epithelial cells. Thus, interference with attachment *in vitro* might also reduce the colonization rate *in vivo* with these bacteria. The aim of this study was to analyze the inhibitory effect of human milk on attachment.

Four highly adhering strains of both *S. pneumoniae* and *H. influenzae* were used. They were tested for adhesive capacity as previously described [1,2]. Adhesion was given as the mean number of attaching bacteria per epithelial cell. In inhibition experiments, adhesion with inhibitor was given as per cent of the adhesion buffer (EID_{50} = Effective inhibitory dose reducing to 50% of the buffer control (50% inhibition). Human milk was obtained during the first week of lactation and was defatted by centrifugation. Some samples were separated in a high molecular weight fraction (HMWF, molecules with more than 12 400 D) and a low molecular weight fraction (LMWF) by filtration through a collodion bag (Sartorius, Göttingen, FRG). Other samples were treated with anti-human immunoglobulin on solid phase to remove the antibodies (Rabbit anti-human immunoglobulin; Dakopatts, Copenhagen, Denmark; coupled to Affigel-10; Bio-Rad, Richmond, CA, USA using the procedure recommended by the manufacturers). Antibodies (total secretory IgA or anti-phosphoryl choline for *S. pneumoniae* and anti-polysaccharide type b or anti-outer membrane preparation for *H. influenzae*) were determined in enzyme-linked immunosorbent assays. Synthetic saccharides and natural saccharides for human milk were

[a] *Telephone:* 46-31-602082

Protein-Carbohydrate Interactions in Biological Systems. ISBN 0 12 436665 1

Table 1. Inhibitory activity of unfractionated and fractionated human milk on adhesion of *S. pneumoniae* and *H. influenzae* to human upper respiratory tract epithelial cells

Milk sample	*Adhesion (% of buffer control)*	*Secretory IgA (g/l)*
S. pneumoniae		
Whole milk (A) strain 1	6	
Whole milk (A) strain 2	0	
Whole milk (A) strain 3	0	
Whole milk (A) strain 4	1	
Whole milk (B)	—	
HMWF (B)	8	
LMWF (B)	44	
HMWF (C)	9	1.090
HMWF (C), immunoabsorbed	31	0.000
H. influenzae		
Whole milk (D) strain 1	7	
Whole milk (D) strain 2	0	
Whole milk (D) strain 3	3	
Whole milk (D) strain 4	23	
Whole milk (E)	15	
HMWF (E)	1	
LMWF (E)	76	
Whole milk (F)	18	0.087
Whole milk (F), immunoabsorbed	49	0.002

Table 2. Inhibition of adhesion of *S. pneumoniae* by oligosaccharides from human milk

Structure	*Source*	EID_{50} *(mg/ml)*
Oligosaccharide preparation	Human milk	6.0
Galβ1→4GlcNAcβ1→3Galβ1→4Glc	Human milk	0.2
Galβ1→3GlcNAcβ1→3Galβ1→4Glc	Human milk	1.3
GlcNAcβ1→3Gal-β-O-Me	Synthetic	3.5
GlcNAcβ1→4Gal-β-O-Me	Synthetic	>10.0

provided by the Swedish Sugar Co, Arlöv, Sweden and Dr D. Zopf, Bethesda, Md, USA respectively.

Table 1 shows that whole breast milk inhibited all strains tested of both organisms. The HMWF inhibited attachment of both bacteria both before and to a lesser extent also after reduction the secretory IgA (specific antibodies were also removed; data not shown). *S. pneumoniae* was also inhibited by the LMWF. This fraction contains no antibodies but oligosaccharides corresponding to a receptor for attaching pneumococci [1]. Purified oligosaccharides from human milk are also inhibitory (Table 2). The oligosaccharide preparation (first structure in Table 2) contains about 25% of lacto-*N*-tetrarose (the third structure) and has about 22% of its activity.

Breast milk inhibited adhesion of *S. pneumoniae* and *H. influenzae* to human mucosal cells. The anti-adhesive activity against *S. pneumoniae* was found both in the HMWF and in the LMWF. In the HMWF the activity was present also after removal of the antibodies. The activity in the LMWF may be related to its content of oligosaccharides corresponding to the receptor saccharides of *S. pneumoniae* [1].

The anti-adhesive activity against *H. influenzae* was only found in the HMWF and was unrelated to the content of antibodies of these bacteria.

The inhibitory activity of breast milk on adhesion of *S. pneumoniae* and *H. influenzae* may have a protective potential on colonization and infection with these bacteria in breast-fed children.

REFERENCES

1. Andersson, B., J. Dahmén, T. Frejd, H. Leffler, G. Magnusson, G. Noori and C. Svanborg-Edén (1983). *J. Exp. Med.* **158**, 559–570.
2. Porras, O., C. Svanborg-Edén, T. Lagergärd and L. A. Hanson (1985). *Eur. J. Clin. Microbiol.* **4**, 310–315.

Fimbriae of *Rhizobium leguminosarum* and *Rhizobium trifolii*

Gerrit Smit[a], Jan W. Kijne and Ben J. J. Lugtenberg

Department of Plant Molecular Biology, Botanical Laboratory, University of Leiden, Leiden, The Netherlands

The soil bacterium *Rhizobium* attaches to the roots of leguminous plants as a first step in an infection process leading to a nitrogen-fixing symbiosis. The role of fimbriae in the attachment process is currently under study.

Fimbriae of *Rhizobium leguminosarum* strain 248, cultivated in TY-medium [1], were found to be 5-6 nm in diameter and up to 15 μm long. Another type, 3-4 nm in diameter, was rarely observed. Fimbriae were formed in batch culture and standing culture, as well as on solid media, and were also formed at low temperatures (10°C). Most fimbriated cells were found at the late logarithmic phase of growth in batch culture (up to 80% of the cells). *R. trifolii* strain 5523 produces a similar type of fimbriae in abundance during early to mid-log phase. Also *sym*-plasmid cured *R. leguminosarum* and *R. trifolli* strains were able to produce fimbriae. Haemagglutination (human, guinea pig and calf erythrocytes) by rhizobial cells was found to be correlated with the degree of fimbriation of the culture, and could not be inhibited by mannose and galactose.

Attachment of the rhizobia to pea root hairs was correlated with the amount of fimbriated cells in batch culture. *Sym*-plasmid cured strains of both *R. leguminosarum* and *R. trifolii* were able to attach to the roots, which indicates that *sym*-plasmid borne nodulation genes are probably not involved in the attachment process. Addition of different monosaccharides (e.g., mannose, 3-*O*-methylglucose: pea lectin haptens) did not inhibit attachment of *R. leguminosarum*, which contradicts the lectin-recognition hypothesis for rhizobial adherence to host plant cells [2].

[a] *Telephone:* 31-71-148333 ext. 6659

Protein-Carbohydrate Interactions in Biological Systems. ISBN 0 12 436665 1

REFERENCES

1. Beringer, J. E. (1974) R factor transfer in *Rhizobium leguminosarum*. *J. Gen. Microbiol.* **84**, 188–198.
2. Dazzo, F. B., C. A. Napoli and D. H. Hubbell (1970). Adsorption of bacteria to roots as related to host specificity in the Rhizobium-clover symbiosis. *Appl. Env. Microbiol.* **32**, 166–171.

Adhesion of an ETEC Strain Mediated by a Non-Fimbrial Adhesin

C. Forestier[a], A. Darfeuille-Michaud, B. Joly and R. Cluzel

Service de Bactériologie, Faculté de Pharmacie, 63001 Clermont Ferrand Cedex, France

INTRODUCTION

Enterotoxigenic *Escherichia coli* (ETEC) cause acute diarrhoea in both humans and animals. Two steps are involved in the pathogenesis of ETEC: colonization of the intestinal epithelium and production of toxins. Colonization is mediated by pilus-like antigens such as K88, K99, CFA/I and CFA/II, which are responsible for the adherence of ETEC strains to the intestinal epithelium.

We examined many ETEC strains isolated from cases of human diarrhoea in Senegal (Africa) for their ability to adhere to the human intestinal epithelial cells. Among adhesive strains, we found one of them (strain 2230) to possess a cell surface non-fimbrial protein which mediates adherence of bacteria to the brush border of human enterocyte.

Table 1.

	Erythrocytes			
Strains	*Human (A+)*	*Bovine*	*Chicken*	*Guinea pig*
H 10407 (CFA/I)	R	R	R	N/S
Pb 176 (CFA/II)	N	R	R	N/S
2230	N	N	N	N/S

R = mannose-resistant HA.
S = mannose-sensitive HA.
N = no HA.

[a] *Telephone:* 33-73-265675

Protein-Carbohydrate Interactions in Biological Systems. ISBN 0 12 436665 1

Table 2.

Strains	*Pili 1*	*Adhesion*
H 10407	+	+ +
(CFA/I$^+$)	−	+ +
H 10407p	+	−
(CFA/I$^-$)	−	−
Pb 176	+	+ +
(CFA/II$^+$)	−	+ +
Pb 176p	+	−
(CFA/II$^-$)	−	−
2230	−	+ +
	−	+ +
2230 p	+	−
	−	−

+ + = adhesion index superior to 20.
− = adhesion index inferior to 5.

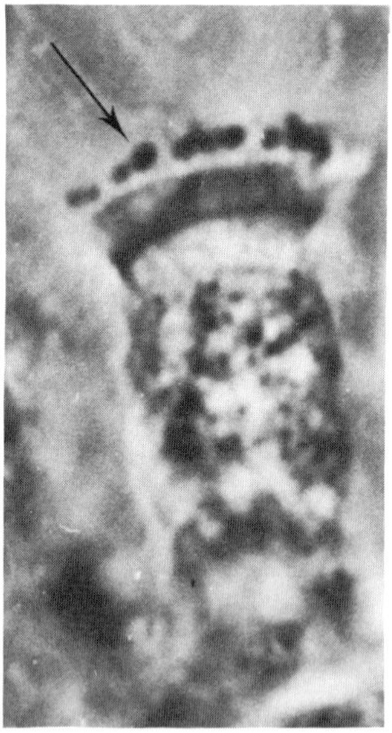

Figure 1.

MATERIALS AND METHODS

Preparation of antisera. Immunization of rabbits with purified pili infected intravenously. Each antiserum was absorbed against negative adhesive variant to obtain specific antiserum.

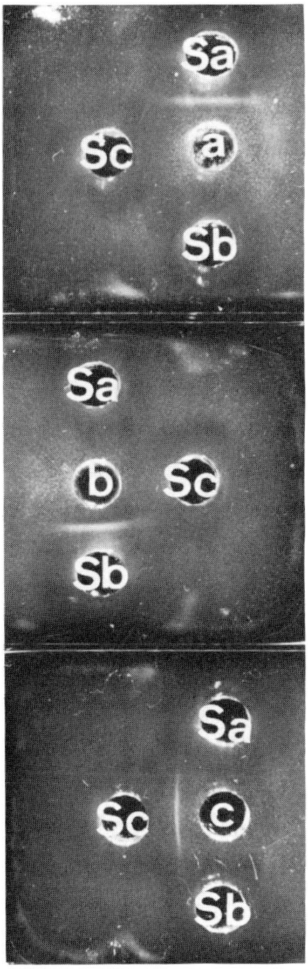

Figure 2.

Double immunodiffusion test. A saline extract of 2230 strain was tested against each CFA antiserum by the Outcherlony gel immunodiffusion technique.

Bacterial strains. E. coli 2230 strain isolated from a case of diarrhoea in Dakar (Senegal) and its spontaneous negative variant 2230p. CFA positive strains for comparative studies: H10407 (CFA/I), Pb 176 (CFA/II) and their negative variants H10407p, Pb 176p.

In vitro *adhesion assay.* With human duodenal enterocytes.

Haemagglutination tests. Slide agglutination at room temperature and at 4°C with human (A+), bovine, chicken and guinea pig erythrocytes.

Electron microscopy. Observation of bacteria after staining with 1% phosphotungstic acid, pH 6.8.

RESULTS

Haemagglutination (HA) pattern. Results are given in Table 1.

In vitro *enterocytes adhesion capacity (Fig. 1).* Results are given in Table 2.

Electron microscopy. E. coli 2230 strain did not show any filamentous structures on the bacterial surface. 2230 antigen extract did not contain any identified filamentous fimbria-like structures, whereas CFA/I and CFA/II extracts consisted in fimbrial antigens that tend to aggregate.

Antigenicity of E. coli *2230 antigen (Fig. 2).* Immunodiffusion of antigen extract. Well (a), antigen 2230; well (b), antigen Pb 176 (CFA/II); well (c), antigen H 10407 (CFA/I). Other wells contain specific antisera prepared against strains 2230 (Sa), Pb 176 (Sb) and H 10407 (Sc).

CONCLUSION

The ETEC strain 2230 adheres to the brush border of human enterocytes and does not cause any MRHA with human (A^+), bovine, chicken or guinea pig erythrocytes. The adhesive factor of this strain does not show filamentous structures under the electron microscope and does not cross-react with antisera against classical adhesive factors CFA/I (H 10407) and CFA/II (Pb 176) by the immunodiffusion test. 2230 antigen may represent another colonization factor antigen. Further experiments are in progress to characterize biochemically this adhesin and to determine its genetic control.

REFERENCES

1. Evans, D. G. and D. J. Evans (1978). New surface-associated heat labile colonization factor antigen (CFA/II) produced by enterotoxigenic *Escherichia coli* of serogroups 06 and 08. *Infect. Immun.* **21**, 638-647.
2. Evans, D. G., R. P. Silver, D. J. Evans, D. G. Chase and S. L. Gorbach (1975). Plasmid controlled colonization factor associated with virulence in *Escherichia Infect. Immun.* **12**, 656-667.

Fimbriae of *Bordetella pertussis*

L. A. E. Ashworth[a], A. Robinson, L. I. Irons,
A. B. Dowsett, A. Gorringe and P. Wilton-Smith

Experimental Pathology and Vaccine Research and Production Laboratories, PHLS Centre for Applied Microbiology and Research, Porton, Salisbury, UK

INTRODUCTION

Adherence of *Bordetella pertussis* to the ciliated mucosa is necessary for colonization of the upper respiratory tract by this non-invasive organism. To prevent infection, and thereby provide herd immunity, vaccines should contain those surface antigen(s) which act as adhesins. From knowledge of other organisms, fimbriae are potentially important adhesins and we have explored the properties of fimbriae of *B. pertussis* in our work on a subcellular pertussis vaccine.

FIMBRIAL ANTIGENS

B. pertussis organisms bear fimbriae of diameter about 4 nm as seen in the electron microscope. The length and number of fimbriae vary with growth conditions and with serotype; serotype 1,3 < serotype 1,2 < serotype 1,2,3.

We have previously shown by immunoelectron microscopy using ferritin-tagged polyclonal agglutinin 2 [1], that serotype antigen (agglutinogen) 2 is borne by fimbriae. This work has been extended using monoclonal antibodies (mAbs) to agglutinogens 1,2 and 3 (IgM McAbs AG1, AG2 and AG3, respectively). These McAbs, either untagged or labelled with colloidal gold, reacted with fimbriae of *B. pertussis* as follows: AG2 labelled 90% of fimbriae on cells of serotype 1,2 but only 4% on cells of serotype 1,3; the reverse was found with AG3. AG1 did not react with the fimbriae of any strain, although it agglutinated all Phase 1 strains of *B. pertussis*.

In an attempt to determine whether individual fimbriae can bear both antigens 2 and 3, AG3-15 nm gold and an IgG McAb to agglutinogen 2 tagged with 3 nm

[a] *Telephone:* 44-980-610391

Protein-Carbohydrate Interactions in Biological Systems. ISBN 0 12 436665 1

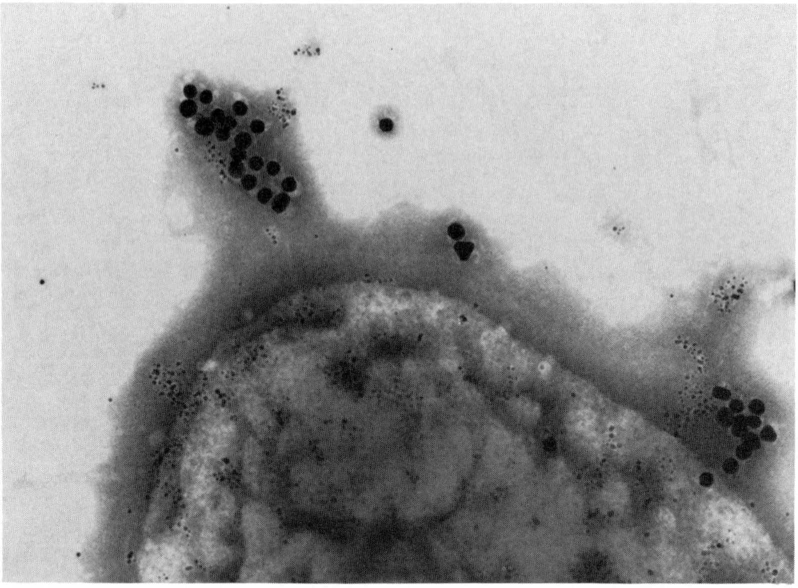

Figure 1. Serotype 1,2,3 *Bordetella pertussis* double labelled with anti-2 and anti-3 monoclonal agglutinins tagged with 3 nm and 15 nm diameter gold particles, respectively.

gold were used to label a serotype 1,2,3 strain. Individual cells bore both labels (Fig. 1) but the question of double labelling of individual fimbriae is still open.

ISOLATION OF FIMBRIAE

A competitive ELISA utilizing McAb AG2 [2] and an assay based on the inhibition of bacterial agglutination by polyclonal anti-2 were used to monitor the release of fimbriae from organisms of 1,2,3 serotype by several methods. Homogenization for 15 min released more than 50% of the cell-bound agglutinogen 2 into solution. Cell disintegration (Braun homogenizer) or sonication released almost all cell-bound agglutinogen 2, but with a marked loss of activity in ELISA. Heating at 60°C gave little release, while extraction with 3M KSCN or heating at 60°C in 2M urea released up to 75% of the agglutinogen with little loss of antigenicity.

The agglutinogen successfully released was treated with DNase and RNase, concentrated 20-fold and run on columns of Sepharose CL 6B. Agglutinogen 2 activity eluted as a single peak near the void volume.

PROPERTIES OF FIMBRIAE

Fimbriae isolated from organisms of serotypes 1,2 and 1,2,3 had similar dimensions to those observed on organisms in the electron microscope. There was a tendency to associated side by side to form bundles, fimbriae from serotype 1,3 organisms showing this property to a much greater extent than those from serotypes 1,2 and 1,2,3. Isolated fimbriae of serotypes 2 and 3 could be labelled in immunoelectron microscopy with McAbs AG2 and AG3, respectively.

The fimbriae of serotypes 2 and 3 gave single bands at 22.5kD and 22kD, respectively, on SDS-PAGE. Neither polyclonal anti-2 antibody nor McAbs AG2 and AG3 reacted with Western blots of these gels. Fimbriae purified from a 1,2,3 serotype did not agglutinate erythrocytes of man, horse, dog, ox, sheep, or goose.

FIMBRIAE AS ADHESINS

In an *in vitro* assay of adherence of X-mode *B. pertussis* to Vero cells, monoclonal anti-2 strongly inhibited the adherence of strains bearing type 2 fimbriae. Serotype-specific inhibition by monoclonal anti-3 was weaker but significant [3].

CONCLUSIONS

The above findings show that serotype agglutinogens 2 and 3 of *B. pertussis* are borne by fimbriae and that these can act as adhesins. The evidence for this also indicates that inhibition of adherence by antibody is a realistic goal for vaccination, although the need for this antibody to be available at the site of colonization should be recognized. It is hoped to identify the receptor for *B. pertussis* fimbriae on Vero cells and determine its expression on cells of the upper respiratory tract.

Whole cell pertussis vaccine contains agglutinogens and serum agglutinins have been widely utilized as the prime indicator of individual anti-pertussis immunity. Despite this, the acellular vaccines now under trial consist of pertussis toxin (toxoided) and filamentous haemagglutinin (FHA) with low levels of agglutinogen 2 present only as a contaminant. This is remarkable in that FHA was originally considered as a protective antigen because it was thought to be fimbrial. However, present evidence [1] shows that FHA is not a fimbrial antigen.

These studies strengthen the case for inclusion of fimbrial agglutinogens in acellular pertussis vaccines.

REFERENCES

1. Ashworth, L. A. E., L. I. Irons and A. B. Dowsett (1982). *Infect. Immun.* **37**, 1278.
2. Ashworth, L. A. E., A. Robinson, L. I. Irons, C. P. Morgan and D. Isaacs (1983). *Lancet* **ii**, 878.
3. Gorringe, A., L. A. E. Ashworth, L. I. Irons and A. Robinson (1984). *FEMS Microbiol. Lett.* **26**, 5.

Structure-Function Relationships in Diphtheria Toxin as Deduced from the Sequence of Three Non-Toxic Mutants

R. Rappuoli[a], G. Ratti, G. Giannini, M. Perugini and J. R. Murphy*

*Sclavo Research Centre, Siena, Italy, and *Section of Biomolecular Medicine, University Hospital, Boston University Medical Centre, Boston, MA, USA*

Diphtheria toxin is a protein of 58 350 D (535 amino acid residues), produced by *Corynebacterium diphtheriae* lysogenic for one of the phages which carry the toxin gene (tox). The toxin can be cleaved by mild trypsin digestion into two functionally distinct subunits: fragment A and fragment B [1,2]. Fragment A catalyses the NAD^+-dependent ADP-ribosylation of elongation factor 2 (EF2), thereby inhibiting protein synthesis in eukaryotic cells. Fragment B is thought to mediate the entry of fragment A into the cytosol of eukaryotic cells. This fragment can be divided into three functionally distinct regions: a $-COOH$ terminal region which binds to toxin receptors on the surface of sensitive cells, and two $-NH_2$ regions which are highly hydrophobic and are involved in the translocation of fragment A across the cell membrane. The toxin contains two disulphide bridges: the first is between cys-186 and cys-201 and defines the junction between fragment A and fragment B; the second is between cys-461 and cys-471. Following nitrosoguanidine mutagenesis of phage beta, which carries the *tox* gene, several mutants have been obtained which encode for proteins immunologically related to diphtheria (Cross Reacting Materials, CRMs).

We have determined the nucleotide sequence of CRM 197 (which has a non-functional fragment A), CRM 1001 (which is unable to mediate the translocation of fragment A into the cytoplasm) and CRM 45 (which is unable to bind the

[a]*Telephone:* 39-577-293111

Protein-Carbohydrate Interactions in Biological Systems. ISBN 0 12 436665 1

receptors). This analysis identifies the amino acids and/or the regions of toxin which are essential for the intoxication of intact eukaryotic cells.

CRM 197 is a non-toxic protein of 58 350 D which has a functional B fragment, but an enzymatically inactive fragment A. Sequence analysis has shown that a single G to A transition which changes glycine-52 to glutamic acid, abolishes the ability of fragment A to bind NAD [3]. Glycine-52 is likely to be part of the NAD-binding site as Tryptophan-50 [4] has also been shown to be involved in the binding of NAD.

Since CRM 197 is non-toxic and immunologically indistinguished from diphtheria toxin, it could be used as a new vaccine against diphtheria. Indeed, it has been shown to induce protective levels of antibodies in animal model systems and to be an effective carrier for bacterial-polysaccharide antigens. Compared to the diphtheria vaccine currently used, CRM 197 would have the advantage of absolute safety and higher purity.

CRM 1001 is a non-toxic protein of 58 350 D which has an enzymatically active fragment A and a fragment B which binds to the toxin receptors on the cell surface, but is not able to mediate the translocation of fragment A into the cytosol. The tox-1001 allele has been shown to have a point mutation which changes cysteine-471 into tyrosine, thus affecting the formation of the disulphide bridge in the *C*-terminal portion of the molecule.

Since CRM 1001 can compete with diphtheria toxin for the receptors on the surface of the cells, this disulphide bridge is not necessary for the binding of toxin receptors. However, CRM 1001 is non-toxic for CHO cells up to 10^{-7} M (i.e. at least 10^3 times less toxic than wild-type diphtheria toxin), therefore the mutation described above must block the process which, after binding to the cell surface, leads to the transfer of the *N*-terminal fragment A into the cytosol. Furthermore, since 17 out of 27 monoclonal antibodies which react with wild-type fragment B do not react with CRM 1001 [5], the loop determined by the 461-471 disulphide bridge is likely to be a major antigenic determinant.

CRM 45 is a non-toxic protein of 41 800 D which has a functional fragment A, but its fragment B is truncated and is unable to bind the toxin receptor on the cell surface. This alteration is due to a single C to T transition which changes glutamic acid-387 into a stop codon and causes premature termination of the protein [3]. From this we can conclude that the receptor binding domain of diphtheria toxin is contained within the *C*-terminal 149 amino acids.

REFERENCES

1. Pappenheimer, A. M. Jr (1977). *Ann. Rev. Biochem.* **46**, 69–94.
2. Ratti, G., R. Rappuoli and G. Giannini (1983). *Nucleic Acids Res.* **11**, 6589–6595.
3. Giannini, G., R. Rappuoli and G. Ratti (1984). *Nucleic Acids Res.* **12**, 4063–4069.
4. Michel, A. and J. Dirkx (1977). *Biochim. Biophys. Acta* **491**, 286–295.
5. Zucker, D. R. and J. R. Murphy (1984). *Molecular Immunol.* **21**, 785–793.

Conformation of Small Peptides in Solution, Determined by N.M.R. Spectroscopy and Computer Simulation

Morten Meldal[a]

The Technical University of Denmark, Lyngby, Denmark

The solution conformations in dimethylsuphoxide of the peptides Boc-Asn-Tyr-OMe (1) and Phe-Asn-Glu-Asn-Met-Ala-Tyr-OMe (2), the latter a proposed antigen from *Eschericia coli* K88 and protein fimbriae [1], have been determined using a modified version of the minimization program "GESA" generally used for energy minimization of conformations of oligosaccharides [2,3].

The energy calculations include VDW, hydrogen bond, electrostatic and torsional interactions and minimization is carried out under variance of ϕ ψ, ω and ξ. The number of starting conformations for the program is greatly reduced by experimentally determined short proton-proton distances found by one- and two-dimensional experiments. Thus the number of starting conformations for (1) could be limited from 442 possible starting conformations to only 8 allowed conformations, and there was excellent agreement between the distances in the calculated minimum energy conformation (cf. fig. 1) and the short proton-proton distances found by N.O.E. measurements.

For compound (2) the allowed conformations could be reduced from 0.6×10^9 to 6900 possible conformations by N.O.E. experiments. These were reduced further by 260 minimizations of successive tripeptide fragments allowing only 25 different conformations of the heptapeptide. The minimum energy conformation of the heptapeptide (2), cf. Fig. 2, which linked to BSA was found to be a potent antigen, suggests that the peptide is positioned on the surface of the protein with the polar groups pointing towards the surroundings.

[a] *Telephone:* 45-2-883236

Protein-Carbohydrate Interactions in Biological Systems. ISBN 0 12 436665 1

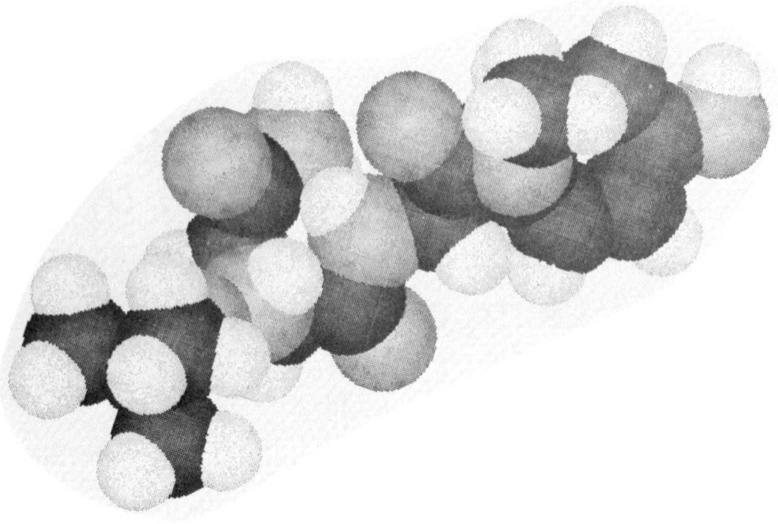

Figure 1. Minimum energy conformation of the peptide Boc-Asn-Tyr-OMe.

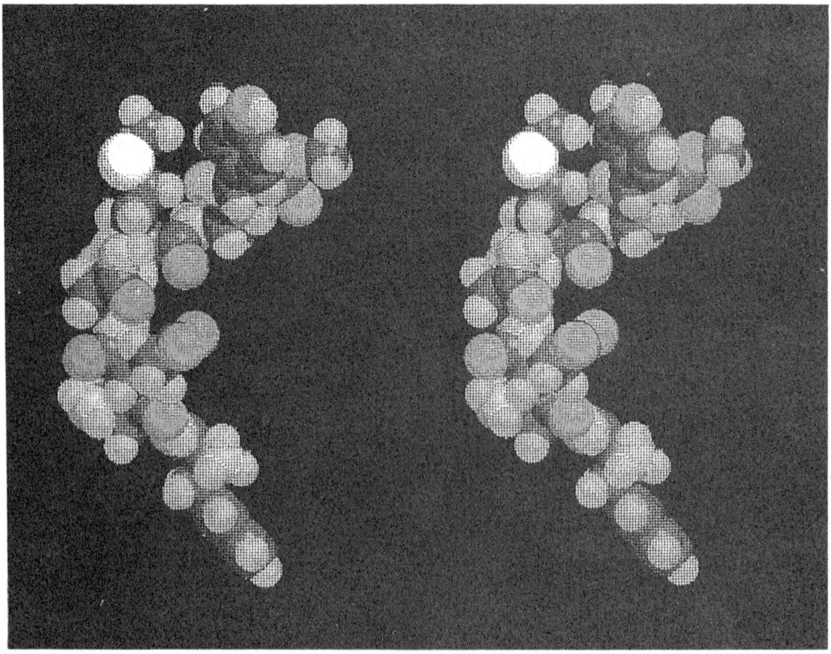

Figure 2. Minimum energy conformation of the peptide Phe-Asn-Glu-Asn-Met-Ala-Tyr-OMe in a stereo projection.

REFERENCES

1. Klemm, P. and L. Mikkelsen (1982). *Infect. Immun.* **38**, 41-45.
2. Bock, K. (1983). *Pure Appl. Chem.* **55**, 605-622.
3. Meyer, B. (1982). XIth International Carbohydrate Symposium, Vancouver 1982, Abstract II/25.

VI. Bacterial Invasion

Co-Chairpersons: Helena Mäkelä
Hans Wolf-Watz

Invasion of Eukaryotic Cells by *Shigella*: A Genetic Approach

P. J. Sansonetti[a], A. T. Maurelli, A. Ryter*, B. Baudry and P. Clerc

Service des Entérobactéries, and
**Unité de Microscopie Electronique, Département de Biologie Moléculaire, Institut Pasteur, Paris, France*

INTRODUCTION

The main step in the pathogenesis of bacillary dysentery is the invasion of the human colonic epithelial cells by the pathogen [1]. This invasive process encompasses complex features including penetration into epithelial cells, intracellular multiplication, spreading to adjacent cells and to the conjunctive tissue of the intestinal villus. Repetition of this process causes death of the epithelium and allows diffusion of microorganisms in the lamina propria in which they evoke an intense inflammatory reaction. This reaction kills bacteria which rarely reach the submucosa but also causes abscesses and ulcers which generate the dysenteric symptoms.

Several laboratory models are available to assay for the organism's capacity to invade epithelial cells. These include the Sereny test [2], which detects an organism's ability to produce a keratoconjunctivitis in guinea pig by invading corneal epithelial cells; tests employing mammalian cells monolayers, HeLa or others, which explore the steps of intracellular penetration and multiplication [3,4]; and the histologic or electron microscopic examination of the intestine of experimentally infected animals [3]. Another virulence factor is the Shiga toxin which is both enterotoxic, thus likely to be responsible for the jejunal fluid loss in shigellosis [5], and cytotoxic through inhibition of eukaryotic cell protein synthesis [6]. The actual role of this toxin in the pathogenic process is still unclear.

[a] *Telephone:* 33-1-3061919

Protein-Carbohydrate Interactions in Biological Systems. ISBN 0 12 436665 1

PLASMID MEDIATED INVASIVENESS OF EUKARYOTIC CELLS

Plasmids in the 120 to 140 MD size range (190–220 kb) are associated with the invasive phenotype of the four *Shigella* species [1,7,8]. Variants which have lost this plasmid are no longer invasive in the experimental models. They regain invasiveness through transfer of this plasmid M90T, a *S. flexneri* serotype 5 carries the 220 kb virulence plasmid pWR100. This plasmid is absent in its non-invasive derivative BS176.

Although chromosomal loci are also required for expression of a complete virulence phenotype [9], the role of this plasmid in expression of virulence is certainly a critical one and deserves further genetic and molecular study. Moreover, the high level of homology observed among virulence plasmids belonging to different *Shigella* species [7] should allow us to generalize any particular result to these other dysentery-producing organisms.

LOCALIZATION OF PLASMID SEQUENCES INVOLVED IN HeLa CELL INVASION

The kanamycin resistance transposon Tn*5* was used in strain M90T to generate a bank of avirulent mutants through transposition onto the virulence plasmid pWR100. F'_{ts}114-lac::Tn*5* served as the transposon donor as previously described [1,8]. One thousand mutants were screened for ability to invade HeLa cells. Among 120 non-invasive clones, 112 had either lost pWR100 or sustained deletions in this plasmid. In the eight remaining mutants, no deletion could be detected and transposition of Tn*5* onto pWR100 was demonstrated by hybridization with a ^{32}p-labelled Tn*5* probe. On the basis of their *Eco*R1 restriction patterns, these plasmids fell within three groups since Tn*5* insertions mapped in three fragments (i.e. 7.6, 11.5 and 17 kb). The corresponding fragments of the unmutagenized pWR100 were cloned into the *Eco*R1 site of the cloning vector pBR325. These recombinant plasmids were selected for further study: pHS3188 (7.6 kb insert), pHS4011 (11.5 kb insert) and pHS4033 (17 kb insert). After transformation into the avirulent plasmidless strain BS176, none of them restored invasiveness into HeLa cells, indicating that a larger sequence was certainly necessary for expression of the invasive phenotype.

COSMID CLONING OF pWR100 VIRULENCE SEQUENCES

Cosmid vectors permit cloning of large fragments of DNA which should improve the likelihood of cloning sequences large enough to encode the functions necessary for invasion of HeLa cells. This approach required construction of a λ sensitive *Shigella* recipient which could be screened for expression of the invasive phenotype in HeLa cells. A spontaneous, non-invasive derivative of *S. flexneri* 2a was first isolated. This strain had lost the 220 kb virulence plasmid. A spontaneous Mal^+ derivative was then isolated as Mal^+ *Shigellae* are capable of expressing the λ receptor and are sensitive to host range mutants of λ. Introduction of the *gal*U::Tn*10* mutation into a Mal^+ *Shigella* renders the strain sensitive to wild-type λ and such a strain, BS169, can be used for cosmid cloning. An additional advantage of the *gal*U::Tn*10* mutation is that, in a virulent background, it alters the virulence properties of the bacteria such that invasion of HeLa cells causes the monolayer to detach 1-2 hours after infection. Isogenic *gal*U::Tn*10* *Shigella* which does not possess the 220 kb plasmid and cannot invade the cells, has no effect on the monolayer. The cosmid cloning technique which has been utilized in these experiments can be summarized as follows: pWR100 plasmid DNA was partially restricted by *Sau*3a and sized in a NaCl gradient. Cosmid vector pJB8 [10] was cleaved by *Bam*HI and dephosphorylated. Ligation was performed at high concentration of DNA. Recombinant molecules were then packaged into λ and transduced into BS169. A bank of 800 transductants was screened for the ability of clones to cause monolayer detachment. Six stable recombinant plasmids capable of restoring invasiveness on BS169 were identified. They were subsequently transformed into BS176, the non-invasive plasmidless variant of M90T. Relevant properties of these invasive transformants are summarized in Table 1. Temperature-dependent invasiveness which was recently described in *S. flexneri* [11] was conserved by the newly invasive strains. Although they attained a high rate of HeLa cell infection, they appeared negative in the Sereny test. Completion of a keratoconjunctivitis is likely to be dependent on the bacteria being able to multiply freely within the cytoplasm of infected epithelial cells, invade adjacent cells and lyse the cells. This cycle of invasion-release-invasion produces infectious foci. The negative response of the recombinants in the Sereny test may be due either to a partial cloning of the sequences necessary for expression of the complete invasive phenotype or to insufficient expression of the cloned genes by the recombinant plasmids.

Table 1. Characterization of clones containing cosmid recombinant molecules

Strains	*% of HeLa cells invaded*		*Sereny test*	*Size of recombinant plasmids (kb)*
	Bacteria grown at 30°C	*Bacteria grown at 37°C*		
M90T	<1	95	+	
BS176	0	0	−	
BS176(pHS4108)	5	49.2	−	49.2
BS176(pHS4181)	5	26	−	42
BS176(pHS4195)	3	42	−	48.5
BS176(pHS4685)	1	8.5	−	42.9
BS176(pHS4707)	2	16	−	45
BS176(pHS4717)	1	16	−	42.9

CHARACTERIZATION OF RECOMBINANT PLASMIDS

The three *Eco*RI fragments of pWR100 which were characterized as containing sequences involved in virulence through Tn*5* mutagenesis and were subsequently cloned into pBR325 (i.e. pHS3188, pHS4011, and pHS4033) hybridized strongly

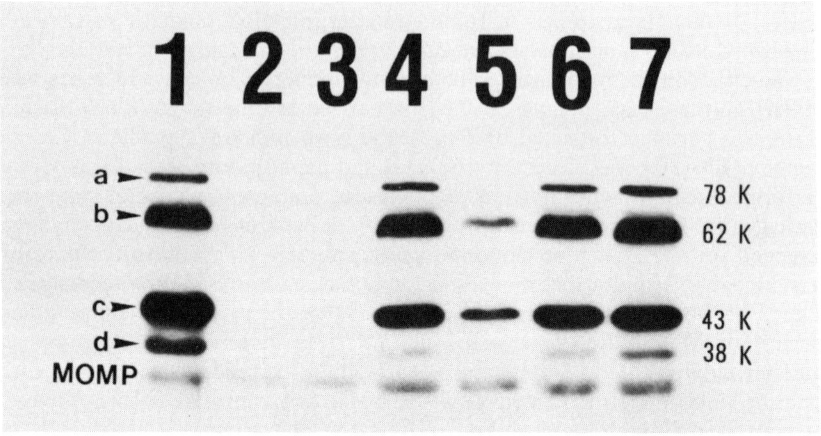

Figure 1. Western blot hybridization analysis of peptides expressed by recombinant clones. Lane 1, M90T, 37°C; lane 2, BS176; lane 3, M90T, 30°C; Lane 4, BS176(pHS4108), 37°C; lane 5, BS176(pHS4108), 30°C; lane 6, BS176(pHS4181), 37°C; lane 7, BS176(pHS4195), 37°C

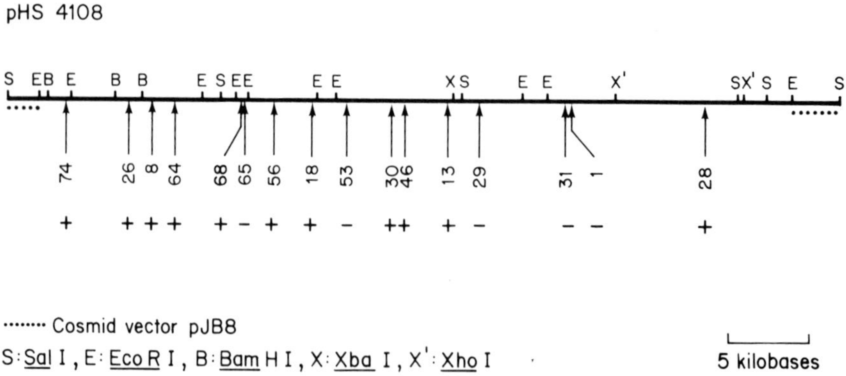

Figure 2. Map of recombinant plasmid pHS4108 with location of Tn*5* insertions inhibiting invasiveness (−) and preserving invasiveness (+).

with fragments of the recombinant plasmids, thus indicating that these sequences were either totally (i.e. 7.6 and 11.5 kb) or partially (i.e. 17 kb) represented within cloned inserts. They are therefore critical for the invasive process. A sequence of 37 kb, which appears common to all inserts, can be defined as the provisional minimum plasmid sequence necessary for HeLa cell invasion.

Recombinant plasmid pHS4108 was selected for further analysis. In order to demonstrate that BS176 carrying this plasmid produced polypeptides which were associated with expression of the invasive phenotype by the parent M90T, Western blot analysis of whole bacterial extracts was performed. The antiserum used had been shown to specifically recognize four major peptides in extracts of bacteria containing pWR100 (T. L. Hale *et al*, submitted). As shown in Fig. 1, peptides of the same molecular weight as a,b,c, and d of M90T were expressed by recombinant clones pHS4108 as well as pHS4181 and pHS4195. Peptide d was reduced in each clone. M90T grown at 30°C produced little detectable amounts of the four peptides. In contrast, when grown at 30°C, the recombinant clone pHS4108 expressed more peptide b and c than M90T at the same temperature. Invasion of HeLa cells by BS176(pHS4108) was also slightly depressed at 30°C (Table 1). This residual expression at 30°C may reflect a higher gene dosage of the cloned insert since cosmid vector pJB8 is a multicopy plasmid. A map of pHS4108 using restriction and endonucleases *Eco*RI, *Sal*I, *Bam*HI, *Xba*I and *Xho*I is shown in Fig. 2. A battery of Tn*5* insertions has been obtained in this plasmid, some of which inhibit invasiveness of the recombinant clone. The location of these insertions is also shown in Fig. 2. Sequences involved in the invasive process are scattered on the insert. The effect of these mutations on peptide expression is currently being studied.

INCOMPLETE RESTORATION OF THE VIRULENCE PHENOTYPE: AN EXPLANATION

Although recombinant plasmids restore the ability of BS176 to penetrate into HeLa cells, the bacteria cannot be considered as fully invasive since they are negative in the Sereny test. In an attempt to explain this discrepancy, we compared intracellular growth capacities of M90T and BS176 (pHS4108) as well as their respective patterns of HeLa cell invasion in electron microscopy. Results are shown in Figs 3 and 4. M90T expresses highly efficient growth ability unlike BS176(pHS4108) which grows very poorly. Fifteen minutes after infection, M90T lies freely within the cytoplasm (A) whereas 2 h after infection, BS176(pHS4108) is still entrapped within its membrane-bound phagocytic vacuole (B). Moreover, as shown in C and D, membrane-limited vacuoles end up fusing together in a large, irregular, intracytoplasmic pocket. Therefore, M90T seems to express plasmid-mediated ability to lyse vacuole membranes and multiply freely in the cytoplasm whereas BS176(pHS4108) appears unable to express this property and remains trapped within vacuoles in probably poor growth conditions. Cloning therefore allowed differentiation of two steps in plasmid-mediated HeLa cell invasiveness: induction of phagocytosis and ability to lyse phagocytic membranes, thus allowing free and highly efficient intracellular multiplication.

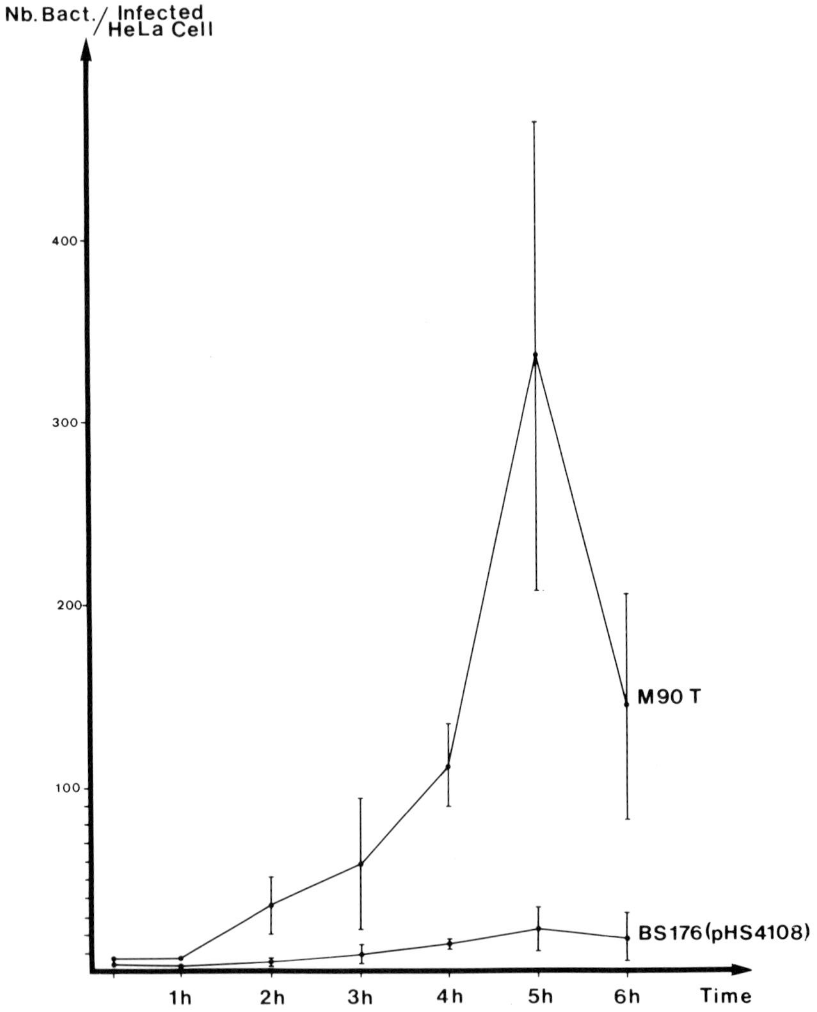

Figure 3. Kinetics of intracellular growth of M90T and BS176(pHS4108) within HeLa cells. All relevant data are reported in the figure. Each point represents the mean and standard deviation of five experiments.

CONCLUSION

Shigellae induce their own phagocytosis by non-professional phagocytes. Genes necessary to achieve this critical step of their pathogenic process are located on a 220 kb plasmid in *S. flexneri*. A 37 kb sequence of this plasmid has been cloned which is necessary to induce intracellular penetration. According to preliminary

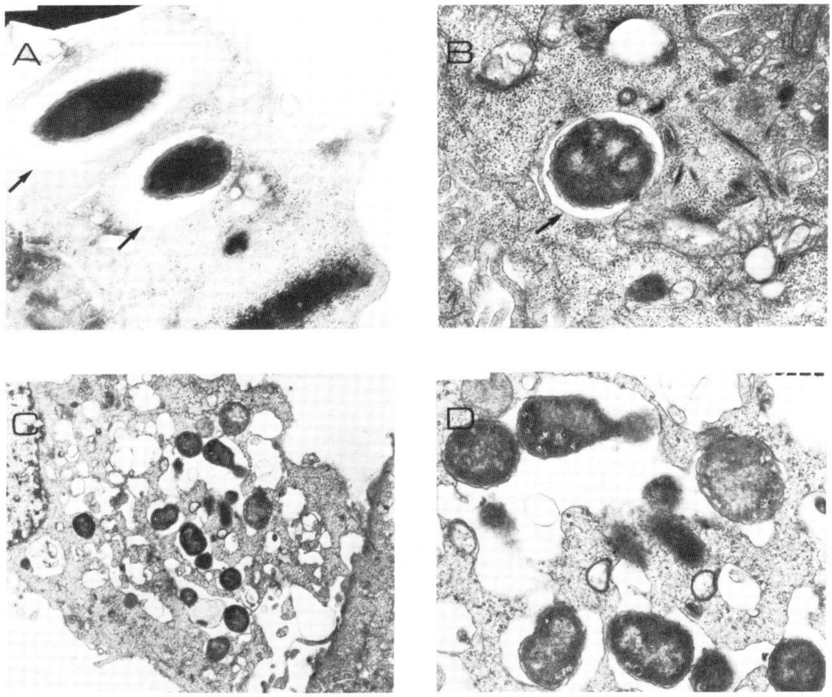

Figure 4. HeLa cells infected by M90T and BS176(pHS4108). Observation by transmission electron microscopy was made at various times. (A) M90T, 15 min after infection. (B) BS176 (pHS4108), 60 min after infection. (C) and (D) BS176(pHS4108), 4 h after infection.

experiments, the genes involved are scattered on the cloned sequence. They express a temperature-regulated set of four polypeptides which are certainly essential in the invasive process. Early lysis of the phagocytic vacuole appears critical for intracellular multiplication which is essential for continuation of the infectious process.

SUMMARY

Bacteria belonging to the genus *Shigella* cause a dysenteric syndrome in human beings. The main step of the pathogenic process is invasion of the colonic mucosa. Invasion encompasses complex features such as induction of bacterial phagocytosis by epithelial cells, intracellular multiplication, invasion of adjacent cells as well as of the conjunctive tissue of the lamina propria, elicitation of a strong inflammatory reaction which is responsible for abscesses and ulcerations.

This presentation summarizes a genetic, molecular and ultrastructural study of this invasive process by *S. flexneri*. A 200 kb plasmid is required for the very steps of induction of phagocytosis and intracellular multiplication. The genes involved in this process have been cloned. They reside on a 37 kb sequence which encodes several proteins that can be revealed either by [^{35}S] methionine labelling in minicells or by Western blotting using the serum of a monkey which has been orally immunized by *S. flexneri*. The respective role of these proteins in the process of epithelial cell invasion is currently being investigated. Plasmid-bearing strains also exhibit fast replication within HeLa cells. This crucial property is related to the ability of these strains to achieve early and efficient lysis of the phagocytic vacuole.

ACKNOWLEDGEMENTS

We wish to thank T. L. Hale for performing Western blot analysis, H. d'Hauteville and J. Mounier for excellent technical assistance, and Z. Lebri for typing this manuscript.

REFERENCES

1. Sansonetti, P. J., D. J. Kopecko and S. B. Formal (1981). *Shigella sonnei* plasmids: evidence that a large plasmid is necessary for virulence. *Infect. Immun.* **34**, 75-83.
2. Sereny, B. (1957). Experimental keratoconjunctivitis shigellosa. *Acta Microbiol. Hung.* **4**, 367-376.
3. La Brec, E. H., H. Schneider, T. T. Magnani and S. B. Formal (1964). Epithelial cell penetration as an essential step in the pathogenesis of bacillary dysentery. *J. Bacteriol.* **88**, 1503-1518.
4. Ogawa, H., A. Nakamura and R. Nakaya (1967). Virulence and epithelial cell invasiveness of dysentery bacilli. *Jap. J. Med. Sci. Biol.* **20**, 315-318.
5. Keusch, G. T., G. F. Grady, L. J. Mata and J. McIver (1972). Pathogenesis of *Shigella* diarrhoea. I. Enterotoxin production by *Shigella dysenteriae* I. *J. Clin. Invest.* **51**, 1212-1218.
6. Brown, J. E., S. W. Rothman and B. P. Doctor (1980). Inhibition of protein synthesis in intact HeLa cells by *Shigella dysenteriae* I toxin. *Infect. Immun.* **32**, 137-144.
7. Sansonetti, P. J., H. d'Hauteville, C. Ecobichon and C. Pourcel (1983). Molecular comparison of virulence plasmids in *Shigella* and enteroinvasive *Escherichia coli*. *Ann. Microbiol. (Inst. Pasteur)* **134A**, 295-318.
8. Sansonetti, P. J., D. J. Kopecko and S. B. Formal (1982). Involvement of a plasmid in the invasive ability of *Shigella flexneri*. *Infect. Immun.* **35**, 852-860.
9. Sansonetti, P. J., T. L. Hale, G. J. Dammin, C. Kapfer, H. H. Collins, Jr and S. B. Formal (1983). Alterations in the pathogenicity of *Escherichia coli* K-12 after transfer of plasmid and chromosomal genes from *Shigella flexneri*. *Infect. Immun.* **39**, 1392-1402.
10. Maniatis, T., E. F. Fritsch and J. Sambrook (1982). *In* "Molecular Cloning". Cold Spring Harbor Edn, New York.
11. Maurelli, A. T., B. Blackmon and R. Curtiss III (1984). Temperature dependent expression of virulence genes in *Shigella* species. *Infect. Immun.* **43**, 195-201.

Virulence-Associated Characteristics of *Shigella flexneri* and the Immune Response to Shigella Infection

Thomas Larry Hale[a], Edwin V. Oaks, Gabriel Dinari and Samuel B. Formal

Department of Bacterial Diseases, Walter Reed Army Institute of Research, Washington, DC 20307, USA

The enteroinvasive phenotype allows enteric bacteria to escape the competition of the colonic lumen, which is a fully occupied environmental niche, and to multiply in an unoccupied niche represented by the cytosol of colonic epithelial cells. By occupying the latter environment, enteroinvasive organisms such as *Shigellae* and *Escherichia coli* are able to escape toxic bile and fatty acids in the lumen. In the process, they also gain access to nutrients drawn from the bloodstream of the host rather than from the colon with its established bacterial flora.

INTERACTIONS OF *SHIGELLAE* WITH MAMMALIAN GLYCOPROTEINS

Two processes which are crucial for the escape of enteroinvasive pathogens from the lumen of the bowel are the penetration of the colonic mucus layer and invasion of colonic epithelial cells. Both of these processes probably involve interactions with mammalian glycoproteins. For example, *S. flexneri* binds to the mucus layer on guinea pig colonic epithelial cells, and this adherence is inhibited by the presence of fucose, glucose, or *S. flexneri* lipopolysaccharide [1]. Crude guinea pig mucus also agglutinates some strains of *S. flexneri* and inhibits the invasion

[a] *Telephone:* 1-202-576-3344

Protein-Carbohydrate Interactions in Biological Systems. ISBN 0 12 436665 1

of HeLa cells by either agglutinating or non-agglutinating strains. In contrast, crude mucus preparations from the colon of a Rhesus monkey neither agglutinate shigellae nor inhibit HeLa cell invasion (G. Dinari, in prep.). Thus there may be one or more species of glycoprotein present in rodent colonic mucus which is protective because it causes some strains to be trapped in the mucus layer and to be removed by peristalsis. A similar glycoprotein may also inhibit the invasion of colonic epithelial cells by the shigellae which reach the epithelial surface. In

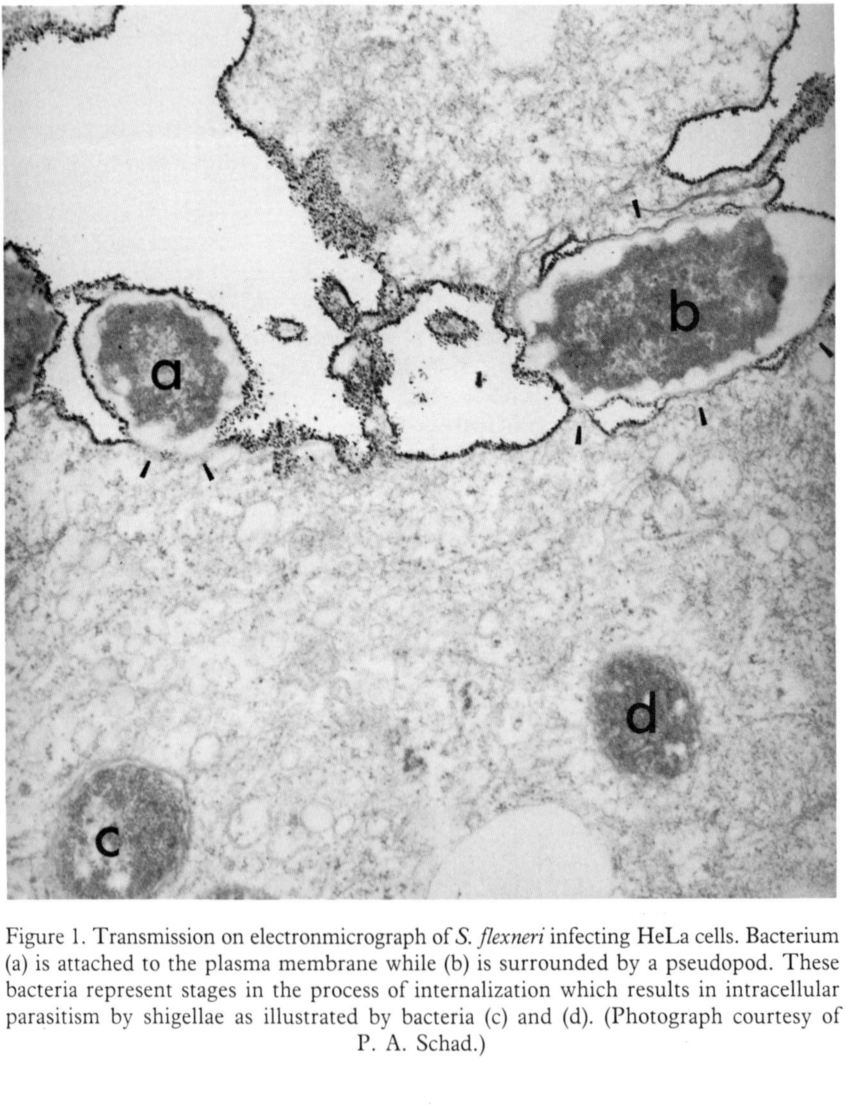

Figure 1. Transmission on electronmicrograph of *S. flexneri* infecting HeLa cells. Bacterium (a) is attached to the plasma membrane while (b) is surrounded by a pseudopod. These bacteria represent stages in the process of internalization which results in intracellular parasitism by shigellae as illustrated by bacteria (c) and (d). (Photograph courtesy of P. A. Schad.)

contrast, primate mucus apparently lacks these protective glycoproteins, and this may be a factor in the selection of primates as a natural host for shigella infection.

Even though shigellae avoid or overcome interactions with primate mucus glycoproteins, there is some evidence that they must interact with the glycocalyx of epithelial cells. This suggestion is based on the finding that HeLa cells treated for 48 hours with tunicamycin, an inhibitor of *N*-linked oligosaccharide chain addition to nascent polypeptides [2], causes a significant reduction of HeLa cell invasion by *S. flexneri*. This inhibitor of glycosylation may render HeLa cells resistant to infection by preventing the synthesis of membrane glycoprotein receptors for shigellae. Functionally similar receptors may reside on the surface of colonic epithelial cells, and the interaction of lectin-like proteins on the surface of invasive bacterial cells with these receptors may induce the uptake of the bacteria by sequential ligand binding.

The latter hypothesis is based on morphological, physiological, and genetic data. Morphologically, the invasion of colonic epithelial cells by enteric pathogens appears to be an endocytic event [3], and the invasion of cultured epithelial cells by *S. flexneri* shown in Fig. 1 has the appearance of a receptor-mediated phenomenon. In the experiments which generated the latter electronmicrograph, an *en bloc* procedure was employed to label the acid mucosaccharides of fixed HeLa cell membranes with colloidal thorium dioxide. This staining process demonstrates areas of close apposition between attached bacteria and the host cell membrane (see arrows in Fig. 1). The exclusion of thorotrast particles suggests that these areas of close apposition represent receptor–ligand binding between the two membranes. Sequential binding of the bacterial and host cell membranes appears to facilitate the uptake of the bacteria by a "zipper mechanism" similar to that described in phagocytic cells [4]. Invasion of HeLa cells by shigellae is inhibited by cytochalasin B, iodoacetate, and dinitrophenol [5]. Treatment with these inhibitors of microfilament function, glycolysis, or oxidative phosphorylation indicates that systems which are required by professional phagocytes for the ingestion of particulate matter are also required for the ingestion of shigellae by HeLa cells. Apparently, there are physiological similarities between phagocytosis and the ingestion of shigella by non-professional phagocytes.

BACTERIAL DETERMINANTS ASSOCIATED WITH THE INVASIVE PHENOTYPE

Since a large plasmid is necessary for expression of the invasive phenotype in either shigella or enteroinvasive *E. coli* [6], the hypothetical lectin-like bacterial receptor which interacts with the mammalian plasma membrane is probably encoded by this large plasmid. Since organisms which lack the 140 MD virulence plasmid are rarely isolated from primates or from the environment, the products of the plasmid virulence genes must favor survival of the species. Consequently, these genes are conserved in all enteroinvasive strains of shigella and *E. coli* [6]. Seven virulence-associated, plasmid-coded polypeptides have been identified by radio-labelling newly synthesized polypeptides in anucleate minicells containing the *S. flexneri* 140 mdal virulence plasmid. These plasmid-coded polypeptides, which

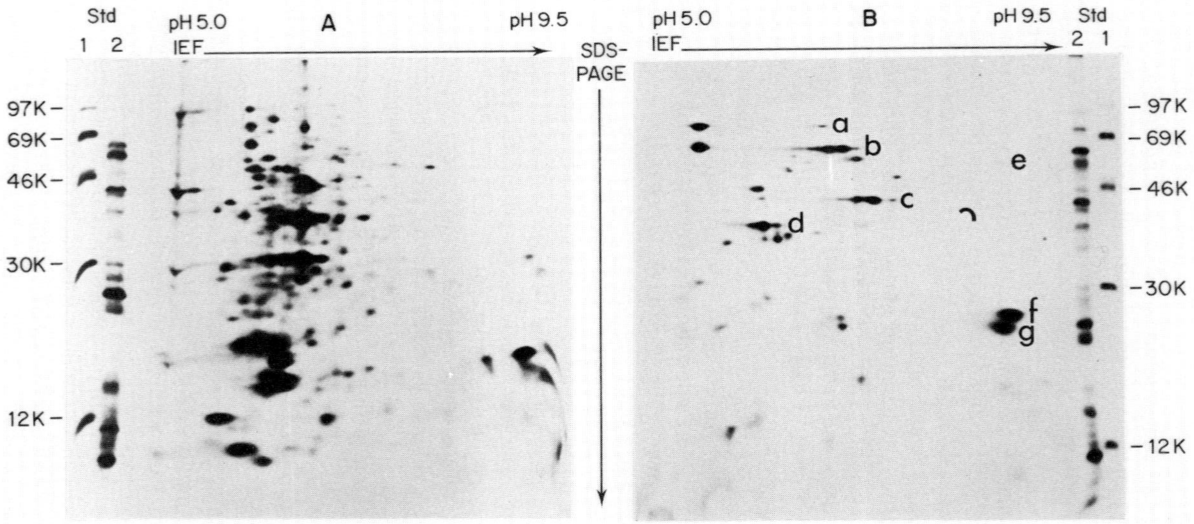

Figure 2. Two-dimensional fluorograms of polypeptides radiolabelled with [^{35}S] methionine in a mixture of vegetative cells and minicells from *S. flexneri* serotype 5 (A) or in isolated minicells (B). Only plasmid-coded polypeptides were labelled in the later anucleate cells, and these are present in such minute quantities in vegetative cells that most of them are not detectable in panel (A). The first dimension isoelectric focusing step was carried out in a pH 5.0–9.5 gradient, and two standards were run in the second dimension SDS polyacrylamide gel electrophoresis step: (1) ^{14}C-labelled molecular weight standards and (2) a radiolabelled minicell extract.

are shown in Fig. 2B, are minor components of the total complement of chromosomal and plasmid-encoded bacterial proteins as shown in Fig. 2A. Nonetheless, four of them (a–d) are the major proteinaceous antigens inducing serum antibody in response to natural infections in either monkeys or humans. In addition, antisera from monkeys which have been infected with *S. flexneri* recognize virulence-associated plasmid-coded polypeptides in all enteroinvasive strains of shigella and *E. coli* thus far tested in ELISA or immunoblots (T. L. Hale *et al.*, submitted; E. V. Oaks *et al.*, in prep.). These data indicate that the polypeptides shown in Fig. 2B are the key antigens which allow the primate immune system to differentiate enteroinvasive pathogens from the normal intestinal flora.

Preliminary experiments indicate that water extracts of virulent shigellae contain quantities of plasmid-coded polypeptides b and c which are easily detectable in immunoblots (Western blots) using convalescent antisera from Rhesus monkeys which have been infected with *S. flexneri* 2a. When water extracts of *S. flexneri* 5 are added to HeLa cell monolayers, it can be demonstrated, with Western blots of the detergent-solubilized monolayers, that polypeptides b and c adhere to the HeLa cells. Indirect fluorescent antibody staining of these polypeptides shows that they initially adhere in a diffuse pattern over the entire surface of the cell. Within three hours, however, they cap into large aggregates on the surface of the cell. Adherent antigen is not removed by treatment with Triton X-100, and the antigen does not adhere to HeLa cells which have been rendered resistant to invasion by treatment with tunicamycin. Thus, it could be speculated that virulence-associated polypeptides specifically bind to glycoproteins on the surface of HeLa cells. If a similar process occurs when these polypeptides are expressed on the surface of a bacterial cell, they could conceivably induce the uptake of the entire bacterium by sequentially binding to aggregated plasma membrane ligands, resulting in the invasion of the mammalian cell.

REFERENCES

1. Izhar, M., Y. Nuchamowitz and D. Mirelman (1982). Adherence of *Shigella flexneri* to guinea pig intestinal cells is mediated by a mucosal adhesin. *Infect. Immun.* **35**, 1110–1118.
2. Ronnett, G. V. and M. D. Lane (1981). Post-translational gylcosylation-induced activation of aglycoinsulin receptor accumulated during tunicamycin treatment. *J. Biol. Chem.* **256**, 4704–4707.
3. Takeuchi, A., H. Spring, E. H. Le Brec and S. B. Formal (1965). Experimental bacillary dysentery: an electron microscopic study of the response of the intestinal mucosa to bacterial invasion. *Am. J. Pathol.* **47**, 1011–1044.
4. Griffin, F. M., Jr, J. A. Griffin and S. C. Silverstein (1976). Studies on the mechanism of phagocytosis II. The interaction of macrophages with anti-immunoglobulin IgG-coated bone marrow-derived lymphocytes. *J. Exp. Med.* **144**, 788–809.
5. Hale, T. L., R. E. Morris and P. F. Bonventre (1979). Shigella infection of Henle Intestinal Epithelial Cells: Role of the host cell. *Infect. Immun.* **24**, 887–894.
6. Hale, T. L., P. J. Sansonnetti, P. J. Schad, S. Austin and S. B. Formal (1983). Characterization of virulence plasmids and plasmid-associated outer membrane proteins in *Shigella flexneri*, *Shigella sonnei* and *Escherichia coli*. *Infect. Immun.* **40**, 340–350.

Plasmids and Virulence of *Yersinia enterocolitica*

P. Gemski[a], G. R. Fanning, M. A. Sodd and J. A. Wohlhieter*

*Departments of Biological Chemistry and *Bacterial Immunology, Walter Reed Army Institute of Research Washington, DC 20307, USA*

Of several species now classified in the genus Yersinia, three are well established as provoking significant diseases in humans. *Yersinia pestis* and *Yersinia pseudotuberculosis* have long been appreciated as pathogens causing highly invasive infections in man and other mammals. It is only during the past decade, however, that *Yersinia enterocolitica* has emerged as a diarrhoeal pathogen receiving wide attention from clinicians, food microbiologists and researchers. As a consequence, *Y. enterocolitica* is now recognized as an important invasive pathogen which is widely distributed in the environment. In addition to causing gastroenteritis (particularly among children), this organism is associated with infections characterized as acute mesenteric lymphadenitis, terminal ileitis, septicaemia and reactive arthritis [1]. *Y. enterocolitica* represent a heterogeneous group of organisms which have been differentiated into over 20 serotypes and 5 biotypes. The most common serotypes associated with human disease appear to be 0:3, 0:5, 0:8 and 0:9 [1].

The central focus of many recent studies of the virulence of *Yersiniae* has been a class of plasmids referred to as Vwa. It is now clear that *Y. enterocolitica*, *Y. pseudotuberculosis* and *Y. pestis* share virulence determinants, coded by Vwa plasmids, which function by an as yet unresolved mechanism in invasive pathogenesis. It is our purpose here to discuss briefly the plasmid characteristics and virulence phenotypes of *Y. enterocolitica*. Other chapters in this volume address the virulence of *Y. pestis* and *Y. pseudotuberculosis*.

[a] *Telephone:* 1-202-576-2594

Protein-Carbohydrate Interactions in Biological Systems. ISBN 0 12 436665 1

Table 1. Plasmids of *Yersiniae*

Species	*Plasmid size* (MD)	*Functions*	*Reference*
Y. enterocolitica	33	lac⁺; Tn 951	[23]
	36	unknown	[4, 10, 11]
	40–48	Vwa	[4, 6, 8–11]
	82	unknown	[7]
Y. pseudotuberculosis	40–45	Vwa	[5, 16]
Y. pestis	6	pst	[3]
	45	Vwa	[3, 18]
	65	unknown	[3]
Y. ruckerii	2–3	unknown	[2]
	35	tet^R	[2]
	70	virulence (?)	[2]

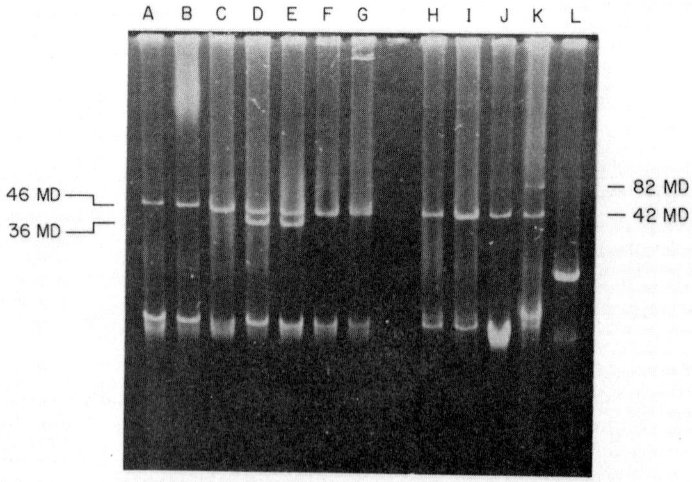

Figure 1. Agarose gel electrophoresis of plasmid DNA from several serotypes on *Y. enterocolitica*. Plasmid DNA was prepared by rapid alkaline extraction and subjected to electrophoresis in Tris-borate-EDTA-buffered agarose gels. (A) 0:3 strain 348; (B) 0:5 strain 1746; (C) 0:7,8 strain 1137; (D) 0:8 strain 2383; (E) 0:8 strain Y7P-2; (F) 0:8 strain Y7P-1; (G) 0:8 strain WA; (H) 0:13 strain 1209; (I) 0:20 strain 874; (J) 0:20 strain 1223–74; (K) 0:40 strain 3973; (L) NTR strain 3979.

PLASMIDS OF Y. ENTEROCOLITICA

Numerous surveys of plasmid DNA content of several species of Yersinia have revealed a broad range of plasmid sizes [2-11]. As shown in Table 1, certain phenotypes can be associated with some of these plasmids; however many remain cryptic with respect to function.

Considerable attention has been placed on the Vwa class of plasmids because of its association with pathogenesis. Figure 1 illustrates plasmid profiles of a few serotypes of *Y. enterocolitica* isolated from human infections. Vwa plasmids, ranging in mass from 40-48 MD, have been associated with numerous phenotypic properties which may relate to expression of virulence. These include: calcium dependency [4], production of V-W antigens [12], colony morphology [13] autoagglutination [14], serum resistance [15], novel outer cell membrane proteins [10, 16], adherence [17] and invasiveness [4-6, 10, 11, 18]. Figure 1 also illustrates additional plasmid types that have been detected in some *Y. enterocolitica* serotypes. Functions of 82 MD, 36 MD and a 12-15 MD plasmid species remain obscure.

MOLECULAR DNA HOMOLOGY OF Vwa PLASMIDS

We have examined the molecular homology of Vwa plasmids through use of solution hybridization techniques [19]. Purified plasmid DNA, after being

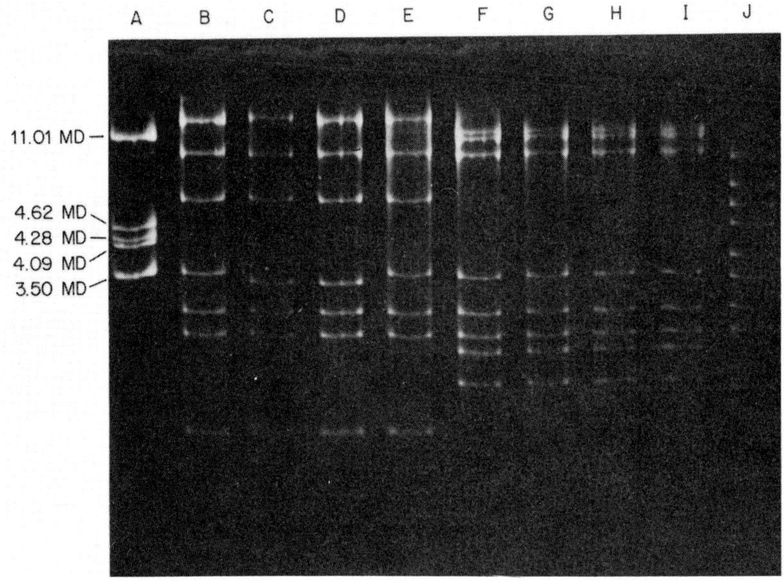

Figure 2. Agarose gel electrophoresis of Vwa plasmids digested by *Bam* HI. (A) λ DNA; (B) *Y. enterocolitica* 0:8 WA; (C) *Y. enterocolitica* 0:8 Ye WL; (D) *Y. enterocolitica* 0:8 E2708; (E) *Y. enterocolitica* 0:8 Ye DT; (F) *Y. enterocolitica* 0:3 MCH 628; (G) *Y. enterocolitica* 0:3 79A1; (H) *Y. enterocolitica* 0:3 56A1; (I) *Y. enterocolitica* 0:3 64A1; (J) *Y. pseudotuberculosis* III.

Table 2. DNA relatedness of plasmids of *Y. enterocolitica*

				A		B	
Species	*Strain*	*Serotype*	*Plasmid size (MD)*	*% DNA relatedness*	*% divergence*	*% DNA relatedness*	*% divergence*
Y. enterocolitica	MCH628	0:3	46	67	2.2	100	0.0
	79A1	0:3	46	98	2.2	100	1.2
	63A1	0:3	46	85	1.8	95	0.0
	56A1	0:3	46	81	2.2	97	0.0
	477	0:3	46	80	2.0	88	0.0
	Ye835	0:3	46	78	2.0	86	0.0
	76A1	0:3	46	72	1.6	82	0.0
	64A1	0:3	46	93	2.0	100	0.0
Y. enterocolitica	WA	0:8	42	100	0.6	64	2.9
	YeWL	0:8	42	103	0.4	65	2.5
	E2708	0:8	42	90	0.0	54	2.5
	YeDT	0:8	42	77	0.6	55	2.5
Y. enterocolitica	1223-75-2	0:20	82	8	---	4	---
Y. pseudotuberculosis	YIII	III	42	64	2.4	64	3.2

A, the Vwa plasmid from WA used as the reference plasmid.
B, the Vwa plasmid from MCH 628 used as the reference plasmid.

sheared by sonication and denatured, was reacted with ^{32}P-labelled reference plasmid DNA (sheared, denatured) under conditions allowing heteroduplex formation (0.001–0.01 μg reference DNA + 15 μg unlabelled DNA in 0.28M phosphate buffer, 60°C, 16 h). Hybridized DNA was separated from unhybridized DNA using hydroxyapatite with phosphate buffer. Table 2 summarizes the levels of DNA homology that we detected among various plasmids. Hybridization of Vwa reference plasmids from serotype 0:8 (Column A) and serotype 0:3 (Column B) with Vwa plasmid DNA from several virulent strains revealed a high degree of genetic relatedness. These plasmids share over 50% DNA sequence homology at optimum reassociation temperatures (60°C).

We have also examined the similarity of Vwa plasmids from 0:3 and 0:8 strains of *Y. enterocolitica* by comparing DNA fragmentation patterns after digestion by restriction endonucleases [9]. After treatment with several different restriction enzymes, Vwa plasmids from strains within each serotype were found to share many common fragments. Common restriction fragments were also detected when Vwa plasmids of different serotypes were compared (Fig. 2). However, distinct banding patterns were evident in such comparisons indicating some divergence in the microevolution of Vwa plasmids. It is clear from such studies that Vwa plasmids represent a class of related plasmids that have conserved regions of substantial genetic homology [6, 8–10, 18].

CHARACTERIZATION OF THE 82 MD PLASMID

We have reported previously on a plasmid, about 82 MD in size, which is found in conjunction with Vwa plasmids in some serotypes of *Y. enterocolitica* [7, 13]. Because these strains cause lethal infections in mice, it had been suggested that functions encoded by 82 plasmid DNA may contribute to virulence. The 82 MD plasmid of *Y. enterocolitica* serotype 0:20 strain 1223-75 has now been further characterized. As shown in Fig. 3, segregants that retain their 82 MD plasmid but lose their Vwa plasmid can be readily recovered. This plasmid, designated 82-(0:20), was purified from such segregants and its molecular homology compared with Vwa plasmids from both 0:8 and 0:3 isolates.

Less than 10% DNA relatedness with Vwa plasmids was detected, indicating that the 82-(0:20) plasmid belongs to a DNA hybridization group distinct from the Vwa class (Table 2). This conclusion is further supported by results of restriction endonuclease analyses (Fig. 4). *Hin*dIII digests of 82-(0:20) show a fragmentation pattern distinct from the Vwa plasmid of serotype 0:8 (Fig. 4A) and 0:3 (Fig. 4B). Moreover, digestion of a plasmid preparation from strain 1223-75, which has both the 82 and Vwa plasmid, reveals a composite pattern (42/82 columns, Fig. 4).

The DNA homology of the 82-(0:20) plasmid was next compared with that of an 82 MD plasmid species observed in *Y. enterocolitica* serotype 0:8 strain WA, using nitrocellulose filter blot hybridizations. Plasmid DNA from 0:8 strain WA, 0:20 segregant 1223-75-2 and 0:20 strain 1223-75 was separated via agarose gel electrophoresis, transferred to a nitrocellulose filter and hybridized with either a ^{32}P-labelled 82-(0:20) DNA or a ^{32}P-labelled Vwa DNA probe. The results of

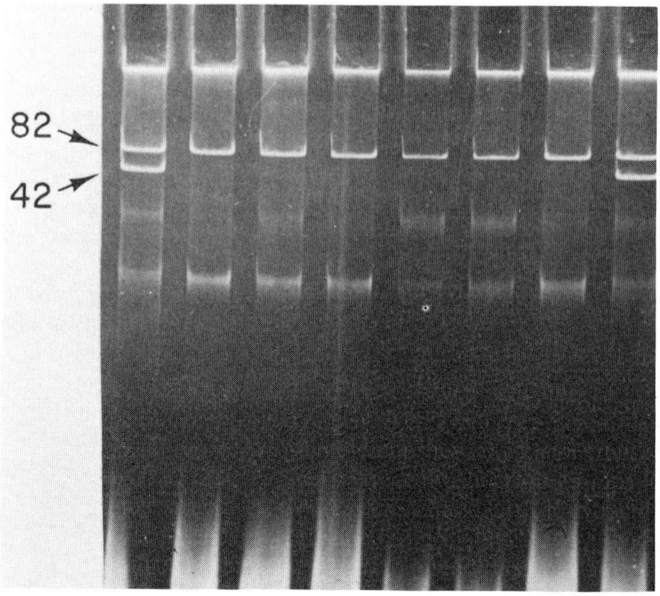

Figure 3. Agarose gel electrophoresis of plasmid DNA from virulent *Y. enterocolitica* 0:20 strain 1223-75 and avirulent segregants 1223-75-2.

autoradiographic analysis using the 82-(0:20) probe (Fig. 5B, bottom) indicate little, if any, DNA homology between the 82-(0:20) plasmid and the 82-(0:8) plasmid. The distinctions between these two 82 MD plasmid species was further emphasized when we used Vwa plasmid as a reference probe (Fig. 5B, top). As expected from the solution hybridization results (Table 2), Vwa and the 82-(0:20) were found to lack significant homology (lanes 2, 3). In contrast, hybridization was detected between the Vwa plasmid and the 82-(0:8) plasmid (lane 1) of strain WA.

These results indicate that the 82-(0:20) plasmid isolated from *Y. enterocolitica* behaves as a unique entity distinct from the 82 MD plasmid species detected in serotype 0:8 and perhaps other serotypes. Indeed, the strong homology between Vwa plasmids and the 82-(0:8) plasmid is consistent with the interpretation that some 82 MD species represent molecular dimers of Vwa plasmids. Several other findings are consistent with this interpretation:

(a) Although extensive efforts have been made, we have failed to recover a segregant of serotype 0:8 strain which possesses only the 82 MD plasmid, a sharp contrast to the ease with which such segregants can be recovered from strain 1223-75 (Fig. 3).

(b) Restriction endonuclease analysis of plasmid preparations from 0:8 strain WA which contain both Vwa and 82 plasmid species reveal a pattern identical to that of preparations containing only the Vwa plasmid; hence no unique plasmid entity can be demonstrated.

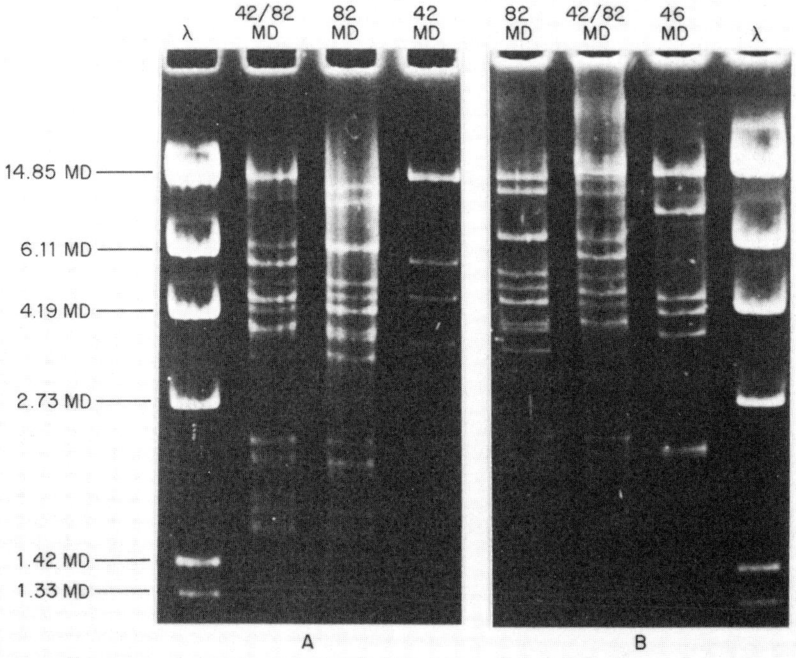

Figure 4. Agarose gel electrophoresis of *Hin*dIII digests of *Y. enterocolitica* plasmids. Comparison of *Y. enterocolitica* serotype 0:8 Vwa plasmid (A) and serotype 0:3 Vwa plasmid (B) with the 82-(0:20) plasmid of serotype 0:20 segregant 1223-75-2. Lanes 42/82 represent a plasmid preparation from serotype 0:20 1223-75 [contains 82-(0:20) and Vwa].

(c) The detection of 82 MD species in 0:8 strains can be variable, often of low intensity and affected by whether phenol and chloroform are employed in the plasmid extraction procedure.

(d) A dimeric form of the Vwa plasmid is consistent with the high association that we have observed between mouse virulence and presence of 82 MD plasmids in conjunction with Vwa plasmids [7].

VIRULENCE PHENOTYPES OF *Y. enterocolitica*

Although initial efforts to study the virulence of *Y. enterocolitica* strains in small animal models had yielded variable results [1], there are now several reliable experimental models for assessing the pathogenicity of this organism. Through the use of a natural oral infection model, in which Swiss albino mice are deprived

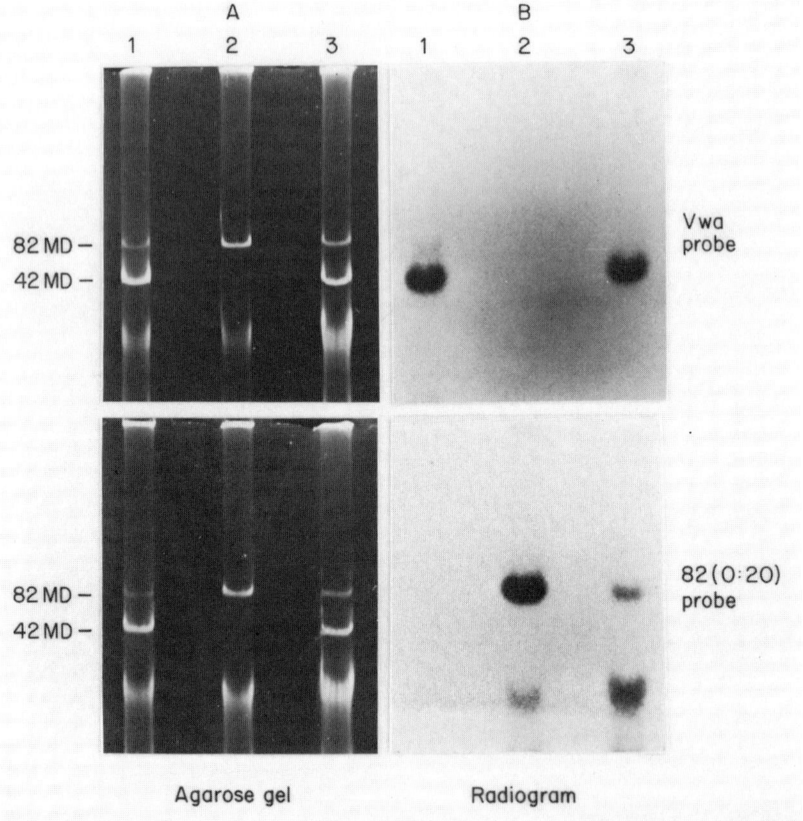

Figure 5. Blot hybridization of 82 MD plasmids of *Y. enterocolitica*. (A) Agarose gel electrophoresis of plasmid DNA preparations. (B) Autoradiograph of nitrocellulose hybridization blot (1) plasmid preparation from serotype 0:8 WA [contains 82-(0:8) and Vsw plasmid]. (2) plasmid preparation from serotype 0:20 segregant 1223-75-2 [contains 82-(0:20) plasmid]. (3) plasmid preparation from 0:20 parent strain 1223-75 [contains 82-(0:20) and Vwa plasmids].

of water overnight and then are allowed to drink *ad libitum* from a distilled water suspension of the pathogen (10^9 cells/ml), we have delineated two distinct virulence phenotypes of *Y. enterocolitica* which are dependent on possession of a Vwa plasmid by the pathogen [13]. A summary (Table 3) of results, obtained with just a few of the strains that we have examined, illustrates these virulence phenotypes. Generally, we observe that some strains (typified by Vwa^+ serotype 0:3) provoke a mild diarrhoea in mice, which is readily seen within 48 to 72 h post infection and which can persist approximately a week. Bacteriological examination of faeces reveals that such infected mice can shed the challenge strain for periods in excess of 21 days. Despite such colonization, these animals rarely

Table 3. Virulence phenotypes of *Y. enterocolitica*

			Oral infection of mice					
Strain	*Serotype*	*Plasmids (MD)*	*Diarrhoea*[a]	*Colonization*[b]	*Dissemination*[c]	*Death*[d]	LD_{50} *(i.p.) Mice*[e]	*Conjunctivitis*[f]
WA	0:8	42(82)	+	+	+	+	10^2	+
WA	0:8	---	−	−	−	−	$>10^8$	−
1223-75	0:20	42,82	+	+	+	+	10^2	+
1223-75-2	0:20	82	−	−	−	−	$>10^8$	−
MCH 628	0:3	46	+	+	−	−	$>10^8$	−
MCH 628	0:3	---	−	−	−	−	$>10^8$	−

[a]Faecal consistency was wet; diarrhoea observed on days 4, 5, 6 post infection.
[b]The presence of detectable Yersiniae in faeces beyond 96 h post infection.
[c]Detection of *Yersiniae* in blood, liver and spleen of moribund animals.
[d]Animals were monitored for 21 days.
[e]Number of bacteria, intraperitoneal injection.
[f]1×10^{10} cells deposited into conjunctival sac of guinea pigs and mice.

die from the infection. Strains with this type of virulence property generally fail to evoke a conjunctivitis reaction and have an LD_{50} (i.p.) in mice in excess of 10^8 organisms.

The other type of virulence phenotype that we observe in mice is exemplified by the highly invasive, extraintestinal, lethal disease caused by Vwa^+ serotype 0:8 strains. Infected mice are colonized and develop diarrhoea within 48 to 72 h. By days 7 to 14 following infection, animals become moribund and evidence of extraintestinal dissemination of the pathogen to various vital organs can be established bacteriologically. Such sepsis usually leads to death of the infected host within 14 days. *Y. enterocolitica* strains which exhibit this highly invasive virulence phenotype generally cause a conjunctivitis reaction and have an LD_{50} (i.p.) in mice of less than 10^3 cells.

The nature of Vwa plasmid-controlled factors that function in colonization or persistence of the pathogen in mouse intestine remains undefined. In addition to adhesion to the mucosal surface, one must consider survival and intracellular growth of the pathogen within the mucosa and submucosa of the gut as a basis for colonization [20]. Our understanding of the reasons why colonization with some serotypes leads solely to mild diarrhoea, whereas with other serotypes it leads to disseminating lethal infections, is obscure. Genes controlling extra-intestinal invasive properties could be chromosomal and/or plasmid associated [6,13,21]. It has also been suggested that the enhanced invasiveness of serotype 0:8 strains could be due to their low requirement for iron as compared to serotypes 0:3 and 0:9 [22]. The role of smooth O-antigen in contributing to extraintestinal invasiveness is currently being examined. We have isolated rough LPS mutants of Vwa^+ *Y. enterocolitica* 0:8 strains and found that some of these retain a full lethal virulence phenotype (P. Gemski and A. A. Lindberg, in prep.).

SUMMARY

(1) Molecular characteristics of *Y. enterocolitica* plasmids and their role in virulence has been reviewed. Vwa plasmids are a class of closely related plasmids that have conserved significant genetic homology. 82 MD plasmids from serotype 0:20 are unique and share little homology with Vwa plasmids and 82 MD plasmids of serotype 0:8. The 82 MD plasmids seen in 0:8 strains share significant homology with Vwa plasmids and could represent a dimeric form of Vwa.

(2) Through use of a mouse oral infection model, two distinct virulence phenotypes have been delineated; some serotypes produce merely a mild diarrhoeal infection, whereas others cause a diarrhoea which progresses to become a lethal, disseminating infection.

ACKNOWLEDGEMENTS

We are indebted to J. R. Lazere, T. Casey, D. Wells, J. N. Coulby, B. Kay and other colleagues who provided support, expertise and the strains.

REFERENCES

1. Swaminathan, B., M. Harmon and I. J. Mehlman (1982). A review: *Yersinia enterocolitica. J. Appl. Bact.* **52**, 151-183.
2. Cook, T. and P. Gemski (1982). Studies of plasmids in the fish pathogen, *Yersinia ruckeri*. Abst. XIIIth Int. Cong. of Microbiol, p. 26.5.
3. Ferber, D. M. and R. R. Brubacker (1981). Plasmids in *Yersinia pestis. Infect. Immun.* **31**, 839-841.
4. Gemski, P., J. R. Lazere and T. Casey (1980). Plasmid associated with pathogenicity and calcium dependency of *Yersinia enterocolitica. Infect. Immun.* **27**, 682-685.
5. Gemski, P., J. R. Lazere, T. Casey and J. A. Wohlhieter (1980). Presence of virulence-associated plasmid in *Yersinia pseudotuberculosis. Infect. Immun.* **28**, 1044-1047.
6. Heesemann, J., C. Keller, R. Morawa, N. Schmidt, H. J. Seimans and R. Laufs (1983). Plasmids of human strains of *Y. enterocolitica*: molecular relatedness and possible importance for pathogenesis. *J. Infect. Dis.* **147**, 107-115.
7. Kay, B. A., K. Wachsmuth and P. Gemski (1982). New virulence-associated plasmid in *Yersinia enterocolitica. J. Clin. Microbiol.* **15**, 1161-1163.
8. Laroche, Y., M. van Bouchaute and G. Cornelis (1984). A restriction map of virulence plasmid pVYE 439-80 from a serogroup 9 *Y. enterocolitica* strain. *Plasmid* **12**, 67-70.
9. Lazere, J. R., J. A. Wohlhieter and P. Gemski (1982). Vwa plasmids of *Yersinia enterocolitica* serotype 0.3. *Am. Soc. Microbiol.* **B42**.
10. Portnoy, D. A., S. L. Moseley and S. Falkow (1981). Characterization of plasmid-associated determinants of *Yersinia enterocolitica* pathogenesis. *Infect. Immun.* **31**, 775-782.
11. Zink, D. L., J. C. Feeley, J. G. Wells, C. Vanderzant, J. C. Vickery, W. D. Roof and G. A. O'Donovan (1980). Plasmid-mediated tissue invasiveness in *Yersinia enterocolitic. Nature (Lond.)* **283**, 224-226.
12. Carter, P. B., R. J. Zahorchak, and R. R. Brubacker (1980). Plague virulence antigens from *Yersinia enterocolitica. Infect. Immun.* **28**, 638-640.
13. Lazere, J. and P. Gemski (1983). Association of colony morphology with virulence of *Yersinia enterocolotica. FEMS Microbiol. Lett.* **17**, 121-126.
14. Laird, W. J. and D. C. Cavanaugh (1980). Correlation of autoagglutination and virulence of *Yersiniae. J. Clin. Microbiol.* **11**, 430-437.
15. Pai, C. H. and L. De Stephano (1982). Serum resistance associated with virulence in *Yersinia enterocolitica. Infect. Immun.* **35**, 605-611.
16. Bolin, I., I. Norlander and H. Wolf-Watz (1982). Temperature inducible outer membrane protein of *Yersinia pseudotuberculosis* and *Yersinia enterocolitica* is associated with the virulence plasmid. *Infect. Immun.* **37**, 506-512.
17. Vesikari, T., T. Nurmi, M. Maki, M. Skurnik, C. Sundqvist and P. Granfors (1981). Plasmids in *Yersinia enterocolitica* serotypes 0:3 and 0:9: Correlation with epithelial cell adherence *in vitro*. *Infect. Immun.* **33**, 870-876.
18. Portnoy, D. A., H. F. Blank, D. T. Kingsbury and S. Falklow (1983). Genetic analysis of essential plasmid determinants of pathogenicity of *Y. pestis. J. Infect. Dis.* **148**, 297-304.
19. Fanning, G. R., J. R. Lazere, J. N. Coulby, J. A. Wolhieter and P. Gemski (1983). Molecular homology among plasmids of *Yersinia enterocolitica. Am. Soc. Microbiol.* Abst. B78.
20. Une, T. (1977). Studies on the pathogenicity of *Yersinia enterocolitica*. I. Experimental infection in rabbits. *Microbiol. Immunol.* **21**, 349-363.
21. Heesemann, J. and R. Laufs (1983). Construction of a mobilizable *Y. enterocolitica* virulence plasmid. *J. Bacteriol.* **155**, 761-767.

22. Robins-Brown, R. M. and J. K. Prpic (1985). Effects of iron and desferrioxamine on infections with *Y. enterocolitica*. *Infect. Immun.* **47**, 774–779.
23. Cornelis, G., D. Ghosal and H. Saedler (1978). Tn951: A new transposon carrying a lactose operon. *Mol. Gen. Genet.* **160**, 215–222.
24. Kay, B. A., K. Wachsmuth, P. Gemski, J. C. Feeley, T. J. Quan and D. J. Brenner (1983). Virulence and phenotypic characterization of *Yersinia enterocolitica* isolated from humans in the United States. *J. Clin. Microbiol.* **17**, 128–138.

Possible Determinants of Virulence of *Yersiniae*

Hans Wolf-Watz[a], Ingrid Bölin, Åke Forsberg and Lena Norlander

Division of Microbiology, National Defense Research Institute, Umeå, Sweden

All three pathogenic species of *Yersinae, Y. pestis, Y. pseudotuberculosis* and *Y. enterocolitica*, possess a related group of plasmids (Fig. 1) which are essential for virulence as well as Ca^{2+}-dependent growth at 37°C *in vitro* [1]. At 26°C plasmid containing bacteria can form colonies on agar-medium lacking Ca^{2+}, whereas at 37°C they are unable to grow. However, when Ca^{2+} is added to the agar, colonies can be formed (Ca^{2+}dependency). Only bacteria devoid of the plasmid or having a mutation within the plasmid can grow at 37°C in the absence of Ca^{2+} (Ca^{2+} independency).

The region of the plasmid containing the gene locus of the plasmid YVO19 involved in Ca^{2+} dependency has been established by Portnoy *et al.* [2]. By the use of transposon Tn*5* and selection for Ca^{2+}-independent mutants at 37°C of the *Y. pestis* strain EV76, they were able to obtain a large number of Ca^{2+}-independent plasmid mutants. Upon analysis of these mutants it was found that a 20 kb region of plasmid DNA seemed to be involved in the unknown mechanism giving rise to the Ca^{2+}-dependent behaviour of strain EV76 (see Fig. 1). Furthermore when these plasmid mutants were tested for their ability to cause a fatal infection in mice, it was found that all these Ca^{2+}-independent mutants were avirulent, in contrast to the wild-type strain [2]. Thus, there is a high correlation between Ca^{2+} dependency and virulence. We have repeated these experiments for virulent *Y. pseudotuberculosis* and similar results have been obtained.

When the individual virulence plasmids of *Y. enterocolitica, Y. pseudotuberculosis*

[a] *Telephone:* 46-90-189230

Protein–Carbohydrate Interactions in Biological Systems. ISBN 0 12 436665 1

and *Y. pestis* were compared it was found that the plasmids of *Y. pestis* and *Y. pseudotuberculosis* were almost identical while the plasmid of *Y. enterocolitica* showed about 50% homology [1]. Interestingly, the region found to be involved in the Ca^{2+}-response was conserved in all three species (Fig. 1), showing that this region has been maintained during evolution, indicating its possible importance in the process of virulence.

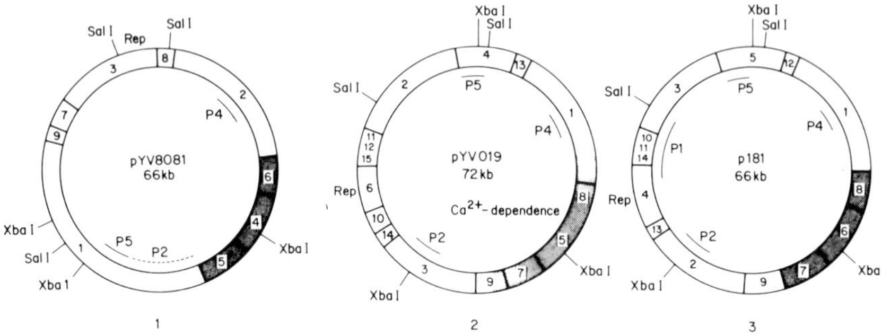

Figure 1. Endonuclease restriction maps of plasmids pYV8081, *Y. enterocolitica*: PYV 019, *Y. pestis*; and pIBl, *Y. pseudotuberculosis*. The numbered fragments refer to the *Bam* HI map. The stippled areas show the region of homology of the plasmids and also indicate the extension of the Ca^{2+} region (pYV019). P1, P2, P4 and P5 indicate the map position of the temperature-inducible proteins YOP1, YOP2, YOP4 and YOP5, respectively.

When a virulent strain YPIII (pIB1) of *Y. pseudotuberculosis* is grown in a rich media devoid of Ca^{2+} at 37°C the synthesis of five proteins is induced [1]. These proteins can be recovered in the outer membrane fraction (Fig. 2). When the same strain is incubated at 26°C or if a plasmid-cured avirulent strain YPIII or a Ca^{2+}-independent plasmid mutant is analysed, these proteins are not detected (Fig. 2). Similar results are obtained when virulent and avirulent strains of *Y. enterocolitica* are studied [1]. We name these proteins YOP1-YOP5 (*Y*ersina *o*uter *m*embrane proteins) with the following molecular weights: YOP1, 150 000, YOP2, 44 000; YOP3, 40 000; YOP4, 34 000; and YOP5, 26 000.

For a long time it was the general opinion that strains of *Y. pestis* did not have the capability to express the YOPs when these were studied *in vitro*. This was puzzling since it had been shown by *Escherichia coli* minicell experiments that all three plasmids harboured the structural genes of these proteins [1] and since it also could be shown by us that all three species expressed these proteins during the process of infection [3]. However, Heesemann *et al.* reported in 1985 that plasmid containing strains of *Y. enterocolitica*, exported to the culture medium a number of polypeptides [4]. The appearance of these proteins in the supernatant was found to be thermoregulated and only occurred after incubation

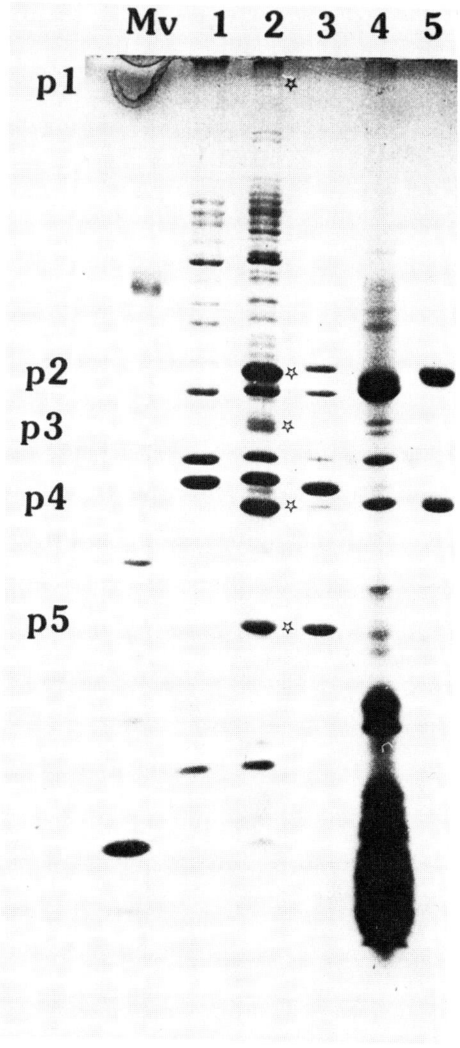

Figure 2. Outer membrane protein profiles of two strains of *Y. pseudotuberculosis* and protein profiles of the exported proteins of virulent strains of *Y. enterocolitica, Y. pestis* and *Y. pseudotuberculosis*. Strains YPIII and YPIII (P1B1) were grown under conditions allowing expression of the plasmid coded temperature inducible proteins (YOPs).

Outer membrane protein profile of strain YPIII, (1); and YPIII (pIB1), (2). Note the induced proteins denoted P1-P5 marked with the symbol*, representing YOP1-YOP5.

Lanes 3-5: the three different species of *Yersiniae* represented by strains 8081, EV76 and YPIII (pIB1) were grown under conditions allowing expression of the exported proteins. Protein profile of proteins recovered from the culture supernatant. Lane 3, *Y. enterocolitica;* lane 4, *Y. pestis;* and lane 5, *Y. pseudotuberculosis.*

at 37°C. When the same strains were grown at 26°C or if plasmid-cured strains were used, these exported proteins could not be recovered in the culture supernatant. Moreover, the appearance of these proteins was also suppressed by the addition of Ca^{2+} to the growth medium [4].

We have been able to show that similar exported proteins also are made when *Y. pseudotuberculosis* is studied (Fig. 2). Because these exported proteins show identical molecular weights to the earlier described YOPs (2–5) (Fig. 2) it is suggested that these proteins are the same. This was further reinforced when we were able to show, by Western blotting analysis, that the YOPs and the exported proteins are immunologically related.

These results inspired us to search for these exported proteins in *Y. pestis* and we were indeed able to find that also a plasmid-containing strain of *Y. pestis* had the capability to synthesize a number of polypeptides which could be recovered in the culture supernatant (Fig. 2). These proteins showed corresponding molecular weights and were also immunologically related to the corresponding exported proteins of *Y. pseudotuberculosis* (Fig. 2). Thus, it must be concluded that *Y. pestis* also has the capability to synthesize the exported proteins *in vitro* and that these proteins are very similar to those of *Y. pseudotuberculosis* and *Y. enterocolitica*.

During these studies it was observed that when *Y. pestis* strain EV76 was used a large amount of low molecular weight polypeptides appeared in the culture supernatant (Fig. 2). These low molecular weight peptides, particularly two peptides of molecular weights about 11 000 and 12 000, also showed cross-reactivity with the exported proteins of *Y. pseudotuberculosis*. Upon prolonged incubation at 37°C of the culture supernatant it was observed that the amount of the exported proteins was greatly reduced and only the low molecular weight polypeptides could be observed after analysis. We conclude from this that there is proteolytic activity in the culture supernatant and that the exported proteins are degraded. We suggest that this is the reason why the YOPs or the exported proteins were not observed in earlier studies of *Y. pestis*. Consequently, this proteolytic activity is higher for *Y. pestis* than for *Y. pseudotuberculosis* and *Y. enterocolitica*. In line with this we have observed a similar proteolytic activity during our studies using *Y. pseudotuberculosis*, however much reduced compared to that of *Y. pestis* strain EV76.

One of the temperature plasmid-induced proteins, YOP1, is however, not exported to the culture medium and is therefore a true outer membrane protein (Fig. 2). This protein is only seen in strains of *Y. enterocolitica* and *Y. pseudotuberculosis* [5]. By molecular cloning and site-specific mutagenesis we have been able to show that this high molecular weight protein is involved in mannose-resistant haemagglutination as well as autoagglutination. However, structural gene mutants of this protein do not show reduced virulence, indicating that this protein is not a major virulence determinant [6].

We have also been able to clone from the virulence plasmid pIB1 of *Y. pseudotuberculosis*, the structural genes of YOP2, YOP4 and YOP5, respectively. The cloning strategy was "shot gun" cloning into pBR322 followed by expression of the hybrid molecules in a minicell producing strain of *E. coli* K 12. The expressed peptides were thereafter precipitated with YOP-specific antisera and analysed by SDS-PAGE.

By subcloning the genetic location of these three genes of plasmid pIB1 was first established and it was found that YOP2 was located close to the single *Xba*I site of *Bam*HI fragment 2. YOP4 overlapped a single *Eco*RI site of *Bam*HI fragment 1 and YOP5 was located between the left *Bam*HI site and the *Xba*I site of *Bam*HI fragment 5 (Fig. 1). The different identified subclones harbouring the respective gene were thereafter used as probes to identify the location of these structural genes on plasmid pYV019 of *Y. pestis* and plasmid pYV8081 of *Y. enterocolitica*.

With respect to plasmid pYV019 it was found that the genes mapped at the same sites as was found for plasmid pIB1 (Fig. 1). This result is not surprising because it was shown earlier that the two plasmids are almost identical [1, 7]. Upon analysis of plasmid pYV8081 we found that the gene of YOP4 seemed to be at a conserved position when compared to plasmid pIB1 (Fig. 1). However, genes of YOP2 and YOP5 were found at other positions (Fig. 1). Both were located on *Bam*I fragment 1 of plasmid pYV8081. Moreover it was also observed that the YOP2 gene of pYV8081 did not show full homology with the corresponding gene of pIB1.

Thus it is apparent that the virulence plasmids of *Yersiniae* have been subjected to evolutionary rearrangements but have maintained the capability to synthesize the inducible plasmid-coded proteins. This can be applied as an argument for the importance of these proteins in the process of virulence, especially because *Y. pestis* has an ecological niche different from the that of *Y. enterocolitica* and *Y. pseudotuberculosis*.

However, further work is needed to establish a possible role of these proteins in the process of virulence. Such work is in progress in our laboratory.

ACKNOWLEDGEMENTS

This work was supported by the Swedish Medical Research Council.

REFERENCES

1. Portnoy, D., H. Wolf-Watz, I. Bölin, A. Beeder and S. Falkow (1984). Characterization of common virulence plasmids in *Yersiniae* species and their role in the expression of outer membrane proteins. *Infect. Immun.* **43**, 108-114.
2. Portnoy, D., H. Blank, D. Kingsbury and S. Falkow (1982). Genetic analysis of essential plasmid determinants of pathogenicity in *Yersinia pestis*. *J. Infect. Dis.* **148**, 297-304.
3. Bölin, I., D. Portnoy and H. Wolf-Watz (1985) Expression of the temperature-inducible outer membrane protein of *Yersiniae. Infect. Immun.* **48**, 234-240.
4. Heesemann, J., B. Algermissen and R. Laufs (1984). Genetically manipulated virulence of *Yersinia enterocolitica. Infect. Immun.* **46**, 105-110.

5. Bölin, I., L. Norlander and H. Wolf-Watz (1982). Temperature-inducible outer membrane protein of *Yersinia pseudotuberculosis* and *Yersinia enterocolitica* is associated with the virulence plasmid. *Infect. Immun.* **37**, 506–512.
6. Bölin, I. and H. Wolf-Watz (1984). Molecular cloning of the temperature-inducible outer membrane protein 1 of *Yersinia pseudotuberculosis*. *Infect. Immun.* **43**, 72–78.
7. Wolf-Watz, H., D. Portnoy, I. Bölin and S. Falkow (1984). Transfer of the virulence plasmid of *Y. pestis* to *Y. pseudotuberculosis*. *Infect. Immun.* **48**, 241–243.

Virulence of Enteropathogenic *Yersinia* Studied by Genetic and Immunochemical Methods

Jürgen Heesemann[a], Uwe Gross, Jörg Schröder and Rainer Laufs

Institute of Medical Microbiology and Immunology, University of Hamburg, Hamburg, FRG

The most important species clinically of the genus *Yersinia* are *Y. pestis, Y. pseudotuberculosis* and *Y. enterocolitica* [1]. These species are genetically closely related, particularly in respect of harbouring similar virulence plasmids [2-6]. In spite of this relationship, it is surprising that only *Y. pseudotuberculosis* and *Y. enterocolitica* are recognized as enteric pathogens [7]. Human infections with enteropathogenic *Yersiniae* comprise a spectrum of diseases, ranging from intestinal manifestations (e.g. enteritis, colitis) to extraintestinal manifestations (e.g. erythema nodosum, arthritis, uveitis, myocarditis). The most important prerequisite to virulence of these species is the presence of a plasmid of about 40 megadalton (MD) in size.

Our experimental strategy for the analysis of the pathogenicity of *Yersinia* is based essentially upon three approaches:

(1) genetic manipulation of the virulence plasmids;
(2) analysis of virulence expression using different test systems including animals, cell and organ cultures;
(3) analysis of the immune response of artificially infected rabbits and naturally infected humans.

[a] *Telephone:* 49-40-4683147

Protein-Carbohydrate Interactions in Biological Systems. ISBN 0 12 436665 1

GENETIC MANIPULATION OF *YERSINIA* PLASMIDS

Yersinia virulence plasmids are non-selftransferable and do not have a selection marker, e.g. antibiotic resistance genes [8]. To overcome these problems, we have developed a two-step method (Fig. 1) which allows us to construct mobilizeable *Yersinia* plasmids using the mobilizeable vector, pRK290, of which replicon is of wide host range [8-10]. Any *Bam*HI fragment of the virulence plasmid can be inserted into the vector, resulting in a hybrid molecule (*in vitro* recombination). Then the hybrid molecule is mobilized into a plasmid-bearing *Yersinia* using the helper plasmid pRK2013. By homologous recombination well-defined cointegrates are generated by the *Yersinia* recipient. Subsequently these cointegrates can be efficiently mobilized into different species. Using *Escherichia coli* minicell producers, we were able to identify plasmid-encoded proteins. Transposon Tn*5* mutagenesis of the cointegrate was carried out in *E. coli*, having a chromosomal Tn*5* insertion. Chromosomal and extrachromosomal virulence determinants could be identified by using *Yersinia* recipients of different chromosomal virulence determinants.

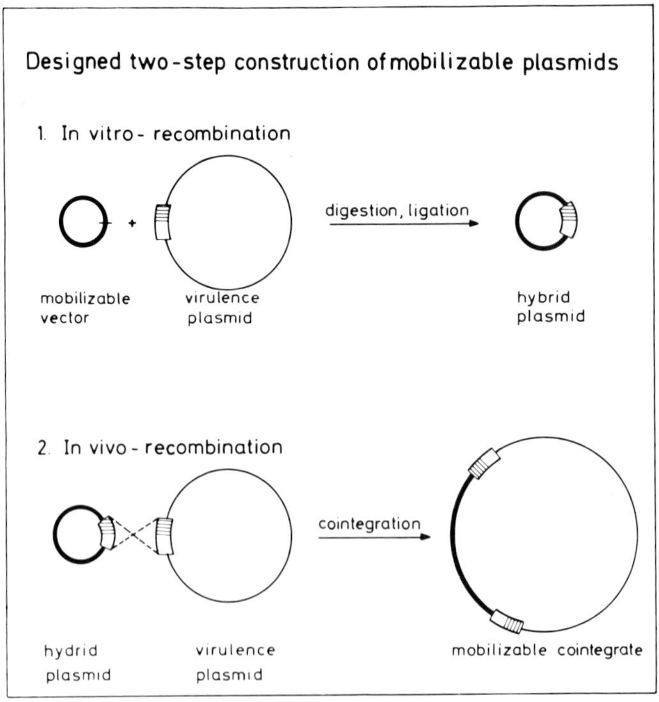

Figure 1. Schematic drawing of the two-step procedure of constructing mobilizeable *Yersinia* plasmids, using the mobilizeable vector pRK290.

IDENTIFICATION OF TRANSFERABLE VIRULENCE FUNCTIONS

Transferable virulence-associated characteristics were identified by mobilization experiments. Plasmidless *E. coli*, *Y. enterocolitica* of different serotypes and *Y. pseudotuberculosis* were chosen as recipient strains. After introduction of the virulence plasmids the obtained exconjugants were phenotypically compared with the isogenic plasmidless recipient. The most striking results of these transfer experiments can be summarized as follows [8, 10]:

(1) *Yersinia* plasmids are not functionally expressed by *E. coli*.

(2) The original virulence characteristics can be re-established in the plasmid-cured host strain by introduction of the isogenic plasmid.

(3) Plasmid-associated expression of animal virulence (mouse lethality, guinea pig lethality and conjunctivitis provocation) are plasmid- and strain-specific, indicating that extrachromosomal and chromosomal determinants are involved in virulence.

(4) Surface agglutinogens and cell adhesion functions are transferable traits within the genus *Yersinia*.

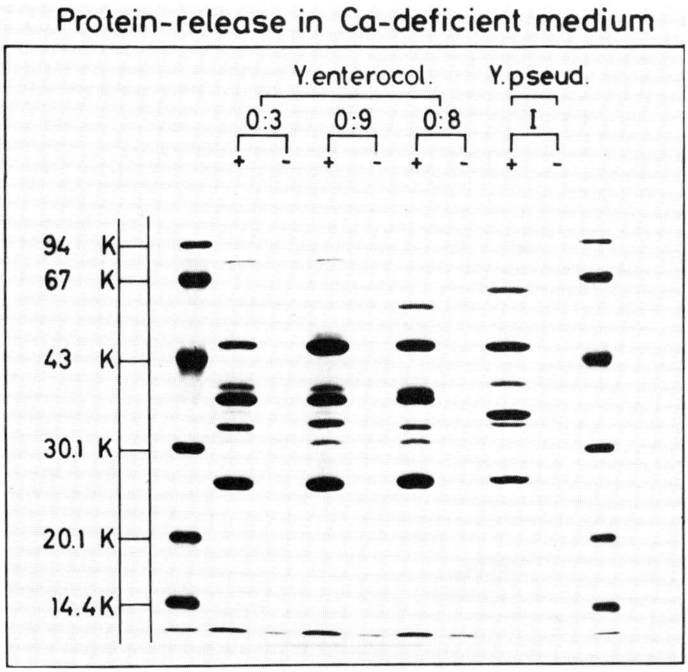

Figure 2. SDS-polyacrylamid gel (11%) electrophoresis patterns of released proteins obtained by ammonium sulphate precipitation from supernatants of calcium-deficient brain-heart infusion medium after cultivation of plasmid-positive (+) and plasmid-negative (−) *Yersinia*, respectively. Strains: *Y. enterocolitica* of serotype 0:3, 0:9, 0:8, and *Y. pseudotuberculosis* of serotype I.

(5) Synthesis and release of proteins into the medium are plasmid-controlled functions (see Fig. 2).

MODULATION OF BACTERIAL ADHERENCE BY GENETIC MANIPULATION

The common human pathogenic *Yersiniae* (*Y. enterocolitica* serotypes 0:3, 0:8 and 0:9, *Y. pseudotuberculosis* serotype I) interact strongly with HEp-2 cells and embryonal intestinal (CCL6) cells in culture. Using a double-immunofluorescence technique, we were able to demonstrate striking differences of cell interaction between isogenic pairs. The plasmid-positive strains remained extracellularly cell-attached, whereas the plasmidless derivatives were rapidly ingested by HEp-2 cells [11]. We wondered whether plasmid-encoded functions are involved in cell interaction. A plasmidless serotype 0:5 *Y. enterocolitica* strain, which interacted with neither cells nor glass surfaces, was chosen to study plasmid-mediated cell interaction (Fig. 3A). After receiving a *Y. enterocolitica* plasmid, the serotype 0:5 exconjugant expressed cell-adherence functions (Fig. 3B). The cell-associated bacteria were found to be exclusively extracellularly located (immunofluorescence results not shown) and exhibited a diffuse adherence pattern.

This adherence pattern changed dramatically concomitantly with the loss of plasmid-mediated surface agglutinogens and a large membrane protein (200–250 kD) after introduction of a virulence plasmid with a Tn*5* insertion within a 2.9 MD *Bam*HI fragment. As shown in Fig. 3C, the serotype 0:5 strain harbouring the mutagenized plasmid interacts only with the border of the cells and weakly with the glass surface. A similar adherence pattern is obtained with the serotype 0:5 strain harbouring the non-mutagenized plasmid in presence of diluted (up to 1:500) rabbit antisera to released proteins (Fig. 3D). By immunoblot experiments using cross-absorbed antisera we were able to provide evidence that a plasmid-encoded 200 kD membrane protein mediates agglutination with specific antisera and cell adherence. This large protein is suggested to be released into calcium-deficient media as dissociated subunits of 25 kD and 47–58 kD.

THE ABILITY TO RELEASE PLASMID-ENCODED PROTEINS IS AN ESSENTIAL PREREQUISITE FOR VIRULENCE

Functions controlling protein release can be selectively inactivated by Tn*5* insertions at different regions of the virulence plasmid. As revealed by immunoblots of whole cell lysates, such mutants were found to be still able to synthesize the protein species of the released fractions (see Fig. 2) and the plasmid-encoded agglutinogen. The most striking consequence of the inactivation of protein release functions was monitored by serotype 0:8 exconjugants. In the HEp-2 cell assay these mutants were found to be internalized (20–30% of cell associated bacteria were found intracellular); obviously they had lost phagocytosis resistance functions. Furthermore these mutants were not lethal for mice.

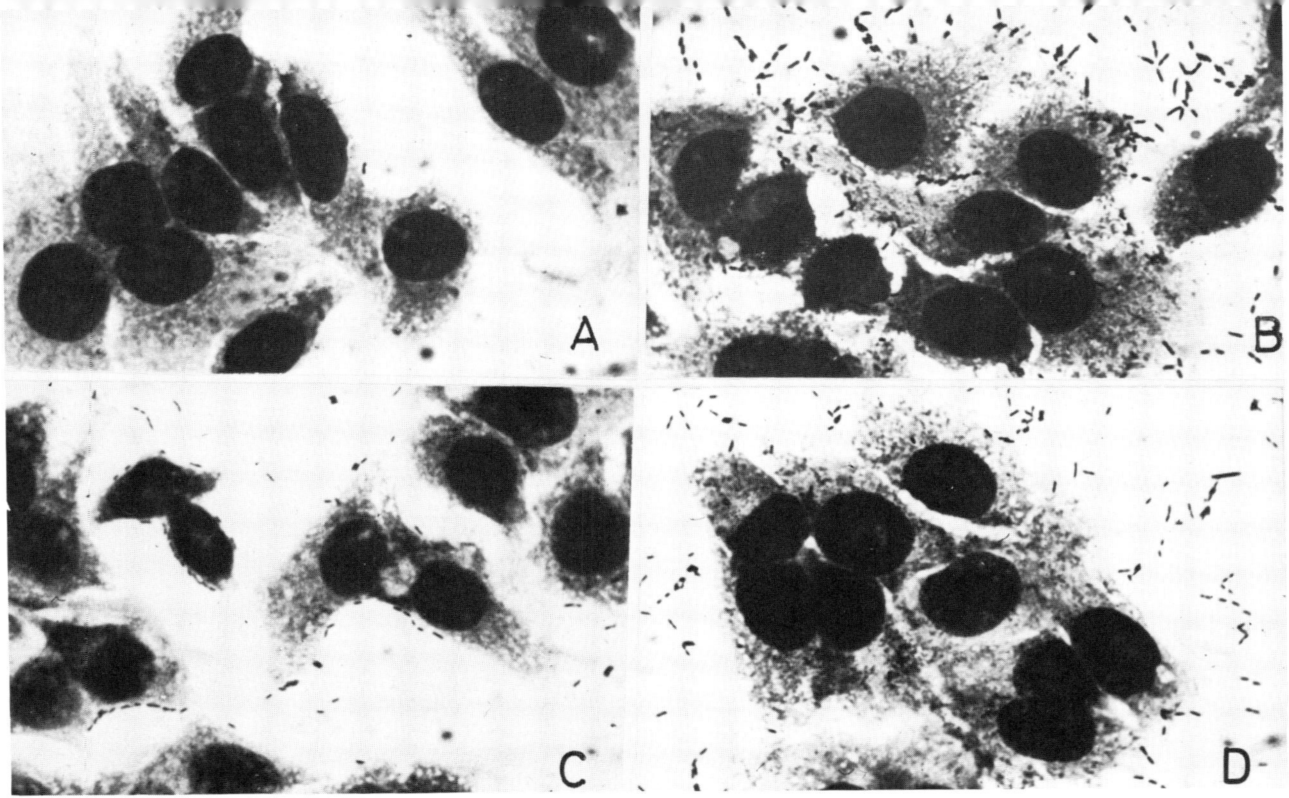

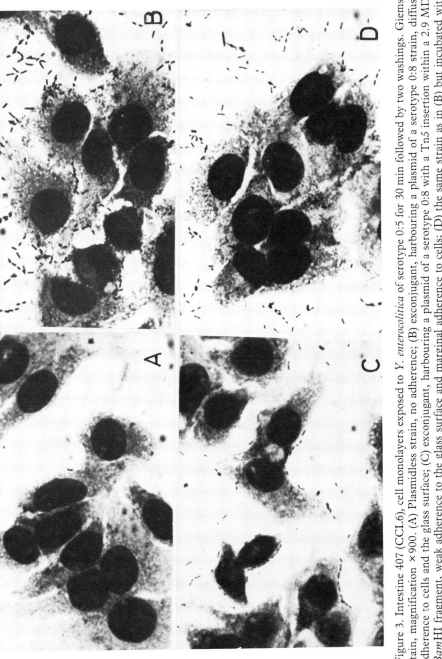

Figure 3. Intestine 407 (CCL6), cell monolayers exposed to *Y. enterocolitica* of serotype 0:5 for 30 min followed by two washings. Giemsa stain, magnification ×900. (A) Plasmidless strain, no adherence; (B) exconjugant, harbouring a plasmid of a serotype 0:8 strain, diffuse adherence to cells and the glass surface; (C) exconjugant, harbouring a plasmid of a serotype 0:8 with a Tn*5* insertion within a 2.9 MDl *Bam*HI fragment, weak adherence to the glass surface and marginal adherence to cells; (D) the same strain as in (B) but incubated with diluted rabbit antisera against proteins of the released fraction (dilution 1:200), efficient blocking of bacterial attachment to cells and the glass surface.

THE IMMUNE RESPONSE TO PROTEINS OF THE RELEASED FRACTION (RP)

Rabbits were orogastrically infected with virulent (plasmid-positive) and avirulent (plasmid-negative) *Y. enterocolitica* strains of serotype 0:3 (inoculum size was 10^{10} bacteria suspended in 10 ml of 5% $NaHCO_3$). Rectal swabs obtained thereafter were positive for about two weeks in the case of rabbits infected with the avirulent strain (av-rabbits), whereas rabbits challenged with the virulent strain (v-rabbits) excreted *Yersinia* for a period of 4–5 weeks. Sera were obtained for a period of three months and tested for agglutinating antibodies to the avirulent strain and for antibodies against plasmid-encoded released proteins. (The plasmid-encoded proteins released by different serotypes show strong cross-reactions. Therefore we used the released protein fraction of a genetically manipulated serotype 0:8 super-producer.) Agglutination titres of 1:50 up to 1:200 were found from day 7 to day 21 with av-rabbits and from day 10 to day 42 post infection with v-rabbits.

The immune response to plasmid-encoded proteins is shown by the immunoblots in Fig. 4. As expected, seroconversion is not found with rabbits challenged

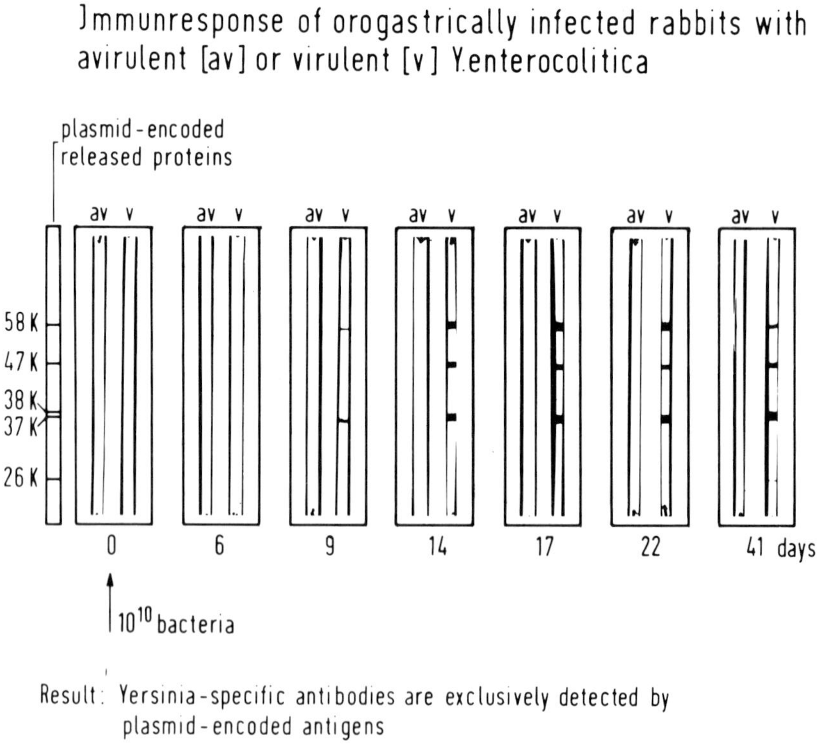

Figure 4. Immune response of rabbits to plasmid-encoded antigens of *Y. enterocolitica* (released protein fraction).

with plasmidless strains. However a strong immune response to plasmid-encoded proteins is demonstrable with rabbits infected with plasmid-positive strains, beginning with antibodies to the 37 kD and 58 kD protein at day 9 followed by additional antibodies to the 26 kD and 47 kD proteins at day 14 post infection.

Similar serological data are found with sera of patients with culturally proven infection due to *Y. enterocolitica*. For example, the course of the serological response of a female patient has been followed up for a period of two years. At the beginning of clinical symptoms seroconversion was observed by the agglutination reaction and the immunoblot method. A striking feature is that specific IgG-class antibodies are demonstrable from the beginning of the disease up to the end of this study. The acute phase of the disease obviously correlates with the detection of IgM-class antibodies against released proteins and of agglutinating antibodies. Finally, the IgA-response during the reconvalescent phase may be of predictive value for serological diagnosis of Yersiniosis, particularly in view of the fact that at this time stool cultures are negative and agglutination titres are weak or negative.

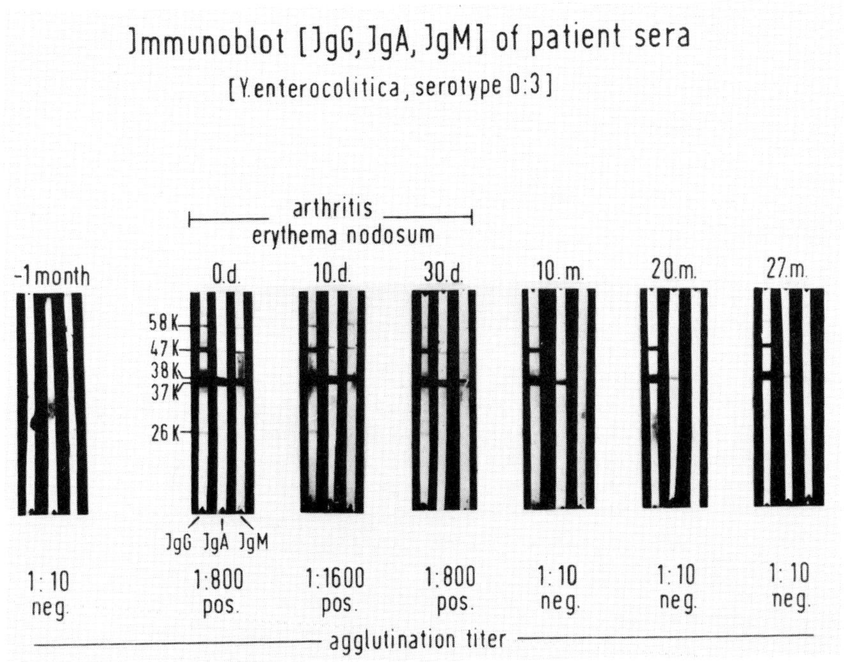

Figure 5. Immune response of a female patient with complicated Yersiniosis to plasmid-encoded antigens of *Y. enterocolitica* and to surface agglutinogens of plasmidless *Y. enterocolitica* of serotype 0:3.

CONCLUSION

(1) Animal virulence, serum resistance, phagocytosis resistance, cell adherence and cytotoxicity of enteropathogenic *Yersinia* are controlled by plasmids of 42–46 MD in size.

(2) Plasmid-encoded proteins which were localized within the outer membrane and in the culture supernatant are suggested to be closely associated with virulence.

(3) Inactivation of protein release by Tn*5* insertions abolishes animal virulence and phagocytosis resistance.

(4) The released protein fraction consists of the V-antigen and subunits of an outer membrane protein which are probably involved in cell adherence.

(5) The released proteins are excellent antigens for serological diagnosis of *Yersinia* infections.

ACKNOWLEDGEMENTS

Thanks are due to R. Brubaker for providing us with anti-V serum. We are particularly grateful to A. Grote and A. Koppe for excellent assistance. This work was supported by the Deutsche Forschungsgemeinschaft.

REFERENCES

1. Brubaker, R. R. (1972). The genus Yersinia: biochemistry and genetics of virulence. *Curr. Top. Microbiol.* **57**, 111–158.
2. Zink, D. L., J. C. Feeley, J. G. Wells, C. Vanderzant, J. C. Vickery, W. D. Roof and A. O. O'Donovan (1980). Plasmid-mediated tissue invasiveness in *Yersinia enterocolitica. Nature (Lond.)* **283**, 224–226.
3. Gemski, P., J. R. Lazere and T. Casey (1980). Plasmid associated with pathogenicity and calcium dependency of *Yersinia enterocolitica. Infect. Immun.* **27**, 682–685.
4. Portnoy, D. A. and S. Falkow (1981). Virulence-associated plasmids from *Yersinia enterocolitica* and *Yersinia pestis. J. Bacteriol.* **148**, 877–883.
5. Heesemann, J., C. Keller, R. Morawa, N. Schmidt, H. J. Siemens and R. Laufs (1983). Plasmids of human strains of *Yersinia enterocolitica*: molecular relatedness and possible importance of pathogenesis. *J. Infect. Dis.* **147**, 107–115.
6. Portnoy, D. A., H. Wolf-Watz, J. Bölin, A. B. Beeder and S. Falkow (1984). Characterization of common virulence plasmids in *Yersinia* species and their role in the expression of outer membrane proteins. *Infect. Immun.* **43**, 108–114.
7. Bottone, E. J. (1977). *Yersinia enterocolitica*: a panoramic view of a charismatic microorganism. *Crit. Rev. Microbiol.* **5**, 211–214.
8. Heesemann, J. and R. Laufs (1983). Construction of a mobilizable *Yersinia enterocolitica* virulence plasmid. *J. Bacteriol.* **155**, 761–767.
9. Ditta, G., S. Stanfield, D. Corbin and D. R. Helinski (1980). Broad host range DNA cloning system for Gram-negative bacteria: construction of a gene bank of *Rhizobium meliloti. Proc. Natl. Acad. Sci. USA* **77**, 7347–7351.
10. Heesemann, J., B. Algermissen and R. Laufs (1984). Genetically manipulated virulence of *Yersinia enterocolitica. Infect. Immun.* **46**, 105–110.
11. Heesemann, J. and R. Laufs (1985). Double immunofluorescence microscopic technique for accurate differentiation of extracellularly and intracellularly located bacteria in cell culture. *J. Clin. Microbiol.* **22**, 168–175.

Two Plasmid-Borne Loci Controlling the Response of *Yersinia pestis* to Ca^{2+} and Temperature

Jon D. Goguen[a] and Janet Yother*

Department of Microbiology and Immunology,
University of Tennessee Center for the Health Sciences, Memphis, TN 38163, USA
*and *Department of Microbiology and Immunology,*
University of Alabama in Birmingham, Birmingham, AL 35294, USA

The three established members of the genus *Yersinia* share an unusual phenotype which we call the low calcium response (LCR). When exponential phase cultures of these organisms growing in Ca^{2+}-free medium are shifted from 26°C to 37°C, growth ceases over a period of about two generations and, between the time of the shift and cessation of growth, a protein known as V antigen is produced. If 2.5 mM Ca^{2+} is present in the medium, growth continues normally and V is not detected [1-3]. Thus, expression of the LCR is controlled by both Ca^{2+} and temperature. In all three species, the genes which confer expression of the LCR are carried by plasmids [4-7]. Either loss of these plasmids or plasmid mutations which block expression of the LCR result in loss of virulence [4-6, 8,9]. In *Y. pestis* strain KIM5, the LCR is determined by the 75 kb plasmid pCD1. We have previously established [8] that at least three distinct units of transcription clustered within a 17 kb region of this plasmid containing LCR genes, although this region does not contain the structural gene encoding V antigen. In parallel with our efforts to determine the mechanism(s) by which these genes contribute to pathogenesis, we have begun to explore their role in regulation of the LCR. Our initial analysis [8] employed *lcr::lac* operon fusions and revealed that transcription in two of these units, *lcrB* and *lcrC*, is induced significantly at 37°C as compared with 26°C, while transcription of the third unit, *lcrA*, is affected only slightly by temperature. Ca^{2+} concentration has no effect on

[a] *Telephone:* 1-901-528-5500

Protein-Carbohydrate Interactions in Biological Systems. ISBN 0 12 436665 1

transcription at these loci. In the following sections, we describe the isolation and analysis of pCD1 mutants specifically altered in their responses to Ca^{2+} and temperature.

ISOLATION AND MAPPING OF Ca^{2+}-BLIND MUTANTS

Mutants that fail to express the LCR even in Ca^{2+}-free medium at 37°C may or may not have a defect in the mechanism by which the low Ca^{2+} condition is detected and the response initiated: a defect in any gene required for expression of the LCR, for example one involved in regulation of the response by temperature, could result in this phenotype. In contrast, mutants which express the LCR at 37°C even in the presence of Ca^{2+} are likely to have a defect in the Ca^{2+} detector, since other components of the mechanism must remain functional. To isolate mutants of this type, which we refer to as Ca^{2+}-blind or LCR-constitutive (LCR^c), cultures of *Y. pestis* strain UTP1000 mutagenized with ethylmethanesulphonate (EMS) were subjected to repeated rounds of selection in ampicillin-containing Ca^{2+}-free medium at 37°C alternated with periods of growth at 26°C in medium free of ampicillin. URP1000 is a derivative of KIM5 in which LCR is encoded by pUT1000, a pCD1 derivative carrying a kanamycin resistance determinant. Five independent mutants were isolated by this method, and a sixth was obtained without EMS mutagenesis. The mutants were designated UTP1001-1006 and their pCD1 derivatives pUT1001-1006.

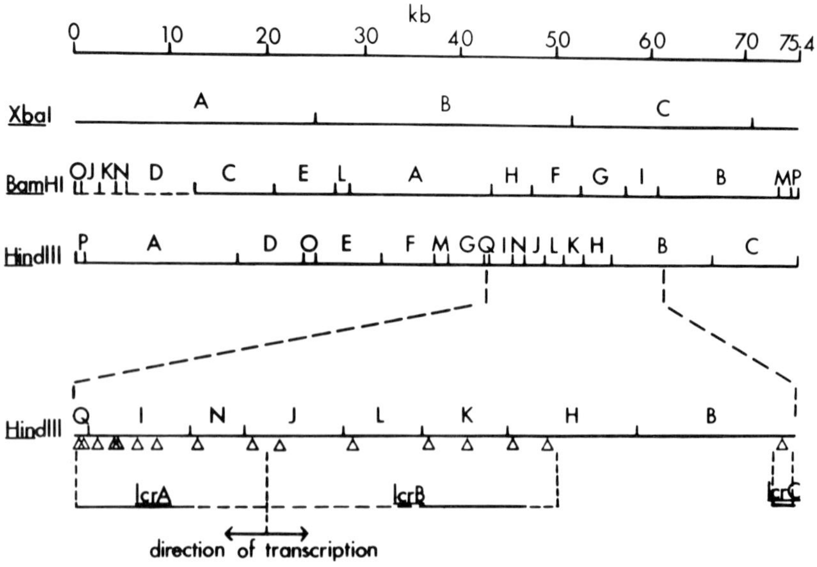

Figure 1. Restriction maps of pCD1. The enlarged section shows the LCR region. Arrows indicate the positions of Mu d1(Ap *lac*) insertions which result in the LCR-phenotype.

When tested in chemically defined medium, all six mutants ceased growth and produced V antigen at 37°C regardless of Ca^{2+} concentration. The lesions responsible for this phenotype were clearly plasmid-borne since replacement of pUT1001-1006 by pUT1000 restored normal expression of LCR and transduction of pUT1001-1006 to another *Y. pestis* strain conferred the Ca^{2+}-blind phenotype.

The location of the mutation carried by pUT1001 was determined by a combination of techniques. The first of these employed a collection of one LCR^+ and 16 LCR^- Mu d1(Ap *lac*) insertion mutants of pCD1. The locations of these insertions, construction and mapping of which have been described previously [8], are shown in Fig. 1. The ability of each of these insertion mutants to complement the lesion in pUT1001 was determined by constructing 17 strains each carrying both pUT1001 and one of the pCD – ::Mu d1(Ap *lac*) cointegrates and determining the plating efficiencies of these strains on Ca^{2+}-supplemented medium at both 30 and 37°C. Results of these experiments are shown in Fig. 2. Only insertions mapping between 45.4 and 48.6 kb, a portion of the *lcrA* locus, fail to improve plating efficiency at 37°C, indicating that the mutation in

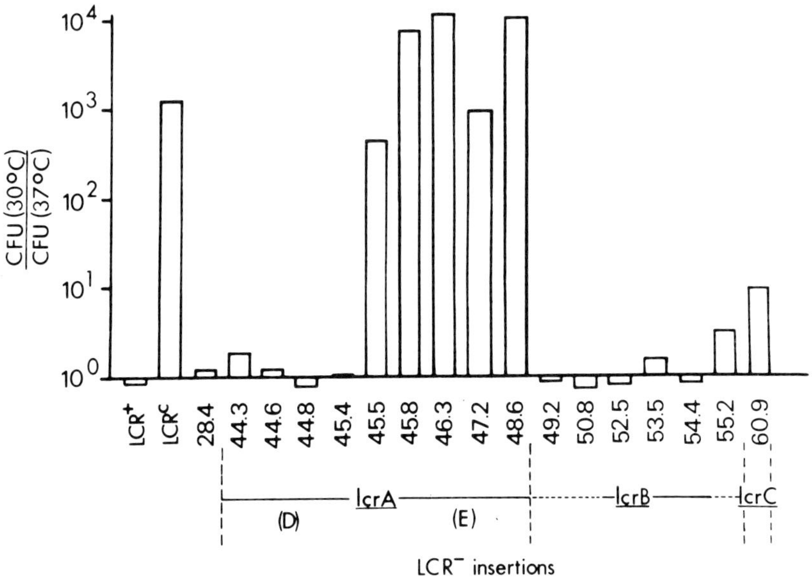

Figure 2. Complementation of an LCR^c mutation by LCR^- pCD1 derivatives. The LCR^- pCD1::Mu d1 plasmids shown in Fig. 1 were used in complementation experiments to map the LCR^c mutation of pUT1001. LCR^+ is UTP1000. LCR^c is UTP1001. Numbers represent map positions, in kilobases, of the Mu d1 insertions. The Mu d1 insertion at position 28.4 does not affect LCR expression. Results are expressed as the ratio of the plating efficiency obtained at 30°C to that obtained at 37°C on Ca^{2+}-supplemented medium.

pUT1001 lies in this region. In keeping with these results, the non-complementing portion of *lcrA* has been designated *lcrE*, the complementing portion designated *lcrD*, and the designation *lcrA* abolished. The mutation carried by pUT1001 has been designated *lcrE1*.

These complementation experiments are potentially misleading because they (a) required the use of antibiotic selection to maintain strains carrying incompatible plasmids, and (b) the complementation observed was in some way incomplete because the expected plating efficiencies were not obtained on Ca^{2+}-deficient medium (data not shown). To confirm independently the location of *lcrE1*, the ability of deletion derivatives of pCD1 and cloned pCD1 restriction fragments to correct this mutation by recombination was determined. Experiments with deletion derivatives indicated that only those retaining the *Xba*I B fragment of pCD1 had this ability. In experiments with cloned *Hin*dIII and *Bam*HI fragments, only *Bam*HI H and *Hin*dIII J could correct *lcrE1*. These results are entirely consistent with those obtained in the complementation experiments, and place *lcrE1* between 47.8 and 48.8 kb on the pCD1 map.

Previous work has shown that insertion mutations in the locus now designated *lcrE* prevent expression of the LCR [8]. The data described above establish that point mutations in this locus prevent Ca^{2+} from inhibiting LCR expression, and indicate that some product of this locus is directly involved in mediating the response of *Y. pestis* to Ca^{2+} concentration.

A THERMALLY CONTROLLED pCD1 REGULON

In addition to *lcrB* and *lcrC*, two other pCD1 loci at which transcription is induced at 37°C have been identified by mutagenesis of pCD1 with Mu d1(Ap *lac*). These loci, *trtA* and *trtB* (*t*emperature *r*egulated *t*ranscription) are shown on the map in Fig. 3.

The properties of *trt::lac* transcriptional fusions with respect to growth, ß-galactosidase production, and production of V antigen, are given in Table 1. Note that all for all three fusions shown, synthesis of ß-galactosidase is increased at least 11-fold at 37°C as compared with 26°C. Moreover, the amount of ß-galactosidase synthesized two- to four-fold higher than is achieved in fully induced *Escherichia coli* and constitutes a substantial fraction of total protein synthesis. These high levels of ß-galactosidase production may account for the failure of strains carrying these fusions to grow at 37°C, although two of the strains also carry the *lcrE1* mutation which should also prevent their growth at this temperature (see above). Although these factors make it impossible to assess the LCR phenotype of the *trt* fusions by observing their growth as a function of Ca^{2+} concentration at 37°C, the other feature of LCR induction, production of V antigen, can be used for this purpose. By this criterion, *trtA* mutants are LCR^+ and *trtB* mutants are LCR^-, indicating that *trtB* function is probably required for expression of LCR.

If high levels of ß-galactosidase production are responsible for the failure of the *trt::lac* fusions to grow at 37°C, mutants in which induction of transcription is blocked should be easily selected by requiring growth at this temperature and

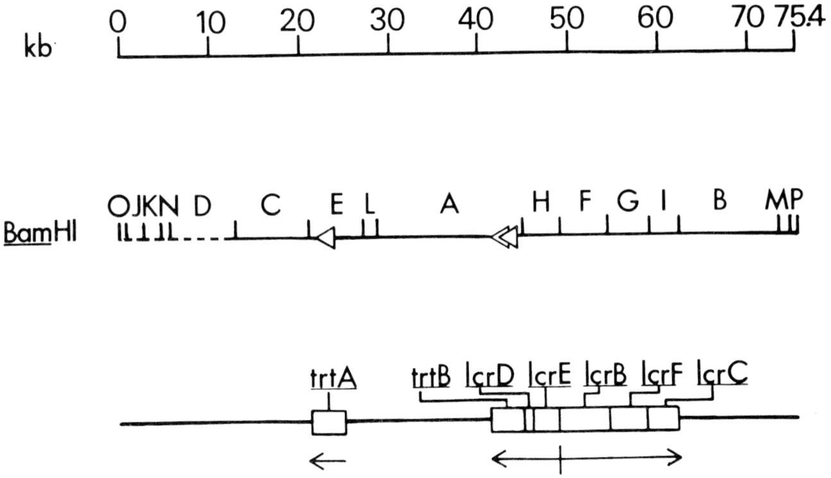

Figure 3. Locations of Mu d1 insertions in the *trt* loci on pCD1 are indicated by the arrowheads in the *Bam*HI map. These arrowheads point in the direction ZYA relative to the orientation of the *lac* genes carried by Mu d1. Loci in the LCR region are designated *lcr*. Arrows below the map indicate directions of transcription.

Table 1. Properties of *Y. pestis* strains harbouring Mu d1(Ap *lac*) insertions in temperature regulated loci of pCD1

Strain	*Plasmid*[a]	*Locus*	*Growth*[b] *ß-gal. units*[c] *30°C*	*37°C* *(−Ca^{2+})*	*37°C* *(+Ca^{2+})*	*V antigen*[d]
UTP1422	pUT1007::Mu d1 22	*trtA*	+/560	−/6500	−/7300	+
UTP1429	pUT1007::Mu d1 42	*trtB*	+/250	−/4100	−/3600	−
UTP2A1	pCD1::Mu d1 42	*trtB*	+/250	−/3300	−/3200	−

[a] Numbers denote position of Mu d1 integration.
[b] A (−) indicates cessation of growth within two generations when exponentially growing cultures are shifted from 30 to 37°C.
[c] Units of ß-galactosidase activity are as defined by Miller [10].
[d] V antigen production was determined at 37°C in Ca^{2+}-deficient medium.

could be useful in identification of genes controlling induction. When the three fusion strains listed in Table 1 were mutagenized with Tn*5*, mutants able to grow at 37°C were readily obtained. These mutants plated with equal efficiency at 30 and 37°C on Ca^{2+}-deficient medium and hence were LCR$^-$. As shown in Fig. 4, induction of ß-galactosidase synthesis at 37°C in all of the mutants was substantially reduced.

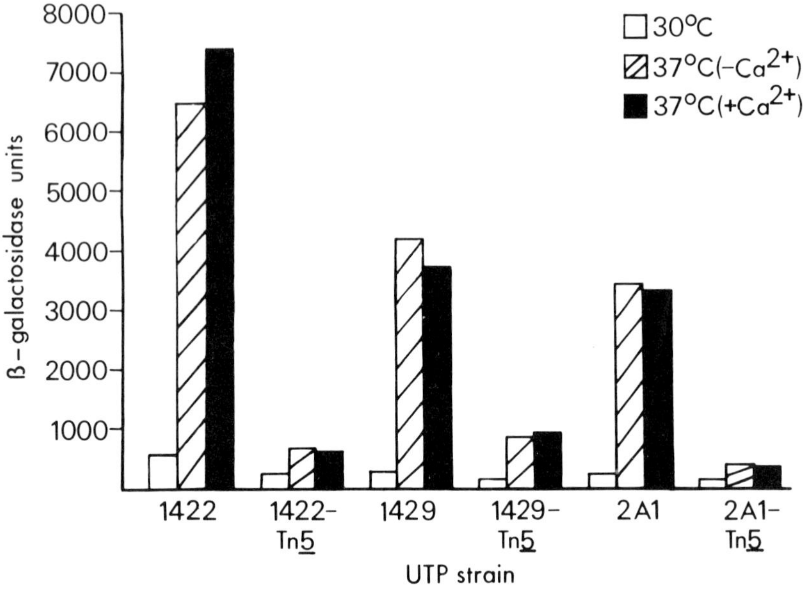

Figure 4. ß-galactosidase production by *trt*::Mu d1 fusions. Levels of ß-galactosidase are shown for UTP1422, 1429, and 2A1 and for LCR^- mutants of these strains obtained following mutagenesis with Tn*5*. Values for the Tn*5* mutants are the averages for at least 5 independent mutants.

The locations of Tn*5* insertions in 32 of these mutants were determined. Thirty of them (9 affecting *trtA* and 21 affecting *trtB*) mapped within a 1.5 kb region between 54.9 and 56.4 kb on the pCD1 map. We have designated this region, which lies between the *lcrB* and *C* loci, as *lcrF* (see Fig. 3). Six *lcrF*::Tn*5* mutants of the native pCD1 plasmid were also isolated and all were found to be LCR^-, indicating that *lcrF* function is also required for expression of the LCR.

These results show that pCD1 contains a thermally controlled regulon which includes at least two loci, *trtA* and *trtB*, in which transcription is induced at 37°C as compared with 26°C. Although these loci are widely separated from each other, induction of transcription at both of them is controlled by *lcrF*, indicating that this locus encodes a diffusible positive regulator. A product of *trtB* is apparently required for expression of LCR (see above). Thus, the LCR^- phenotype of *lcrF* mutants might be due to insufficient synthesis of this product at 37°C. However, we think it likely that *lcrF* is also responsible for the induction of transcription observed at two other pCD1 loci involved in LCR expression, *lcrB* and *C*. Experiments designed to test this hypothesis are in progress.

SUMMARY

(1) Ca^{2+}-blind mutants of *Y. pestis* which express the LCR (both growth restriction and synthesis of V antigen) at 37°C regardless of Ca^{2+} concentration were isolated. By complementation and recombination analysis, these mutants were mapped to a locus on the plasmid pCD1 designated *lcrE*. A product(s) of this locus is directly involved in mediating the response of *Y. pestis* to Ca^{2+} concentration and is required for expression of LCR.

(2) The *lcrF* locus of pCD1 was found to encode a positive regulator that mediates induction of transcription in response to temperature at two remote pCD1 loci, *trtA* and *trtB*. Mutations in either *lcrF* or *trtB* can prevent expression of LCR.

ACKNOWLEDGEMENTS

This work was supported by faculty development funds provided to J.D.G. by the Department of Microbiology and Immunology, University of Tennessee Center for the Health Sciences and by a pre-doctoral fellowship awarded to J.Y. by the Department of Microbiology and Immunology, University of Alabama in Birmingham. Initial stages of one project also received support from Public Health Research Grant AI 19451.

REFERENCES

1. Brubaker, R. R. and M. J. Surgalla (1964). The effect of Ca^{2+} and Mg^{2+} on lysis, growth, and production of virulence antigens by *Pasturella pestis*. *J. Infect. Dis.* **114**, 13-25.
2. Higuchi, K. and H. L. Smith (1961). Studies on nutrition and physiology of *Pasturella pestis*. IV. A differential plating medium for the estimation of the mutation rate to avirulence. *J. Bacteriol.* **81**, 605-608.
3. Zahorchak, R. J. and R. R. Brubaker (1982). Effect of exogenous nucleotides on Ca^{2+} dependence and V antigen synthesis in *Yersinia pestis*. *Infect. Immun.* **38**, 953-959.
4. Ferber, D. M. and R. R. Brubaker (1981). Plasmids in *Yersinia pestis*. *Infect. Immun.* **31**, 839-841.
5. Gemski, P., J. R. Lazere and T. Casey (1980). Plasmids associated with pathogenicity and Ca^{2+} dependency of *Yersinia enterocolitica*. *Infect. Immun.* **27**, 682-685.
6. Gemski, P., J. R. Lazere, T. Casey and J. A. Wohlheiter (1980). Presence of a virulence-associated plasmid in *Yersinia pseudotuberculosis*. *Infect. Immun.* **28**, 1044-1047.
7. Portnoy, D. A., H. Wolf-Watz, I. Bölin, A. B. Breeder and S. Falkow (1984). Characterization of common virulence plasmids in *Yersinia* species and their role in expression of outer membrane proteins. *Infect. Immun.* **43**, 108-114.
8. Goguen, J. D., J. Yother and S. C. Straley (1984). Genetic analysis of the low calcium response in *Yersinia pestis* Mu d1(Ap *lac*) insertion mutants. *J. Bacteriol.* **160**, 842-848.
9. Portnoy, D. A., H. F. Blank, D. T. Kingsbury and S. Falkow (1983). Genetic analysis of essential plasmid determinants of pathogenicity in *Yersinia pestis*. *J. Infect. Dis.* **148**, 297-304.
10. Miller, J. H. (1972). Experiments in molecular genetics. Cold Spring Harbor Laboratory, New York.

Cytotoxic Effect of Virulent *Yersinia pseudotuberculosis* on HeLa Cells

Roland Rosqvist[a], and Hans Wolf-Watz*,

*Division of Experimental Medicine and *Division of Microbiology, National Defence Research Institute, Umeå, Sweden*

INTRODUCTION

It is generally accepted that *Yersinia pseudotuberculosis* infects a host via the oral route. The first step in the infection process is probably adherence to and penetration of intestinal epithelial cells followed by infection of the mesenteric lymph nodes [1]. Occasionally the invader breakthrough causes sepsis and infections of organs where growth of the pathogen leads to tissue damage and necrosis. To study these pathogenic events model systems have been employed, for example the interaction between the pathogen and HeLa cells. In this study we have investigated the plasmid-involved functions concerning the process of invasion of HeLa cells and intracellular behaviour of the virulent strain YPIII(p1B1) and its avirulent plasmid-free isogenic derivative YPIII.

EXPERIMENTAL DETAILS

HeLa cells were seeded at a density of about 2×10^5 cells per well in a 6-well tissue culture plate (35 mm diameter, Nunc Denmark) using Leibovitz L-15 medium with 10% heat-inactivated newborn calf serum. Cultured HeLa cells were infected with the virulent plasmid-containing strain YPIII(p1B1) or its isogenic avirulent plasmid-free strain YPIII of *Y. pseudotuberculosis* at a cell to bacteria ratio of 1 to 10. The infected cell cultures were incubated at 37°C for 1, 2, 4 and 5 h respectively. The cell cultures were washed and the HeLa cells

[a] *Telephone:* 46-90-189230

Protein-Carbohydrate Interactions in Biological Systems. ISBN 0 12 436665 1

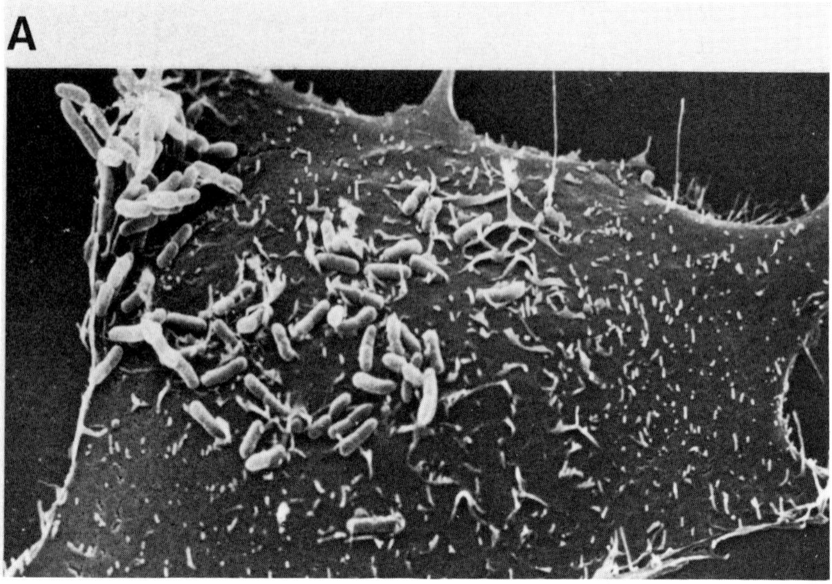

Figure 1A.

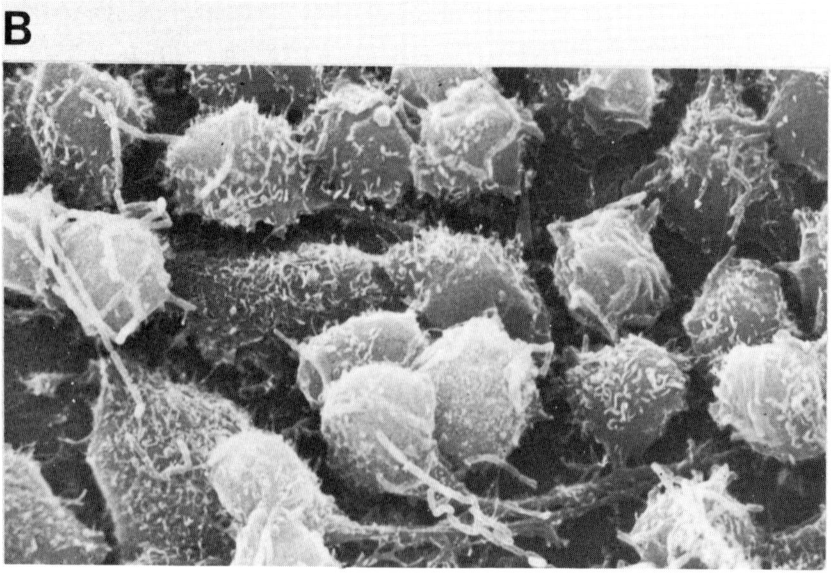

Figure 1B.

cells were lysed with 0.1% Triton X-100. The number of viable bacteria associated with the HeLa cells was determinated by viable count on blood agar base plates. To establish the number of intracellular bacteria, gentamicin was added to a final concentration of 50 μg/ml to the infected cell cultures at various times after infection. The incubation was continued for at least one hour in the presence of gentamicin whereafter the cells were washed, lysed and viable count was performed as described above. To study possible intracellular multiplication of *Y. pseudotuberculosis* within the HeLa cells gentamicin was added to infected HeLa cells one hour after infection and the incubation was continued for up to 24 h. After various times the infected cells were washed and lysed and viable count was performed.

RESULTS

Adherence, invasion and intracellular behaviour of Y. pseudotuberculosis

No significant difference was observed when the virulent and avirulent strains were compared with respect to the infection rate and both strains invaded HeLa cells to the same extent. No increase in the intracellular concentration of bacteria could be demonstrated either when the virulent strain YPIII(p1B1) or the avirulent strain YPIII was studied.

A cytotoxic effect on HeLa cells induced by virulent Y. pseudotuberculosis

The virulent strain YPIII(p1B1) induced a cytotoxic effect on the HeLa cells whereas the avirulent derivative strain YPIII did not. The cytotoxic effect was not induced when the incubation temperature was lowered to 26°C. This result suggests that plasmid-mediated temperature-inducible functions [2] might be involved in the expression of the cytotoxic effects. Therefore, a set of transposon Tn*5*-derived plasmid p1B1 insertion mutants of *Y. pseudotuberculosis* strain YPIII(p1B1) and strains of YPIII, that carried different transposon Tn*5*-derived insertion mutants of the virulence plasmid pYV019 of *Y. pestis* strain EV76 were analysed with respect to their cytotoxic interaction with HeLa cells.

One strain YPIII-P3, is Ca^{2+}-independent but virulent and expresses the YOPs and the V-antigen. This strain was also found to give a cytotoxic response indicating that Ca^{2+}-dependency *per se* is not an absolute requirement for the expression of the cytotoxicity. One of the transposon Tn*5*-induced mutants used in this study was Ca^{2+}-dependent but unable to express YOP1. This mutant was cytotoxic, indicating that protein YOP1 is not an essential determinant for expression of the cytotoxicity of *Y. pseudotuberculosis*. The cytotoxic effect correlated well with the virulence of the tested strains with one exception—strain YPIII-P2. This strain shows after 4 h of incubation a reduced viability at 37°C. This phenomenon may explain the avirulent behaviour of this strain.

CONCLUSIONS

The virulent plasmid-containing strain YPIII(p1B1) induces a cytotoxic effect on cultured HeLa cells whereas its isogenic plasmid-free strain YPIII does not. The plasmid-dependent ability to express virulence is strongly correlated with the cytoxic properties of *Y. pseudotuberculosis*.

REFERENCES

1. Braude, A. I. (1981). "Medical Microbiology and Infectious Diseases." W. B. Saunders, Philadelphia.
2. Portnoy, D. A., H. Wolf-Watz, I. Bölin, A. B. Beeder and S. Falkow (1984). Characterization of common virulence plasmids in *Yersinia* species and their role in the expression of outer membrane proteins. *Infect. Immun.* **43**, 104–114.

Poster Session

Monoclonal Antibody to the Autoagglutination Protein P1 of *Yersinia*

Mikael Skurnik[a], and Kari Poikonen

Department of Medical Microbiology, University of Oulu, Finland

A hybridoma cell line producing a monoclonal antibody, D66, to the autoagglutination protein P1 of *Yersinia enterocolitica* 6471/76 (serotype 0:3) was prepared. D66 recognized both the high molecular weight form and the subunit form of P1 (Fig. 1).

Virulent *Y. enterocolitica* and *Y. pseudotuberculosis* strains representing various serotypes and bearing the virulence plasmid (in which mutations had probably caused differences during evolution) were received from around the world. They were tested with D66 using SDS-PAGE and the immunoblotting technique. D66 recognized P1 of every strain which was positive in autoagglutination test and expressed P1 detectable in SDS-PAGE (Table 1). This result indicates that P1 has not changed much during evolution, and this also means that the presence of P1 must be important to the proper function of the virulence plasmid. There were, however, differences in the molecular weights of the subunit form of P1 of different strains, indicating that some mutations have occurred in its gene during evolution.

D66 was used to detect bacteria expressing P1 *in vivo*. *Y. enterocolitica* 6471/76 was grown at room temperature so that the temperature-inducible P1 was not expressed. Washed bacteria in 7.5% sodium bicarbonate solution were inoculated i.g. in rats using a balloon catheter. At fixed time points rats were killed, pieces of the small intestine were removed, as well as enlarged mesenteric lymph nodes of the 144 h rat, and sections were stained with D66 using the peroxidase anti-peroxidase method. In this experiment it was observed that P1 was expressed very soon after the challenge, and that this already occurred in the lumen of the intestine.

[a]*Telephone:* 358-81-332133

Protein-Carbohydrate Interactions in Biological Systems. ISBN 0 12 436665 1

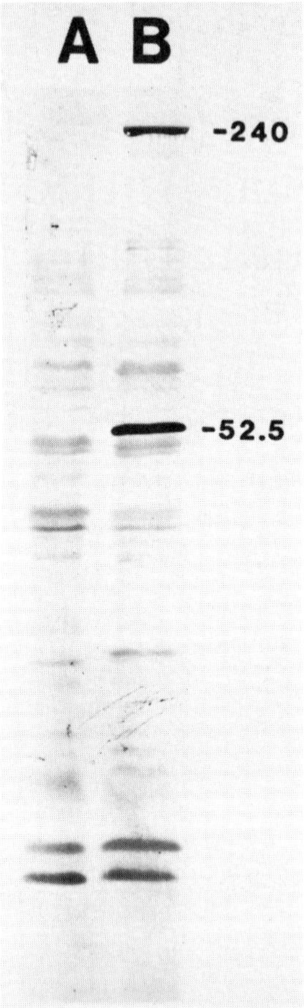

Figure 1. Immunoblot of the total proteins of *Y. enterocolitica* 6471/76c (A) and 6471/76 (B), a plasmid-cured and plasmid-bearing pair of strains, respectively. The bound monoclonal antibody D66 was detected by the peroxidase-conjugated rabbit antibodies to mouse immunoglobulins. The numbers indicate the apparent molecular weights of the bands in kD. The light staining of the other bands was due to prolonged (overnight) incubation with the antibodies.

The biological function of P1 is still obscure. Because of its fibrillar structure, and agglutinating properties, it is most probably involved in adherence. Another possible function could be the formation of a protecting layer on the bacterial cell surface. D66 will be useful in experiments in which the biological role of P1 is studied.

Table 1. *Yersinia* strains tested with D66, *n* = number of strains tested, AA = autoagglutination, P1 = the autoagglutination protein P1 detected in SDS-PAGE stained with Coomassie Brilliant Blue, D66 = P1 bands stained in immunoblots by the immunoperoxidase method and D66

0-serotype	*n*	*AA* (+/−)	*P1* (+/−)	*D66* (+/−)
Yersinia enterocolitica				
1	2	2/0	2/0	2/0
1,2,3	1	1/0	1/0	1/0
2	2	2/0	2/0	2/0
3	20	16/4	16/4	16/4
4	1	1/0	1/0	1/0
4,32	1	1/0	1/0	1/0
5	2	2/0	2/0	2/0
5,27	6	6/0	6/0	6/0
8	10	8/2	8/2	8/2
9	11	11/0	11/0	11/0
13	1	1/0	1/0	1/0
13a,13b	2	2/0	2/0	2/0
13,18	1	1/0	1/0	1/0
15	2	1/0	1/0	1/0
20	2	1/1	1/1	1/1
21	3	3/0	3/0	3/0
34	1	1/0	1/0	1/0
Yersinia pseudotuberculosis				
I B	1	1/0	1/0	1/0
II	1	1/0	1/0	1/0
III	1	1/0	1/0	1/0

VII. Bacterial Surface Components and their Importance for Virulence

Co-Chairpersons: Frederick Sparling
Maggie So

Genetic Basis of Virulence and Type b Capsule Expression in *Haemophilus influenzae*

E. R. Moxon, S. Ely, J. S. Kroll, I. Allan, S. Zamze, J. Tippett, S. Fulford and *S. K. Hoiseth

Infectious Diseases Unit, University Department of Paediatrics, John Radcliffe Hospital, Oxford, UK, and
**Office of Biologics, National Center for Drugs and Biologics, Food and Drug Administration, Bethesda, MD, USA*

Haemophilus influenzae is prevalent in the upper respiratory tract of humans, the only known natural host and reservoir for this bacterium. Most carriers are asymptomatic and balanced parasitism is thus the rule. Nonetheless, *H. influenzae* is responsible for a wide spectrum of serious infections, especially in childhood. These include meningitis, epiglottitis, cellulitis, septic arthritis, pneumonia and otitis media. (For a general review, see [1]). Two distinct pathogenetic mechanisms have been noted; one involves invasive, bloodstream dissemination (as in meningitis), while the other involves contiguous spread of organisms within the respiratory tract (as in otitis media). The former kind of infection is usually caused by encapsulated strains, and the latter by strains lacking the ability to express capsule. Until recently, knowledge of the microbial and host determinants of *H. influenzae* infections was confined almost exclusively to the capsular antigen, especially that of type b, a ribosyl-ribitol phosphate polymer designated PRP. Serum antibodies specific for PRP are critical in host defence against type b infections. Very little is known about the pathogenesis of infection caused by non-capsulated *H. influenzae* strains and the mechanisms of tissue damage. The role of somatic antigens [e.g. outer membrane proteins (OMP), fimbriae, lipopolysaccharide (LPS)] or other products such as IgA1 proteases is not yet clear, although an understanding of their contribution to pathogenicity would

[a]*Telephone:* 44-865-65292

Protein–Carbohydrate Interactions in Biological Systems. ISBN 0 12 436665 1

provide a logical basis for developing strategies aimed at reducing or even preventing the morbidity and mortality caused by *H. influenzae.*

This communication presents a molecular approach to the analysis of *H. influenzae* pathogenicity, focussed on the role of capsule. Using chromosomal donor DNA from strains representative of each of the six different capsular types, Zwahlen *et al.* showed that type b transformants were significantly more virulent than each of the others [2]. This suggested a unique contribution of genes for PRP in virulence, but the possibility of a critical role of genes linked to those for PRP could not be addressed using transformation with uncharacterized high molecular weight chromosomal DNA as donor. Furthermore, when independently derived b^+ transformants expressing similar amounts of capsule, but differing in their LPS phenotype, were compared, significant differences in virulence were found [3].

The biosynthesis, transport, polymerization and surface assembly of each capsular antigen must require the coordinated functions of several genes. These genes could be organized in a linear arrangement, physically linked—but interrupted in one or more loci by sequences unrelated to capsule expression—or unlinked. The latter possibility is unlikely since transformation experiments have shown that all of the genes necessary and sufficient for type b capsule expression can be carried on a single molecule of DNA of about 50 kb [4]. This region is designated *cap b.*

The approach was therefore as follows; a capsule-deficient mutant was isolated from a virulent, wild-type b strain [5]. This mutant, designated Sec-1, was made competent for DNA transformation and used as recipient for the uptake of cloned DNA. A Charon 4 lambda library was screened for clones that could restore the type b phenotype to Sec-1. Encapsulated transformants were identified by their iridescent phenotype. This resulted in the identification of a lambda clone (Charon 4:48) containing a 13.4 kb insert of haemophilus DNA which contains some of the genes for type b expression. A restriction endonuclease map of this insert is shown in Fig. 1. The *Eco*RI fragments (9 and 4.4 kb) of the original 13.4 kb insert were sub-cloned into pBR328 and used as transforming DNA, or as probes for Southern hybridization, to analyse a variety of clinical and laboratory isolates. Both *Eco*RI fragments had DNA necessary but not sufficient for type b capsule expression based on transformation experiments with various capsule-deficient mutants (Table 1).

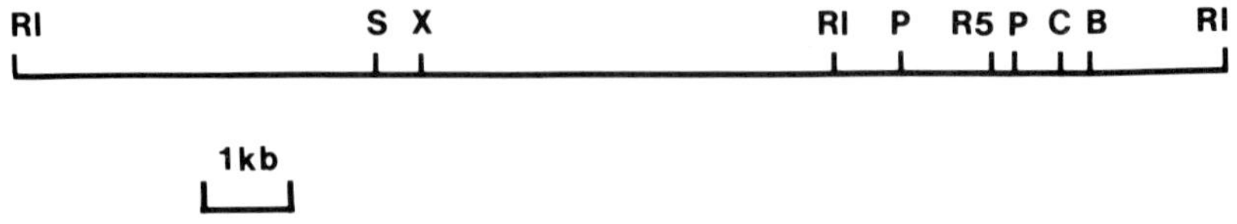

Figure 1. Restriction endonuclease map of the 13.4 kb haemophilus insert in Charon 4:48 showing selected sites only. Restriction endonuclease recognition sites shown are *Eco*RI (R1), *Sal*I (S), *XHo*I (X), *Pst*I (P), *Eco*RV (R5), *Cla*I (C) and *Bst*EII (B).

Table 1. Transformation of capsule-deficient mutants of *Haemophilus influenzae*

	Sub-cloned Eco*RI fragment*	
Strain	*4.4*	*9*
Sec-1	+	−
S-2	+	+
Rt20	−	−
RM5013	−	+

Transformation was performed with the 4.4 kb and 9 kb *Eco*RI fragments (sub-cloned into pBR328). +/− = necessary/unnecessary for transformation. Type b transformants were identified by their characteristic iridescence and confirmed serologically.

Table 2. Virulence of *H. influenzae* in infant rats

Strain	*Incidence of bacteraemia*	*Incidence of meningitis*
Sec-1	0/4	0/4
Sec-1/b^+ transformant	8/9	8/9

Five-day old rats were inoculated intranasally with either 10^6 cfu of Sec-1 or type b transformants of Sec-1. Blood and CSF were cultured 5 days after challenge.

Three out of four capsule-deficient mutants derived from strain Eagan required one, the other, or both of the sub-clones. These transformants also regained virulence potential based on experimental infection of infant rats [6]. A typical experiment is shown in Table 2.

Southern hybridization analysis of the parent b^+ and of the b^- variants with the complete 13.4 kb probe showed that S-2, Rt20 and RM5013 lacked hybridization to a 9 kb *Eco*RI fragment; Sec-1 retained a similar pattern to the parent strain. Further analysis of the mutation in Sec-1 was performed by cutting the plasmid containing the sub-cloned 4.4 kb *Eco*RI fragment with different enzymes, thus generating a variety of fragments as donors for transformation. The mutation in Sec-1 mapped within a 200 bp fragment between the *Eco*RV and *Pst*I sites (Fig. 1). Sec-1 appears to be a point mutant since it reverts to the b^+ phenotype under strong selection and shows an identical Southern hybridization pattern to its b^+ parent, strain Eagan. The remaining 3 variants have evidently undergone more complex changes.

Probing representative strains of the 6 different capsular serotypes (a–f) of *H. influenzae* with the 4.4 kb *Eco*RI fragment has shown that there are sequences homologous to one *Eco*RI restriction fragment in each [5]. Further analysis indicates that the left-hand *Eco*RI-*Eco*RV 1.7 kb subfragment of the 4.4 kb fragment hybridizes with DNA from all capsular serotypes, whereas the right-hand 2.7 kb subfragment hybridizes only with the chromosomal DNA of type a and type b strains [7]. Sequences in this 2.7 kb subfragment may be involved in the biosynthesis of ribitol-5-phosphate since this capsule constituent is found only in serotypes a and b. Furthermore, since there are no other compositional features common to all six capsular serotypes, the existence of homology with the 1.7 kb probe suggests that some regulatory or transport step(s) may be shared by all capsular serotypes.

The 9 kb *Eco*R1 probe has homology with two to four *Eco*R1 restriction fragments in various type b strains, indicating sequence repetitions. The 4.4 kb probe identifies a single DNA fragment in the *Eco*RI digested genome, but when the chromosomes of several type b strains are digested with the restriction endonuclease *Bam*HI and hybridized with the 4.4 kb probe, two fragments of about 18 and 20 kb in size are seen. When the DNA of b^+ strains and their spontaneous b^- derivatives is examined, the *Eco*RI digests of the latter commonly show loss of a 9 kb fragment homologous with the 9 kb probe, while in the *Bam*HI digests of the DNA of these b^- strains, probed with the 4.4 kb probe, the 18 kb fragment is no longer found. Such observations suggest that the genetic basis for the loss of the capsular phenotype involves a complex rearrangement of genes of the *cap b* region. This high frequency of spontaneous loss of capsule production (0.1–0.3%), is too high to be explained by simple base-change mutation. Recent experiments using cosmid clones of the *cap b* region suggest that this may be due to *rec*-dependent recombination between two copies of an 18 kb tandem repeat (Hoiseth, Moxon and Silver, submitted).

Repeating DNA sequences have proved useful in defining *Eco*RI restriction enzyme polymorphisms, allowing, in principle, the recognition of different "clonotypes". In practice, analysis of more than 100 strains isolated from the USA, UK, Australia, Papua New Guinea, Africa, Malaysia, Holland, Sweden and France indicate that three polymorphisms describe more than 95% of type b clinical isolates despite the geographical and temporal (1960–1980) diversity of the strains (Fig. 2a). These polymorphisms correlate with distinctive OMP sub-types and indicate the existence of three major clonotypes (Fig. 2b). To date no direct correlation between clonotype and LPS sub-type has been observed.

Such an an analysis indicates that one of these polymorphisms has been prevalent in the USA for decades but is extremely rare in continental Europe. An examination of a random sample of 15 strains from Holland spanning the years 1977–1983 showed that all type b strains causing meningitis during this period possessed an identical *Eco*RI polymorphism (Fig. 3).

In contrast, all three major polymorphisms were found among a smaller number of isolates from children with meningitis in Baltimore during a four month period in 1983. The relative prevalence of particular clonotypes in different geographical regions may provide insight into the differing patterns of type b disease characteristic of different countries. It would appear that a limited number of highly virulent clonotypes of type b strains have become disseminated throughout the world and presumably account for the vast majority of cases of meningitis and other systemic *H. influenzae* infections. However, recent reports suggest that

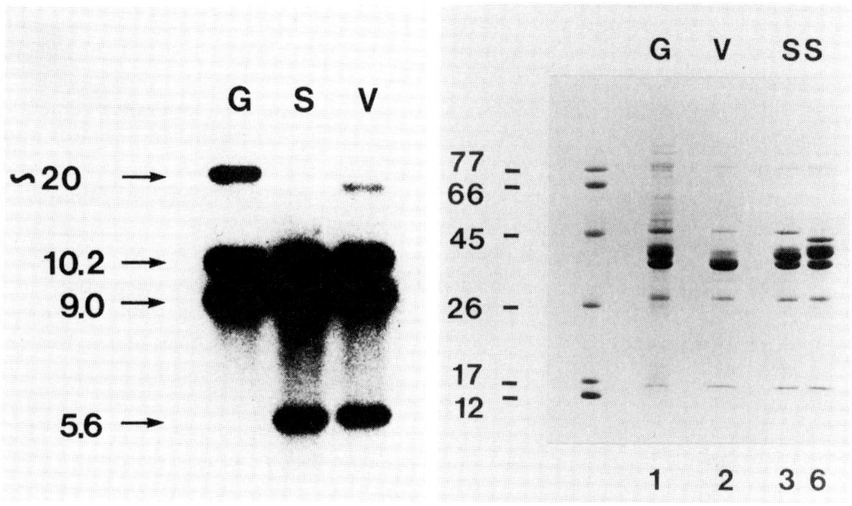

(a) (b)

Figure 2. Correlation of restriction enzyme polymorphisms and outer-membrane protein profiles of strains *H. influenzae* type b. (a) *Eco*RI restriction enzyme polymorphisms of type b *H. influenzae* DNA. Chromosomal DNA from type b strains Gast (G), Sterm (S) and Vallee (V) was digested with *Eco*RI. DNA fragments were then separated by agarose gel electrophoresis and transferred to a nitrocellulose filter prior to hybridization with the ^{32}P-labelled 9 kb *Eco*RI fragment subcloned from Charon 4:48. Molecular lengths in kilobase pairs are indicated on the left. (b) Outer membrane protein profiles of strains of *H. influenzae* type b. Outer membrane protein preparations, prepared as described by Barenkamp *et al.* [9], from strains representing the three commonly occurring polymorphisms are shown. Number below the lanes indicate OMP subtype, namely 1H, 2L, 3L and 6U respectively. Of the 21 OMP subtypes described to date [9] only those which account for the majority of strains having a particular polymorphism are shown. The apparent molecular weights of standard proteins are indicated on the left in kilodaltons.

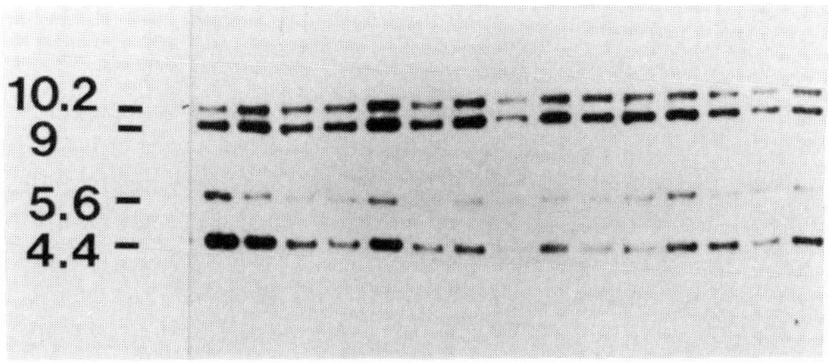

Figure 3. Type b *H. influenzae* DNA from Dutch clinical isolates. Chromosomal DNA was digested with restriction endonuclease *Eco*RI. The DNA fragments were separated by agarose gel electrophoresis and transferred to a nitrocellulose filter prior to hybridization with ^{32}P-labelled Charon 4:48 DNA. Molecular lengths in kilobase pairs are indicated on the left.

type a strains may account for a more substantial fraction of cases of *Haemophilus influenzae* meningitis and pneumonia in developing countries than has been previously recognized [8,9].

In addition to the well known importance of unencapsulated strains of *H. influenzae* as a cause of otitis media, several reports have emphasized the occurrence of rare instances of invasive *H. influenzae* disease caused by such strains in neonates, infants and adults, and it therefore seemed of interest to determine whether these strains showed evidence of clonality. Thirty-two unencapsulated *H. influenzae* strains isolated from blood, CSF or lung tissue (obtained by percutaneous aspiration) and associated with systemic infection, were analysed. These strains, with few exceptions, lack any homology to the Charon 4:48 probe and, based on biotype, LPS, OMP and the use of another DNA probe containing the structural gene for IgA1 protease [10], appear genetically distinct and heterogeneous.

This diversity indicates the likely complexity involved in searching for an effective vaccine for the prevention of infections caused by unencapsulated strains of *Haemophilus influenzae*.

ACKNOWLEDGEMENTS

We would like to acknowledge the collaboration of Dr Martin Kleiman (Indiana), Dr Richard Wallace, Dr Eric Hansen and Dr Mary Ann Jackson (Texas), Mr Mike Gratten, Ms Janet Montgomery and Dr Michael Alpers (Papua New Guinea), Dr Loek van Alphen (Amsterdam) and Dr Paul Zadik (Sheffield). Dr Jim Bricker (Boston) kindly provided the IgA1 protease probe. We thank Mrs Sheila Hayes for help in preparing the manuscript.

This work is supported by grants from the Medical Research Council and National Institutes of Health Postdoctoral Fellowship 1F32AI06978-02 (S.K.H.).

REFERENCES

1. Sell, S. H. and P. F. Wright (Eds) (1982). "*Haemophilus Influenzae*, Epidemiology, Immunology and Prevention of Disease". Elsevier Biomedical, Amsterdam.
2. Zwahlen, A. J., J. A. Winkelstein and E. R. Moxon (1983). Surface determinants of *Haemophilus influenzae* pathogenicity: comparative virulence of capsular transformants in normal and complement-deficient host. *J. Infect. Dis.* **148**, 385–394.
3. Zwahlen, A., L. G. Rubin and E. R. Moxon (1983). Contribution of lipopolysaccharide to *Haemophilus influenzae* pathogenicity: Comparative virulence of genetically related strains. Abstract No. 942: Presented at the 23rd Interscience Conference on Antimicrobial Agents and Chemotherapy at Las Vegas, Nevada, 24–26 October 1983.
4. Catlin, B. W., J. W. Bendler III and S. H. Goodgal (1972). The type b capsulation locus of *Haemophilus influenzae*: map, location and size. *J. Gen. Microbiol.* **70**, 411–422.
5. Moxon, E. R., R. A. Deich and Connelly, C. J. (1984). Cloning of chromosomal DNA from *Haemophilus influenzae*: its use for studying the expression of type b capsule and virulence. *J. Clin. Invest.* **73**, 298–306.
6. Moxon, E. R., A. L. Smith, D. R. Averill and D. H. Smith (1974). *Haemophilus influenzae* meningitis in infant rats after intranasal inoculation. *J. Infect. Dis.* **129**, 154–162.

7. Hoiseth, S. K., C. J. Connelly and E. R. Moxon (1985). Genetics of spontaneous, high frequency loss of b capsule expression in *Haemophilus influenzae*. *Infect. Immun.* **49**, 389-395.
8. Barenkamp, S. J., R. S. Munson Jr and D. M. Granoff (1981). Subtyping isolates of *Haemophilus influenzae* type b by outer-membrane protein profiles. *J. Infect. Dis.* **143**, 668-676.
9. Shann, F., M. Gratten, S. Germer, V. Linneman, D. Hazlett and R. Payne (1984). Aetiology of pneumonia in children in Goroka Hospital, Papua New Guinea. *Lancet* **ii**, 537-541.
10. Bricker, J., M. H. Mulks, A. G. Plaut, E. R. Moxon and A. Wright (1983). IgA1 proteases of *Haemophilus influenzae*. Cloning and characterization in *Escherichia coli* K.12. *Proc. Natl. Acad. Sci. USA* **80**, 2681-2685.

Identification of *Mycoplasma hyopneumoniae* Proteins from an *Escherichia coli* Expression Library and Analysis of Transcription and Translation Signals

Mo-Quen Klinkert[a], Christoph Taschke,
Heinz Schaller and Richard Herrmann

Mikrobiologie, University of Heidelberg, Heidelberg, FRG

Mycoplasma hyopneumoniae is a pathogen of the respiratory tract of the pig, which produces a chronic non-fatal pneumonia, arising whenever large groups of pigs are housed within a confined air space [1,2]. Enzootic pneumonia in pigs is the most important swine disease in the world; about 45–70% of slaughtered market pigs contained pneumonic lung lesions [3]. Studying the structure, organization and expression of the mycoplasma genome is important in advancing our understanding of mycoplasma pathogenesis. With the molecular biology of the mycoplasma being in its initial phase of development, there is still very little known about the organism and the mechanisms leading to the disease.

We have devised a strategy for identifying mycoplasma-specific coding sequences by using antibodies raised against authentic mycoplasma proteins [4]. We were interested in localizing the genes within the chromosome, in analysing the control signals regulating gene expression and in the characterization of the gene products.

IDENTIFICATION OF MYCOPLASMA-SPECIFIC PROTEINS

Mycoplasma DNA treated with DNase I to fragments averaging 300 base pairs long, were ligated to the fusion expression vector [5]. By expressing mycoplasma

[a]*Telephone:* 49-6221-541

Protein-Carbohydrate Interactions in Biological Systems. ISBN 0 12 436665 1

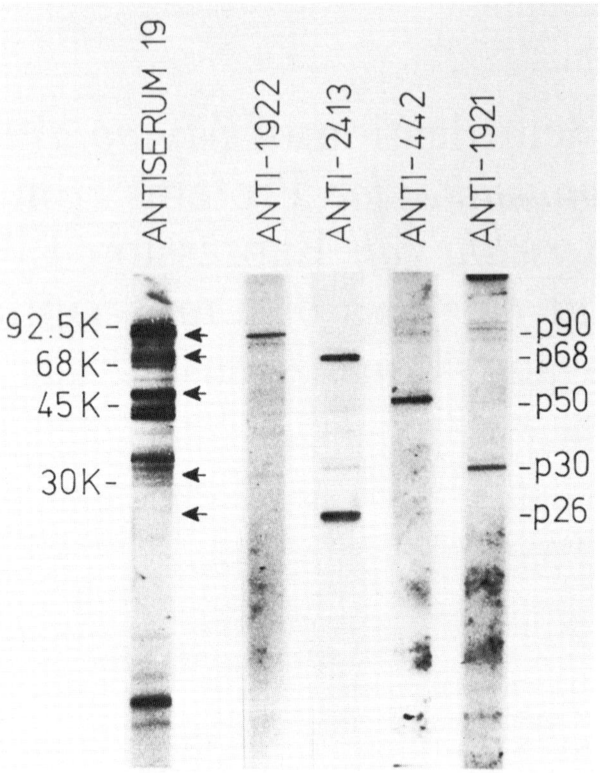

Figure 1. Western blots of *M. hyopneumoniae* proteins using specific antisera. Mycoplasma proteins were separated by PAGE, transferred to nitrocellulose filter and treated with various antisera. The mycoplasma profile stained by a pig antiserum (antiserum 19), raised by the inhalation of viable mycoplasmas, is used as a reference. Mycoplasma proteins recognized by the anti-fusion protein antisera are indicated by arrows. The molecular weight standards are shown on the left.

sequences in *Escherichia coli* as proteins fused to the phage MS2 polymerase, and by using specific antibodies made against the fusion proteins as probes in Western analysis, we have identified antigenically-related mycoplasma proteins.

Anti-fusion protein antisera were found to react specifically with mycoplasma-specific proteins in a Western blot (Fig. 1). Preimmune antisera from rabbits immunized with the various fusion proteins did not stain any mycoplasma proteins (not shown). Anti-1922 antiserum reacted specifically with a protein designated P90, according to its molecular weight of 90 kD, anti-442 recognized P50 (molecular weight 50 kD), while anti-1921 and anti-1925 picked up a protein each of molecular weight 30 kD, termed P30. Anti-2413 recognized two mycoplasma proteins, termed P68 and P26, according to their molecular weights of 68 kD and 26 kD. It is not clear why an antiserum directed against one fusion protein should react with two protein bands. These could either be two different proteins sharing a common antigenic determinant, or P68 could be a precursor of P26. The second possibility is supported by Southern blot analysis [6], in which the clone hybridizes to only one genomic fragment (data not shown).

ISOLATION AND CHARACTERIZATION OF THE MYCOPLASMA GENE CORRESPONDING TO pME2413

The recombinant plasmid pME2413 was chosen for further analysis [7]. To localize the complete gene corresponding to P68/P26 on the mycoplasma genome, the DNA insert isolated from the plasmid pME2413 was used as a probe to screen a genomic library set up in phage lambda. Mycoplasma DNA sequences represented in the expression plasmid were mapped within a single 4.5 kb *Eco*RI fragment of mycoplasma DNA. From restriction analysis of this fragment and partial sequence analysis data, the *Eco*RI fragment is expected to contain the complete coding sequence for a 68 kD protein.

The initiation of transcription was mapped by S1-nuclease protection [8] and primer extension experiments [9,10] approximately 50 nucleotides upstream of the pME2413 probe, which is shown by the hatched box in Fig. 2. By DNA sequencing [7] and comparing this region with the characteristic features of *E. coli* [11] and *Bacillus subtilis* [12] expression units, a promoter region for *M. hyopneumoniae* was defined: the −10 promoter sequence was in full agreement with the consensus sequence of *E. coli* being TATAAT [11], whereas the −35 region showed hardly any homology to *E. coli*. Characteristic of this region however, is an AT-rich stretch. It is conceivable that the proposed −35 recognition sequence for the RNA polymerase would be replaced here by the AT-rich stretches observed or by factors not yet known [11,13].

Assuming the codon TGA does not function in mycoplasmas as a stop codon, but codes for tryptophan, as shown for *M. capricolum* [14], there is an open reading frame with the first initiation codon 9 nucleotides away from the 5′-end of the mRNA, and the second 39 nucleotides away. The observed minimal distance between the start of the message and the initiation triplet for *E. coli* is 26. Moreover, upstream from the second ATG is a sequence GGAAAT, which could act as a ribosome binding site (GGAGGT for *E. coli*), supported by the finding that the 3′-end of the 16S rRNA of mycoplasma is also conserved [7], as in *M. capricolum* [15] and in *E. coli* [11]. Such a sequence is not present for the first start codon. The second ATG in the open reading frame is therefore postulated to function as the translation initiation codon.

DISCUSSION

We have devised a strategy for mapping specific coding sequences within the genome of *M. hyopneumoniae*. The procedure involved "shot gun expression" of mycoplasma DNA coding sequences using an *E. coli* expression vector. By screening the expression plasmid library with antiserum specific for mycoplasma

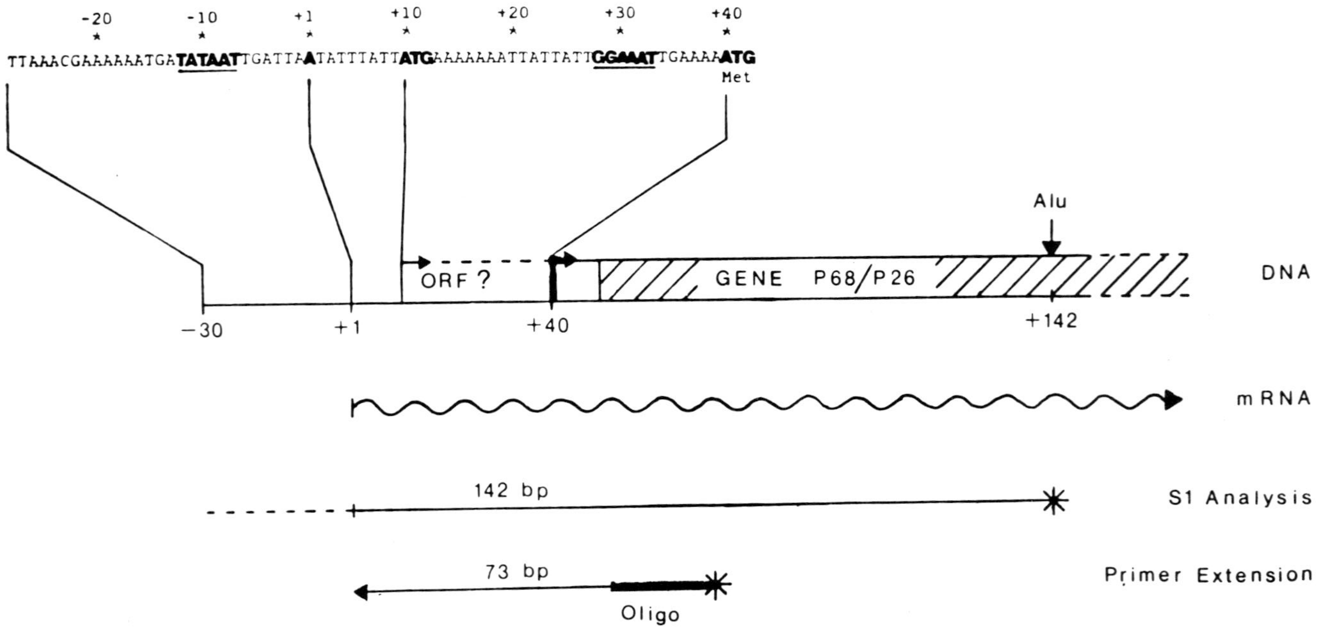

Figure 2. Putative gene expression unit of *M. hyopneumoniae*. The P68/P26 open reading frame is shown by the boxed region. The hatched box depicts the 242 bp mycoplasma DNA insert found in recombinant plasmid pME2413. The thin arrow indicates the beginning of the open reading frame; the heavy arrow the position from which translation presumably initiates. The 5′-end of the mRNA is mapped by S1 protection and primer extension analysis. The dashed line shows the non-protected part of the DNA probe used for S1 analysis (probe labelled at the 5′-end of the AluI site, marked by an asterisk), the solid line the protected region. The solid box indicates the oligonucleotide also labelled at the 5′-end, shown by an asterisk, used for primer extension. The blow up shows the −30 to +40 sequence. The −10 region TATAAT is marked in heavy print; there is no −35 consensus region. The start of the message is given as +1. The ribosome binding sequence and the initiation translation codon are shown, also marked in heavy print.

surface proteins (raised in a pig by the intranasal inoculation of viable mycoplasmas), clones encoding *E. coli*-mycoplasma hybrid proteins were identified. Specific antibodies against mycoplasma antigens generated from fusion proteins were found capable of reacting with discrete mycoplasma surface components in a Western blot.

The technique of inserting small random DNA fragments into *E. coli* expression plasmids has advantages over cloning very large fragments or even whole genes, since it is likely that the complete gene products or large parts of them may be lethal to the cell. Clones encoding bacterial fusion proteins containing one or more epitopes can then be identified, using antibodies of defined specificities. Recombinant plasmids in turn are used to localize genes of unknown map location. This method is generally useful for detecting any gene of interest. While the immunological screening of a complete genomic library of 5×10^4 colonies (for approximately 300 bp long fragments covering a genome of 1×10^6 bp) is certainly laborious, the success of the system lies in the quality of the antiserum.

A gene identified in this way was examined extensively for signals functioning in gene expression. Such a study is expected to contribute towards understanding the mechanism of gene regulation in mycoplasma. In connection with these studies, we have also located two other promoter regions belonging to the single rRNA operon of *M. hyopneumoniae* [7]. All sequences have in common a −10 region, while the −35 sequences are replaced by AT-rich stretches. However, no general conclusions about the features of mycoplasma promoters can be drawn, until a more heterogeneous group of control-sequence regions is analysed. Additional studies on RNA polymerase binding and comparing the biological activities of mutated promoter sequences will also be required for the characterization of putative transcription signals.

It is of further interest to determine whether *M. hyopneumoniae* promoters are also functional in other mycoplasma species. For this purpose it is necessary that promoters of other species be characterized and systems be set up tha enable the transfer of genetic information from one species to another to take place.

ACKNOWLEDGEMENTS

This work is supported by Biogen, S.A. (M.K) and the Deutsche Forschungsgemeinschaft He 780/4-1.

REFERENCES

1. Muirhead, M. R. (1979). Respiratory diseases of pigs. *Br. Vet. J.* **135**, 497-508.
2. Switzer, W. P. and R. F. Ross (1975). Mycoplasmal diseases. *In* "Diseases of Swine" (Eds H. W. Dunne and A. D. Leman), 4th ed, pp. 741-764. The Iowa State University Press, Ames.
3. Whittlestone, P. (1973). Enzootic pneumonia of pigs (EPP). *Adv. Vet. Sci. Comp. Med.* **17**, 1-55.
4. Klinkert, M. Q., R. Herrmann and H. Schaller (1985). Surface proteins of *Mycoplasma hyopneumoniae* identified from an *Escherichia coli* expression plasmid library. *Infect. Immun.* **49**, 329-335.

5. Remaut, E., P. Stanssens and W. Fiers (1981). Plasmid vectors for high-efficiency expression controlled by the PL promoter of coliphage lambda. *Gene* **15**, 81-93.
6. Maniatis, T., E. F. Fritsch and J. Sambrook (1982). "Molecular Cloning: A Laboratory Manual", pp. 90-91. Cold Spring Harbor Laboratory, Cold Spring Harbor, NY.
7. Taschke, C. (1985). Genexpression bei Micoplasma hyopneumoniae. Diplomarbeit, Ruprecht-Karls Universität Heidelberg.
8. Berk, A. J. and P. A. Sharp (1977). Sizing and mapping of early Adenovirus mRNAs by gel electrophoresis of S1 endonuclease-digested hybrids. *Cell* **12**, 721-732.
9. Kavsan, V. M. (1978). Reverse transcription of eucaryotic and viral RNAs. Conditions for obtaining long product. Translated from *Molekulyarnaya Biologiya* **13**, 266-280.
10. Retzel, E. F., M. S. Collett and A. J. Faras (1980). Enzymatic synthesis of deoxyribonucleic acid by the Avian Retrovirus reverse transcriptase *in vitro*: optimum conditions required for transcription of large ribonucleic acid templates. *Biochem.* **19**, 513-518.
11. Rosenberg, M. and D. Court (1979). Regulatory sequences involved in the promotion and termination of RNA transcription. *Ann. Rev. Genet.* **13**, 319-353.
12. Ogasawara, N., S. Moriya and H. Yoshikawa (1983). Structure and organization of rRNA operons in the region of the replication origin of the *Bacillus subtilis* chromosome. *Nucleic Acids Res.* **11**, 6301-6318.
13. Young, R. A. and J. A. Steitz (1978). Complementary sequences 1700 nucleotides apart from a ribonuclease III cleavage site in *Escherichia coli* ribosomal precursor RNA. *Proc. Natl. Acad. Sci. USA* **75**, 3593-3597.
14. Yamao, F., A. Muto, Y. Kawauchi, M. Iwami, S. Iwagami, Y. Azumi and S. Osawa (1985). UGA is read as tryptophan in *Mycoplasma capricolum*. *Proc. Natl. Acad. Sci. USA* **82**, 2306-2309.
15. Iwami, M., A. Muto, F. Yamao and S. Osawa (1984). Nucleotide sequence of the rrnB 16S ribosomal RNA gene from *Mycoplasma capricolum*. *Mol. Gen. Genet.* **196**, 317-322.
16. Steitz, J. A. (1979). Genetic signals and nucleotide sequences in messenger RNA. *In* "Biological Regulation and Development", Vol. I. Gene Expression (Ed. R. F. Goldberg). Plenum Press, New York.
17. Towbin, H., T. Staehelin and J. Gordon (1979). Electrophoretic transfer of proteins from polyacrylamide gels to nitrocellulose sheets: procedure and some applications. *Proc. Natl. Acad. Sci. USA* **76**, 4350-4354.

Nutritional Character, O Antigen, Cryptic Plasmid and Mannose-Resistant Adhesin—Relevance to Virulence of *Salmonella* sp.

B. A. D. Stocker[a], M. F. Edwards,
N. A. Nnalue and M. C. Halula

Department of Medical Microbiology, Stanford University School of Medicine, Stanford, CA 94305, USA

For several years work in our laboratory has concerned the genetics of various properties of *Salmonella* sp. relevant to virulence. I, and my then student, Susan Hoiseth, studied the effect of certain auxotrophic characters on mouse-virulence of *S. typhimurium*. Most *Salmonella* sp. are nutritionally non-exacting (and the auxotrophic characters of some species, such as nicotinic requirement of *S. dublin*, probably have no relevance to virulence, because the required metabolite is available in host tissues). However, it has long been known that some blocks in biosynthetic pathways cause loss of virulence in respect of invasive infection, apparently because they make the bacteria dependent on metabolites which are not available in mammalian tissues (or not available in concentrations sufficient for bacterial multiplication). *p*-aminobenzoate (pAB), a precursor of folic acid in bacteria and plants, is not a mammalian metabolite (mammals acquire folic acid preformed, as a dietary component or vitamin). *Escherichia coli* and *Salmonella* sp. cannot assimilate folic acid and must instead make it themselves, from pAB, a pteridine and glutamic acid. Enterochelin (enterobactin), required by *Salmonella* sp. to capture iron from host iron-binding proteins, likewise is not available in mammalian tissues and bacteria must synthesize it from 2,3-dihydroxybenzoate (DHB). pAB and DHB, like the aromatic amino acids and some minor aromatic compounds, are derived from chorismic acid, the final product of the aromatic

[a] *Telephone:* 1-415-497-2006

Protein-Carbohydrate Interactions in Biological Systems. ISBN 0 12 436665 1

biosynthesis pathway. A complete block in this pathway therefore makes *Salmonella* sp. dependent on two metabolites which are not available in host tissues and would therefore be expected to cause loss of virulence.

The availability of transposon insertions in various genes of *S. typhimurium* allowed a test of this prediction. Gene *aroA* (which specifies an enzyme for a step in the aromatic biosynthesis pathway) inactivated by insertion into it of the tetracycline-resistance transposon Tn*10* was transduced into mouse-virulent strains of *S. typhimurium*, by selection for tetracycline resistance. The transductants obtained (grown in broth with tetracycline, to maintain selection for the transposon) were essentially non-virulent for BALB/c mice; e.g., LD_{50} by i.p. injection $>10^6$, compared to <20 for the parent strain. Strains with a gene inactivated by transposon insertion can revert, at low frequency, by "clean excision" of the transposon. Secondary mutations, causing tetracycline sensitivity and failure to revert (in tests able to detect reversion at a frequency of 10^{-11}/bacterium/generation) were isolated from the *AroA*::Tn*10* strains. They proved highly effective as live vaccines in mice [1]: thus, a single i.p. dose of 2×10^5 CFU protected BALB/c mice against 10^4 LD_{50} given i.p., or 100 LD_{50} given by mouth some weeks later, and a single oral dose of 2×10^8 CFU likewise protected against oral challenge. Of three such *aroA* (non-reverting) strains of *S. typhimurium* tested as parenteral route live vaccine in calves, only one gave satisfactory protection against oral challenge [2]; the effective strain, SL1479, has also been found to be highly efficient when given to calves by feeding, three doses, by our veterinary collaborators at Uppsala [3].

Two students, first Ronald Brown, subsequently Mary Frances Edwards, have applied the same principles for preparation of candidate oral-route live vaccine strains of *S. typhi*. The first constructs (R. Brown and B. A. D. Stocker, unpubl.), derived from strain Ty2 and from a strain of phage type A, had transposon-generated non-reverting mutations at *aroA* and in the *purJHD* operon, each causing a very considerable reduction in virulence as tested in a mouse model infection. These strains would probably have been efficient as live vaccines, but were not proposed for test because it became feasible to introduce more precisely defined genetic lesions, and a more appropriate block in purine biosynthesis.

Two newly constructed strains (M. F. Edwards and B. A. D. Stocker, unpubl.), now being tested in volunteers, were derived from a Vi-positive strain of phage type A. This strain was first given, by two steps of transduction, a deletion of part of gene *aroA*, previously characterized in *S. typhimurium*; then, by two steps of transduction, a point mutation in gene *hisG* (as a "marker" character, not expected to affect virulence); and then, by cotransduction with an adjacent silent Tn*10* insertion, a proven deletion of part of gene *purA*, causing a specific requirement for adenine. [This last type of mutation was used because while purine auxotrophy has long been known to cause attenuation [4], recent experiments in my laboratory (W. McFarland and B. A. D. Stocker, unpubl.) showed that a block between IMP and AMP, causing adenine requirement, in a mouse-virulent Vi-positive strain of *S. dublin*, resulted in more complete loss of virulence than blocks either before IMP, or between IMP and GMP, causing requirements satisfied, respectively, by any purine or guanine.] The tetracycline-resistance caused by the silent Tn*10* insertion was eliminated by selection for tetracycline-sensitivity [5]. Each of the attenuating mutations, tested by itself, was shown to cause a very considerable loss of virulence, as measured in the mouse/i.p. injection/with hog gastric mucin test. The Vi-negative mutant was obtained by selection with a Vi-specific phage.

Investigation by N. A. Nnalue of the effect of *galE*, *aroA* and other mutations on the mouse-virulence of *S. choleraesuis* has revealed some interesting differences from the situation in *S. typhimurium*, all of them probably attributable to the O antigen, 06,7, of strains of group C1. The general transducing phage P22 adsorbs only to *Salmonella* of O groups B, D or A (i.e., those with O factor 12). Bacterial genes can, however, be transduced from *S. typhimurium* to *S. choleraesuis*, or vice versa, by phage P1, if each strain makes LPS of the appropriate chemotype, Rc (galactose-deficient), in consequence of mutation at *galE*, preventing synthesis of UDP-galactose. Some *galE* mutants of mouse-virulent *S. choleraesuis* strains resembled *galE* mutants of *S. typhimurium* by greatly reduced mouse-virulence, e.g., LD_{50}, i.p., for BALB/c mice, *c.* 10^6, compared to *c.* 400 for parent strain. However, a *galE* but galactose-resistant mutant of one strain was nearly as virulent as its wild-type parent (LD_{50}, *c.* 500, compared to *c.* 200 for parent strain) and a galactose-resistant mutant of a non-virulent *galE* mutant of another strain had LD_{50} *c.* 100, the same as that of its wild-type ancestor. We think that the greater mouse-virulence of at least some *galE* mutants of *S. choleraesuis* strains reflects the fact that the O oligosaccharide repeat unit of type 06,7 contains no galactose, so that uptake of only two molecules of galactose, for formation of the two galactose units of the LPS core, suffices for conversion of an Rc LPS molecule to a complete ("smooth") LPS molecule of type 06,7; by contrast, *Salmonella* of groups B and D, whose O repeat unit includes a galactose, require around 30 galactose residues to make an LPS molecule of normal O chain length.

S. choleraesuis made *aroA*$^-$ by P1 transduction from *S. typhimurium* resembled *aroA* derivatives of *S. typhimurium* by essentially complete loss of virulence. However, they differed in that when used as live vaccine given by i.p. route to BALB/c mice they gave virtually no protection against later challenge with a virulent strain of *S. choleraesuis*. The probable explanation for this was discovered by counts of colony-forming units in liver and spleen of mice killed at intervals after live-vaccine injection, i.p. route. The number of viable *S. typhimurium* found was 20% or more of the number inoculated and showed little decline for several days after injection, but the number of viable *S. choleraesuis* recovered from liver and spleen even as early as 24 h after administration was only 1/100 to 1/1000 of the number inoculated. An effective live-vaccine strain of *S. typhimurium* was used as parent of a derivative differing from it only by O antigen, 06,7, instead of 01,4,12. The low numbers of the 06,7 strain recovered from liver and spleen indicate that the failure of *S. choleraesuis* made aromatic-deficient to persist after i.p. injection is indeed a consequence of the chemical composition of its O repeat unit, determined at the *rfb* gene cluster. It seems likely that the greater ability of 06,7 bacteria to bind C3b opsonizes them effectively for killing by peritoneal phagocytic cells. *S. choleraesuis* made *aro*$^-$ and tested as live vaccine, i.p. route, in Ity-r mice, or by i.v. route in BALB/c mice, showed some protecting ability, though much less than that seen with *aro*$^-$ live-vaccine strains of *S. typhimurium*. It thus appears that the immunizing ability of *Salmonella* bacteria with auxotrophic mutations preventing multiplication in host tissues varies according to the chemical nature of their O repeat unit and also with route of administration and mouse genotype, probably because all these factors affect the degree of persistence as live bacteria in liver and spleen.

Another investigation concerned the so-called cryptic plasmid, of molecular weight *c.* 60 MD, present in many, though not all, isolates of *S. typhimurium*. M. F. Edwards transduced a Tn*10* insertion in the cryptic plasmid of strain LT2 (received from John Roth, University of Utah) into the 60 MD cryptic plasmid of several mouse-virulent strains of *S. typhimurium*; the strains with the Tn*10*-tagged plasmids were found unaltered in virulence. The presence of Tn*10* in the cryptic plasmid in a subline of strain TML facilitated the isolation from it of a plasmid-free variant; this was achieved by growth in broth with novobiocin, followed by selection for tetracycline-sensitivity on Bochner medium [5]. Another clone "cured" of the 60 MD plasmid was obtained by conjugational introduction of an incompatible R factor conferring kanamycin resistance. Both the "cured" isolates in the TML line were of reduced virulence, LD_{50}, i.p., for BALB/c mice, *c.* 10^4, compared to <40 for the parent strain. (Re-isolates from mice which had died from infections caused by large inocula of the cured strains remained plasmid-free, as indicated by absence of plasmid DNA bands on gels). Similar increases in LD_{50}, i.p. route, were found for cured (by the novobiocin/Bochner method) derivatives of two other mouse-virulent strains, one *S. typhimurium* and one *S. enteritidis*. Reintroduction of the Tn*10*-labelled plasmid into one of the cured clones in the TML line reduced its LD_{50} to the original level, or nearly so. The cured TML derivatives were also of reduced virulence when given to BALB/c mice by the oral route; no deaths from in gestion of 4×10^8 CFU, to be compared to oral LD_{50} of $<5\times10^5$ CFU for the parent strain (and also for a cured derivative re-infected with the Tn*10*-tagged plasmid). Strain Q1, an old laboratory strain which is non-virulent despite being smooth and prototrophic, was found not to possess any large plasmid; introduction of the labelled plasmid into strain Q1 by DNA-mediated transformation did not increase its virulence to a detectable extent.

Tests of *in vitro* adhesion to, or penetration into, mammalian cells showed no difference between the TML-line strain and its cured derivative. Furthermore, the numbers of live bacteria recovered from liver and spleen of mice killed at intervals after intraperitoneal challenge were about the same, for the first three days, for the plasmid-bearing and plasmid-free TML-line strains; thereafter the numbers increased exponentially for the plasmid-bearing strain but remained constant for the cured strain. The evidence so far thus suggests that the 60 MD plasmid of *S. typhimurium* TML is required for high virulence because it improves the ability of the bacteria to multiply in the liver and spleen, presumably within phagocytic cells, rather than by increasing the ability of the bacteria to gain entrance from the gut lumen or to adhere to or be taken up by phagocytic cells.

Madelon Halula has identified a mannose-resistant haemagglutinating ability (for fixed goat or sheep red cells) in most tested isolates of *S. typhimurium*, and has cloned the chromosomal region responsible for this property, inactivated relevant gene(s) by transposon insertion and introduced the resulting adhesin-negative chromosomal segment into the chromosome of haemagglutination-positive strains. For a fuller account, see her poster in Part I, this volume.

ACKNOWLEDGEMENTS

This work was supported by Public Health Service Research Grants AI07168 and AI18872 from the National Institute of Allergy and Infectious Diseases and by Research Contracts with U.S. Army Medical Research and Development Command, SmithKline Animal Health and SmithKline-RIT and a donation from Johnson and Johnson Inc.

REFERENCES

1. Hoiseth, S. K. and B. A. D. Stocker (1981). Aromatic-dependent *Salmonella typhimurium* are non-virulent and are effective as live vaccines. *Nature* **291**, 238–239.
2. Smith, B. P., M. Reina-Guerra, S. K. Hoiseth, B. A. D. Stocker, F. Habasha, E. Johnson and F. Meritt (1983). Safety and efficacy of aromatic-dependent *Salmonella typhimurium* as live vaccine for calves. *Am. J. Vet. Res.* **45**, 59–66.
3. Robertsson, J. A., A. A. Lindberg, S. K. Hoiseth and B. A. D. Stocker (1983). *Salmonella typhimurium* infection in calves: evaluation of protection and survival of virulent *S. typhimurium* challenge bacteria after immunization with live or inactivated *S. typhimurium* vaccines. *Infect. Immun.* **41**, 742–750.
4. Bacon, G. A., T. W. Burrows and M. Yates (1951). The effects of biochemical mutation on the virulence of *Bacterium typhosum*: The loss of virulence of certain mutants. *Br. J. Exp. Path.* **32**, 85–96.
5. Bochner, B. R., H. Huang, G. L. Schieven and B. N. Ames (1980). Positive selection for loss of tetracycline resistance. *J. Bacteriol.* **143**, 926–933.

The Protective Effect of Monoclonal Antibodies Directed Against Gonococcal Outer Membrane Protein IB

M. Virji[a], K. Zak, J. N. Fletcher and J. E. Heckels

Department of Microbiology, University of Southampton Medical School, Southampton, UK

Gonococcal isolates can be divided into two general categories depending on their expression of major outer membrane protein IA or IB. Monoclonal antibodies have been produced following immunization with group IB strains. Of five antibodies investigated, none reacted with group IA strain, meningococci or other Gram-negative bacteria. Four antibodies demonstrated type specificity but one antibody, SM24, cross-reacted with all the P.IB-expressing strains tested. The position of the epitope recognized by this antibody was located by Western blotting, following cleavage of the protein with trypsin and chymotrypsin, and was found to reside in a peptide of molecular weight 7 kD at the centre of the surface exposed area of the transmembrane protein. The potential efficacy of this antibody against gonococcal infection was investigated using model systems [1,2].

A protective effect of SM24 against invasion of epithelial cells was demonstrated by reduction in the cytotoxic effect of several P.IB-expressing strains for monolayers of Chang conjunctiva cells. The antibody was also effective in promoting complement-mediated killing of such strains and its opsonic effect was demonstrated by an increase in luminol enhanced chemiluminescence of human PMN (see Table 1).

The identification of conserved epitopes on the gonococcal surface which may elicit protective antibodies is of considerable interest in the search for a gonococcal vaccine. Antibody SM24 recognizes such an epitope on the approximately 50%

[a]*Telephone:* 44-703-777222 ext. 3918

Protein-Carbohydrate Interactions in Biological Systems. ISBN 0 12 436665 1

Table 1. Biological properties of antibody SM24

	Strain P9		*Strain SU72*	
	−	+	−	+
Protection of epithelial cells (% tissue culture cells killed)	95	5	74	7
C′-mediated bactericidal killing (% gonococci killed)	2	83	3	37
Opsonization (relative chemiluminescence of PMN)	1	7.7	1	1.5

−,+ indicates absence or presence of antibody SM24, respectively.

of gonococcal strains which express protein IB. If a similar epitope exists on the remaining IA strains, then a combined peptide vaccine should protect against all strains.

REFERENCES

1. Virji, M. and J. E. Heckels (1984). The role of common and type-specific pilus antigenic domains in adhesion and virulence of gonococci for human epithelial cells. *J. Gen. Microbiol.* **130**, 1089–1095.
2. Virji, M. and J. E. Heckels (1985). Role of anti-pilus antibodies in host defence against gonococcal infection studied with monoclonal anti-pilus antibodies. *Infect. Immun.* **49**, 621–628.

Poster Session

Virulence Factors in Avian *Escherichia coli*

Maryvonne Dho[a], Jean-Pierre Lafont and Annie Brée

Institut National de la Recherche Agronomique, Station de Pathologie Aviaire et de Parasitologie, Monnaie, France

Escherichia coli is the most frequent Gram-negative organism in avian respiratory infections. It generally develops as a secondary invader after mycoplasmal and/or viral infections of the respiratory tract and is responsible for severe lesions and mortality. Virulent strains are lethal for day-old chicks, and invasive disease can be reproduced in chickens with lethal *E. coli* strains in the presence of concurrent respiratory virus or mycoplasma infections.

Several bacterial properties which could be involved in virulence were studied in 60 independent *E. coli* strains originating from poultry. On the basis of lethality for day-old chicks, by subcutaneous injection, three classes were distinguished among the strains studied: lethal class (LC; 5 deaths out of 5 inoculated chicks, LD_{50} ranging fro 10^2 to 10^7 organisms/chick), intermediate lethality class (ILC; 2, 3 or 4 deaths) and non-lethal class (NLC: 0 or 1 death). The frequency of the major poultry pathogenic 0-types 01, 02 and 078 was much higher in LC strains (17/26) than in ILC (3/21) or NLC strains (1/13).

Adhesive ability was studied by bacterial adhesion assays on isolated pharyngeal epithelial cells of axenic chickens as described [1]. Adhesive properties were found in 24 strains, of which 19 belonged to LC, whereas 36 non-adhesive strains included only 7 LC isolates. No particular haemagglutination (HA) pattern was correlated to adhesiveness. All adhesive strains exhibited mannose-sensitive HA of chicken and guinea pig erythrocytes.

Ten strains were studied by electron microscopy after negative staining. Two morphologically distinct types of pili were observed on adhesive strains whereas non-adhesive strains either were non-piliated or possessed pili of only one type. The purification of these pili is presently being undertaken.

[a] *Telephone:* 33-47-64 55 65

Protein-Carbohydrate Interactions in Biological Systems. ISBN 0 12 436665 1

Production of the iron siderophore aerobactin enables *E. coli* to grow in the iron-restricted conditions of the internal organs and body fluids. Such iron-uptake abilities represent virulence determinants in several bacterial species [2]. The growth of avian strains under iron-limiting conditions in the presence of transferrin was studied as described [1]. The ability to grow in these conditions was strongly correlated with lethality. The aerobactin iron supply system is frequently encoded by col V plasmids. Among the group of strains studied, we demonstrated a highly significant correlation between colicin V production and growth in the presence of iron-sequestering transferrin. Agarose gel electrophoresis showed the presence of one 70 MD plasmid in antibiotic-sensitive col V-producing strains. The incidence of adhesiveness and iron-uptake ability is summarized in Fig. 1.

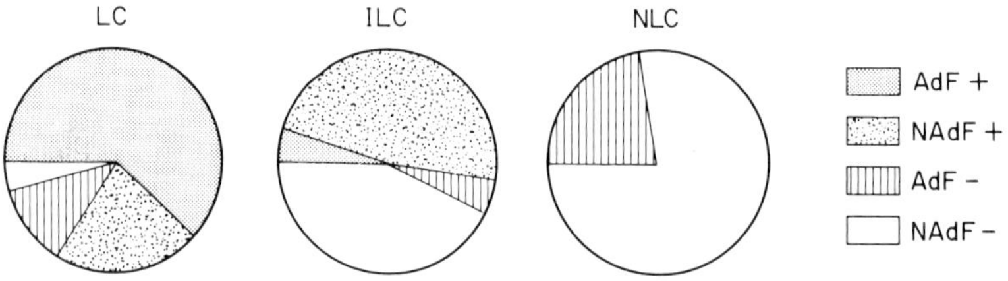

Figure 1. Adhesiveness (Ad) or non-adhesiveness (NAd) and presence (F+) or absence (F−) of iron-captation ability in *E. coli* strains of the different lethality classes.

An experimental model using the natural respiratory route of infection was developed. Groups of axenic or SPF 13-day-old chickens raised in germ-free isolators were inoculated with infectious bronchitis virus (IBV) and two days later with various *E. coli* strains. The lesions recorded 9 days after infection were significantly more important in chickens inoculated with AdF^+ lethal strains than with $NadF^-$ non-lethal strains (Table 1).

Table 1. Pathogenicity of *E. coli* strains in SPF and axenic chickens

	IBV alone		*IBV+NadF⁻* E. coli		*IBV+ADF⁺* E. coli	
% of animals with:	*ax*	*spf*	*ax*	*spf*	*ax*	*spf*
Thoracic air sacs lesions	32.6	64.3	58.6	65.5	84.1	100
Fibrin on air sacs	2.2	0	3.4	6.9	54.5	81.5
Other lesions	2.2	0	0	0	25.0	37.0
Reisolation of *E. coli*	ND	0	24.0	6.9	94.8	96.1
Mortality	0	0	0	0	2.2	6.9

None of the 60 strains was haemolytic. Although the resistance to the bactericidal power of normal adult hen serum was found in a majority of LC strains and a minority of NLC strains, this property was not significantly correlated with lethality.

Virulence of avian *E. coli* is probably a multifactoral property in which adhesion is involved in a first step of mucosal colonization, the ability to grow under conditions of iron deprivation being required in a further invasive step of infection. However, the possibility of other additional properties being correlated with virulence cannot be excluded [3].

REFERENCES

1. Dho, M. and J. P. Lafont (1984). *Avian Dis.* **28**, 1016-1025.
2. Finkelstein, R. A., C. V. Sciortino and M. A. McIntosh (1983). *Rev. Infect. Dis.* **5**, suppl. 4, S759-S777.
3. Smith, H. (1984). *J. Appl. Bacteriol.* **47**, 395-404.

Virulence and Congo Red Binding Ability Encoded by the 140 MD Plasmid of *Shigella flexneri* 2a

Masanosuke Yoshikawa[a], Chihiro Sasakawa, Kunio Kamata, Somay Yamagata-Murayama, Takashi Sakai and Souichi Makino

Institute of Medical Science, University of Tokyo, Tokyo, Japan

A 140 MD plasmid has been implicated to be associated with the virulence of *Shigella flexneri* [1]. *S. flexneri* 2a strain No. 50 was chosen for this study from 87 strains because of its simple plasmid composition. Upon subculturing fresh isolates of this strain taken from guinea pig eyes in liquid media, it has been shown that the loss of virulence as determined by the mouse Sereny test, the loss of Congo red binding ability (Pcr^+: [2]) as observed on a Congo red-containing agar plate, and the faster growth rate than the virulent wild-type, occur coordinately, together with a detectable large-scale molecular alteration of the 140 MD plasmid. This coordinate phenotypic change is ascribed to the fact that once an avirulent Pcr^+ cell has emerged it overgrows as a consequence of its selective advantage over the virulent wild-type in artificial media. Forty derivatives with molecular alterations of the plasmid thus obtained were analysed by digesting with *Sal*I and a deletion map was constructed. The molecular alterations have been classified into 13 types, 12 deletions and one insertion of an IS*1*-like element. Sixteen out of 39 deletions belong to a single type and 6 to another, suggesting a contribution of deletion hot spots. The wild-type plasmid DNA was partially digested with *Sal*I and cloned into pBR322. Of 359 recombinant clones thus obtained those containing 2 or more *Sal*I fragments were used to make a *Sal*I restriction cleavage map. The circular map thus made was exactly the same as that made by the deletion analysis. Using Tn*5* for insertional inactivation the DNA sequences required for the virulencc has been found to extend over more than 100 kb, either scattering or in cluster. One of the Tn*5* insertion derivative

[a]*Telephone:* 81-3-443-8111

Protein-Carbohydrate Interactions in Biological Systems. ISBN 0 12 436665 1

has been found to be avirulent but still Pcr^+, indicating that these two phenotypes are separable. By the Southern hybridization method using IS*1*L from Tn*9* as the probe, this plasmid has been shown to contain three copies of IS*1*-like elements. One spontaneously isolated insertion mutant described above contains four copies of them, suggesting that one copy has been inserted, thus resulting in an insertion mutant lacking the virulence and Pcr^+ phenotype. The wild-type plasmid DNA was partially digested with *Sau*3A, ligated to pBR322 and used to transform *Escherichia coli* MC1061 by selecting Pcr^+ phenotype. When this clone is introduced into an avirulent derivative of *S. flexneri* with a small deletion on the 140 MD plasmid it is converted to fully virulent and Pcr^+. A plasmid-cured *S. flexneri* transformed with this clone is avirulent and Pcr^-. These observations indicate that this clone contains the genetic information essential but not sufficient for full virulence and Pcr^+ phenotype. A DNA sequence of 1.2 kb of this clone has been identified to be responsible for these phenotypes. A region required for plasmid replication has been self-cloned, which may be useful in examining whether or not the apparent instability or the tendency to suffer large-scale molecular alterations of the 140 MD plasmid is ascribed to the nature of the replication machinery itself.

REFERENCES

1. Sansonetti, P., D. J. Kopecko and S. B. Formal (1982). *Infect. Immun.* **35**, 852-860.
2. Maurelli, A. T., B. Blackmon and R. Curtiss III (1984). *Infect. Immun.* **43**, 397-401.

Poster Session

Virulence Factors of Uropathogenic *Escherichia coli*

L. V. Wood[a], B. W. Davis and L. L. Lance

Anti-infective Research Department, Norwich Eaton Pharmaceuticals, Inc., Norwich, NY 13815, USA

We examined 67 strains of *Escherichia coli*, isolated from patients with urinary tract infection (UTI), for possession of various virulence factors: haemolysis and haemagglutination (HA) were tested by standard methods; antibiotic sensitivity was tested by the Kirby-Bauer method and adherence to cultured cell lines was examined using a modification of the method of Svanborg-Edén [1]. Vero, HEp-2 and 5637 cells were used and adherence measured by Nomarski differential phase contrast microscopy. Adherence counts were compared to counts of a positive control organism and expressed as the adherence index (AI). The UTI strains were compared to 11 faecal strains and 12 environmental isolates.

UTI isolates were significantly more likely to produce alpha or beta haemolysin for sheep erythrocytes than faecal or environmental strains ($p < 0.001$). UTI isolates were also significantly more likely to possess mannose-sensitive or mannose-resistant haemagglutinins for human and/or monkey erythrocytes than faecal ($p = 0.001$) or environmental ($p = 0.001$) strains. There were no clear differences in antibiotic resistance patterns between the three groups, although environmental strains tended to be more resistant than faecal strains. UTI isolates were most often resistant to ampicillin (29%), cephalothin (29%), sulfisoxazole (27%) and tetracycline (20%). Drugs showing the lowest overall resistance were nalidixic acid (0%), nitrofurantoin (1%), chloramphenicol (1%), trimethoprim (2%) and TMP/SMX (2%). Faecal strains did not adhere well to any of the three cell lines (mean A.I. values for Vero, HEp-2 and 5637 were 11.8, 3.8 and 8.0, respectively). Corresponding AI values for environmental strains were 28.8, 16.2 and 19.3 and for UTI strains 20.1, 14.5 and 28.9. We compared possession of three virulence characteristics (HA of human and/or monkey erythrocytes;

[a] *Telephone:* 1-607-335-2555

Protein–Carbohydrate Interactions in Biological Systems. ISBN 0 12 436665 1

haemolysis of sheep erythrocytes; haemolysis of sheep erythrocytes; and AI >20 for 5637 cells) in the three groups. Zero per cent of foecal strains possessed all three virulence factors and 66% possessed none. Eight per cent of environmental strains possessed all three and 50% were negative. However, 41% of UTI strains had all three factors and only 1% were negative. In terms of virulence characteristics this clearly shows the "clonality" of UTI isolates.

REFERENCE

1. Svanborg-Edén, C. (1978). *Scand. J. Infect. Dis.* **15(S)**, 1-69.

Capsular Polysaccharide Structures of *Pasteurella haemolytica* and their Potential as Virulence Factors

C. Adlam[a], J. M. Knights, A. Mugridge, J. C. Lindon, J. M. Williams and J. E. Beesley

Wellcome Biotechnology Ltd, Beckenham, Kent, UK

INTRODUCTION

Pasteurella haemolytica organisms may be separated by antibiotic resistance, biochemical fermentation and colonial morphology into two biotypes, A (Arabinose fermenting) and T (Trehalose fermenting) consisting of 15 serotypes. These organisms are Gram-negative and encapsulated.

A strains cause enzootic pneumonia in all ages of sheep and septicaemia/pneumonia in very young lambs. The major ovine serotype is A2. Serotype A1 strains cause pneumonia in cattle (shipping fever/transit fever), while T strains cause "systemic" pasteurellosis in older post weaned lambs.

Although stress and pre-infection with viruses and other agents are known to pre-dispose to infection, the mechanisms whereby *P. haemolytica* organisms bring about primary lodgement and invasion of the host are obscure. As part of a programme of work to understand further these mechanisms, we have purified and characterized the capsular polysaccharides of several of the disease-causing serotypes.

RESULTS

Using chemical analysis and n.m.r. spectroscopy the structures of five of the serotype capsular polysaccharides have been elucidated:

[a] *Telephone:* 44-1-658-2211

Protein–Carbohydrate Interactions in Biological Systems. ISBN 0 12 436665 1

A1 →3)-β-*N*-acetylaminomannuronic-(1→4)-β-
N-acetylmannosamine-(1→
(with 4 position of the uronic acid being *O*--acetylated)

A2 →2)-α-D-*N*-acetylneuraminic acid-(8→
(colaminic acid)
(and a dextran polymer)

A7 →3-*O*-(2-*N*-acetyl-6-*O*-acetylα-D-glucosamine-1)→
(phosphate)→3-*O*-(2-*N*-acetylβ-D galactosamine-1)→

T4 →(2-glycerol-1)→(phosphate)→(6-α-D-galactose-1)→
(2-*O*-acetyl, 3-*O*-acetyl)

T15 →(2-glycerol-1)→(phosphate)→(4-α-D-galactose-1)→
(2-*O*-acetyl, 3-*O*-acetyl)

DISCUSSION

Several of these structures are similar to or identical with bacterial polysaccharides already described:

(1) A1. Some similarity with enterobacterial common antigen which is *N*-acetylaminomannuronic acid linked 1, 4 to *N*-acetyl-D-glucosamine.

(2) A2. Colaminic acid is identical with capsular polysaccharides of *Neisseria meningitidis* B, *E. coli* K1 and *Moraxella nonliquefaciens*.

(3) A7. Similar to the capsular polysaccharides from *N. meningitidis* L and *Haemophilus influenzae* type f.

(4) T4 and T15. These polymers are unusual in being glycerol teichoic acids which are normally associated with Gram-positive rather than Gram-negative bacteria. Although structurally very similar, they do not cross-react immunologically.

(5) T15. This polymer is identical to the capsular polysaccharides of *N. meningitidis* H and *E. coli* K62 (a human urinary tract pathogen).

CONCLUSIONS

The finding that the capsular polysaccharides of several serotypes of *P. haemolytica* are similar or identical to those of other unrelated human pathogens suggest that these organisms may have a similar host, target cell receptors and possible possess common genetic determinants.

The recognition that *P. haemolytica* serotype A2 has a colaminic acid capsule (and produces a dextran polymer) explains the difficulty which workers have experienced in producing protective extract vaccines against this serotype.

Colaminic acid is a particularly difficult antigen to raise specific antisera to and this has caused problems in producing effective human vaccines against *N. meningitidis* B.

Further details concerning polymers A1, T4 and T15 have been published [1-3].

REFERENCES

1. Adlam, C. *et al.* (1984). *J. Gen. Microbiol.* **130**, 2415-2416.
2. Adlam, C. *et al.* (1985). *J. Gen. Microbiol.* **131**, 387-394.
3. Adlam, C. *et al.* (1985). *J. Gen. Microbiol.* **131**, 1963-1972.

Monoclonal Antibodies to Weak Immunogenic *Escherichia coli* and Meningococcal Capsular Polysaccharides

D. Bitter-Suermann[a], I. Goergen and M. Frosch

Institute of Medical Microbiology, Johannes-Gutenberg University, Mainz, FRG

The poor immunogenicity of the capsular polysaccharides of pathogenic *Escherichia coli* K1 and K5 or meningococci group B and C is well known and until now antibodies of the IgG class could not be produced by current methodology. In our approach we made use of the immunological dysregulation of NZB autoimmune mice and produced some unique monoclonal antibodies to capsular polysaccharides as compared to the inefficiency or failure to produce antibodies to these antigens in the conventional BALB/c mouse model. Therefore, we now routinely immunize NZB mice and BALB/c mice in parallel for production of monoclonal antibodies against polysaccharide antigens and thus take advantage of the different immune reactivities of the two strains. Some of these monoclonal antibodies, listed in Table 1, which react with polysaccharides K1, K5, meningococci A, B, and C, enterobacterial common antigen (ECA) and LPS-core structures, now offer the possibility of highly sensitive diagnostic approaches. With the anti-MBPS antibody free capsular polysaccharide is detectable in CSF at the 1 ng level, and even such K1 *E. coli* can be typed by latex agglutination or ELISA which are K1-phage negative or have genetic defects in the extracellular assembly of their K1 capsular material. Analysis of antigenic epitopes on the LPS outer core of the Gram-negatives or the sharing of antigenic determinants (e.g. between between K5 and ECA) is now possible. In addition, cross-reactivities with self-antigens can be demonstrated (e.g. between K1 and

[a]*Telephone:* 49-6131-173128

Protein-Carbohydrate Interactions in Biological Systems. ISBN 0 12 436665 1

the embryonal N-CAM) and in this case poor immunogenicity as a factor of bacterial pathogenicity is, in part, explained on the basis of the hosts self-protection capacity in order to avoid autoimmunity. The NZB mouse is a particularly suitable model for such situations.

Table 1. Monoclonal antibodies against polysaccharide antigens of *E. coli* and meningococci obtained from NZB or BALB/C mice

Monoclonal antibodies	*Ig-isotype*	*Antigenic determinant*		*Mouse strain*
mAb 735	γ 2a	$\alpha(2\rightarrow8)_n$Neu NAc	(MBPS/K1)	NZB
mAb 1125	γ 2a	$\alpha(2\rightarrow9)_n$Neu NAc	(MCPS)	NZB
mAb 932	γ 2a	α-Man NAc (1-P-6) 3 ↓ OAc	(MAPS)	BALB/C
mAb 1087	γ 2a	see mAb 932	(MAPS)	BALB/C
mAb 1091	μ	-4)-β-GlcUA (1→4)-α-GlcNAc(1-	(K5)	NZB
mAb 1094	μ	see mAb 1091	(K5)	NZB
mAb 865	μ	-4)-β-4-ManNAcUA (1→4)-α-GlcNAc (1→3)-α-Fuc-4-NAc(1-	(ECA/K5)	BALB/C
mAB 898	γ 2a	see mAb 865	(ECA)	BALB/C
mAb 786	γ 3	-α-D-Hex (1→2)-α-D-Glc(1→3)-α-D-Glc(1→3)Hep-	(LPS outer core)	BALB/C

VIII. Bacterial Toxins and Receptors

Co-Chairmen: R. John Collier
Torkel Wadström

Photoaffinity Labelling and Site-Directed Mutagenesis of an Active Site Residue of Diphtheria Toxin

R. John Collier[a]

Microbiology and Molecular Genetics, Harvard Medical School and The Shipley Institute of Medicine, Boston, MA 02181, USA

The actions of several bacterial exotoxins depend on their ability to catalyse transfer of the ADP-ribose moiety of NAD to specific target proteins in animal cells. The target for both diphtheria toxin (DT) and exotoxin A from *Pseudomonas aeruginosa* is elongation factor 2 (EF-2) [1]. ADP-ribosylation of EF-2 inactivates it, which leads to inhibition of protein synthesis and cell death. Two regulatory subunits of adenylate cyclase are the targets for, respectively, pertussigen and the cholera toxin class of enterotoxins. ADP-ribosylation of either subunit alters cyclase activity and produces various physiological changes, which depend on the type of cell affected.

Little detailed information has emerged, until recently [2], about the precise location of the active site in any of the ADP-ribosylating toxins. In DT the active site resides within fragment A, a 193-residue amino-terminal fragment, that may be isolated after mild tryptic digestion and reduction of the intact toxin. Fragment A contains a single NAD binding site ($K_d = 8\ \mu M$), which is responsible for its ADP-ribosyl transferase and NAD-glycohydrolase activities [3]. In the studies outlined below we used photoaffinity labelling with native NAD to identify a residue within the nicotinamide subsite of the NAD site. Evidence has been obtained that this residue, glutamic acid-148, is at or near the catalytic centre and may play a direct role in catalysis of the ADP-ribosylation reaction [2].

[a] *Telephone:* 1-617-732-1930

Protein-Carbohydrate Interactions in Biological Systems. ISBN 0 12 436665 1

PHOTOAFFINITY LABELLING WITH RADIOLABELLED NAD

Figure 1 shows results obtained when mixtures of fragment A and NAD radiolabelled in the adenine moiety, the adenylate phosphate or the nicotinamide moiety, were irradiated at 254 nm for various periods, and trichloroacetic acid-precipitable label was measured [2]. With [*carbonyl*-^{14}C]NAD the rate and extent of incorporation were markedly higher than with the other preparations. The photolabelled fragment A was essentially devoid of ADP-ribosyl transferase activity. The high efficiency of incorporation from [*carbonyl*-^{14}C]NAD (approaching 1 mol/mol protein) was atypical and led us to study this photolabelling reaction in greater detail.

Incorporation of label from [*carbonyl*-^{14}C]NAD into other proteins was also tested [2]. Besides DT fragment A, nucleotide-free DT, CRM-45, and activated *Pseudomonas* exotoxin A showed levels of incorporation much higher than other proteins tested, including three NAD-linked dehydrogenases. These results

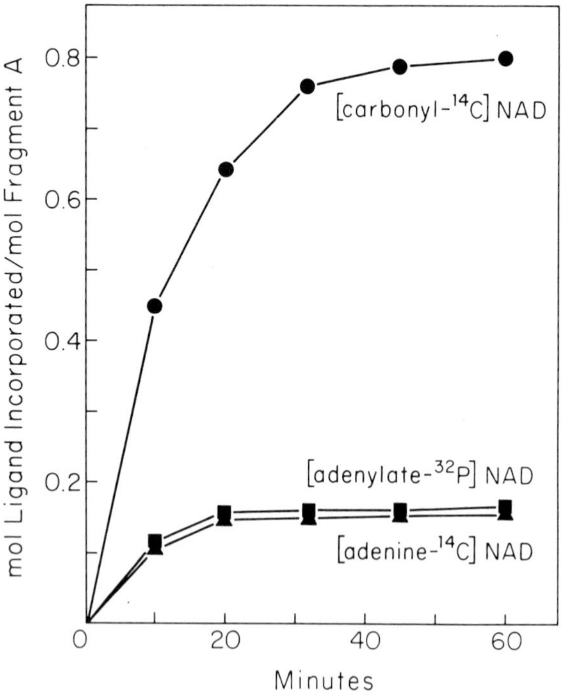

Figure 1. Photoaffinity labelling of DT fragment A with native NAD preparations. Reaction mixtures contained 20 μM fragment A and 40 μM NAD radiolabelled in the nicotinamide (●), adenylate phosphate (■), or adenine (▲) moiety were irradiated with 254 nm UV at 4°C. At intervals aliquots were removed, and trichloroacetic acid-precipitable material was determined.

suggested that the efficient photolabelling might be characteristic of ADP-ribosylating toxins, or a subset thereof.

ISOLATION AND SEQUENCING OF RADIOLABELLED PEPTIDES

When photolabelled fragment A was cleaved with CNBr and the products were fractionated on Sephadex G-75, all of the label ran with a single peptide, CNBr-3, except for that migrating with uncleaved or partially cleaved protein. CNBr-3 containing the photolabel was digested with thermolysin, and the products were analysed by HPLC. Again most of the label was associated with a single peak, which eluted at 45 min. The rest of the label migrated with two other thermolytic peptides, corresponding to incomplete digestion products.

The peptide eluting at 45 min was hydrolysed in acid, and the products were dansylated and separated by thin-layer chromatography. Three derivatives were found, in equimolar quantities: Dns-Val, Dns-Tyr, and an unidentified residue migrating between Dns-Thr and Dns-Ser on polyamide. Valine was amino-terminal, and carboxypeptidase A liberated only tyrosine. The only tripeptide sequence within fragment A containing valine in the first position and tyrosine in the third, is Val-Glu-Tyr, corresponding to residues 147–149. When the major labelled thermolytic peptide was sequenced manually by dansyl-Edman and Edman methods, 99% of the label was released at the second position. In addition, we found that radiolabelled thermolytic or chymotryptic peptides showed an absorption peak at 264 nm, with an extinction coefficient of 6.1/mM cm at 260 nm. These data suggested that nicotinamide, or a derivative thereof, was transferred to position 148 of fragment A.

NATURE OF THE PHOTOPRODUCT AT POSITION 148

Subsequent results from chemical methods, fast atom bombardment mass spectrometry, and nuclear magnetic resonance confirmed this prediction. Radiolabel from [N4-^{3}H]NAD was incorporated into fragment A as rapidly and efficiently as that from carbonyl-labelled NAD, and label from [N4-^{13}C]NAD or [N1-^{15}N]NAD was also efficiently incorporated. The thermolytic tripeptide, Val-X-Tyr, was isolated from fragment A irradiated in the presence of native NAD or NAD containing ^{13}C at position 4 of the nicotinamide moiety, and mass spectra were determined by fast atom bombardment. The mass of MH^+ at high resolution corresponded precisely with a calculated composition of $C_{24}H_{32}N_5O_6$. This represented a net loss of a single oxygen atom and addition of $C_5H_4N_2$, compared with the unmodified Val-Glu-Tyr. These alterations are equivalent to decarboxylation of glutamic acid and addition of nicotinamide, either via the carbonyl nitrogen or a ring carbon. N.m.r. spectra of the photolabelled tripeptide confirmed that only the nicotinamide and tripeptide moieties were present and established that the nicotinamide moiety was attached to the decarboxylated γ-methylene group of glu-148 via the number 6 ring carbon.

From the results obtained we infer that irradiation of the NAD:fragment A complex brings about formation of a new carbon–carbon bond between the γ-methylene group of Glu-148 and the number 6 carbon of the nicotinamide ring and that this is accompanied by decarboxylation of the Glu-148 side chain and hydrolysis of the bond linking ADP-ribose to nicotinamide (Fig. 2) ([4], Tweton *et al.*, unpublished observation). The resulting product at position 148 may be termed α-amino-γ-(6-nicotinamidyl)butyric acid. The other products have not been identified, but are formally equivalent to ADP-ribose and formic acid.

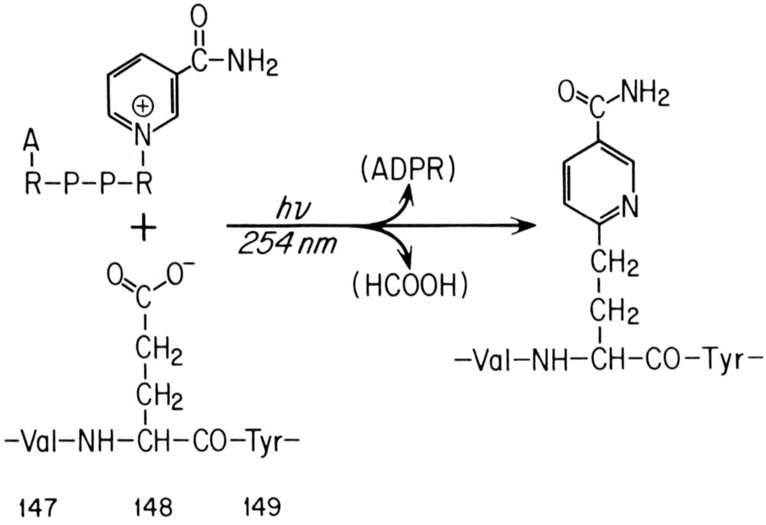

Figure 2. Photoaffinity labelling of Glu-148 by the nicotinamide moiety of NAD.

SITE-DIRECTED MUTAGENESIS OF Glu-148

To evaluate the role of Glu-148 in the ADP-ribosylation of EF-2, we replaced this residue with aspartic acid by *in vitro* mutagenesis of a toxin gene fragment cloned in *E. coli*. Gene fragment F2, which encodes all of the A fragment and 189 residues of B, was subjected to oligonucleotide-directed mutagenesis, to change the codon for residue 148 from GAA to GAT.

Periplasmic fractions from strains carrying a plasmid containing either the wild-type or the Asp-148 form of F2 were analysed. Wild-type and mutant periplasmic fractions gave essentially identical band patterns on Western blots, and the same concentrations of trypsin were effective in converting higher molecular weight peptides to fragment A. Wild-type extracts showed greatly enhanced ADP-ribosylation activity after trypsin-treatment, and the activity comigrated with fragment A on SDS-polyacrylamide gels. The activity of the trypsin-treated Asp-148 extracts was nil (within experimental error, *c.* 0.6%), regardless of trypsin

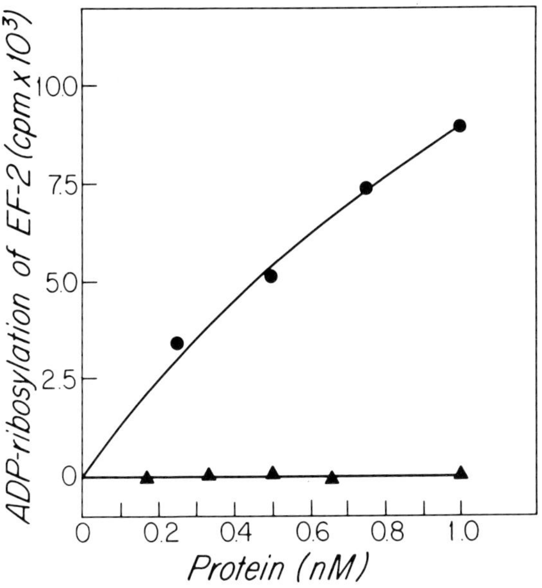

Figure 3. Titration of trypsin-treated Glu-148 (wild-type) and Asp-148 (mutant) peptides in ADP-ribosylation assay. Periplasmic fractions from isolates containing wild-type or mutant F2 gene fragments were digested with 1 μg/ml trypsin, to generate fragment A, and then analysed for ADP-ribosyl transferase activity. Glu-148, ●——●; Asp-148, ▲——▲.

treatment (Fig. 3). However, similar concentrations of NAD protected the Asp-148 and wild-type forms of fragment A against proteolytic degradation, implying that the mutant fragment A retained similar affinity for NAD. Emerick *et al.* [5] have reported that deletion of Glu-148 gave an enzymically inactive molecule, but NAD binding was not measured.

DISCUSSION

The results obtained imply that Glu-148 is essential for the ADP-ribosylation activity of DT and suggest that this residue may participate directly in catalysis. Substitution of Asp for Glu would be expected to produce minimal structural perturbation, and this prediction is supported by the findings that little or no change in NAD affinity or in sensitivity to trypsin was seen in the mutant. Since the Asp side chain differs from that of Glu only in being shorter by one methylene group, our results suggest that the precise location of the side-chain carboxylate at position 148 is crucial for enzymic activity.

The position of the new carbon–carbon bond in the photoproduct implies close contact in the fragment A:NAD complex between the γ-methylene carbon of Glu-148 and carbon number 6 of the nicotinamide ring. Such contact would constrain the Glu-148 carboxylate oxygens to within a radius of approximately

0.2–0.3 nm of the nicotinamide C-6, and therefore within range of contact with the *N*-glycosidic linkage and of ionic interaction with the cationic pyridine ring nitrogen. The Glu-148 carboxylate must therefore be at or near the active site, and conceivably might play a direct role in catalysis. Such a role might be highly sensitive to the precise position of the side chain carboxylate group, and therefore explain the dramatic reduction in activity upon substitution of Asp at position 148.

Exotoxin A from *Pseudomonas aeruginosa* is also efficiently photolabelled with nicotinamide-radiolabelled NAD, and we have preliminary evidence that labelling occurs at a single site (unpubl. res.). Studies are currently under way to determine if other ADP-ribosylating toxins, such as cholera and pertussis toxins, and other enzymes involving NAD as substrate, may be photolabelled by this method. If the results demonstrated here for diphtheria toxin are found to be general among the aDP-ribosylating toxins, the methods employed may prove useful in devising novel vaccines against diseases involving such toxins, as well as in defining and characterizing active sites of this class, and perhaps other classes, of enzymes.

SUMMARY

(1) When DT NAD is mixed with fragment A and the mixture is irradiated at 254 nm, the nicotinamide moiety is efficiently transferred to a single site on the protein, glutamic acid at position 148. A new carbon–carbon bond is formed between the C-6 ring carbon of the nicotinamide and the γ-methylene group of Glu, the Glu side chain is decarboxylated, and the ADP-ribose moiety of NAD is cleaved from the nicotinamide.

(2) Glu-148 was replaced by Asp in a cloned fragment of DT, and the Asp-148 mutant fragment A was found to have little or no ADP-ribosyl transferase activity, although it retained high affinity for NAD.

(3) The results imply that Glu-148 is at or near the catalytic centre of the toxin and suggest that this residue may play a direct role in catalysis.

ACKNOWLEDGEMENTS

The seminal photochemical results and the protein chemistry were carried out by Stephen Carroll. Rodney Tweten and Joseph Barbieri were responsible for the studies in which Asp was substituted for Glu-148. The mass spectrometric measurements were performed by James McCloskey and coworkers, at the University of Utah, and the n.m.r. measurements were by Norman Oppenheimer and coworkers, at the University of California, San Francisco. A portion of this work was supported by Public Health Research Grant AI-22021 from the National Institute of Allergy and Infectious Diseases and Grant CA-39217 from the National Cancer Institute.

REFERENCES

1. Collier, R. J. (1975). *Bact. Rev.* **39**, 54–85.
2. Carroll, S. F. and R. J. Collier (1984). *Proc. Natl. Acad. Sci. USA* **81**, 3307–3311.

3. Kandel, J., R. J. Collier and D. W. Chung (1974). *J. Biol. Chem.* **249**, 2088-2097.
4. Carroll, S. F., J. A. McCloskey, P. F. Crain, N. J. Oppenheimer, T. M. Marschner and R. J. Collier (1985). *Proc. Natl. Acad. Sci. USA* **82**, 7237-7241.
5. Emerick, A., L. Greenfield and C. Gates (1985). *DNA* **4**, 78.

Genetics of Cholera Toxin

John Mekalanos[a], Virginia Miller,
Ronald Taylor and Ina Goldberg

Department of Microbiology and Molecular Genetics, Harvard Medical School, Boston, MA 02115, USA

The cholera toxin A and B subunit genes are organized in a single transcription unit, the *ctx* operon, which is located on a genetic element that undergoes duplication and amplification events in *Vibrio cholerae* [1-4]. The promoter of the *ctx* operon requires for its function a positive regulatory gene called *toxR* [4]. Strains of *V. cholerae* vary widely in the capacity to produce cholera toxin in laboratory media. In this paper we will summarize our most recent results concerning the role of positive regulation and gene amplification in the control of cholera toxin gene expression. We have found that variation in at least three different genetic parameters is responsible for the toxin production phenotypes of different strains of *V. cholerae*.

STRUCTURAL AND FUNCTIONAL HETEROGENEITY OF *toxR* LOCI

The *toxR* gene of the highly toxinogenic strain 569B has been cloned and used as a probe for homologous sequences in other less toxinogenic and non-toxinogenic *V. cholerae* [3]. This analysis demonstrated that all *V. cholerae* tested as well as several other *Vibrio* species (e.g. *V. harveyi* and *V. parahemolyticus*) contain sequences homologous to the *toxR* gene. We determined whether these *toxR*-homologous sequences represented active *toxR* genes by cloning these sequences from two El Tor strains of *V. cholerae*. The cloned segments were then tested for their ability to activate a *ctx* promoter/*lacZ* fusion in *Escherichia coli*. Table 1 shows the *toxR* genes of strain E7946 and RV79 activate the *ctx*/*lacZ* fusion but not as well as the *toxR* gene of 569B. Construction of hybrid *toxR* genes by recombinant DNA methods has shown that this difference is probably

[a]*Telephone:* 1-617-732-1937

Protein-Carbohydrate Interactions in Biological Systems. ISBN 0 12 436665 1

Table 1. Activation of a *ctx-lacZ* operon fusion by *toxR* clones from three different strains of *V. cholerae*

Plasmid	*Origin of* toxR *gene*	*Activation*
pVM7	*V. cholerae* 569B	14
pE5	*V. cholerae* E7946	3.5
pR1	*V. cholerae* RV79	3.5

not due to a difference in the *toxR* structural genes *per se* but rather reflects a difference in transcriptional control of these genes.

Southern blot analysis of 569B, RV79 and E7946 indicates that strain 569B contains a deletion near the *C*-terminus of the *toxR* gene relative to other *V. cholerae* (Fig. 1). We have shown that this deletion mutation is at least partially responsible for the difference in the activities between the *toxR* genes of 569B, E7946, and RV79. For example, *V. cholerae* 569B *toxR* recombinants that have substituted the *toxR* locus of RV79 produce 13-fold less toxin. Reassortment of restriction fragments suggest that the DNA deleted in the *toxR* region of strain 569B encodes an additional control system that affects the expression of *toxR*. This deletion mutation may have been selected by intestinal passage in infant rabbits, a process that resulted in the genesis of the highly toxinogenic 569B strain from its avirulent parent [5].

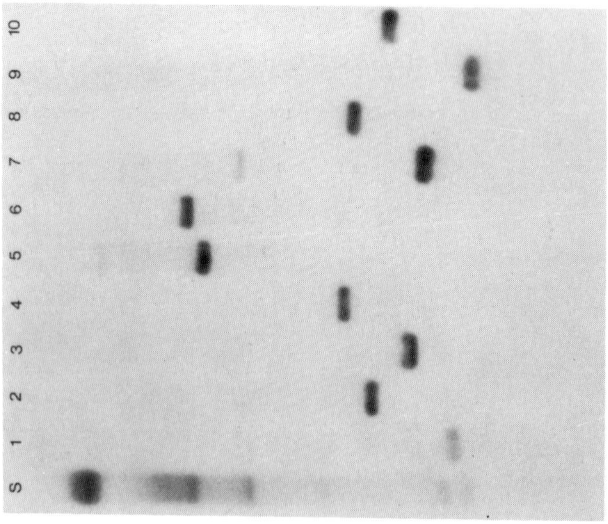

Figure 1. Southern blot analysis of the *toxR* genes of strains 569B (odd numbered lanes) and RV79 (even numbered lanes). After the standard (S), each pair of lanes is digested with a different restriction enzyme.

MECHANISM OF *toxR*-MEDIATED ACTIVATION OF THE *ctx* PROMOTER

A maxicell analysis of the protein products encoded by the cloned *toxR* gene of 569B has shown that a 33 kD protein appears to be required for *toxR* activity (Fig. 2). DNA sequence analysis has confirmed that the *toxR* gene is composed of an open reading frame sufficient to encode a protein of this size. Introduction of mutations in this open reading frame eliminate *toxR* activity and the production of the 33 kD protein in maxicells (Fig. 2), confirming the role of this protein in activation of *ctx* transcription.

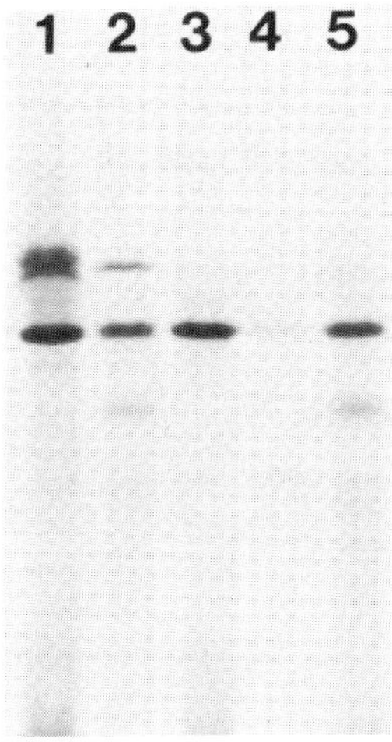

Figure 2. Identification of the *toxR* gene product. Maxicell analysis of *toxR*$^+$ (lane 2) or *toxR*$^-$ (lanes 3-5) plasmids in *E. coli*. *ToxR*$^-$ plasmids carry *in vitro* constructed deletion and insertion mutations in an open reading frame corresponding to the *toxR* gene. The uppermost band in lane 2 is approximately 33 kD in size.

Binding studies suggest that a protein present in extracts of *E. coli* cells carrying the *toxR* gene is capable of binding to the *ctx* promoter region (data not shown). In an effort to obtain additional information about the *toxR* binding site, we have performed a deletion analysis of the *ctx* promoter. A set of deletion mutations in the *ctx* promoter were constructed with nuclease *Bal*31 and the effect of these mutations on *ctxB* expression were scored in *E. coli* in either the presence or absence of the *toxR* gene. As shown in Fig. 3, deletions that remove five or more copies of the repetitive sequence TTTTGAT eliminate the ability of the *ctx* promoter to be activated above basal levels by *toxR*. The high constitutive levels seen with deletions 9 and 6-2 are due to fusion of the *ctx* operon with a strong upstream promoter. Subcloning experiments have shown that this upstream promoter is not invovled in the transcriptional activation of *ctx* by *toxR*.

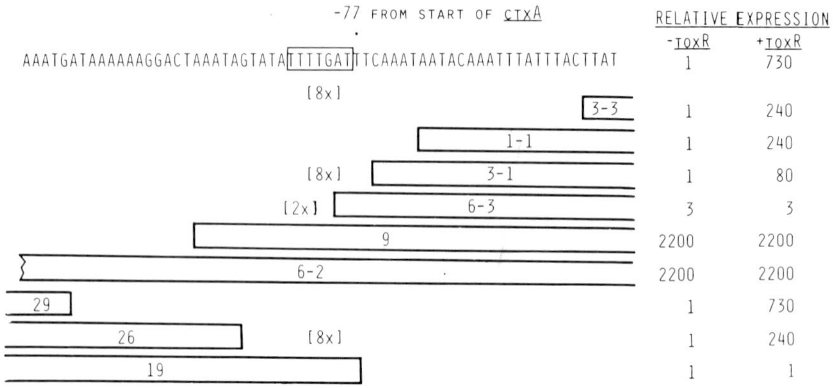

Figure 3. Deletion analysis of the *ctx* promoter. The DNA sequence of a portion of the *ctx* promoter is shown together with the extent of several deletions introduced into the region with nuclease *Bal*31. The promoter used in this analysis is from strain 569B and contained initially 8 tandem copies of the sequence TTTTGAT [1,4]. The number of these repeats remaining after a given deletion is shown in brackets under this sequence. The effect of these various deletions on basal (−ToxR) and activated (+ToxR) promoter activity is shown on the left.

The TTTTGAT sequence is tandemly repeated 3, 4 or 8 times 77 base pairs upstream of the start of *ctxA* in different *ctx* promoters and the copy number of this repeat appears to correlate with the amount of toxin produced by the different strains [1,4]. Together these data suggest that the *toxR* gene product activates *ctx* gene expression by binding to the TTTTGAT repeated sequence. Perhaps cooperative binding of *toxR* monomers to multiple copies of this seven base pair sequence is involved in the influence of repeat copy number on *ctx* expression.

RecA-DEPENDENT *ctx* GENE AMPLIFICATION

Strains of *V. cholerae* can carry multiple copies of the *ctx* operon located in either widely separated sites on the chromosome or in tandem repeats [2,3]. Intestinal passage of *V. cholerae* selects for variants that produce more toxin and some of these show amplification of the *ctx* genetic element [2]. Thus, gene amplification apparently plays a role in cholera toxin regulation. The *ctx* genetic element is composed of a central core region containing the *ctx* operon flanked by two or more copies of a repetitive sequence called RS1. Amplification of the *ctx* element generates direct repeats of the core region hyphenated with one or two copies of RS1 [2].

The structure of the amplified derivatives suggest that homologous recombination might be involved in the amplification phenomenon. We have tested this possibility by constructing a *recA* mutant of *V. cholerae* and then testing the effect of this mutation on cholera toxin gene amplification. The *recA* gene

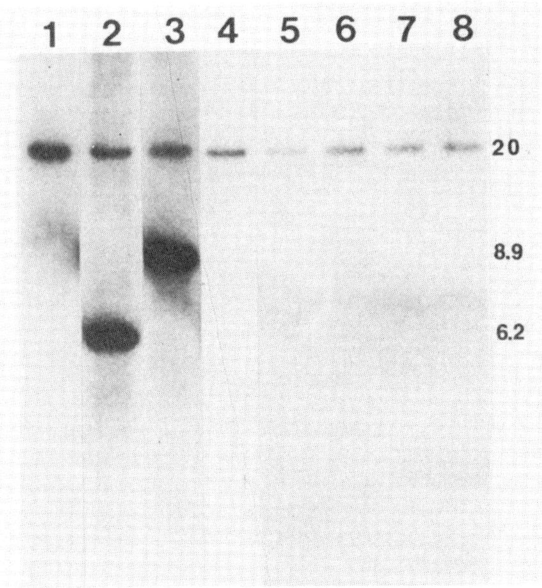

Figure 4. Southern blot analysis of *ctx* amplification events in *recA*$^+$ and *recA*$^-$ *V. cholerae*. Strain SM116 *recA*$^+$ (lane 1) was grown on media containing 3 mg/ml of kanamycin to give two amplified derivatives shown in lanes 2 and 3. Strain SM117 *recA*$^-$ (lane 4) was treated in a similar manner and gave a much lower frequency derivatives shown in lanes 5-8. Amplification of the *ctx* region is indicated by the appearance of the 8.9 or 6.2 tandem repeats seen in lanes 2 and 3.

of *V. cholerae* was cloned by the method of Better and Helinski [4] and a mutation was introduced into it by *in vitro* methods. This mutation was crossed onto the *V. cholerae* chromosome and the resultant strain (SM117) was compared to its parent (SM116) in a *ctx* amplification assay.

This assay involves the use of a kanamycin resistance marker introduced into the *ctx* operon of SM116. We have shown that selection of variants of strain SM116 that are resistant to 3 mg/ml of kanamycin results in amplification of the *ctx* genetic element (strain SM116 is usually resistant to only 150 μg/ml of kanamycin). Two such amplified derivatives are shown in Fig. 4, lanes 2 and 3, which have respectively, one and two copies of RS1 between their directly repeated *ctx* core regions. These amplification events generate fragments of 6.2 and 8.9 kilobase pairs, respectively, when the DNA of these derivatives is hybridized to a *ctx* specific probe in this Southern blot assay. In contrast, strain SM117 (*recA*$^-$) threw off variants resistant to 3 mg/ml of kanamycin at a frequency at least 100-fold less than SM116 (*recA*$^+$). Moreover, these variants showed no evidence of amplification of the *ctx* element as indicated by the absence of bands of 6.2 and 8.9 kilobase pairs in size (Fig. 4, lanes 5-8). Thus, homologous recombination mediated by the *recA* gene product is involved in the amplification

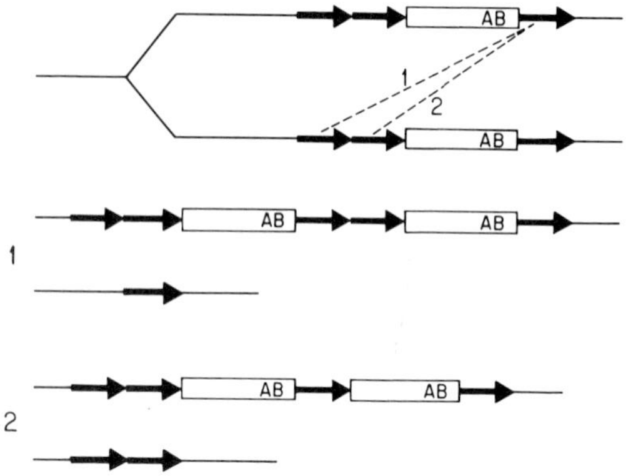

Figure 5. Proposed model for the generation of *ctx* duplication, amplification and deletion events by unequal crossing over (1 and 2) between RS1 copies (arrows) flanking the *ctx* core (open box) region. The structure of the *ctx* genetic element is analogous to that present in strains SM116 and SM117.

of the *ctx* genetic element in *V. cholerae*. Presumably, unequal cross-over events between directly repeated copies of RS1 that flank the core region generate these *ctx* amplification events (Fig. 5). The reciprocal product of such events would be deletion mutations of the core region of the *ctx* genetic element (Fig. 5). We have been able to isolate these mutants by virtue of their kanamycin-sensitive phenotype compared to their parental strains SM116 and SM117. These deletion mutations also so a considerable dependency on the *recA* gene of *V. cholerae*.

SUMMARY

Several genetic factors influence the level of cholera toxin production by different strains of *V. cholerae*. First, the positive regulatory gene *toxR*, whose product stimulates *ctx* transcription, exists in different allelic states that affect its relative activity. Second, the *ctx* promoters of various strains vary in the number of copies of the TTTTGAT repetitive sequence. We have presented evidence that the *toxR* gene product acts at the tandem repetitions of this seven base pair sequence and that the copy number of these repeats effect the relative activity of the *ctx* promoter. Third, a second regulatory locus, *htx*, not discussed here, plays an as yet unknown role in *ctx* expression [6]. Finally, the copy number of the *ctx* genetic element, which can undergo duplication and amplification events in *V. cholerae*, can change the level of toxin production in a given strain by presumably a gene dosage mechanism. Homologous recombination, at least in part mediated by the *recA* gene of *V. cholerae*, is required for the latter rearrangements. This may be the first of a variety of examples where *recA*-dependent homologous recombination plays a direct role in the generation of genome rearrangements that enhance the virulence of pathogenic micro-organisms.

ACKNOWLEDGEMENTS

This work was supported by grant AI-18045 from the National Institute of Allergy and Infectious Disease.

REFERENCES

1. Mekalanos, J. J., D. J. Swartz, G. D. N. Pearson, N. Harford, F. Groyne and M. deWilde (1983). Cholera toxin genes: nucleotide sequence, deletion analysis, and vaccine development. *Nature* **306**, 551-557.
2. Mekalanos, J. J. (1983). Duplication and amplification of toxin genes in *Vibrio cholerae*. *Cell* **35**, 253-263.
3. Sporecke, I., D. Castro and J. J. Mekalanos (1984). Genetic mapping of *Vibrio cholerae* enterotoxin structural genes. *J. Bacteriol.* **157**, 253-261.
4. Miller, V. and J. J. Mekalanos (1984). Synthesis of cholera toxin is positively regulated at the transcriptional level by *toxR*. *Proc. Natl. Acad. Sci. USA* **81**, 3471-3475.
5. Dutta, N. K. and H. K. Habbu (1955). Experimental cholera in infant rabbits: a method of chemotherapeutic investigation. *Br. J. Pharmacol. Chemother.* **10**, 153-159.
6. Mekalanos, J. J. and J. R. Murphy (1980). Regulation of cholera toxin production in *Vibrio cholerae*: genetic analysis of phenotypic instability in hypertoxinogenic mutants. *J. Bacteriol.* **141**, 570-576.

Mechanisms of Enterotoxin Secretion from *Escherichia coli* and *Vibrio cholerae*

T. R. Hirst[a]

Department of Medical Microbiology, University of Göteborg, Göteborg, Sweden

All living cells have developed efficient mechanisms for correctly localizing proteins in different cellular compartments. Gram-negative bacteria, for example, export specific polypeptides to the various compartments of the cell envelope and to the extracellular milieu. These export processes are of pervading significance for pathogenic microbes, because the important determinants of pathogenicity, e.g., surface-associated adhesins or extracellular toxins, are exported proteins.

In this paper I shall focus on the export and secretion of enterotoxins from *Vibrio cholerae* and *Escherichia coli* which are responsible for causing severe diarrhoeal disease in man and farm animals. *V. cholerae* secretes a potent cholera enterotoxin comprised of five identical B subunits which bind to G_{M1}-ganglioside receptors and a single A subunit that catalyses ADP-ribosylation of adenylate cyclase. Enterotoxinogenic *E. coli* produce a heat-labile enterotoxin (LT) which is strikingly similar in structure and function to cholera toxin. However, in contrast to cholera toxin, LT is only exported as far as the periplasmic space of the *E. coli* cell envelope. This raises two important questions: (1) how and under what conditions is LT released from *E. coli*; and (2) what mechanisms or processes does *V. cholerae* possess that endow it with the capacity to secrete cholera toxin completely?

ENTEROTOXIN EXPORT FROM *E. COLI*

Significant progress in our understanding of the early stages of LT export, assembly and localization in *E. coli* have been made. Biochemical and genetic

[a]*Telephone:* 46-31-602070

Protein–Carbohydrate Interactions in Biological Systems. ISBN 0 12 436665 1

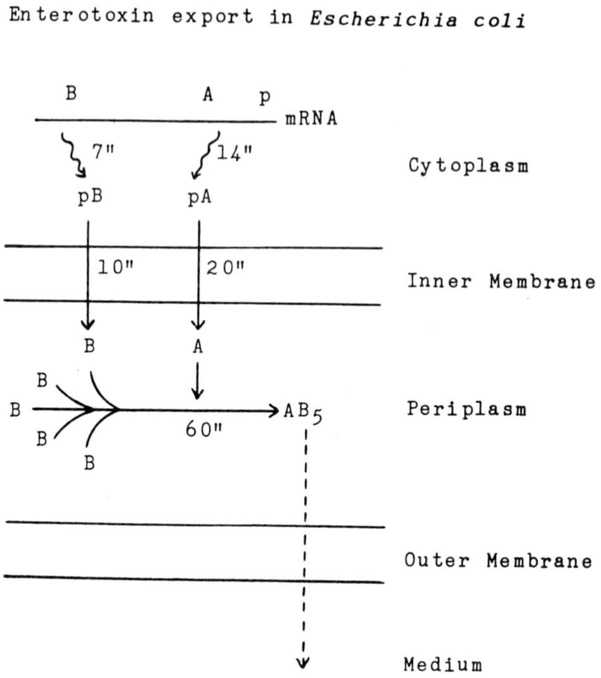

Figure 1. A diagrammatic representation of the events during enterotoxin export in *E. coli.* The A and B subunits are synthesized as precursors (pA and pB) from a polycistronic mRNA, and are then translocated and processed to mature polypeptides (A and B) which are released into the periplasmic space where they assemble into holotoxin (AB_5). The approximate half-time for synthesis, translocation and assembly are indicated.

studies have established that the A and B subunits are initially synthesized as precursors containing typical "signal sequences" of 18 and 21 amino acids, respectively [1-3]. The precursors are translocated through the cytoplasmic membrane, proteolytically processed, and the mature polypeptides released into the periplasmic space (Fig. 1). Kinetic studies have revealed that translocation and processing are rapid events, with a half-time for the release of mature subunits into the periplasm of approximately 20 and 10 seconds for the A and B subunits respectively [4,5]. The mature monomers enter a pool of assembling subunits which gives rise to holotoxin (AB_5 in Fig. 1). Details of this assembly are at present complicated and difficult to interpret since the A subunits influence the rate of removal of B subunits from the monomer pool, but also have a negligible effect on the kinetics of B subunit pentamer formation compared with strains synthesizing only the B subunits of LT (S. Hardy, pers. comm.). These data suggest that the A subunit normally participates in the formation of an "A-B" intermediate during the assembly of the B subunit pentamer but that formation of the intermediate is not rate-limiting or obligatory for B subunit pentamerization.

The assembled holotoxin is located within the periplasmic space. This conclusion was arrived at after nearly a decade of controversy about the location of LT in which numerous reports suggested different locations for the toxin within the subcompartments of the *E. coli* cell envelope [6–8]. The confusion arose because: (1) procedures used to release periplasmic proteins gave conflicting results—that is lysozyme/EDTA and polymixin B treatments completely released LT, whereas osmotic shock released only negligible amounts of LT; and (2) isolated membranes from enterotoxinogenic *E. coli* contained as much as two-thirds of the total LT, most of which was associated with the outer membrane fraction [8].

These inconsistencies were recently resolved after studies on the remarkable solubility characteristics of LT during cell fractionation [6] in which it was found that the apparent location of LT shifted depending on the ionic conditions used. Buffers of low ionic strength caused LT to aggregate and thereby pellet with spheroplasted cells, and to cosediment with membrane preparations. However, in buffers of physiological ionic strengths LT was soluble and found to be periplasmically located [6]. This latter finding raises the crucial question of how LT is able to mediate disease if it is shielded from the intestinal epithelium by being located within the periplasm of the *E. coli* cell? Until recently we assumed that the laboratory conditions used to study the location of LT in *E. coli* failed to permit the release of periplasmic LT but that growth *in vivo*, in an environment containing bile salts and other stringent conditions, would promote its release. To test this, *E. coli* strains $286C_2$ and G6pEWD299 were cultured in a 1:1 mixture of L-broth and intestinal fluid aspirates taken from humans. To our surprise only a negligible amount ($< 1\%$) of LT and β-lactamase, a periplasmic marker enzyme, were released. This suggests to us that the likely corollary of the growth of enterotoxinogenic *E. coli in vivo* is the release of only a small fraction of the total LT which is produced.

ENTEROTOXIN SECRETION FROM *V. CHOLERAE*

The capacity of *V. cholerae* to secrete an extracellular toxin was established nearly 40 years ago, but the mechanism by which this organism accomplishes this biological feat is still poorly understood. In a recent article [7], we proposed that the secretion of enterotoxins from *V. cholerae* occurs via an export pathway analogous to LT export in *E. coli*, but with an additional secretory step which releases the toxin into the medium. This hypothesis predicted that the toxin subunits transiently enter the periplasmic space of the *V. cholerae* cell envelope where they assemble into holotoxin prior to either their leakage or secretion through the outer membrane.

In order to test these predictions the following studies have been initiated: (1) the transfer into *V. cholerae* of various plasmids encoding the A and B subunits of LT to determine the role of the subunits in the secretion process; (2) experiments on the kinetics and pathway of toxin secretion in *V. cholerae*;

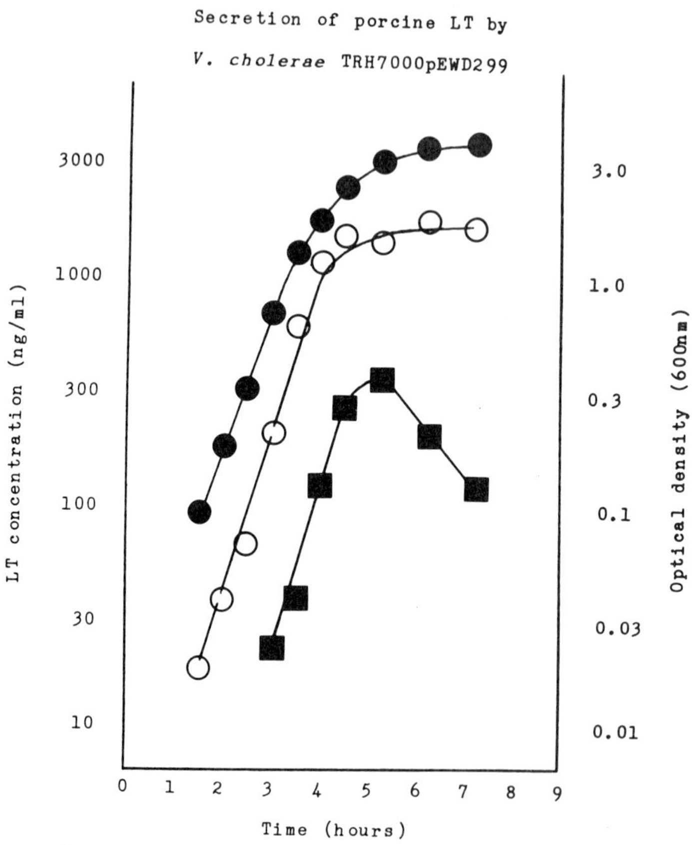

Figure 2. *V. cholerae* TRH7000pEWD299 was grown in syncase broth containing 50 μg/ml of thymine and 200 μg/ml ampicillin. Samples were removed during growth (●) and the concentration of LT in the cells (■) and media (○) determined by G_{M1}-ELISA.

and (3) an investigation into the location of cholera toxin in a mutant defective in toxin secretion.

These studies have shown that *V. cholerae* not only has the capacity to secrete cholera toxin, but also LT (Fig. 2). This was demonstrated by using a *V. cholerae* strain (TRH7000) from which the cholera toxin genes had been genetically deleted, as the host for the expression of a recombinant plasmid encoding LT. Approximately 80-90% of the LT produced during exponential growth of this bacterium was secreted into the medium. Concomitantly, TEM β-lactamase (a periplasmic enzyme) failed to be released (Fig. 3). This provides persuasive evidence that enterotoxin secretion from *V. cholerae* is not due to the generalized leakage of periplasmic constituents across the outer membrane.

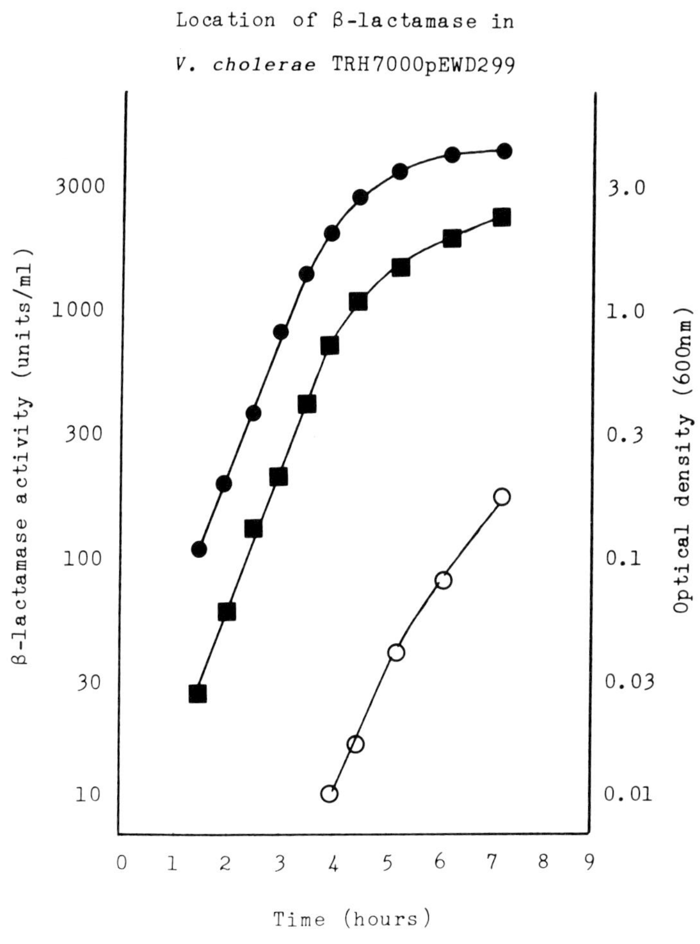

Figure 3. *V. cholerae* TRH7000pEWD299 was cultured as described in the legend to Fig. 2. Samples were removed during growth (●) and the activity of β-lactamase in the cells (■) and media (○) was determined by measuring the rate of hydrolysis of nitrocefin.

The role of the individual toxin subunits in the secretion process was investigated by transferring into *V. cholerae* TRH7000 various plasmids which encoded either for the A or B subunits of LT (Table 1). The B subunits were secreted in the absence of A subunit synthesis but not vice versa. This implies that the B subunits contain the structural information necessary for the secretion of LT and that the A subunits are only secreted by virtue of their association with B subunits. If this is the case then the A and B subunits should normally assemble prior to the entry of the B subunits into the secretion step of the export

Table 1. Cellular location of LT and its subunits in *V. cholerae*

V. cholerae *strain*[a]	*Toxin*	*Toxin concentration (ng/ml)* Extracellular	*Cell-associated*
TRH7000pWD600[b]	LT holotoxin	1110	<67
TRH7000pWD605[c]	LT-A subunit	0	1200
TRH7000pWD615[b]	LT-B subunit	1380	170

[a] The *V. cholerae* strains were grown in a rich syncase broth supplemented with 50 μg/ml thymine and 2 μg/ml tetracycline. As a control the strain TRH7000 lacking any plasmids was shown to be negative in both toxin assays.
[b] Assayed by G_{M1}-ELISA.
[c] Assayed in the pigeon erythrocyte lysate assay.

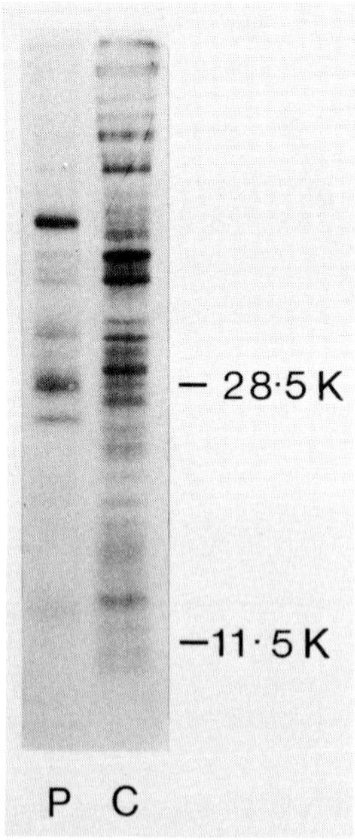

Figure 4. Polymixin B-induced release of periplasmic proteins from *V. cholerae* TRH7000pMMB25/pWD600. Cells were cultured in M9 minimal salts medium supplemented with 0.4% glucose, 50 μg/ml thymine, 200 μg/ml ampicillin, 2 μg/ml tetracycline, and a mixture of 18 amino acids (50 μg/ml) which excluded methionine and tryptophan. At an Abs (600 nm) of *c.* 0.6, 1 ml of cells was radioactively labelled with 200 μCi of ^{35}S-methionine (~ 1000 Ci/mmol) for 30 min at 37°C and then the cells separated from the media and washed with PBS. These cells were resuspended in 1 ml of ice-cold 0.25 M sucrose/0.1 M sodium phosphate pH 7.0/2000 U/ml polymixin B and incubated on ice for 15 min. The cells were pelleted (10 000 × *g*, 2 min) and the supernatant (P fraction) and the cell pellet (C fraction) analysed by SDS-PAGE and autoradiography. A ratio of 10:1 of the P:C fraction was loaded on the gel.

pathway. The most likely location for toxin assembly would be the periplasmic space.

In order to study this subcellular compartment, a method was developed for fractionating *V. cholerae* using polymixin B. Strain TRH7000 harbouring plasmid pMMB25 was used because this strain synthesized easily assayable marker enzymes for the periplasm (β-lactamase) and the cytoplasm (catecol oxygenase). Conditions were established in which 95% of the β-lactamase, but less than 5% of the catecol oxygenase, was released. This procedure resulted in the release of a characteristic set of polypeptides (Fig. 4, lane P) which represented approximately 5% of the total cellular protein.

Pulse-chase experiments, performed on *V. cholerae* strains secreting LT, revealed that the half-time for toxin secretion (measured as the appearance of newly synthesized, labelled B subunits in the medium) was approximately 6 min (Fig. 5). Fractionation of pulse-labelled and chased cells taken at various times after the addition of the chase, showed that the polymixin B-released periplasmic fraction contained labelled toxin B subunits which disappeared from this fraction concomitantly with the appearance of B subunits in the medium. The periplasmically located toxin intermediates had assembled by the earliest chase time of 30 s. I therefore conclude that enterotoxins produced by *V. cholerae* transiently enter the periplasm of the cell envelope prior to their secretion through the outer membrane.

This latter intriguing step remains as a focus of future research, which will no doubt benefit from the isolation of mutants defective in toxin secretion. One such mutant is currently being investigated, which was isolated by NTG-mutagenesis of *V. cholerae* 569B by Dr R. Holmes in 1973. Approximately 90%

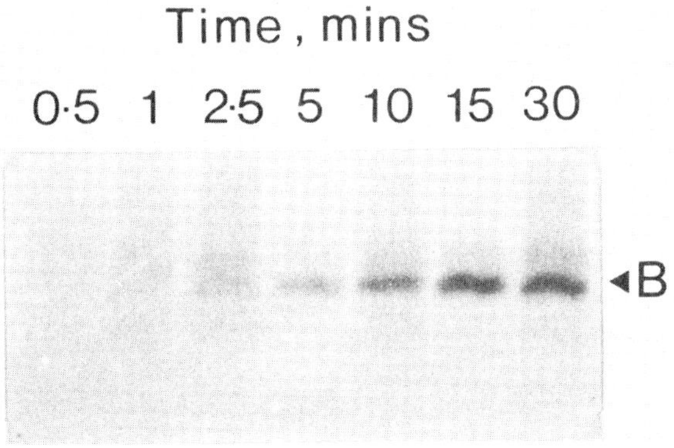

Figure 5. Kinetics of LT secretion from *V. cholerae*. Strain TRH7000pMMB25/pWD600 was cultured as described in the legend to Fig. 4. Cells were pulsed-labelled with ^{35}S-methionine (200 μCi/ml of cells) for 15 s, followed by the addition of 1 mM L-methionine. Samples were removed at the various times indicated and rapidly chilled on ice, and the cells and media separated by centrifugation. The media was analysed by SDS-PAGE and autoradiography, and the region of the autoradiogram containing the B subunit of LT is shown. No B subunit band appeared in control strains which did not synthesize LT.

of the cholera toxin produced by the mutant strain remains within the periplasm. It therefore seems likely that outer membrane components are crucial for the secretion of holotoxin which mechanistically may function in a facilitated fashion by recognizing B subunit domains of the toxin.

ACKNOWLEDGEMENTS

I would like to thank S. Johanson for her skilled technical assistance, and J. Hirst for helping prepare this manuscript. I would also like to extend a special thanks to my coworkers past and present, Drs J. Holmgren, S. Hardy, L. L. Randall, J. Sanchez, M. Bagdasarian and B.-E. Uhlin for their help and stimulating discussions.

This work was made possible by a long-term fellowship from the European Molecular Biology Organization and through financial support from the World Health Organization and the Swedish Board of Technical Development.

REFERENCES

1. Dallas, W. S. and S. Falkow (1980). Amino acid sequence homology between cholera toxin and *Escherichia coli* heat-labile enterotoxin. *Nature (Lond.)* **288**, 499–501.
2. Palva, E. T. *et al.* (1981). Synthesis of a precursor to the B subunit of heat-labile enterotoxin in *Escherichia coli*. *J. Bacteriol.* **146**, 325–330.
3. Spicer, E. K. and J. A. Noble (1982). *Escherichia coli* heat-labile enterotoxin, nucleotide sequence of the A subunit gene. *J. Biol. Chem.* **257**, 5716–5721.
4. Hirst, T. R. *et al.* (1983). Assembly *in vivo* of enterotoxin from *Escherichia coli*: formation of the B subunit oligomer. *J. Bacteriol.* **153**, 21–26.
5. Hofstra, H. and B. Witholt (1984). Kinetics of synthesis, processing and membrane transport of heat-labile enterotoxin, a periplasmic protein in *Escherichia coli*. *J. Biol. Chem.* **259**, 15 182–15 187.
6. Hirst, T. R. *et al.* (1984). Cellular location of heat-labile enterotoxin in *Escherichia coli*. *J. Bacteriol.* **157**, 637–642.
7. Kunkel, S. L. and D. C. Robertson (1979). Factors affecting release of heat-labile enterotoxin by enterotoxinogenic *Escherichia coli*. *Infect. Immun.* **23**, 652–659.
8. Wensink, J. *et al.* (1978). Isolation of membranes of an enterotoxigenic strain of *Escherichia coli* and distribution of enterotoxin activity in different subcellular fractions. *Biochim. Biophys. Acta* **514**, 128–136.
9. Hirst, T. R. *et al.* (1984). Mechanism of toxin secretion by *Vibrio cholerae* investigated in strains harboring plasmids that encode heat-labile enterotoxins of *Escherichia coli*. *Proc. Natl. Acad. Sci. USA* **81**, 7752–7756.

Identification of the *E. coli* Heat-Stable Enterotoxin Receptor on Rat Intestinal Brush Border Membranes

Jean Gariépy[a] and Gary K. Schoolnik*

*Departments of Medicine and *Medical Microbiology, Stanford University School of Medicine, Stanford, CA 94305, USA*

Diarrhoea is a life-threatening dehydrating illness of malnourished children in developing countries. Enterotoxigenic *Escherichia coli* strains (ETEC) are a major cause of diarrhoea and elaborate at least three classes of enterotoxins which differ widely in their mechanism of action. One of these enterotoxins, abbreviated ST1 (or ST_a), is a heat-stable low molecular weight peptide [1-3]. ST1 activates a membrane-bound intestinal guanylate cyclase and the subsequent elevation of intracellular cGMP leads to a profound secretory diarrhoea [4-8]. Its activity is assayed in suckling mice [9]. ST1 enterotoxins from human, bovine, and porcine ETEC strains have been purified and sequenced [1-3,10-12]. Although their *N*-terminal amino acid sequences are heterogeneous, the enterotoxic domain is confined to a highly conserved *C*-terminal segment of 13 amino acids [13, unpubl. res.]. Recently, a detergent-solubilized 100 000 D brush border protein has been reported to bind radiolabelled ST1 [14]. We have synthesized an enterotoxic analogue of ST1, termed ST1b(1-19] [2,3] (Fig. 2). We report in this communication the design of a photoreactive radiolabelled derivative of this toxin that was used to cross-link specifically two molecules on rat intestinal brush border membranes that may constitute the ST toxin receptor.

[a]*Telephone:* 1-415-497-3083

Protein-Carbohydrate Interactions in Biological Systems. ISBN 0 12 436665 1

BINDING OF SYNTHETIC ^{125}I-ST1b(1-19) TO RAT INTESTINAL MEMBRANES IS RAPID, SPECIFIC AND SATURABLE

A 19-amino acid analogue corresponding to the entire sequence of ST1 from an ETEC strain of human origin was prepared by solid-phase peptide synthesis [2,3]. Less than 5 ng of the purified peptide induced diarrhoea in suckling mice (results not shown). The sequence (Fig. 2) includes two tyrosine residues, not involved in the enterotoxicity of the peptide [13, unpubl. res.) that can be iodinated.

The binding of ^{125}I-ST1b(1-19) to intestinal membranes was examined. "Specific binding" was defined as the difference between the counts per minute (cpm) of bound ^{125}I-ST1b(1-19) in the presence and absence of a 400-molar excess of unlabelled ST1b(1-19). ^{125}ST1b(1-19) bound to intestinal membranes rapidly at 37°C; 50% maximal binding was achieved in 2 min (Fig. 1A). Specifically bound ^{125}I-ST1b(1-19) reached an asymptote at a toxin concentration of 30 nM, indicating that saturation of the toxin binding sites had occurred (Fig. 1B).

Competitive inhibition of binding by increasing amounts of unlabelled toxin from 0 to 1 μg was conducted with 100 000 cpm of ^{125}I-ST1b(1-19) and intestinal membranes (Fig. 1C). A linear curve of inhibition is evident between 1.0 and 10 ng of unlabelled toxin; 50% binding inhibition was produced by 2.7 ng of unlabelled ST1b(1-19). ST1b(1-19) and ST1b(6-18), a synthetic toxin analogue comprising residues 6 to 18 of ST1b, were equally active in competition experiments (data not shown), indicating that the receptor binding domain of ST1b is encoded by a peptide as small as 13 amino acids. Unrelated proteins such as ovalbumin (50 μg) were tested as negative controls and did not cause binding inhibition, providing additional evidence that the interaction of synthetic ST1b(1-19) and ST1b(6-18) with rat intestinal membranes is specific. If the intestinal membranes and ^{125}I-ST1b(1-19) were incubated for 1 h at 37°C prior to the addition of a 400-molar excess of unlabelled ST1b(1-19), only one-half of the specifically bound radioactivity was eluted from the membranes. Thus the binding of ^{125}I-ST1b(1-19) to its receptor may not be entirely reversible once the toxin is receptor bound, suggesting that a conformational or covalent modification of the receptor-peptide complex occurs. A mixed disulfide reaction between toxin and receptor has been proposed [15,16].

PREPARATION AND BINDING PROPERTIES OF A PHOTOREACTIVE, LABELLED ANALOGUE OF ST1b(1-19)

N-terminal amino group of ^{125}I-ST1b(1-19) was derivatized with the hetero-bifunctional cross-linker, *N*-hydroxysuccinimidyl *p*-benzoylbenzoate. The structure of the photoreactive analogue, termed ^{125}I-ST1b(1-19)-BB, is depicted in Fig. 2. The binding of ^{125}I-ST1b(1-19)-BB and ^{125}I-ST1b(1-19) to rat intestinal membranes was compared (Table 1) and found to be similar. This result was expected since this *N*-terminal modification of this 19-amino acid analogue is 5 amino acids away from the *C*-terminal toxic domain (residues 6 to 18).

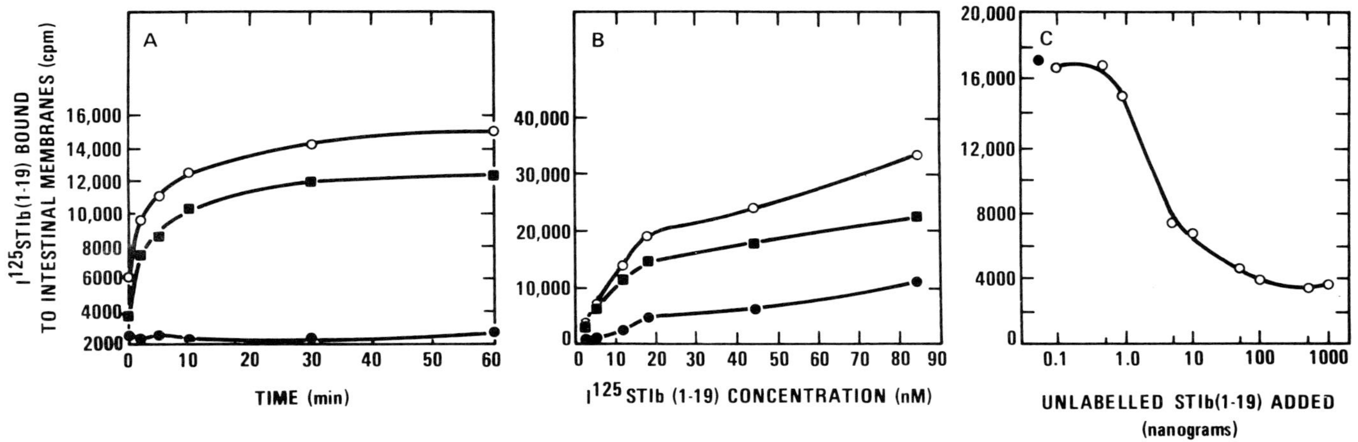

Figure 1. Binding characteristics of ^{125}I-ST1b(1–19) to intestinal membranes. (A) Rate of binding of ^{125}I-ST1b(1–19) (100 000 cpm) to intestinal membranes at 37°C. Specific binding (■) represents the difference in counts per minute (cpm) of membrane-bound radioactivity in the absence (○) and presence (●) of unlabelled ST1b(1–19). (B) Saturation curve for the binding of ^{125}I-ST1b(1–19) to intestinal membranes. The symbol nomenclature is described in (A). (C) Competitive inhibition of binding of ^{125}I-ST1b(1–19) (100 000 cpm) to intestinal membranes in the presence of increasing amounts of unlabelled ST1b(1–19). The closed circle represents the amount of radioactivity bound when no unlabelled ST1b(1–19) was added. Each point represents an average value from experiments performed in triplicate.

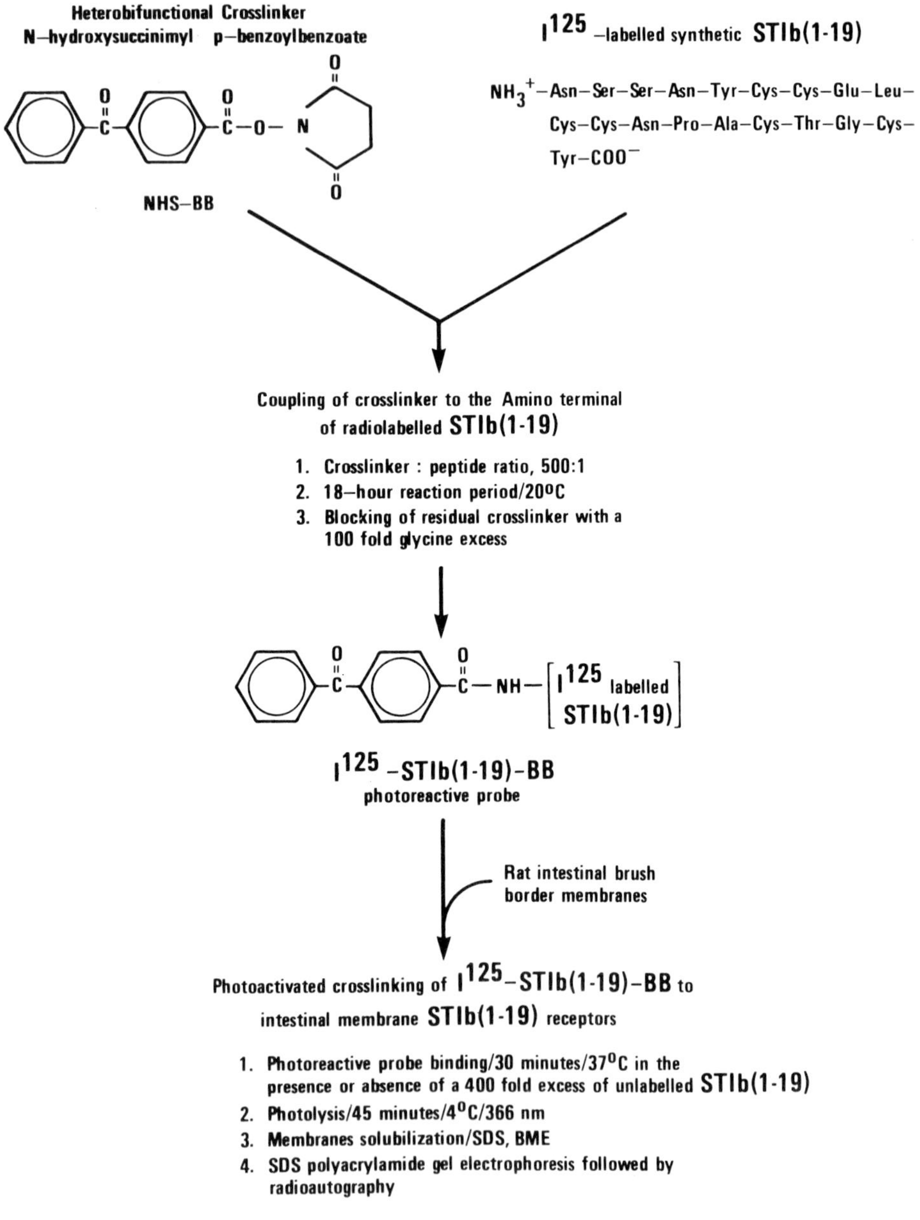

Figure 2. Scheme summarizing the construction of the photoreactive radiolabelled analogue of ST1b(1-19) and its use to identify the toxin receptor on rat brush border membranes. SDS, sodium dodecyl sulphate; BME, 2-mercaptoethanol.

Table 1. Binding of ^{125}I-ST1b(1–19)[a] and ^{125}I-ST1b(1–19)-BB[b] to intestinal membranes

	Membrane-bound radiolabelled analogue (cpm)[c]		
Synthetic toxin analogue	*Total bound*	*Non-specifically bound*	*Specifically bound*
^{125}I-ST1b(1–19)	16 826 (292)[d]	3523 (248)	13 303
^{125}I-ST1b(1–19)-BB	15 502 (1240)	3855 (357)	11 647

[a] Synthetic ^{125}I-ST1b(1–19).
[b] Photoreactive synthetic ^{125}I-ST1b(1–19).
[c] ^{125}I-ST1b(1–19) or ^{125}I-ST1b(1–19)-BB (100 000 cpm) was incubated with intestinal membranes in the presence and absence of a 400-fold molar excess of unlabelled ST1b(1–19).
[d] Values represent the radiolabelled analogue (in cpm) bound to 20 μl of intestinal membranes (7 mg protein/mL). The standard deviation (in parentheses) associated with each value was derived from experiments performed in triplicate.

^{125}I-ST1b(1–19)-BB SPECIFICALLY LABELS TWO INTESTINAL BRUSH BORDER MEMBRANE MOLECULES

^{125}I-ST1b(1–19)-BB and brush border membranes were incubated in the presence or absence of excess unlabelled ST1b(1–19) and subjected to photolysis. The radiolabelled membrane moieties were identified by polyacrylamide gel electrophoresis in the presence of $NaDodSO_4$ and 2-mercaptoethanol followed by radioautography (Fig. 2). Two major molecular species of approximately 57 and 75 kilodaltons (kD) were specifically labelled (Fig. 3, lane 1). Complete inhibition of binding occurred in the presence of excess, unlabelled ST1b(1–19) substantiating the specificity of the reaction (Fig. 3, lane 3). No major component was labelled by photolysis when ^{125}I-ST1b(1–19)-BB was substituted for ^{125}I-ST1b(1–19) indicating that the toxin-receptor complex is dissociated by the reducing conditions under which $NaDodSO_4$ polyacrylamide gel electrophoresis was conducted (Fig. 3, lanes 2 and 4).

Three explanations may account for the specific labelling of at least two molecular species: (1) The ST1b(1–19) receptor may be in close proximity or associated with another membrane molecule; (2) The actual toxin binding site may be formed by closely associated regions contributed by each of these molecules; (3) The 75 kD species may be partially degraded during the preparation of intestinal membranes to a 57 kD species. This last possibility was mitigated by the use of protease inhibitors (EDTA, *p*-toluenesulphonyl chloride and pepstatin) during the isolation and storage of intestinal membranes. In support of the first two possibilities, the putative receptor complex has been solubilized and partially purified by Kuno *et al.* [17] with 0.1% LUBROL-PX. This preparation retains its ability to bind radiolabelled ST1 and when cross-linked with disuccinimidyl suberate, and dissociated in the presence of $NaDodSO_4$ and 2-mercaptoethanol, generates an electrophoretic pattern similar to Fig. 3. The broad 75 kD band (Fig. 3) could be resolved into two bands of about 68 kD and 80 kD.

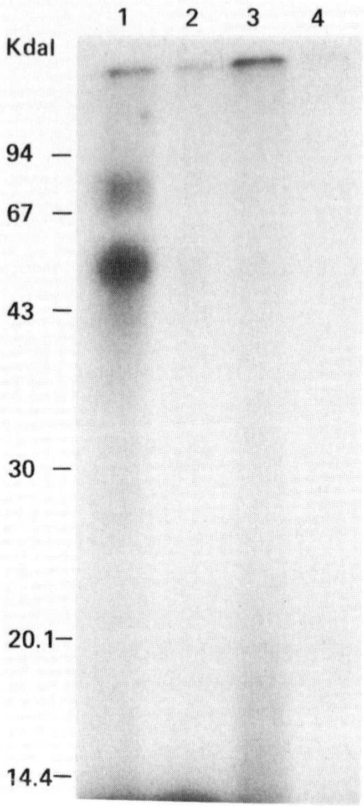

Figure 3. Radioautogram of a 15% polyacrylamide gel depicting the specific binding of ^{125}I-ST1b(1–19)-BB to two rat intestinal membrane molecular species. Brush border membrane samples were incubated at 37°C in the presence of ST1b analogues, photolysed, treated with $NaDodSO_4$ and 2-mercaptoethanol and then boiled prior to the electrophoresis step. The addition of ^{125}I-ST1b(1–19)-BB in the absence (lane 1) and presence (lane 3) of a 400-molar excess of unlabelled ST1b(1–19) indicates that a 57 kD and a 75 kD membrane component are cross-linked specifically by the photoreactive analogue. In contrast, the addition of ^{125}I-ST1b(1–19) in the absence (lane 2) and presence (lane 4) of unlabelled ST1b(1–19) resulted in no labelling of brush border molecules.

SUMMARY

The *E. coli* heat-stable enterotoxin, ST1b was prepared by solid phase peptide synthesis and purified to homogeneity by high pressure liquid chromatography. This analogue, abbreviated as ST1b(1–19), was iodinated and shown to bind specifically to rat intestinal membranes. The amino terminus of the radiolabelled peptide was derivatized using the photoreactive heterobifunctional cross-linker, *N*-hydroxysuccinimidyl *p*-benzoylbenzoate. This photoreactive probe also exhibited binding specificity. It was mixed with rat intestinal brush border membranes and photolysed in the presence or absence of excess unlabelled

ST1b(1-19). Polyacrylamide gel electrophoresis performed in the presence of sodium dodecyl sulphate and 2-mercaptoethanol indicated that the peptide probe was cross-linked specifically to two major molecular species of 57 and 75 kD respectively. One or both of these molecules appears to constitute the enterotoxin receptor or be in close proximity to it.

ACKNOWLEDGEMENTS

We gratefully acknowledge the help of Dr Takayoshi Kuno and Dr Ferid Murad with the membrane filtration assay. We thank Sonia Najjar, Nilda Santiago, and Dr Gary M. Gray for advice on isolation of rat intestinal membranes. J.G. is a fellow of the Medical Research Council of Canada and G.K.S. is a fellow of the John A. Hartford Foundation.

REFERENCES

1. So, M. and B. J. McCarthy (1980). *Proc. Natl. Acad. Sci. USA* **77**, 4011-4015.
2. Moseley, S. L., J. W. Hardy, M. Imdadul Huq, P. Echeverria and S. Falkow (1983). *Infect. Immun.* **39**, 1167-1174.
3. Aimoto, S., T. Takao, Y. Shimonishi, S. Hara, T. Takeda, Y. Takeda and T. Miwatani (1982). *Eur. J. Biochem.* **129**, 257-263.
4. Field, M., L. H. Graf Jr, W. J. Laird and P. L. Smith (1978). *Proc. Natl. Acad. Sci. USA* **75**, 2000-2804.
5. Hughes, J. M., F. Murad, B. Chang and R. L. Guerrant (1978). *Nature (Lond.)* **271**, 755-756.
6. Giannella, R. A. and K. W. Drake (1979). *Infect. Immun.* **24**, 19-23.
7. Guerrant, R. L., J. M. Hughes, B. Chang, D. C. Robertson and F. Murad (1980). *J. Infect. Dis.* **142**, 220-228.
8. Rao, M. C., S. Guandalini, P. L. Smith and M. Field (1980). *Biochim. Biophys. Acta* **632**, 35-46.
9. Dean, A. G., Y. C. Ching, R. G. Williams and L. B. Harden (1972). *J. Infect. Dis.* **125**, 407-411.
10. Lallier, R., F. Bernard, M. Gendreau, C. Lazure, N. G. Seidah, M. Chrétien and S. A. St-Pierre (1982). *Anal. Biochem.* **127**, 267-275.
11. Ronnberg, B., T. Wadstrom and H. Jornvall (1983). *FEBS Lett.* **155**, 183-186.
12. Thompson, M. R. and R. A. Giannella (1985). *Infect. Immun.* **47**, 834-836.
13. Yoshimura, S., H. Ikemura, H. Wanatabe, S. Aimoto, Y. Shimonishi, H. Saburo, T. Takeda, T. Miwatani and Y. Takeda (1985). *FEBS Lett.* **181**, 138-142.
14. Dreyfus, L. A. and D. C. Robertson (1984). *Infect. Immun.* **46**, 537-543.
15. Gariépy, J. P. O'Hanley, S. A. Waldman, F. Murad and G. K. Schoolnik (1984). *J. Exp. Med.* **160**, 1253-1258.
16. Dreyfus, L. A., L. Jaso-Friedmann and D. C. Robertson (1984). *Infect. Immun.* **44**, 493-501.
17. Kuno, T., Y. Kamisaki, S. A. Waldman, J. Gariépy, G. Schoolnik and F. Murad (1985). *Fed. Proc.* **44**, 21-26.

The *Escherichia coli* Haemolysin: Its Gene Organization and Interaction with Neutrophil Receptors

R. A. Welch[a], T. Felmlee, S. Pellett and D. E. Chenoweth*

*Department of Medical Microbiology, University of Wisconsin, Madison, WI, and *Veterans Medical Center, San Diego, CA, USA*

Considerable attention has been paid to the identification of virulence determinants of the opportunistic pathogen, *Escherichia coli*. Epidemiological surveys often reveal properties likely to be involved in pathogenesis because of their more common occurrence among clinical isolates in comparison to normal faecal isolates. These putative virulence factors then become the subject of intensive immunological, biochemical and genetic studies. The goal of such studies is to resolve the "putative" nature of such a property. The *E. coli* haemolysis is illustrative of this phenomenon. Our understanding has progressed to the point where the tentativeness of calling the haemolysin a virulence factor is probably no longer warranted. The purpose of continued investigation is to understand how it contributes to the pathogenesis of extraintestinal *E. coli* disease.

GENE ORGANIZATION

In order to build a foundation for future engineering of the *E. coli* haemolysin, we determined the entire DNA sequence of a chromosomal haemolysin which originated from an O4 serotype, urinary tract isolate [1]. We identified four successive open-reading frames (ORFs) that coincide to four haemolysin genes (HlyC, A, B and D) identified by genetic complementation and analysis of plasmid-encoded polypeptides in *E. coli* minicells [1-3]. Based on features of the DNA sequence and minicell results we proposed two alternative models for the transcriptional organization of the haemolysin determinant [1]. Figure 1 is a

[a] *Telephone:* 1-608-263-2700

Protein-Carbohydrate Interactions in Biological Systems. ISBN 0 12 436665 1

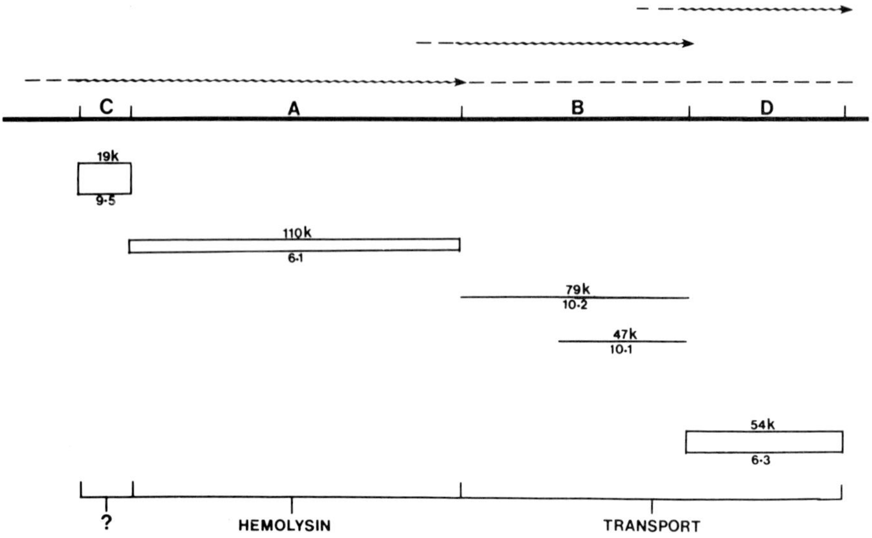

Figure 1. Transcription and gene organization of *E. coli* chromosomal haemolysin. The solid, thick line near the top of the figure shows the sequential 5′ to 3′ order of transcription of the four hly genes C, A, B and D. The wavy lines at the top indicate possible transcripts based on genetic observations as well as analysis of the DNA sequence. At present we are assessing whether or not HlyB is translated from a unique transcript or continuation of the HlyC+A species. HlyD has a unique transcript (Pellet and Welch, in prep.). The boxes beneath the map indicate by their thickness, the relative amounts of each polypeptide detected in *E. coli* minicells. The number above the box indicates in kilodaltons (kD) the molecular mass of that polypeptide. The number below shows the predicted isoelectric point of that polypeptide based on its amino acid composition. At the very bottom of the figure is a legent describing the gene functions. The ? refers to the fact that the HlyC gene product is responsible for activation and transport of the hly by an unknown mechanism.

Table 1. Quantitative analysis of haemolysin transcription

Source of in vivo ^{3}*H-uridine: labelled RNA*	*Relative hybridization of labelled RNA to plus strand DNA of M13mp-subclones specific for:*	
	hlyC+hlyA	*hlyB*
J198 (Hly−)	1[a]	1[a]
J96 (Hly+)	2.2-fold	1.2-fold
WAF270 (J198/pSF4000)	27.6-fold	2.9-fold

[a] The hybridization of labelled RNA from J198 is used as the control of non-specific binding to the M13 DNA probes. The hybridization figures represent the relative CPM above the J198 background. M13 probe DNA [5 μg] was collected on nitrocellulose filters. Labelled RNA [200 000 cpm, 1 μg] was added and the mixture (in 35% formamide -4X SSC) was incubated for 16–24 h at 37°C. The length of the HlyC–HlyA and HlyB probes are similar.

summary of the DNA sequence results and indicates the possibilities of the transcriptional organization.

By subcloning portions of the haemolysin into M13 vectors and utilization of their single-stranded forms for the dideoxy-chain termination method of DNA sequencing, we also acquired hybridization probes of known length and location within the haemolysin. We have begun to use these in a quantitative analysis of transcripts specific for the different haemolysin genes and haemolysins of different clinical origin. Shown in Table 1 are the results of RNA-DNA hybridization studies comparing the amount of *hly*C + A specific tritium-labelled mRNA isolated from the original clinical isolate (J96) to that of a strain (WAF270) harbouring the recombinant haemolysin plasmid (pSF4000). It is apparent that there is an increase in the amount of *hlyC* + A specific transcripts when the chromosomal haemolysin is cloned into the vector pACYC184. This is accompanied by a similar increase in the amount of detectable extracellular haemolysin. Our observations are in contrast to those made by Hartlein *et al.*, where it was suggested the haemolysin is regulated to prevent increases in haemolysin production due to increases in gene copy number [4].

We also demonstrate that relative to the *hly*C+A hybridizations there is approximately four- to eight-fold less labelled mRNA specific for *hly*B. This is consistent with the proposal that there is transcriptional polarity on the downstream *hly*B cistron which is responsible for its lower expression relative to HlyA [1]. We are presently investigating whether or not a rho-independent terminator-like structure predicted to occur between HlyA and HlyB plays any role in the apparent polarity [1].

SECRETION AND STRUCTURE OF THE *E. COLI* HAEMOLYSIN

The determination of the predicted amino acid sequence of the *E. coli* haemolysin permits analyses of its possible secondary structure, hydrophobicity, antigenicity and membrane interactions. Initially, these predictions were tenuous because it was not certain what portion of the haemolysin precursor made up the final extracellular haemolysin protein. There were conflicting reports in the literature [4,5]. We examined the controversy and concluded that the haemolysin was secreted extracellularly as a polypeptide approximately 110 kD in molecular weight without *N-terminal processing* [6]. The possibility remains that some proteolysis at the *C*-terminus may occur and this is involved in the release of the haemolysin from the cell.

The general features of the structural predictions are summarized in Fig. 2. The DNA sequence indicates that the HlyA ORF would encode 1023 amino acids. There are no cysteines and the polypeptide could be characterized as having discrete domains based on isoelectric point and hydropathy [1]. The first 50 amino acids contain a high percentage of basic residues and is hydrophilic. Amino acids 50–230 are less basic and alternate betwen small hydrophilic and hydrophobic regions. Between amino acids 230 and 430 there are two extended hydrophobic regions which are approximately 80 amino acids in length and interrupted by a short hydrophilic region. Within this region, the predicted

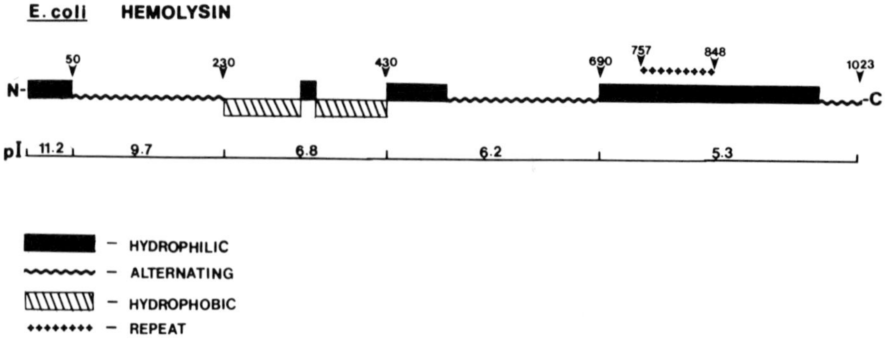

Figure 2.

isoelectric character of the polypeptide is neutral. The remaining half of HlyA appears to be mostly hydrophilic in character and more acidic than the *N*-terminal half. Between amino acids 757 and 848 there are nine repeats of an eight amino acid sequence. The repeated sequence is L-X′-G-G-X²-G-N-D. The L at position 1 and G at positions 3 and 6 occur in all nine repeats. Whereas, the G at position 4, N at position seven and D at position eight occur respectively seven, six and eight times. X′ is a polar amino acid in five of the nine repeats. X^2 is a charged amino acid in five out of nine times.

Based on the predicted structural features, we propose a simple model for HlyA interaction with a target membrane. The positively charged *N*-terminal domain could be envisioned as initiating the binding to the membrane by electrostatic interaction. The 50–230 amino acid region could take on a globular structure. The hydrophobic 230–430 amino acid region could be pictured as being inserted within the target membrane. The remaining acidic hydrophilic half of the HlyA polypeptide could be viewed as remaining extended out from the external side of the target membrane. Divalent cations, especially calcium, are required for lytic activity [7] and the negatively charged *C*-terminal half of HlyA could interact with the cations. Alternatively, the series of amino acid repeats near the *C*-terminus suggests to us an ordered protein structure that may interact with the membrane and/or calcium.

INTERACTION OF THE *E. COLI* HAEMOLYSIN AND HUMAN NEUTROPHILS

Molecular characterization of the *E. coli* haemolysin suggests that certain structural features facilitate this molecule's membrane interactions. Additionally, our recent investigations provide a plausible explanation for the mechanism of action of this virulence factor. Specifically, the observations reported here clearly demonstrate that sublytic concentrations of the *E. coli* haemolysin are sufficient to inhibit binding of chemotactic factors to their human granulocyte receptors.

Human C5a anaphylatoxin is a 74-residue glycopolypeptide that is cleaved from complement component C5 during complement activation. This molecule serves as a critical mediator of the inflammatory response because it promotes both neutrophil and monocyte adherence, chemotaxis, degranulation, and toxic oxygen radical production. Human C5a triggers these types of cellular responses after it binds to specific receptors that have been identified on human peripheral blood granulocytes. These receptors have been characterized by performing ^{125}I-labelled C5a ligand binding assays [8].

In the present study, the effects of haemolysin on C5a receptor function were evaluated by quantitating the binding of 1 nM ^{125}I-C5a to neutrophils after they had been briefly preincubated with supernatants obtained from different strains of *E. coli* that had been grown to late log phase *in vitro*. As shown in Fig. 3, supernatants obtained from a laboratory 04 strain, as well as three different clinical isolates, displayed varying abilities to inhibit binding of ^{125}I-C5a to human neutrophils. Subsequent investigations demonstrated a direct correlation ($r=0.913$, $p<0.001$) between haemolysin production by these clinical isolates and their ability to secrete a factor that abolished C5a receptor function on human neutrophils.

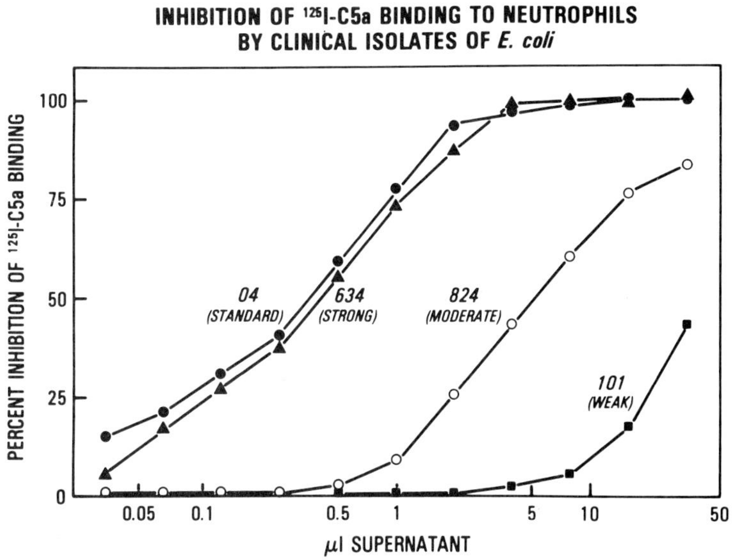

Figure 3. Varying quantities of culture supernatant obtained from a laboratory strain of *E. coli* (04) and three different clinical isolates grown to late log phase *in vitro* were incubated with human PMN for 15 min at 37°C prior to the addition of ^{125}I-C5a (1 nM final concentration). Twenty-two clinical isolates were evaluated in this fashion, 8 (36%) displayed either strong or moderate ability to inhibit PMN binding of ^{125}I-C5a. All 8 of these isolates secreted haemolysin. Of the remaining 14 strains displaying a weak inhibitory effect, only one was haemolytic.

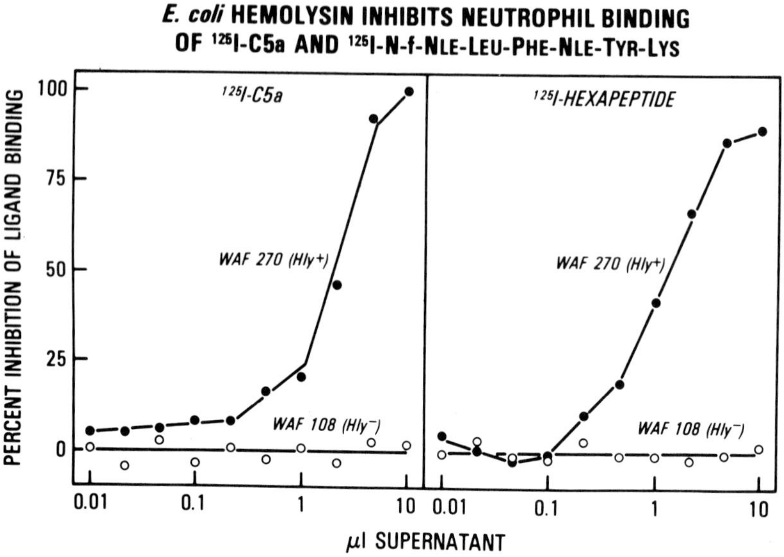

Figure 4. Culture supernatants obtained from the Hly$^+$ recombinant WAF270, but not the Hly$^-$ strain WAF107, inhibited binding of two different chemotactic factors to human PMN. The ampicillin transposon, Tn*1* is inserted within HlyA encoded by pSF4000 present in WAF270 [1,6].

Demonstration that this active factor was in fact the *E. coli* haemolysin was achieved in studies performed with Hly$^+$ and Hly$^-$ recombinant strains [1]. As shown in Fig. 4, the Hly$^+$ WAF270 strain secreted a soluble factor that inhibited neutrophil binding of both C5a and a synthetic hexapeptide chemotactic factor that binds to receptors that are different from the C5a receptors [9]. By contrast, the Hly$^-$ WAF108 mutant did not alter binding of either chemotactic factor. These findings not only suggest that it is the *E. coli* haemolysin that is responsible for the observed phenomena, but also demonstrate that haemolysin exerts a generalized rather than specific perturbation of neutrophil receptor function.

Haemolysin acts to inhibit binding of soluble chemotactic factors to their cellular receptors because of its interaction with the target cells rather than the ligand. As shown in Fig. 5, neutrophil binding of ^{125}I-C5a in the absence of haemolysin is described by a saturable binding plot (upper curve). By contrast, if PMN are preincubated with non-lytic quantities of haemolysin for 15 min at room temperature before ^{125}I-C5a is added, binding of the ligand is totally inhibited (lower curve). Partial inhibition of ^{125}I-C5a binding to neutrophils is observed if haemolysin is either added after C5a has bound to its cellular receptors or added simultaneously with ^{125}I-C5a. In either case, the resulting binding curves demonstrate a reduction in the number of functional C5a receptors on PMN.

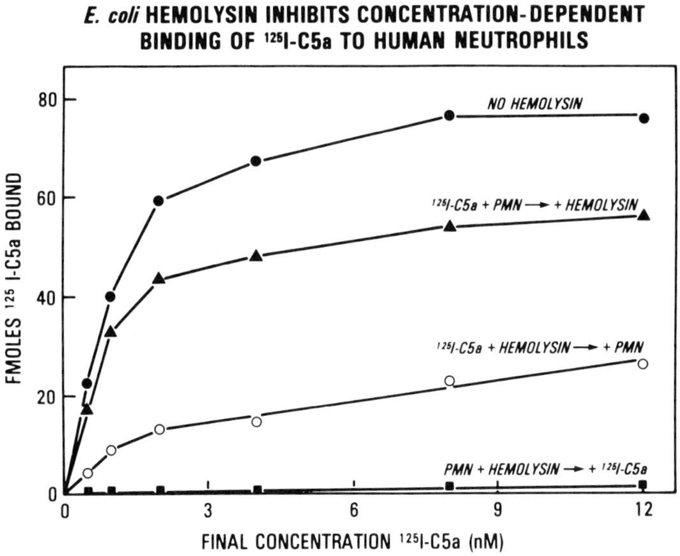

Figure 5. A fixed quantity (5 μl) of *E. coli* 04 supernatant was preincubated with human PMN (■), added simultaneously to PMN with ^{125}I-C5a (○), or added to the ^{125}I-C5a and PMN incubation mixture after ^{125}I-C5a had been allowed to bind to the cells (▲). Normal binding of ^{125}I-C5a to PMN in the absence of haemolysin (●) demonstrates a saturable binding isotherm with half-maximal uptake of the ligand observed at 1 nM concentration.

The results of additional studies also favour the hypothesis that *E. coli* haemolysin acts directly on the intact leukocyte rather than C5a. For example, addition to haemolysin to ^{125}I-C5a-bearing neutrophils produces a time- and temperature-dependent displacement of the bound ^{125}I-C5a. Thus the second curve (triangles) of Fig. 5 represents partial displacement of the pre-bound ligand. Furthermore, incubation of ^{125}I-C5a with haemolysin for prolonged periods of time at 37°C does not produce any detectable biochemical or immunochemical change in the C5a molecule itself. Finally, haemolysin seems to perturb the function of intact cells rather than interact directly with the C5a receptor itself. This statement is predicated on the observation that haemolysin does not inhibit binding of ^{125}I-C5a to the C5a receptor found on isolated plasma membranes.

SUMMARY

Sublytic quantities of *E. coli* haemolysin appear to act directly on intact human leukocytes to produce an impairment of membrane receptor function. As a result, neutrophil binding of soluble chemotactic factors is dramatically reduced and subsequent cellular responses are depressed [10]. Thus, we would propose that *E. coli* haemolysin functions as a true virulence factor because of its ability to inhibit host inflammatory responses.

REFERENCES

1. Felmlee, T., S. Pellett and R. A. Welch (1985). *J. Bacteriol.* **163**, 94–105.
2. Noegel, A., U. Rdest, W. Springer and W. Goebel (1979). *Mol. Gen. Genet.* **175**, 343–350.
3. Wagner, W., M. Vogel and W. Goebel (1983). *J. Bacteriol.* **154**, 200–210.
4. Hartlein, M., S. Schiebl, W. Wagner, U. Rdest, J. Freft and W. Goebel (1983). *J. Cell Biochem.* **22**, 87–97.
5. Mackman, N. and B. Holland (1984). *Mol. Gen. Genet.* **193**, 312–315.
6. Felmlee, T., S. Pellett, E. Y. Lee and R. A. Welch (1985). *J. Bacteriol.* **163**, 88–93.
7. Chenoweth, D. E. and T. E. Hugli (1978). *Proc. Natl. Acad. Sci. USA* **75**, 3943-3947.
9. Niedel, J., S. Wilkinson and P. Cuatrecasas (1979). *J. Biol. Chem.* **254**, 10 700–10 706.
10. Cavalieri, S. J. and I. S. Snyder (1982). *Infect. Immun.* **37**, 966–974.

Identification of the Receptor Glycolipid for Shiga Toxin Produced by *Shigella dysenteriae* Type 1

Alf A. Lindberg[a], Joanne E. Schultz, Marie Westling, J. Edward Brown*, Sara W. Rothman*, Karl-Anders Karlsson[‡] and Nicklas Strömberg[‡]

Karolinska Institutet, Department of Clinical Bacteriology, Huddinge University Hospital, Huddinge, Sweden,
**Division of Biochemistry, Walter Reed Army Institute of Research, Washington, DC 20012, USA, and ‡Department of Medical Biochemistry, University of Göteborg, Göteborg, Sweden*

Bacillary dysentery, caused mainly by strains of the genus *Shigella* and infrequently by certain strains of *Escherichia coli*, is responsible for 10–20% of acute diarrhoeal disease world-wide. The organisms penetrate the epithelial mucosa of the ileum and the large intestine and multiply intracellularly. Only occasionally do they spread beyond the mucosa, causing systemic infection of the host. Beyond invasive properties these bacteria encode for toxin production. The most prominent of these toxins is the Shiga toxin produced by *Shigella dysenteriae* type 1 bacteria.

The Shiga toxin was first described by Shiga in 1896, but still almost 90 years later do we not know its role in the pathogenesis of Shiga dysentery. Recent observations that other genera of *Shigella* and various pathogenic strains of *E. coli*, *Vibrio cholerae* and *V. parahemolyticus* produce a similar toxin, referred to as Shiga-like toxin [6–8], has increased interest in this toxin. These findings suggest that the Shiga and Shiga-like toxin may be a virulence factor in a spectrum of human diseases. The cytotoxic effect of Shiga toxin is through inhibition of protein synthesis [3]. The molecule is composed of one A subunit of MW 30 500 and several B subunits of MW 5000 [10]. The A subunit has the enzymatic

[a] *Telephone:* 46-8-7463475

Protein-Carbohydrate Interactions in Biological Systems. ISBN 0 12 436665 1

activity of the toxin, whereas the B subunits are assumed to carry the binding sites for the obligate receptor-mediated uptake of the toxin molecule.

The chemical nature of the receptor was studied by Keusch and Jacewicz [5]. They concluded, on basis of lysozyme digestion, limited inhibition of toxin binding to cells by oligosaccharides and incomplete receptor blockade by the wheat germ agglutinin, that the receptor was found in a glycoprotein with oligomers of β 1,4 linked GlcNAc residues. In the present study we report on the binding of purified Shiga toxin to glycosphingolipids. Extending a previous study [2] we now show binding to glycolipids with the disaccharide Galα1→4Gal as a common denominator. Inhibition of Shiga toxin binding to HeLa and Vero cells using synthetic receptor analogues has been studied. We also demonstrate that either isolated glycolipids or synthetic receptor analogue can be used for a receptor-binding assay to demonstrate production of Shiga toxin by *S. dysenteriae* type 1 and *E. coli* strains.

SHIGA TOXIN BINDING TO ISOLATED GLYCOLIPIDS

Highly purified Shiga toxin [1] was labelled with ^{125}I and added to glycolipid mixtures where the individual glycolipids first had been separated on a thin-layer chromatogram (Fig. 1). Of the natural glycolipids analysed, the common denominator of positive binders was Galα1→4Gal in a terminal or internal position of the oligosaccharide (Table 1, Fig. 1). However, several glycolipids carrying

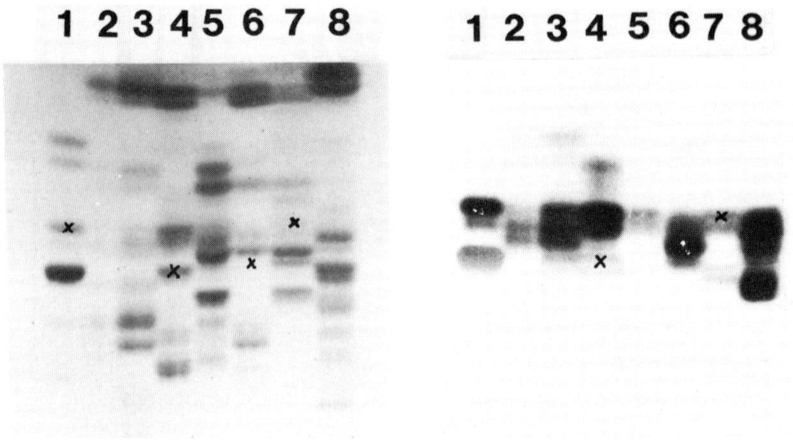

Figure 1. Chromatogram after detection with anisaldehyde (left) and autoradiogram after binding with toxin (right) of total non-acid glycolipids of human erythrocytes (1), human meconium (2,3), monkey intestine (4), dog small intestine (5), rabbit small intestine (6), guinea pig small intestine (7) and rat small intestine (8). The indications with × are for rapid-moving globotriaosylceramide (lanes 1 and 7), slow-moving globotriaosylceramide (lane 6) and globotetraosylceramide (lane 4).

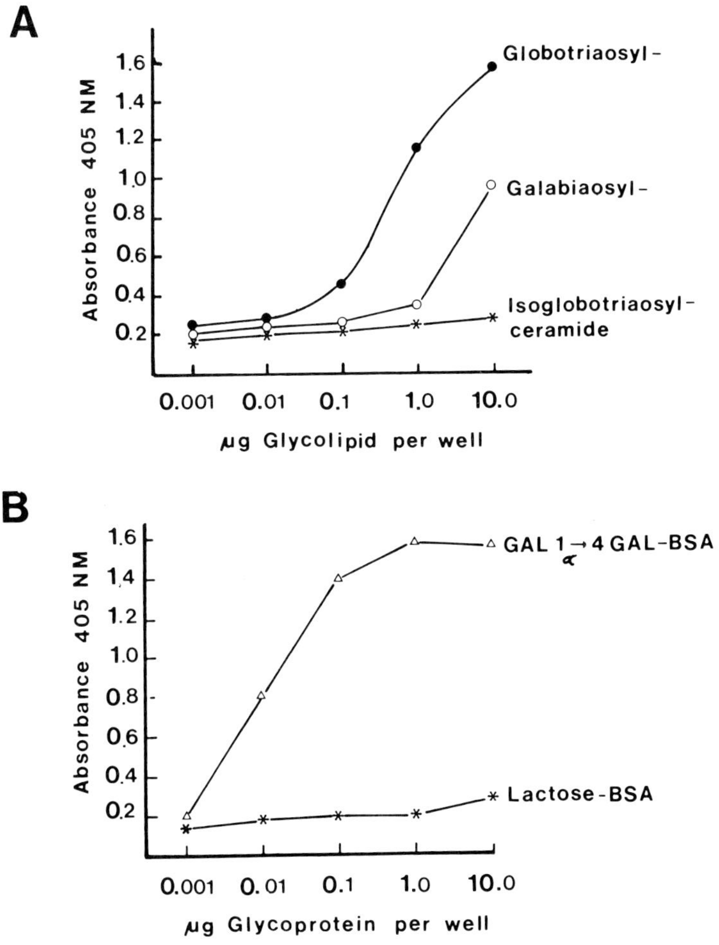

Figure 2. The sensitivity of various isolated glycolipids (A) or α-D-Gal*p*(1→4)-D-Gal*p*-bovine serum albumin (B) as Shiga toxin receptor in a microtitre and solid phase assay. Various concentrations of glycolipid or glycoconjugate were absorbed into microtitre well walls, and 0.5 μg of Shiga toxin was added to each well. Bound toxin was detected with a rabbit Shiga toxin antiserum, followed by a goat anti-rabbit immunoglobulin-alkaline phosphatase conjugate.

the actual disaccharide in an internal position did not bind the toxin. Thus linking GalNAc α1,3 to the terminal non-reducing GalNAc residue in the globotetraosylceramide, making it into a Forssman glycolipid, completely blocked toxin binding (Table 1). Also other extensions in the terminal non-reducing end of a receptor-active saccharide were incompatible with Shiga toxin binding (Table 1).

Table 1. Some glycolipid structures of relevance for toxin binding studies

Glycolipid	*Receptor*	*Name*
Galα1→4GalCer	+	Galabiaosylceramide
Galα1→4Galβ1→4GlcCer	+	Globotriaosylceramide
Galα1→3Galβ1→4GlcCer	−	
GalNAcβ1→3Galα1→4Galβ1→4GlcCer	+	Globotetraosylceramide
GalNAcβ1→3Galα1→3Galβ1→4GlcCer	−	
GalNAcα1→3GalNAcβ1→3Galα1→4Galβ1→4GlcCer	−	Forssman glycolipid
Galβ1→3GalNAcβ1→3Galα1→4Galβ1→4GlcCer	+	
Fucα1→2Galβ1→3GalNAcβ1→3Galα1→Galβ1→4GlcCer	−	
GalNAcβ1→3GalNAcβ1→3Galα1→4Galβ1→4GlcCer	−	
Galα1→4Galβ1→4GlcNAcβ1→3Galβ1→4GlcCer	+	P_1 antigen
Galα1→3Galα1→4Galβ1→4GlcCer	+	
Fucα1→2Galα1→3Galα1→4Galβ1→4GlcCer	−	

Subsequently we studied binding of ^{125}I-labelled Shiga toxin to isolated glycolipids and a synthetic Galα1→4Gal-bovine serum albumin (Gal-Gal-BSA) glycoconjugate coated onto microtitre plate wells (Fig. 2A,B). Of the glycolipids tested Galα1→4Galβ1→4GlcCer, globotriaosylceramide, was the best binder Fig. 2A). Binding of labelled toxin to the Gal-Gal-BSA synthetic glycoconjugate was excellent (Fig. 2B). On a weight basis the glycoprotein appeared 1000-fold more efficient than the galabiosyl glycolipid (Fig. 2). Several reasons may account for this difference such as the multivalency of the glycoprotein (~40 mol Gal-Gal disaccharide per mol BSA) and perhaps a better coating efficiency of the glycoprotein. What is evident from these studies is that the Galα1→4Gal disaccharide functions for binding the Shiga toxin either in the form of a glycolipid or a glycoprotein.

ASSOCIATION OF SHIGA TOXIN TO CULTURED CELLS IN SUSPENSION AND INHIBITION BY RECEPTOR GLYCOCONJUGATE

Cultured HeLa and Vero cells which were in suspension after trypsin treatment were tested for binding of the Shiga toxin. Trypsinized cells were suspended in Eagle minimal essential medium (MEM) and incubated with various concentrations of labelled toxin diluted in phosphate buffered saline (PBS), pH 7.4, supplemented with BSA (1.0 mg/ml), and then incubated for 1 h at 0 or 25°C. Each assay mixture was then layered on a discontinuous Percoll (Pharmacia Fine Chemicals, Uppsala, Sweden) gradient and centrifuged for 15 min at $800 \times g$ at +25°C. The cells layered at the 20–100% Percoll interface. The top Percoll layers containing unbound toxin were removed by Pasteur pipette. Both fractions were counted in a gamma counter. Additions of up to 67 ng/ml ^{125}I-labelled toxin to 4×10^5 HeLa or Vero cells resulted in linear uptake of the toxin (data not shown). This uptake could competitively be inhibited with unlabelled toxin, which indicates that labelled toxin competes with unlabelled toxin for specific binding sites on the cells.

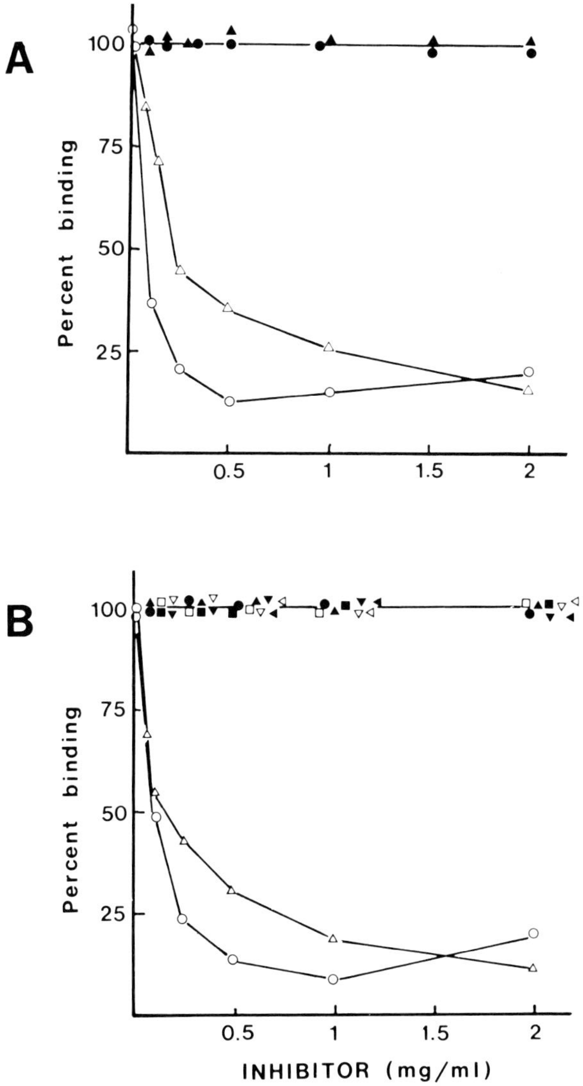

Figure 3. Inhibition of binding of ^{125}I-labelled Shiga toxin to suspensions of HeLa cells (A) or Vero cells (B). Toxin (33 ng/ml) was incubated with inhibitors at either 0°C (○, ●, □, ◁, ◁) or 25°C (△, ▲, ■, ▲, ▼). Aliquots of each mixture was added to 2×10^5 cells, and the suspension was shaken for 1 h at 0 or 25°C. Each assay mixture was then centrifuged through a discontinuous Percoll gradient as described in the text. Symbols: ○, △, Gal-Gal-BSA; ●, ▲, lactose-BSA; □, ■, α-D-Gal*p*(1→4)-D-Gal*p*; ◁, ◀, globotetraose; ▽, ▼, chitotriose.

Various saccharides and glycoconjugates were tested for their ability to inhibit uptake of the ^{125}I-toxin to HeLa and Vero cells at 0 and 25°C (Fig. 3A and B). Inhibitor and toxin were preincubated for 1 h at the indicated temperature before being mixed with the cells. No inhibition of uptake to cells was noted when the Galα1→4Gal disaccharide or the globotetraose tetrasaccharide (≤4 mg/ml) was preincubated with the toxin. However, preincubation of the toxin with the Gal-Gal-BSA glycoconjugate gave significant inhibition of uptake; 50% inhibition was observed with approximately 0.2 mg/ml of the glycoconjugate. No inhibition was observed when Gal-Gal-BSA was added after a 1 h preincubation of toxin with either HeLa or Vero cells. Preincubation of toxin with a lactose-BSA glycoconjugate (≤2 mg/ml) gave no inhibition of toxin binding to HeLa or Vero cells.

Chitobiose and chitotriose were tested for inhibition of toxin uptake since data from Keusch and Jacewicz [5] had suggested that oligomeric β1,4 linked GlcNAc may be the receptor for Shiga toxin. However, no inhibition of ^{125}I-toxin binding was observed in our assays with either chitobiose or chitotriose in concentrations up to 2 mg/ml.

Binding of labelled Shiga toxin to HeLa and Vero cells in monolayer at 0 and 25°C is similar to binding to cells in suspension. Inhibition studies have revealed that the monolayer assay is more sensitive; 50% inhibition of Shiga toxin binding at 0°C is observed with approximately 1.0 μg/ml of the Gal-Gal-BSA glycoconjugate, whereas none of the other saccharides or glycoconjugates tested above showed any inhibition in concentrations up to 2 mg/ml (data not shown). The concentration of Gal-Gal-BSA required for inhibition was equimolar to the concentration of toxin.

The inhibition studies convincingly demonstrate that the Galα1→4Gal disaccharide is a structural element recognized by the Shiga toxin, assumably by the B subunits, and that this disaccharide in multivalent form as in the Gal-Gal-BSA glycoconjugate can inhibit binding/uptake of the toxin to HeLa and Vero cells.

SHIGA TOXIN RECEPTOR BINDING ASSAY

As shown above both galabiosylceramide and the Gal-Gal-BSA glycoconjugate functions as receptors for Shiga toxin when coated onto microtitre plate wells (Fig. 2A,B). We utilized these observations to develop an assay to detect Shiga toxin in solution. A receptor-based microtitre plate assay should be much simpler to use than the cytotoxicity assays commonly used [4].

After coating the microtitre plates with Gal-Gal-BSA (100 μl aliquots of 2.0 μg/ml incubated for 18 h at 20°C), the plates were washed twice with PBS. Fifty μl aliquots of Shiga toxin or culture supernatants were added to each well and the plates incubated at 20°C for 2 h. After washing three times with a PBS-Tween 20 (0.05% Tween 20, v/v) solution, 50 μl of a rabbit Shiga antitoxin antiserum, diluted 1:1000, was added and the plates incubated for 2 h at 20°C. Bound antibodies were detected using a goat anti-rabbit immunoglobulin-alkaline phosphatase conjugate.

The sensitivity of the assay was studied by varying receptor and Shiga toxin concentrations. With a receptor coating concentration of $\geq 1.0\ \mu g$/well an absorbance of ≥ 1.0 after 50 min enzyme-substrate incubation was recorded with a toxin concentration of 0.01 μg/well.

Strains of *S. dysenteriae* type 1 and *E. coli* known to produce Shiga toxin or Shiga-like toxin were grown in iron-depleted Syncase broth for toxin production. Cells were disrupted by sonication and debris centrifuged. The supernatants were saved for determination of cytotoxicity [4] and for testing in the receptor binding assay (Table 2). *S. dysenteriae* type 1 strain 3818-0, which produces Shiga toxin, and *E. coli* strains EDL 931, EDL 932 and H-19, which have been reported to produce Shiga-like toxin [9], were all found to be positive in the receptor binding and cytotoxicity assays. In terms of sensitivity, the cytotoxicity assay is much more sensitive than the receptor assay. The receptor assay was found to be specific; no binding was observed to the Gal–Gal–BSA receptor by supernatants from known enterotoxin-producing strains of *E. coli* (LT and ST toxins), and clinical isolates of *Clostridium difficile*, *Aeromonas hydrophila*, *Campylobacter jejuni*, and *Yersinia enterocolitica*.

Table 2. Receptor binding assay for detection of Shiga and Shiga-like toxin in bacterial sonicate supernatants

Bacterial strain	*Receptor assay activity*[a] *(µg/ml)*	*Cytotoxic activity* *(CD_{50}/ml*[b]*)*
S. dysenteriae 3818-0	16.0	7.8×10^5
E. coli EDL 931	8.0	7.8×10^5
E. coli EDL 932	8.0	7.8×10^5
E. coli H-19	16.0	7.8×10^5
Clostridium difficile VPI	0	<0.01
Campylobacter jejuni 31	0	<0.01
Aeromonas hydrophila	0	<0.01
E. coli SVA	0	<0.01

[a] Amount of Shiga or Shiga-like toxin detected as compared to a standard curve with highly purified Shiga toxin.

[b] Concentration of toxin required to kill 50% of cells (CD_{50}). 1 CD_{50} = 10 pg purified Shiga toxin.

Since binding to the Gal–Gal–BSA receptor occurs with the toxin from three strains producing Shiga-like toxin, and this bound toxin is detected with a rabbit anti-Shiga toxin antibody, the results suggest that the Shiga-like toxin is indeed identical, or very similar, to the true Shiga toxin. Our results also suggest that the receptor binding assay may be developed into a clinical test for detection of Shiga toxin production by bacterial strains or in stool specimens.

SUMMARY

(1) The toxin produced by *Shigella dysenteriae* type 1 bacteria, the Shiga toxin, binds to glycolipids with Galα1→4Gal either in terminal or internal position in the saccharide portion.

(2) Binding of the Shiga toxin to HeLa or Vero cells in suspension, or in monolayer, can be inhibited by preincubation of the toxin with the Galα1→4Gal disaccharide in multivalent form as when covalently linked to bovine serum albumin forming a digalactosyl-BSA glycoconjugate. The Gal-Gal disaccharide, or other digalactosyl-containing saccharides, chitobiose, chitotriose and a lactose-BSA glycoconjugate, were inactive as inhibitors.

(3) The Gal-Gal-BSA glycoconjugate can be used as a receptor analogue in a microtitre plate binding ELISA for detection of Shiga and Shiga-like toxin produced by *S. dysenteriae* type 1 and certain *E. coli* strains.

ACKNOWLEDGEMENTS

This work was partly supported by the Swedish Medical Research Council (grant no. 16x-656).

REFERENCES

1. Brown, J. E., D. E. Griffin, S. W. Rothman and B. P. Doctor (1982). Purification and biological characterization of Shiga toxin from *Shigella dysenteriae* 1. *Infect. Immun.* **36**, 996–1005.
2. Brown, J. E., K.-A. Karlsson, A. A. Lindberg, N. Stromberg and J. Thurin (1983). Identification of the receptor glycolipid for the toxin of *Shigella dysenteriae*. *In* "Glycoconjugates" (Eds M. A. Chester, D. Heinegard, A. Lundblad and S. Svensson), pp. 678–679. Proceedings of the 7th International Symposium on Glycoconjugates, Lund-Ronneby, Sweden.
3. Brown, J. E., S. W. Rothman and B. P. Doctor (1980). Inhibition of protein synthesis in intact HeLa cells by *Shigella dysenteriae* 1 toxin. *Infect. Immun.* **29**, 98–107.
4. Gentry, M. K. and J. M. Dalrymple (1980). Quantitative microtiter cytotoxicity assay for *Shigella* toxin. *J. Clin. Microbiol.* **12**, 361–366.
5. Keusch, G. T. and M. Jacewicz (1977). Pathogenesis of Shigella diarrhea. VII. Evidence for a cell membrane binding of toxin involving β1→4-linked *N*-acetyl-D-glucosamine oligomers. *J. Exp. Med.* **146**, 535–546.
6. Keusch, G. T. and M. Jacewicz (1977). The pathogenesis of *Shigella* diarrhea. VI. Toxin and antitoxin in *Shigella flexneri* and *Shigella sonnei* infections in humans. *J. Infect. Dis.* **135**, 552–556.
7. O'Brien, A. D., M. E. Chen and R. K. Holmes (1984). Environmental and human isolates of *Vibrio cholerae* and *Vibrio parahemolyticus* produce a *Shigella dysenteriae* 1 (Shiga)-like cytotoxin. *Lancet* **i**, 77–78.
8. O'Brien, A. D., G. D. LaVeck, M. R. Thompson and S. B. Formal (1982). Production of *Shigella dysenteriae* type 1-like cytotoxin by *Escherichia coli*. *J. Infect. Dis.* **146**, 763–769.
9. O'Brien, A. D., T. A. Lively, M. E. Chen, S. W. Rothman and S. B. Formal (1983). *Escherichia coli* 0157:H7 strains associated with haemorrhagic colitis in the United States produce a *Shigella dysenteriae* 1 (Shiga) like cytotoxin. *Lancet* **i**, 702.
10. Olsnes, S., R. Reisbig, and K. Eiklid (1981). Subunit structure of *Shigella* cytotoxin. *J. Biol. Chem.* **256**, 8732–8738.

Poster Session

Application of Multilocus Enzyme Gel Electrophoresis to *Haemophilus influenzae*

O. Porras[a], D. Caugant, T. Lagergård and C. Svanborg-Edén

Department of Clinical Immunology and Medical Microbiology, University of Göteborg, Göteborg, Sweden

Studies of epidemiology and pathogenesis of infections caused by *Haemophilus influenzae* have been hampered by the lack of adequate identification schemes. In this study multilocus enzyme electrophoresis was shown to be a highly sensitive tool to discriminate between isolates of *Haemophilus influenzae*. By this technique, each isolate is assigned an electrophoretic type, ET, depending on the electrophoretic mobility of known enzymes in starch gel [1].

Ninety-four isolates were used: 15 non-b capsulated strains, 3 non-capsulated mutants of strains type a, b and f, 29 type b strains and 47 non-typeable (NT) *H. influenzae*. Protein extracts were prepared as follows: bacteria were grown in 100 ml broth overnight, centrifuged, the pellet resuspended in 2 ml of 0.01 M Tris-0.001M EDTA buffer (pH 6.8), sonicated 1 min and centrifuged at $2830 \times g$ for 20 min at 5°C. The supernatants were submitted to horizontal electrophoresis in 11.4% starch gels [1]. The mobility of each isoenzyme was determined after histochemical staining with the specific substrate. Twenty-eight enzymes were tested; 16 of these were not detected (Acid-PHOS, AH, ADH; Alk-PHOS, Ala-DH, CAT, EST, GDH; IDH, LAP, LDH, Leu-DH, MPI, PE3, SDA, TO). Of the 12 enzymes produced, all were polymorphic and migrated toward the anode (MDH, PE2, 6PG, AK, G6P, PGI, FH, GOT, HEX, IPO, G3PDH, P6G). Six enzymes were selected for future ET determinations since they occurred in all the isolates tested (Table 1). The assay conditions are shown in Table 1. *Escherichia coli* strains were used as controls, but they were found to differ in the repertoire of enzymes produced and in the mobility of the proteins.

[a] *Telephone:* 46-31-602082

Protein-Carbohydrate Interactions in Biological Systems. ISBN 0 12 436665 1

Table 1. Electrophoretic conditions and electromorphs for the 6 enzymes used to analyse the *H. influenzae* isolates. The mobility was graded as variants of fast (F++, F+, F), medium (M, M−) or slow (S+, S, S1, S2, S3)

Enzyme (symbol)	*Volts*	*Buffer system*[a]	*MD*[b] *(cm)*	*No. of strains*	*Electromorphs*						
					No.	*1*	*2*	*3*	*4*	*5*	*6*
Malate dehydrogenase (MDH)	130	1	9	94	5	F++	F+	F	M	S	
Phenylalanyl-leucine peptidase (PE2)	130	1	8	94	6	M	M−	S	S1	S2	S3
6-Phosphogluconate dehydrogenase (6PG)	130	1	9	94	4	F++	F+	F	M		
Adenylate kinase (AK)	130	1	8	94	3	F	M	S			
Glucose-6-phosphate dehydrogenase (G6P)	250	1	10	93	6	F+	F	M	S	S1	S2
Phosphoglucose isomerase (PGI)	150	2	10	94	6	F+	F	M	S+	S	S1

[a] Buffer system 1: Tris-Citrate pH 8.0/Tris-Citrate pH 8.0. Buffer system 2: Tris-Citrate pH 6.3/Tris-Citrate pH 6.7.
[b] Migration of dye.

Each isolate was assigned an electrophoretic type (ET) by the combination of the electrophoretical forms (electromorphs) for the six enzymes tested. Strains expressing the same electromorphs for the six enzymes were included in one ET. A total of 49 ETs were detected, with a mean number of 1.98 strain per ET (range 1-28). The ET with the largest number of isolates were No. 14 (28 isolates), No. 35 (4 isolates), Nos. 1 and 13 (3 isolates) and Nos. 3, 6, 7, 15, 24, 27, 28, 38 and 43 with 2 isolates each. The remaining 38 isolates each corresponded to one ET.

ET determinations were subsequently compared to other typing techniques for *H. influenzae*: capsular type and biotype. The type b strains expressed 4 ETs, (Nos. 12, 13, 14, 15) with No. 14 comprising 26/28 of them. ET 14 differed from ET 13 and 15 at one enzyme. ET 12 was expressed by the strain RAB; it differed in 5 enzymes from ETs 13 and 14 and in 4 enzymes from ET 15. No ETs were shared between capsulated type b and non-b capsulated *H. influenzae* or between the non-b capsular types. Thirty-seven ETs were detected among the 47 NT isolates. Five NT strains belonged to the ETs expressed by the type b strains, but no ETs were shared with the non-b strains. The ET variation among NT strains was significantly higher (38/47 strains) than among the type b (4/28). When capsular type and biotype were analysed as a class, a low ratio of ETs per class (0.15) was recorded for type b biotype 1. Most classes comprised one isolate and correspond to one ET. The three more frequent combinations among the NT isolates (NT:1, NT:2, NT:3) showed similar ratios of ET per group (0.80, 0.86, 0.85).

CONCLUSIONS

(1) By multilocus enzyme electrophoresis an ET could be assigned to each isolate; none were NT.

(2) Multilocus enzyme electrophoresis discriminated more variants among *H. influenzae* than other typing techniques including outer membrane protein patterns (data not shown).

(3) Six enzymes (MDH, PE2, 6PG, AK, G6P and PGT) resolved 49ETs. They were polymorphic and occurred in all 94 isolates.

(4) The ET variability was low among type b strains, (26/28 isolates expressed the ET 14) but high among the NTs (37 ETs in 47 isolates).

(5) No ETs were shared between NT and non-b capsulated strains or between type b and non-b capsulated strains. Three ETs were shared between type b and NT strains.

REFERENCE

1. Caugant, D. (1983). Enzyme polymorphism in *Escherichia coli*. Thesis, Department of Clinical Immunology, University of Göteborg, Sweden.

Antibiotics are Necessary for Plasmid Isolation from *Clostridium difficile*

J. Swindlehurst, T. G. Tachovsky and M. Knoppers
Cambridge Research Laboratory, Cambridge, MA, USA

Clostridium difficile is a recognized intestinal pathogen responsible for antibiotic-associated colitis in experimental animals and pseudomembranous colitis in humans. The pathology subsequent to infection with this organism results from the production of a pair of toxins, an enterotoxin-like molecule (toxin A) and a cytotoxin (toxin B), which are responsible for haemorrhage, inflammation and necrosis in a variety of tissues of experimentally infected animals. The toxins are cytotoxic *in vitro* and lethal *in vivo*. Toxin A is enterotoxigenic causing fluid accumulation in rabbit ileal loops; toxin B is cytotoxic for a variety of tissue culture cells and fails to cause oedema when applied to intestinal loops. Because of the association of *C. difficile* with antibiotic-associated diarrhoea, and the fact that many antibiotic resistance genes are located on plasmids, the association of the two toxins of *C. difficile* with the presence or absence of plasmids was examined.

The nine *C. difficile*, four toxin-producing and five non-toxin-producing strains, were obtained from Nancy Taylor (Massachusetts Institute of Technology, Cambridge, MA), American Type Culture Collection (Rockville, MD), and Tracy Wilkins (Virginia Polytechnical Institute, Blacksburg, VA). Cultures were maintained in Brain Heart Infusion Broth (BHI) in anaerobic PRAS II tubes (Scott Laboratory, Fiskeville, RI). Batch cultures were grown in BHI broth supplemented with 0.5 g/l D-L cysteine and 5 mg/l haemin in screw cap flasks. Prewarmed flasks were inoculated with 0.5 to 1.0 ml of an overnight culture, filled with uninoculated medium, capped tightly and incubated overnight (16–18 h) at 37°C. The following antibiotics were added to the medium prior to inoculation; Neomycin, 30 μg; Kanamycin, 30 μg; Streptomycin, 10 μg; and Polymyxin B, 300 Units.

[a]*Telephone:* 1-617-491-2300 ext. 56

Protein-Carbohydrate Interactions in Biological Systems. ISBN 0 12 436665 1

A modification of the plasmid isolation procedure of Just *et al.* was used for large scale plasmid preparation. A ninety minute lysozyme treatment (37°C) was followed by SDS-alkali lysis. Precipitation of the plasmid DNA followed with 0.5 M NaCl and 50% (w/v) PEG (MW 8000). A 42 kilobase plasmid was isolated from *C. difficile* strain 9689 (ATCC) only when the strain was grown in BHI with Kanamycin, Neomycin, Streptomycin and Polymyxin B. Restriction endonuclease activity was seen using various restriction endonucleases such as *Hin*dIII, *Xba*I, and *Eco*RI.

Dot blot hybridization using ^{32}P-labelled CRL-S42 plasmid shows that plasmid DNA sequences exist in all nine toxin-producing and non-toxin-producing strains of *C. difficile.* Curing of the toxin-producing strains was attempted by various methods; however, strain 2-4 was only cured of its neomycin resistance when it was grown at 42°C in the presence of acridine orange. These "cured" strains were still cytotoxic when tested with 3T3 cells and contained plasmid DNA sequences when hybridized with CRL-S42 plasmid.

To correlate the presence of antibiotics with toxin production strain 9689 was grown in BHI with and without antibiotics and aliquots were tested for toxin production by cell rounding (cytotoxicity) of 3T3 cells. Antibiotic-grown cells produced less toxin than those cells grown without antibiotics. The presence of the plasmid DNA sequences in all strains of *C. difficile* grown with and without antibiotics, as compared to the ability to isolate plasmid DNA, suggests that this may be the result of plasmid sequences which shuttle between a plasmid and chromosomal form in the presence and absence of antibiotics. The correlation between this plasmid DNA and toxin production is under study.

REFERENCE

1. Just, L., R. Frankis, W. S. Lowery, R. A. Meyer and G. V. Paddock (1983). *BioTechniques* Sept/Oct, 136-140.

Translocatable Kanamycin Resistance in *Campylobacter*

S. F. Kotarski, T. L. Merriwether,
J. J. Brendle and P. Gemski

Department of Biological Chemistry, Walter Reed Army Institute of Research, Washington, DC 20307-5100, USA

Campylobacter jejuni, recognized as a world-wide cause of gastroenteritis, remains uncharacterized with respect to its invasive disease mechanism. Efforts have focused only recently on development of genetic systems for molecular characterization of *C. jejuni* and its virulence determinants [1,2]. Our characterizations of plasmid-encoded antibiotic resistances of *Campylobacter* have revealed that a kanamycin resistance determinant (Km^r) behaves as a translocatable genetic element.

C. jejuni strains conjugally transferred a 36 MD plasmid encoding tetracycline resistance (Tc^r) and Km^r to *Campylobacter* recipients irrespective of the antibiotic selection that we used to recover transconjugants, however, were resistant to only one of these antibiotics. Analyses showed that the Tc^rKm^s transconjugants inherited a 34 MD plasmid which is smaller than the 36 MD plasmid of the parental strain. Tc^sKm^r transconjugants, however, contained no detectable plasmid DNA.

The segregation of these resistance determinants was also evident in isogenic derivatives recovered after ethidium bromide treatment of the parental strains. Such treatment of Tc^rKm^r strain 3H40 yielded Tc^rKm^s derivatives containing a plasmid about 2 MD smaller than the original 36 MD plasmid. Similar treatment of Tc^rKm^r strain 4B20A yielded segregants without plasmids that retained the Km^r but lost the Tc^r phenotype. Thus, loss of Tc^r is associated with loss of plasmid, whereas loss of the Km^r phenotype appears to be result of a small deletion in the plasmid. Moreover, our finding that Km^r can be retained by strains cured of the 36 MD plasmid suggests that Km^r is translocatable.

[a] *Telephone:* 1-202-576-2594

Protein-Carbohydrate Interactions in Biological Systems. ISBN 0 12 436665 1

We also experimentally demonstrated the translocation of the Km^r determinant from chromosome to plasmid DNA. A self-transmissible, 31 MD plasmid encoding Tc^r was introduced by conjugation into a plasmidless Km^rTc^s segregant of 4B20A. A Tc^rKm^r transconjugant containing the 31 MD plasmid was recovered and then mated with a plasmidless Tc^sKm^s recipient to detect translocation of the Km^r determinant onto the plasmid. Although such matings yielded Tc^rKm^s transconjugants (10^{-5} transconjugants/donor) that had inherited the 31 MD plasmid, this cross also produced Km^rTc^r transconjugants (10^{-9} transconjugants/donor) which possessed a 34 MD plasmid. This cotransfer of Km^r with an associated increase in size from 31 MD to 34 MD of the self-transmissible plasmid supports our conclusion that the Km^r determinant is located within a transposable element. We are currently exploring the use of Km^r translocation as a system for generating insertion mutants of *C. jejuni*.

The Km^r determinant, cloned into the *Cla*I site of pBR322, is fully expressed in *Escherichia coli*. This finding and the discovery of a translocatable genetic element hopefully will provide a basis for the development of DNA shuttle vectors between *E. coli* and *Campylobacter* for use in molecular characterization of this important pathogen.

REFERENCES

1. Taylor, D. E., R. S. Garner and B. J. Allen (1983). *Antimicrob. Agents Chemother.* **24**, 930–935.
2. Tenover, F. C., M. A. Bronson, K. P. Gordon and J. J. Plorde (1983). *Antimicrob. Agents Chemother.* **23**, 320–322.

Genetic Studies on the Production of Shiga-Like Toxin

R. J. Neill, D. H. Wells, N. S. Serrano and P. Gemski

Department of Biological Chemistry, Walter Reed Army Institute of Research, Washington, DC20307, USA

Shigella dysenteriae produces high levels of a cytotoxic protein called Shiga toxin. Recently it has been shown that certain pathogenic *Escherichia coli* produce a toxin structurally and immunologically similar to Shiga toxin. This toxin, commonly referred to as Shiga-like toxin, is produced by *E. coli* isolates associated with hemorrhagic colitis, hemolytic uremic syndrome, and enteropathogenic infantile diarrhoea [1]. In *E. coli* isolate H19, Shiga-like toxin production is bacteriophage associated [2]. Following mitomycin C induction of this strain we isolated a toxinogenic converting phage designated H19M. We have now generated phage mutants altered in toxin-associated genes so as to characterize the role of the phage in toxin production and the role of toxin in diarrhoeal disease.

To determine if this phage encodes a structural gene for the toxin, H19M was mutagenized by exposure to hydroxylamine. Four hypotoxinogenic (Tox^-) phage mutants were further characterized. *E. coli* lysogenized with these Tox^- mutants produced 50- to 1000-fold less Shiga-like cytotoxin as compared to the wild-type phage. Polyacrylamide gel electrophoresis and immunoblot analyses of three of these lysogens revealed reduced levels of one or both of the toxin subunits. The fourth mutant produced no detectable native toxin molecules but contained wild-type levels of toxin subunits A and B. This latter phage thus represents a mutant with a direct change in the physical structure of the toxin and shows that the phage genome encodes the structural gene for at least one of the toxin subunits.

To define phage nucleotide sequences associated with toxin production, Tn*5* transposon mutagenesis was used to insert an identifiable DNA sequence into the genome of H19M. Tox^- derivatives of H19M containing Tn*5* insertions

[a] *Telephone:* 1-202-576-2594

Protein-Carbohydrate Interactions in Biological Systems. ISBN 0 12 436665 1

were obtained. Restriction enzyme analyses of phage DNA indicated that all Tox^- Tn*5* insertions were located in the same 1 kilobase region of the phage genome, which we have designated as *slt*. Such Tox^- insertion mutants were also found to yield phage derivatives containing deletions in the *slt* region. One of these deletion mutants recovered the ability to direct synthesis of the B subunit of Shiga-like toxin but not the A subunit.

We have used a *Pvu*II restriction fragment encompassing a portion of the *slt* region of H19M as a hybridization probe for *slt* sequences. The *slt* probe was reacted with colony blots or with Southern blots of total DNA extracts of individual strains. The results of our initial survey indicate that the *Pvu*II fragment hybridizes to DNA not only from strains of *E. coli* which produce high levels of Shiga-like toxin but also from *S. flexneri*, *S. sonnei*, *S. dysenteriae*. The probe did not hybridize with DNA from *E. coli* K12, *Yersinia*, *Enterobacter*, *Citrobacter*, *Proteus*, and *Serratia*.

We have also addressed the question of phage H19M host range. This phage had been previously shown to lyse rough but not smooth *E. coli* [2]. We have determined the host range of phage H19M to include *Shigellae*. Wild-type H19M or a Tox^+ Tn*5* insertion derivative were used to infect rough isolates of *S. flexneri*, *S. sonnei*, and hypotoxinogenic *S. dysenteriae* and smooth virulent *S. flexneri*. In each case lysogenic derivatives of these strains produced high levels of Shiga-like toxin. Presence of the phage did not inhibit the invasive properties of the virulent *S. flexneri*.

The availability of Tox^- mutant derivatives of phage H19M and the demonstration that *Shigellae* can be infected by H19M allows for construction of host strains of *Shigellae* and *E. coli* which are toxin negative. Through use of such mutants, the relative role of Shiga and Shiga-like toxins in pathogenesis of dysentery and colitis can be delineated. Construction of such mutants is underway.

REFERENCES

1. Cantey, J. R. (1985). *J. Infect. Dis.* **151**, 766–771.
2. Smith, H. W. *et al.* (1983). *J. Gen. Microbiol.* **129**, 3121–3137.

Index

I

K

L

M

N